CHONDRULES and their ORIGINS

CHONDRULES and their ORIGINS

edited by Elbert A. King

Lunar and Planetary Institute
Houston

Typeset by the Lunar and Planetary Institute, printed by Publishers Production International, and bound by Arnold's Bindery in the United States of America.

Library of Congress Cataloging in Publication Data available from the Library of Congress CIP Division, or from the publisher.

Cover design by Sharon Adlis and Donna Chady.

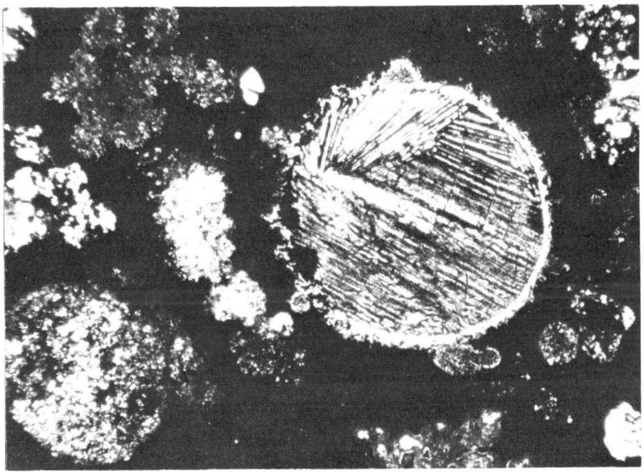

Olivine-rich chondrule in the Allende carbonaceous chondrite. The barred texture and thin rim are typical of many olivine-rich chondrules in virtually all types of chondrites. The texture and circularity of the chondrule indicate that it formed by solidification of a molten drop. Plane polarized light, width of field of view is approximately 3.2 mm.

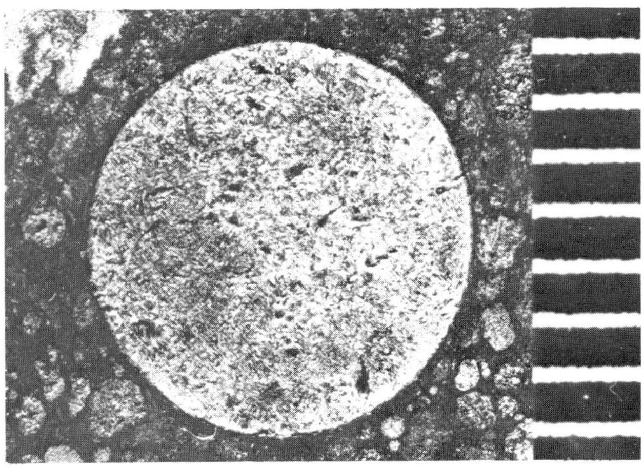

Refractory, calcium- and aluminum-rich chondrule as seen on sawed slab of the Allende carbonaceous chondrite. The major minerals are fassaite, spinel, anorthite and melilite. Both the almost perfectly circular outline and texture of this chondrule indicate an origin by solidification of a molten drop. Reflected light, scale is in mm.

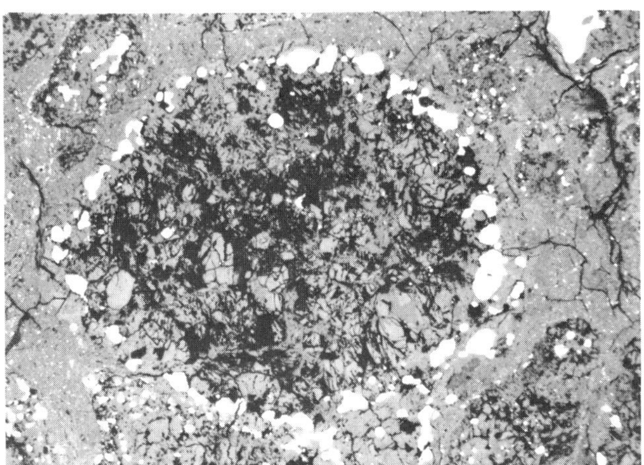

Olivine-rich chondrule with rim of abundant metal (Ni-Fe) blebs and spherules in the Renazzo carbonaceous chondrite. Such metal-rimmed chondrules occur in many different types of chondrites but are particularly large and abundant in Renazzo. Reflected light, width of field of view is approximately 4.0 mm.

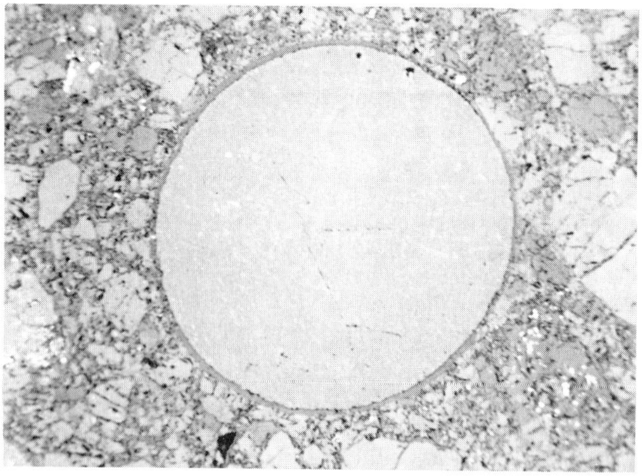

Finely crystalline and glassy chondrule in the Kapoeta howardite. Such chondrules are known to be uncommon components of several "achondritic" meteorites. Reflected light, width of field of view is approximately 3 mm.

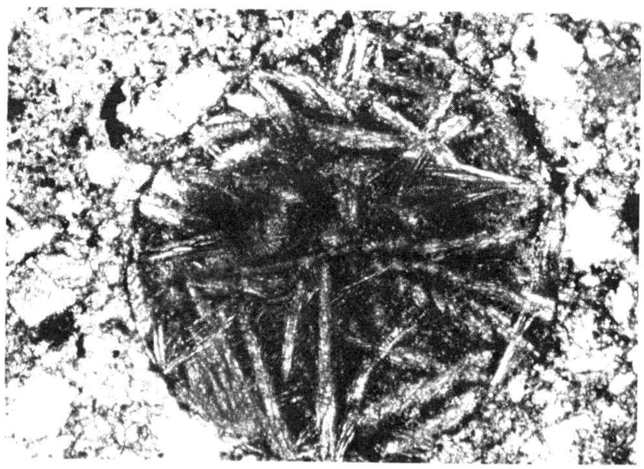

Chondrule in Apollo 14 lunar sample 14318. The lunar chondrule is composed chiefly of plagioclase, pyroxene and turbid glass. Both the outline and texture indicate an origin by solidification of a molten drop. Plane polarized light, width of field of view is approximately 0.8 mm.

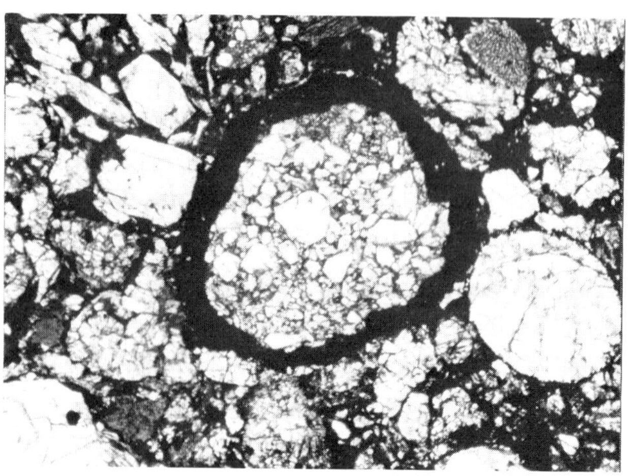

Irregularly shaped olivine-rich chondrule with thick dark rim in the Inman olivine-hypersthene (L3) chondrite. Both the outline and texture of this chondrule do not indicate an origin as a molten drop. Plane polarized light, width of field of view approximately 3.2 mm.

FOREWORD

The origins of chondrules are fundamental problems of most stony meteorites and some planetary surface samples. Yet, until now, no single volume has been devoted entirely to this topic that is so essential to understanding these rocks. Most researchers agree that many chondrules were free molten drops during some portion of their histories, but there the agreement ends. Virtually all other aspects of chondrules are in dispute between competing interpretations and genetic models. The contents of this volume are designed to provide the reader with a broad overview of current ideas in this area of research. Both reviews and original research contributions are included.

All contributions to this volume were technically reviewed under the supervision of an Editorial Board, whose membership is listed on the next page. The concept for the volume grew out of the planning for a Lunar and Planetary Institute Topical Conference of the same name, which was held November 15–17, 1982. However, this is not a proceedings of that conference, as additional papers on the subject were solicited. An abstract volume for the conference is available from the Lunar and Planetary Institute (LPI Contribution 493).

In addition to the authors and editors, I am indebted to a number of LPI staff members for the production of this book. These include Reneé Dotson and Karen Hrametz of the Publications Office; they were assisted by Lisa Bowman, Linda Kofler, Tracey McCasey, and Rosanna Ridings. Design and production of camera-ready pages were done by Sharon Adlis, Donna Chady, Carl Grossman, and Pamela Thompson. Stephen Tellier was responsible for the indexing.

It is obvious that there is still much to be learned about chondrules and their origins, and many contributions are yet to be derived from standard laboratory research methods. It is equally clear that some of the most important answers must be obtained through further solar system exploration, particularly through remote investigations and sample returns from asteroids and comets. We can only urge that these programs not be too long in coming.

Elbert A. King, Editor
Houston, Texas

ASSOCIATE EDITORS

TABLE OF CONTENTS

Condensation of Chondrules

Milton Blander
Argonne National Laboratory, 9700 South Cass Avenue, Argonne, Illinois 60439

By the imposition of well-known kinetic constraints on processes involved in condensation, crystallization, and chemical or crystal reactions, it is shown that the major properties of chondrules could have been produced by the direct condensation of droplets of metastable liquid silicates followed by subsequent crystallization, accretion, and partial comminution of these droplets. Important kinetic factors include nucleation constraints on condensation and crystallization, slow transformations and chemical reactions involving chain silicates and slow diffusion of ions. These factors can lead to the variety of compositions and textures of chondrules, as well as the compositions and content of metals within chondrules. Superimposed on these kinetic factors is the expectation of other effects due to the nature of a cooling nebula. For example, convection in a cooling nebula would impose an oscillatory component on the temperature. Thus, the condensation of chondrules would be followed by reheating of chondrules and accreted "chunks." This would lead to changes in crystallization behavior and "automorphism," as well as the enhancement of reactions with surficial deposits and accretions. It is likely that all observations on chondrules, as well as on calcium-aluminum-rich inclusions (CAI's), can be reconciled with this picture. Non-equilibrium effects make the interpretation of chondrule properties difficult. However, with sufficient information on kinetic factors, they provide us with much more information on chondrule (and meteorite) origins than if chondrules and other meteorite features were formed at equilibrium.

INTRODUCTION

The formation of chondrules directly by condensation from an ancient body of gas remains as a prominent and quite probable mechanism. Critiques of this mechanism have rested on an incorrect or incomplete understanding of the consequences of the process of condensation and of properties of a primordial nebula. It is the purpose of this paper to better define some of these consequences, many of which result from the non-equilibrium effects, which played a likely role in fixing the chemistry and mineralogy of chondrules. A picture of the chondrule formation process will be created based on known physical principles. Because information on kinetic factors is incomplete, not all properties of chondrules can be predicted *a priori*. However, as will be seen later in this paper, the number and importance of the properties of chondrules that are predicted in this picture, based on a small number of physical principles, make this an appealing and probable mechanism.

In order to limit the complexity of the paper, we will define the properties of one possible gaseous source of chondrules, which is considered to be a hot gas with a solar composition (Cameron, 1973) in which the pressure range includes $10^{-4}-10^{-2}$ atmosphere. The physical principles discussed here are not limited to this gas, but will also apply to a broad spectrum of other possible sources.

PROCESSES FOR FORMATION AND DISPOSITION OF CHONDRULES

The process of chondrule formation in a nebula is followed by a series of other processes, which leads to a complex variety of forms and compositions. This complexity is, at least in part, the result of expected departures from equilibrium. Consequently, before discussing the processes involved, a brief discussion of some significant kinetic factors is given.

From known physicochemical principles, we can deduce several important kinetic constraints on chondrule formation. Nucleation constraints must have played an important role in the initial condensation process, as well as in the crystallization of liquid condensates. Because surface-free energies of solids are greater than those of liquids, the critical supersaturation with respect to the condensation of an equilibrium solid is greater than that for the condensation of a metastable liquid. Because entropies of fusion of solids (including silicates) are relatively small, only relatively small differences in such critical supersaturations are needed to condense a metastable liquid, rather than a stable solid, when one is not extremely far below liquidus temperatures. For silicates, it is known from experiments (Blander *et al.*, 1976) that such metastable liquids could be formed hundreds of degrees below liquidus temperatures. Consequently, one predicts that metastable liquids will condense directly from the gas in a manner analogous to the formation of metastable liquid water below 0° C in terrestrial clouds. There probably exists a lower limit below which metastable liquids cannot form. For enstatite chondrites and from experiments on MgO-SiO_2 liquids (Blander *et al.*, 1976), this lower limit appears to be about 1300K. In addition, very large constraints on the formation of metallic iron are predicted from nucleation theory (Blander and Katz, 1967). Measurements by Frurip and Bauer (1977) have confirmed that very large supersaturations (e.g., for liquid iron at 1600K, ln $S_c = 6.3$, where S_c is the critical supersaturation) are necessary to nucleate iron condensation. However, their results are somewhat smaller than predicted by nucleation theory (e.g., ln S_c = 8.5–10). The tendency of the metastable liquid silicates to crystallize will also be blocked by nucleation constraints. Since nucleation is a stochastic (probabilistic) process and rates of liquid to crystal nucleation generally do not have large temperature coefficients, the nucleation of crystals from liquid droplets can take place over a range of temperatures. This will be true whether nucleation is heterogeneous or homogeneous.

Another important kinetic factor is related to diffusion and material transfer between particles in the gas. As the temperature is lowered, diffusion coefficients decrease. Consequently, materials deposited or accreted onto a crystallized chondrule can enter by diffusion only if temperatures are high enough. In silicates, low valent ions such as Na^+ are usually more mobile than high valent ions (Si^{+4}, Al^{+3}), so that there will be selective diffusion and/or exchange of ions of lower valence in certain temperature regimes. On the other hand, transfer of materials between two particles (e.g., chondrules and a later condensate) separated from each other by the gas will tend to be much slower than for particles in contact, since it involves the relatively slow vaporization of the material, transport through the gas, and slow recondensation. The vaporization and recondensation steps are especially slow if the partial pressures of the materials volatilizing are very low (as the result of low abundances or of a majority of it having already condensed).

Another class of kinetic constraints that should be important is related to reaction rates of solids with the nebula. For example, chain silicates have complex structures that are usually difficult to form or decompose at low temperatures. An examination of enstatite chondrites indicates that on the time scale of their formation, reactions involving $MgSiO_3$ such as $3MgSiO_3 + Fe(s) \rightarrow {}^{3}/_{2}\, Mg_2SiO_4 + Si$ (solid soln.) $+ {}^{1}/_{2}\, Fe_2SiO_4$ (solid soln.), "froze in" at temperatures near 1300–1400K. Such constraints would freeze in much of the chemistry and mineralogy of chondrules. This temperature is consistent with the expected temperatures to which $MgO\text{-}SiO_2$ liquids could be subcooled (Blander *et al.*, 1976). Thus, the crystallization of molten chondrules to solid chain silicates would have stopped the further reaction of the major minerals in the chondrule with the nebula. In addition, reactions of non-chain silicates to form crystalline chain silicates would also be stopped because of the kinetic barriers to the formation of the crystalline chain silicates.

With these kinetic factors in mind, we can examine the processes that lead to the formation of chondrules in a large cloud of gas. If the gas is initially hot, it will cool with the consequent formation of metastable liquid silicates as well as some metal. The liquids continue to react with the nebula and become more siliceous as the temperature decreases. The liquids will then crystallize in a range of temperatures defined by crystal nucleation kinetics and, if heterogeneous nucleation is important, by the kinetics of collisions between crystal nuclei and the liquid drops. Further cooling of the gas will lead to the condensation of more volatile solids to form a dust.

Simultaneously with the process of condensation, there will be accretion and comminution of condensed materials. Any metal nucleated will tend to be coated by co-condensing silicates and will be swept up in collisions with liquid droplets. Colliding molten droplets will form larger droplets, and crystallized chondrules that encounter each other will tend to accrete to form "chunks." The rate of formation of such "chunks" at low temperatures will increase if low temperature condensates, which form surficial deposits, act to bind chondrules together. Chondrules accreted at fairly high temperatures will be more "automorphosed" than those accreted at lower temperatures. Lower temperature condensates will condense or accrete onto exposed surfaces of chondrules. Since individual and comminuted chondrules have larger average exposed surfaces than those in chunks, and since the accretion process should occur parallel with nebular cooling and condensation, there will be a distribution of volatile materials that will be the most non-uniform for the most volatile materials and more uniform for the less volatile materials. Incorporation of chunks into meteorites will lead to stones with higher "automorphic" grades and less of the volatiles than when the stones have incorporated a large fraction of individual and comminuted chondrules that have survived to lower temperatures. Automorphic grades should be correlated with volatile contents and the specific volatiles that are more or less depleted should reflect the temperature range of accretion and the average relative exposed surface of the accreted material in that temperature range.

CONSEQUENCES OF KINETIC EFFECTS

If there is a lower limit below which metastable liquids do not form, then the fraction of silicates that can form chondrules will decrease with pressure. In Fig. 1 are plotted the nominal temperatures of first condensation of a molten silicate, as well as a nominal lower limit of 1300K. For a fixed pressure (or even for an adiabat), the length of the line between the temperatures of first condensation and the lower limit reflects the fraction of the silicates that have condensed. If, for example, the "average" silicate crystallizes at 1350K, then the approximate fraction that could have condensed as a liquid is $1-10^{-4.3}/p$, where p is the nebular pressure. (Between $10^{-4.0}$ atm and $10^{-4.7}$ atm, this approximation is somewhat inaccurate.) From this, we can deduce that at pressures of $10^{-2}-10^{-1}$ atm, a major fraction of the magnesiosilicates (and of other chondrule components) could condense as liquids before 1300K, a smaller fraction at 10^{-4} atm, and none at 10^{-5} atm. We have postulated (Blander and Katz, 1967; Blander, 1971, 1979) that E chondrites formed at high pressures ($10^{-2}-10^{-1}$ atm), ordinary chondrites formed near 10^{-3} atm, and C chondrites formed at

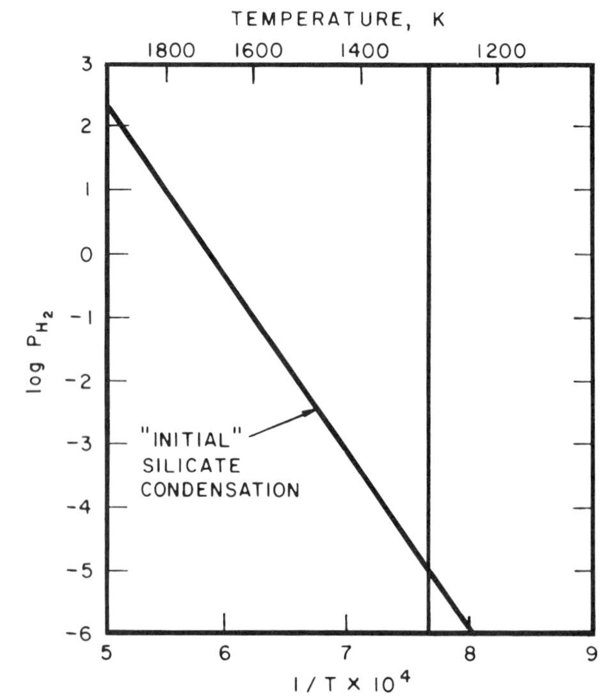

Fig. 1 Representation of the condensation temperatures of liquid magnesium silicates as a function of pressure. The line labeled "initial" silicate condensation represents temperatures at which about 5% of the silica and magnesia have already condensed into a molten mixture, which also contains the more refractory oxides. The vertical line is a representative "lowest" temperature for the formation or maintenance of metastable liquid silicates (see text). A major fraction of droplets are crystallized before cooling to this temperature. This "lowest" temperature would, of course, be a function of the time scale and should decrease with a decrease in time scale. Because crystallization of metastable liquids occurs over a range of temperatures, many droplets will be crystallized before cooling to the "lowest" temperature, and a small fraction could survive to lower temperatures. At high pressures (10^{-2}–10^{-1} atm), liquids can condense for ~200K beyond the initial condensation, and essentially all the silicates will have formed a liquid, whereas at low pressures (10^{-5} atm) the initial condensate forms below 1300K and no liquid can form.

10^{-5}–10^{-4} atm. Thus, we see that this picture is consistent with the fraction of igneous-like material in the different chondrite classes, since E chondrites contain a large proportion of chondrules and other igneous-like materials, whereas many C chondrites (presumably those formed in the lower part of the pressure range for C chondrites) contain little, if any, materials that appear to have been molten.

Of course, the phenomenon is more complex than is indicated above. Because of the stochastic nature of the nucleation process, there is not a sharp lower limiting temperature, but rather a range of temperatures where the probability of condensing liquids is possible, and the temperature range is also a function of the composition of the liquid that tends to form. For example, at high pressures, most of the magnesiosilicates condensed before the limit was approached, and the liquids are consequently more siliceous than at low pressures. Thus, the limit is likely to be lower for the more siliceous materials at high pressures than for the less siliceous materials at low pressures, and the pressure range for the formation of the different classes of chondrites would be narrowed. In addition, the "lower" limit may not be 1300K. However, any change in this temperature merely shifts the pressure range in which the same phenomena should occur.

Each chondrule constitutes a relatively independent chemical system, partly because of slow equilibration through the gas phase. Thus, each liquid drop will have a somewhat different composition, depending on the constituents incorporated by direct condensation and by collisions between drops. Superimposed on this, each droplet will nucleate crystals at a different temperature. The range of nucleation temperatures can be fairly large if homogeneous nucleation is the only mechanism, because of the generally low temperature coefficients of liquid-solid nucleation. The range of nucleation temperatures will be even larger if (as would appear from radiating chondrules) heterogeneous nucleation is also important, since the rates will depend on the rates of collision of crystal nuclei with droplets.

Melts equilibrated with the gas at low temperatures are more siliceous than those at higher temperatures. If reactions of the nebula with the silicates essentially stop upon crystallization (because chain silicates are a reactant or product), then the resultant droplets that had crystallized at different temperatures will have different Si/Mg ratios. This effect can lead to the compositional variability of chondrules that is observed. In addition, the differences in nucleation temperatures, modes of nucleation (homogeneous or heterogeneous), and compositions of droplets could lead to the variability of textures and morphologies that is observed in chondrites.

Because equilibration between solids in contact is generally much more rapid than between materials separated in the gas, accretion and the formation of condensates on surfaces are important precursors of equilibration. Consequently, a collection of crystallized and comminuted chondrules in a cooling gas before accretion will collect later condensates of more volatile materials (perhaps Na^+ and K^+ bearing) on their surfaces. With accretion to chunks, the mobile portions of these surficial deposits, such as Na^+, will equilibrate within the chunks, whereas the individual chondrules will be different in a manner dependent on the quantity of the surficial deposits. In addition, if accretion to chunks occurred at a high enough temperature for sintering of the chondrules, they will "automorphose" to a higher grade. These chunks, having smaller surface areas per unit mass than whole or comminuted chondrules, will collect less of the very low temperature surficial deposits of most volatile materials. The amount of volatiles incorporated will depend on the rates of accretion, which is a continuous process (that can have temperature dependent rates) occurring during the cooling of the nebula. The particular volatiles incorporated or not incorporated into chunks will be related to the relative values of the accretion and condensation temperature ranges.

METAL

Because of non-equilibrium effects, understanding the origins of metal particles in chondrules presents some complexities. Iron-nickel alloys can be considerably supersaturated before condensing because of large surface-free energies and the consequent nucleation constraints (Blander and Katz, 1967). Frurip and Bauer (1977) have shown that measured supersaturations are large, but somewhat smaller than predicted from classical nucleation theory. [For example, at 1600K, the natural logarithms of the critical supersaturation for (metastable) liquid iron is about 6.3, compared to theoretical values of 8.5 to 10 for rates commensurate with the method of measurement.] Consequently, some metal could nucleate during the condensation of chondrule precursors. Since liquid silicates tend to wet metals (i.e., small wetting angles), any metal particles formed would be good nucleation sites and tend to be coated by co-condensing liquids. Metal particles formed during silicate condensation would thus be removed from direct contact with the nebula and cease to be metal condensation nuclei. The co-condensation of liquid silicates provides a removal mechanism for metal nuclei and would serve to increase the supersaturations possible with respect to metal to a higher level than one would have if no liquids had formed. Thus, early-formed metal particles will be effectively isolated from the nebula and will usually be able to react with it only indirectly.

Of the nebular reactions possible with such particles imbedded in silicates, the most significant will involve the relatively oxidizable metals. These can have significant concentrations dissolved in the silicates, which provides a mechanism for diffusive transport. Equilibration of relatively oxidizable metals (e.g., Fe and Ni) between metal particles immersed in a liquid drop should be relatively rapid, but equilibration between metals in separated droplets or between metals in droplets and the gas should be slow. Equilibration of Fe and Ni between crystallized chondrules in contact is possible, as long as the interdiffusion of Fe^{++} and Ni^{++} is rapid enough in the solid silicates. Consequently, different chondrules that remain separated to low temperatures can preserve inherited differences in compositions of even the oxidizable metals, and all chondrules can preserve differences for metals (e.g., platinum metals) that are noble enough so that no good transport mechanism exists by which they can be equilibrated even at high temperatures. Because the metal content has been emplaced by a stochastic process, and because the number of metal particles in chondrules is generally small, one would expect large variations in total metal content and compositions, with larger variations for noble metals, which are expected to co-condense with the silicates in chondrules.

OTHER NEBULAR PROCESSES

Misconceptions concerning the formation of chondrules by direct condensation from a gas stem from the lack of consideration of the many possible post-condensation processes (other than crystallization) that might have influenced the chondrules (or CAI's that could have formed in a manner similar to chondrules). For example, in a large, cooling, hot gas cloud, one expects convection of the gas that would transport gas, dust, chondrules, and chunks in a convective loop. As the gas travels outward toward the nebular periphery, the gas pressure decreases and the gas cools. When the gas loops back, the pressure increases and the gas temperature rises. Thus, local regions in a cooling nebula do not necessarily cool monotonically, but rather have a local oscillating temperature variation superimposed on the overall falling temperature. The magnitude of the oscillation depends on the magnitude of the ratio between the maximum and minimum pressure in the convective loop (or cell), as determined by the equation for adiabatic expansion of gases:

$$\ln(T_1/T_2) = \left(\frac{C_P - C_V}{C_P}\right) \ln \frac{P_1}{P_2}, \quad (1)$$

where T is temperature, P is pressure, and the prefactor of the expression on the righthand side averages 0.27 for hydrogen at temperatures ranging from 1000–1300K. Thus, the final temperatures for different expansion ratios, R, of a gas initially at, e.g., 1300K, are 1165K for $R = 1.5$, 1078K for $R = 2$, and 894K for $R = 4$. This oscillatory temperature component can lead to many important influences on chondrules (and CAI's). For example, droplets condensed, crystallized, and partly accreted during the expansive cooling part of the cycle will be reheated during the compression of the gas. This would enhance the tendency toward automorphism and equilibration of materials (e.g., Na^+-containing condensates) and chondrules in contact. If there are different convective cells with different ranges of pressure and temperature, it would lead to a tendency toward quantization of the range of chondrule compositions in each cell. Since gravitational fields would tend to cause the condensates to fall to the gravitational center, materials from one cell can fall into other cells, especially if they are large aggregations for which gravitational effects are large. Thus, if there are differences in isotopic compositions between inner and outer cells, material condensed in an outer convection cell (let us say one with no ^{17}O or ^{18}O) can fall into another cell (with, say, "normal" oxygen isotopic abundances) and react with the new gas. This could lead to the isotope anomalies found in CAI's by the diffusion process discussed by Blander and Fuchs (1975). This diffusion process is not necessarily limited by the size of aggregations (unless they become extremely large and non-porous), since intergranular and surface diffusion is generally large relative to bulk diffusion into individual mineral grains. Chemical effects could also be possible in the transfer of chondrules and CAI's between cells. One could thus mix condensed materials formed in different cells in different temperature ranges, and materials in a given chondrite are not necessarily formed in a single location. Convection will also hasten accretionary processes, since materials with different kinematic properties (different sizes, different densities) will tend to possess more relative motion than in a stationary gas and collide more often.

We see that a nebular origin should lead to complex thermal histories for chondrules and can lead to more complex chemical histories than envisioned in simplistic models involving a stationary gas. A common reason for not accepting an origin of chondrules or CAI's by condensation in a nebula (e.g., see Lofgren, 1982; Wark and Lovering, 1982) is because of the assumption that there was a process of either condensation *or* reheating. In a large nebula, one could (and probably should) have both condensation *and* reheating in the gas. In addition, there are other complexities in properties imposed on chondrules and CAI's by probable nebular processes. Chondrule textures will reflect oscillations in nebular temperature, which can alter crystal forms and relationships. It is well known that oscillations in temperature can dramatically increase rates of crystallization and recrystallization (Kahlweit, 1975) and, in fact, protein crystals have been grown using oscillatory temperatures when they could not be grown in a reasonable time with monotonically decreasing temperatures. Oscillatory temperatures in convection cells (or in any other place) will have a major influence on crystal textures and forms in chondrules.

CONCLUSIONS

Condensation from a solar-type gas can produce materials that have all the properties of chondrules if kinetic constraints are imposed on the processes involved. The observed variability in compositions, textures, degrees of equilibration of minerals and metals, and the "automorphic" grade of chondrules (or collections of chondrules) can all be reconciled with an origin by direct condensation to liquid droplets. Nebular conditions can affect chondrules. For example, convection in a nebula can lead to oscillatory temperatures and to major effects on automorphism, crystal forms, and equilibration, and can also alter chemical, mineralogical, and physical properties of chondrules. Condensation and reheating of chondrules can both occur in a nebular setting, and indications of either of these two processes in meteorites do not necessarily exclude the other.

The kinetic constraints on chondrule formation are not completely understood, and much more experimentation is needed to fully define their influence. Thus, although all of the most important properties of chondrules can be deduced from what is known of the kinetics, much more information is necessary to quantify our predictions. Non-equilibrium effects present difficulties for those attempting to quantify our understanding of the origins of chondrules and meteorites. The rewards of all these experimental and conceptual difficulties that one must surmount is that a definitive understanding of non-equilibrium effects will ultimately provide us with a great wealth of information on the nebula and/or the processes that produced meteorites 4.6 billion years ago.

Acknowledgment. This work was supported by the National Aeronautics and Space Administration.

REFERENCES

Blander M. (1971) The constrained equilibrium theory: Sulphide phases in meteorites. *Geochim. Cosmochim. Acta* **35**, 61–76.

Blander M. (1979) Non-equilibrium effects on the chemistry of nebular condensates: Implications for the planets and asteroids. In *Asteroids* (T. Gehrels, ed.), pp. 809–821. Univ. of Arizona Press, Tucson.

Blander M. and Fuchs L. H. (1975) Calcium-aluminum-rich inclusions in the Allende meteorite: evidence for a liquid origin. *Geochim. Cosmochim. Acta* **39**, 1605–1619.

Blander M. and Katz J. L. (1967) Condensation of primordial dust. *Geochim. Cosmochim. Acta* **31**, 1025–1034.

Blander M., Planner H., Keil K., Nelson L. S., and Richardson N. L. (1976) The origin of chondrules: experimental investigation of metastable liquids in the system Mg_2SiO_4-SiO_2. *Geochim. Cosmochim. Acta* **40**, 889–896.

Cameron A. G. W. (1973) Abundances of the elements in the solar system. *Space Sci. Rev.* **15**, 121–146.

Frurip D. J. and Bauer S. H. (1977) Homogeneous nucleation in metal vapors. 3. Temperature dependence of the critical supersaturation ratio for iron, lead and bismuth. *J. Phys. Chem.* **81**, 1001–1006.

Kahlweit M. (1975) Ostwald ripening of precipitates. *Adv. Colloid Interface Sci.* **5**, 1–35.

Lofgren G. E. (1982) The importance of heterogeneous nucleation for the formation of microporphyritic chondrules (abstract). In *Papers Presented to the Conference on Chondrules and their Origins*, p. 41. Lunar and Planetary Institute, Houston.

Wark D. A. and Lovering J. F. (1982) The nature and origin of Type B1 and B2 Ca-Al-rich inclusions in the Allende meteorite. *Geochim. Cosmochim. Acta* **46**, 2581–2594.

Meteor Ablation Spherules as Chondrule Analogs

D. E. Brownlee and B. Bates
Department of Astronomy, University of Washington, Seattle, Washington 98195

R. H. Beauchamp
Battelle Northwest Laboratories, Richland, Washington 99352

Meteor ablation spherules are melt products of meteoroids that enter the Earth's atmosphere. They are produced by aerodynamic melting, a process that surely produced chondrule-like objects in the early solar system but that apparently did not play a role in forming chondrules found in chondrites. The properties of ablation spherules do, however, provide insight into the chondrule problem because the spheres are chondrule analogs formed by a known process. Although the spheres have strong similarities to meteoritic chondrules, they also differ in significant respects. The differences between the spheres and chondrules suggest that chondrules could not have formed by flash melting of a primitive fine-grained chondritic precursor.

INTRODUCTION

Meteor ablation spherules are solidified melt droplets from 0.1 mm and larger meteoroids that enter the atmosphere. The spheres in effect are terrestrial chondrules produced in the atmosphere by aerodynamic heating of extraterrestrial materials. Ablation spheres in the 2 μm to 25 μm size range have been collected in the stratosphere (Brownlee *et al.*, 1977) and spheres in the 25 μm to 3 mm size range have been collected from deep-sea sediments (Murray and Renard, 1891; Bruun *et al.*, 1955; Millard and Finkelmann, 1970; Pettersson and Fredriksson, 1958; Brownlee, 1981). These aerodynamic "chondrules" have both strong similarities and distinct differences relative to true chondrules. The comparison of ablation spheres and true chondrules provides insight into the possible origins of meteoritic chondrules because many of the aspects of the formation of the atmospheric spherules are known. The time scale for heating and cooling can be calculated, the ambient pressure at the altitude of melting is known, and to some extent the nature of the unmelted precursor material is known. The ablation spheres are of interest both because they are chondrule analogs formed by a known process and also because they are examples of chondrule-like objects that have not been seen in chondrites but that must be abundantly produced in the atmosphere of all planets.

AERODYNAMIC SPHERULES

Meteor ablation spheres collected in the stratosphere and from the Earth's surface are the only known samples of chondrule-like objects produced aerodynamically, but similar objects must also be common elsewhere in the solar system. Currently they must be produced in the atmospheres of all the planets, and they should also have been produced in the solar nebula whenever high differential velocity existed between solids and gas. Ablation spheres probably constitute 100% of the millimeter-sized metal and silicate particles presently in the atmospheres of the Jovian planets. In the past there may have been times when ablation spheres could have been relatively abundant on the surfaces of terrestrial planets and even on satellites with transient atmospheres.

Aerodynamic melting of small meteoroids occurs when the frictional power input exceeds the power thermally radiated at the object's melting temperature. The entry physics for millimeter-sized meteoroids is relatively simple because the particles have small thermal inertias and they do not generate gas caps during entry (Öpik, 1958). The velocity required to heat a small meteoroid to a temperature T in a gas of density ρ can be expressed by the following equation where ϵ and σ are respectively the emissivity and the Stefan Boltzman constant:

$$V = \left[\frac{8\epsilon\sigma}{\rho} \right]^{1/3} T^{4/3} \text{ (cgs).} \tag{1}$$

Chondritic material melts at ~ 1300° C and the critical relative velocity for chondrite melting at a given air density is:

$$V_C = 1.41 \times 10^{-2} \left[\frac{\epsilon}{\rho} \right]^{1/3} (\text{km s}^{-1}) \tag{2}$$

In the Earth's atmosphere, V_c is 0.6 km s^{-1} at 32 km altitude and 9.4 km s^{-1} at 90 km. For the terrestrial atmosphere and minimum entry velocity, meteoroid melting occurs at altitudes below 100 km at ambient pressures of 10^{-6} atmospheres or more. Melting can occur in any atmosphere as long as the scale height is not so large that the particle significantly decelerates before reaching dense air. In addition to formation in atmospheres, aerodynamic spherules could form directly in the solar nebula gas if velocity gradients existed on the order of 1 s^{-1} (for pressures on the order of 10^{-5} atmospheres and velocities of a few km s^{-1}).

An intriguing aspect of aerodynamic melting in atmospheres is that with properly constrained entry velocities, the process could convert irregular millimeter particles to spheres with very high efficiency. Depending on velocity and entry angle, particles either survive as unmelted micrometeorites or they are converted to spheres and/or vapor. It is conceivable that high densities of spherules could be produced by aerodynamic melting in both planetary atmospheres and transient atmospheres of planetesimals. However, while the aerodynamic production process must have produced large numbers of spherules in the early solar system, it does not appear to have played a significant role in production of chondrules found in chondrites. This is clear because ablation spherules have properties that are distinctly different from those of true chondrules.

ANALOGS TO METEORITIC CHONDRULES

The recovered ablation spherules range in size from a few microns to over 2 mm in diameter and hence cover the same size range as chondrules. The spheres formed in an ambient gas pressure of 10^{-6} to 10^{-3} atm and the time scale for heating and cooling is in the range of 5 to 15 seconds. As will be discussed in the following sections, the precursor material for the spheres is believed to be a fine-grained chondritic composition material with a relatively high carbon content. The spheres occur as stony (S), Fe-Ni sulfide (FSN), and iron (I) types. The S spheres are composed of olivine and glass and have chondrule-like textures (Blanchard *et al.*, 1980). The FSN spheres are composed of magnetite and pyrrhotite and have not been found as particles larger than 25 μm (Brownlee, 1978). The I spheres are composed of Fe-Ni metal, magnetite, and wüstite (Marvin and Einaudi, 1967) and are not analogs to anything in meteorites. The S spheres are the closest analogs to true chondrules and they will be the subject of the following discussion.

THERMAL EFFECTS

A striking difference between the type S ablation spheres and true chondrules is that the spheres show a wide range in degree of peak heating. This ranges from particles that were mildly heated and only partially melted to those that were severely heated to the point where major

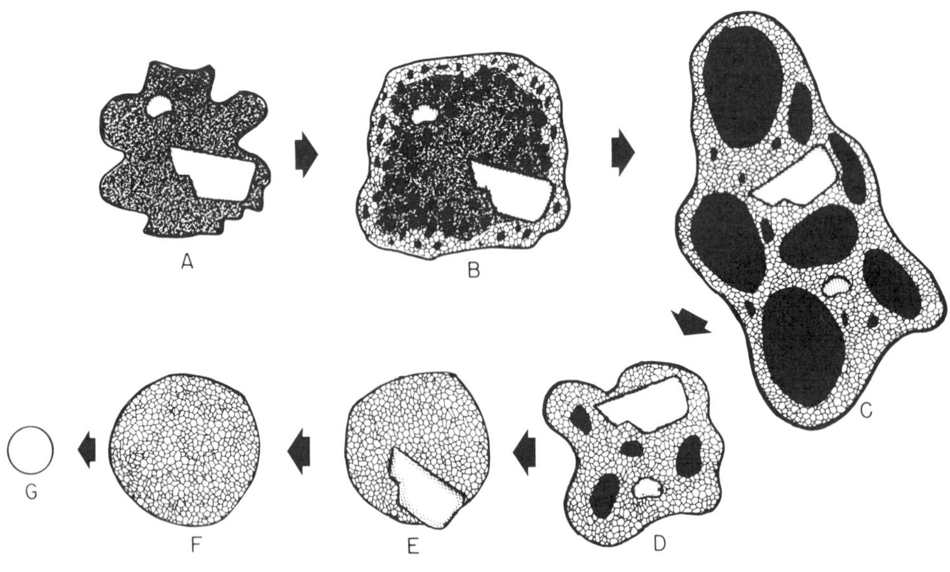

Fig. 1 The thermal alteration of a submillimeter particle depicted in hypothetical polished sections made at successive stages of heating from A to G. At stage F all precursor solids have melted and at stage G volatilization of Fe and Si has occurred. All stages in this figure have been observed in meteor ablation spherules.

elements such as iron and silicon were depleted by volatilization. The range of heating presumably is due to a range in entry velocities and entry angles. The degree of heating is also determined by whether a particle enters singly or is separated from a larger mass deep in the atmosphere. Figure 1 depicts the evolutionary changes that occur in a hypothetical 0.5 millimeter particle as it is subjected to increasing degrees of heating. The drawings A→G show the expected appearance of polished sections of a particle quenched from various stages of increased thermal alteration. All stages of Fig. 1 have been observed in polished sections of collected ablation particles. Part A is the original particle composed of a fine-grained carbonaceous chondrite (CI/CM)-like matrix containing a few large silicate grains. In stage B the matrix has begun to break down, partial melting occurs, and the volume of the particle increases due to vesicle formation. In C the release of volatiles (H_2O, CO_2, or S) produces extreme vesicularity (Fig. 2a). In steps B and C, formation of

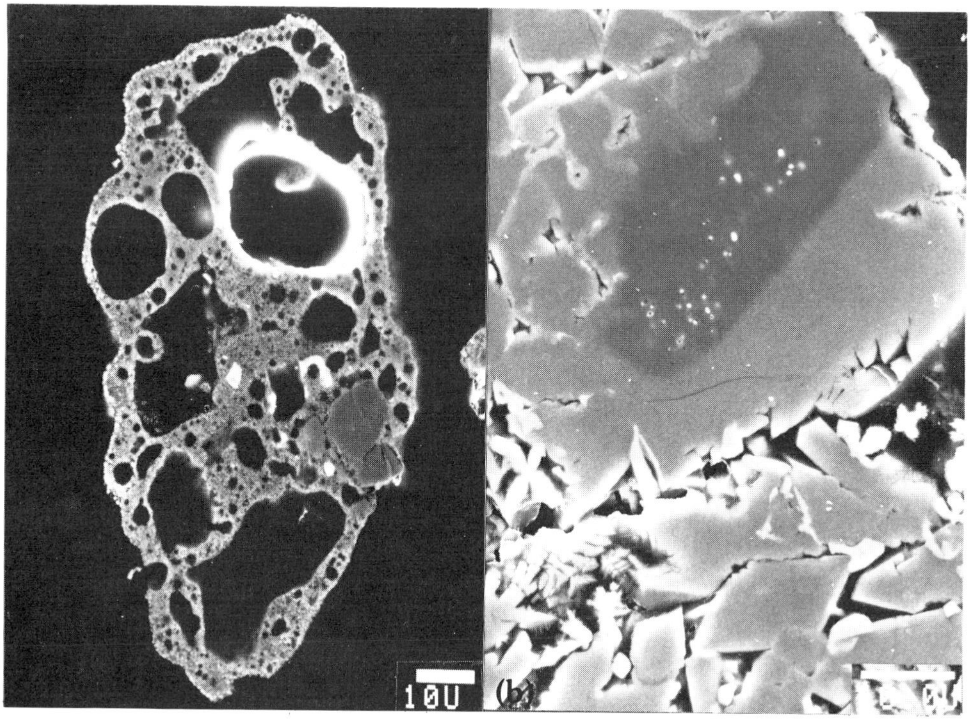

Fig. 2 **(a)** SEM image of a polished section of a vesicular chondritic composition deep-sea "spherule" corresponding to stage C in Fig. 1. The bright grains are sulfides and the large relic grain is forsterite. **(b)** A relic forsterite olivine in a deep-sea stony sphere. The large rectangular relic grain contains bright (dusty) metal grains that are nearly Ni-free, indicating formation by *in situ* reduction. The sharp contact between the dark low Fe relic grain and the surrounding olivine is typical and shows a clear delineation between true relic material and material that has equilibrated with a chondritic composition melt. The sphere containing this grain is shown in Fig. 5. Scale bars = 10 μm.

iron metal begins due to reduction of silicates and/or the decomposition of sulfide. At stage D continued heating has driven out nearly all of the volatiles and all but the two large original mineral grains have been absorbed into the melt. At stage E only one original relic grain remains and it has a distinct reaction rim where it has equilibrated with chondritic composition melt. Continued heating destroys all original grains, depletes all volatiles, and the end product is the non-vesicular fairly spherical object in F. The particle in F is composed of olivine, glass, and magnetite and it has chondritic elemental composition with the exceptions of depletion of siderophiles and strong depletion of Na, Rb, K, and S. In the case of extreme heating (high entry velocity) evolution continues to G, where major elements are depleted by vaporization. The result is a sphere that is depleted in Fe, Si, and sometimes Mg relative to chondritic properties. Particles of this composition have been collected in the stratosphere and have been called CAT spheres (Brownlee *et al.*, 1982) because they are enriched in Ca, Al, and Ti relative to chondritic composition. A stratospheric CAT sphere analyzed by Esat *et al.* (1979) was found to have Mg isotopic fractionation of 1.1% Amu^{-1}, consistent with significant volatilization of Mg.

The wide range of thermal alteration in the ablation spherules is a natural consequence of their mode of formation. Some objects are barely heated while others are heated severely. As is the case with aerodynamic heating, chondrule formation processes such as hypervelocity impact, electrical discharge, and shock waves should also produce particles with thermal modifications ranging from partial melting to significant vaporization.

It appears, however, that peak temperatures reached by chondrules were constrained to a narrower range than that of the ablation spheres. It is evident that stages B through D in Fig. 1 and in particular the highly vesicular stage in Fig. 2a are underrepresented in ferromagnesian chondrules. This difference could be due to differences in the heating time scales and precursor materials, but it probably indicates that the process of chondrule formation always involved temperatures that resulted in significant melting and efficient loss of highly volatile phases. The survival of unmelted relic grains (stage E), usually forsteritic olivine or enstatite, is seen in ~5% of the deep-sea spherules. Relic grains also exist in chondrules as the "dusty" olivine grains identified by Rambaldi and Wasson (1982). A dusty relic grain in a deep-sea spherule is shown in Fig. 2b.

Most of the examined ablation spheres that represent F are depleted in the volatiles Na, K, S, and Rb by factors in the range of $10-10^3$ (Brownlee, 1981; Papanastassiou *et al.*, 1983). The Na depletion is much larger than for typical chondrules (Grossman *et al.*, 1979; Grossman and Wasson, 1982; Lux *et al.*, 1980; Gooding *et al.*, 1980), an indication that many of the spheres have been heated to higher temperatures than chondrules. Investigation of Na loss in chondrules by Tsuchiyama *et al.* (1981) indicates that peak temperatures for chondrules are in the range of 1800°–2200°C. For extreme cases of heating of ablation spheres, the result is the previously mentioned CAT spheres, which have experienced substantial volatilization of Fe, Si, Mg, and more volatile materials. The composition of the CAT spheres is qualitatively similar to residues produced in the laboratory by heating CM matrix to 1800°C for 10 minutes in vacuum (Hashimoto *et al.*, 1979). To produce similar volatilization on the time scale of 5 seconds probably requires a peak temperature above 2000°C. The abundance of CAT spheres in atmospheric ablation spheres has not yet been quantitatively determined because these particles do not have the usual compositional or magnetic properties normally used for identifying extraterrestrial materials in the stratosphere and in deep-sea sediments. In the stratosphere the abundance of CAT spheres is probably more than 10% of the abundance of chondritic composition spheres in the 10 μm size range. While some calcium-aluminum inclusions in carbonaceous chondrites could be evaporative residues, it is clear that the extreme heating required to produce CAT spheres did not play a significant role in producing chondrules in ordinary chondrites. Thus, peak temperatures in the chondrule formation process were both lower and constrained to a narrower range than temperatures produced during meteor ablation.

ELEMENTAL COMPOSITION

We have measured the elemental compositions of the unaltered centers of 260 deep-sea stony spheres in the 0.2 to 1 mm size range by defocused beam microprobe analysis of polished sections. The abundances of Mg, Al, Si, Ca, Ti, and Mn occur with chondritic proportions and, normalized to Si, they cluster around CI/CM values. Histograms for Al/Si and Mn/Si are shown in Figs. 3a and 3b. The relatively small spread of measured values and lack of systematic offsets from CI composition contrasts with measurements of chondrule compositions (McSween, 1977; Lux *et al.*, 1980; Gooding *et al.*, 1980; Hertogen, 1982). Over 75% of the spheres have Mg/Si ratios within 10% of the CI ratio. As can be seen in Fig. 3d, the measured points cluster around a constant Mg/Si line that agrees with CI but is distinct from ordinary chondrites. We interpret these results as an indication that the precursors for the ablation spheres were finer grained than the chondrule precursor materials and that they had a composition similar to CI meteorites.

The other elements that were measured (Na, S, Cr, Fe, Ni) are all depleted relative to CI values. Na and S are usually depleted by a factor of 10 or more and presumably were lost by volatilization. As mentioned previously, the significant depletion of alkalies requires temperatures higher than those that were experienced by chondrules. The depletion of Cr, Fe, and Ni is believed to be due to a loss of a siderophile-rich phase and will be discussed in the following section.

Loss of siderophiles

Ni and Cr, and usually Fe, are depleted by various degrees in nearly all of the stony ablation spherules collected in the stratosphere and in deep-sea sediments. Figure 3d compares sphere compositions with the chondrule compositions of Scott *et al.* (1982). It is seen that the typical Fe abundance is considerably below the CI value but not as low as that in typical chondrules. However, the lowest Fe values are probably underrepresented because the particles were collected magnetically and those with only minor iron contents could not have enough magnetite to permit efficient collection. Magnetite forms in the spherules during heating and oxidation. The spheres in Fig. 3d that have very high iron values probably originally contained a large metal or sulfide grain that was subsequently oxidized and assimilated into the melt.

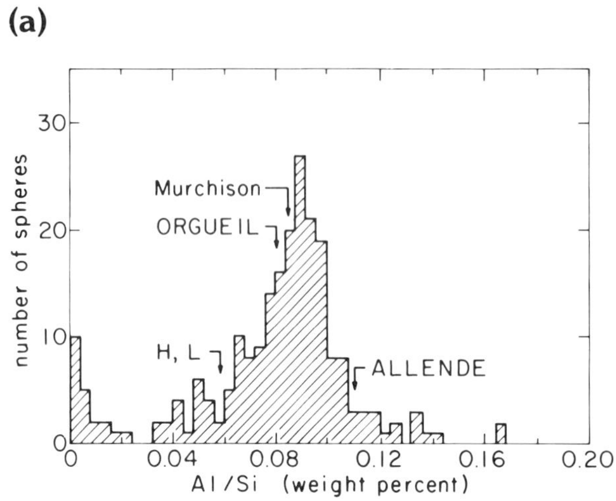

Fig. 3a Results of microprobe analysis of stony spheres using a 30 μm spot on the interiors of spheres unetched by seawater. Distribution of Al/Si (wt.%) histogram is peaked at the Murchison (CM) Al/Si value. The low Al/Si values probably correspond to spheres of monomineralic (olivine) precursors. All spheres were collected from the ocean floor by the cosmic muke rake (Brownlee *et al.*, 1978).

A large range of coupled Ni and Cr depletion is seen in the spheres and the depletions of these elements are correlated (Fig. 3c). The obvious implication is that the spheres lost a siderophile-rich (including Cr) phase during atmospheric entry. The depletion of Ni and Cr in nearly all of the spheres implies that this process was quite common. Rare stony spheres have been found that contain a metal bead. In these cases Ni is concentrated into the metal bead. A dense metal bead could be lost in the majority of cases from a molten silicate sphere by deceleration out of the Earth side of the sphere or by centrifugal spin out.

The remarkable aspect of the siderophile depletion is that it appears to have occurred by loss of a metal bead from a material that otherwise has CI/CM composition (metal-free meteorites). Although oxidation clearly takes place late in the entry process, it can be preceded by a phase of brief but strong reduction. CI meteorites have C/Fe atom ratios near unity, and rapid heating of

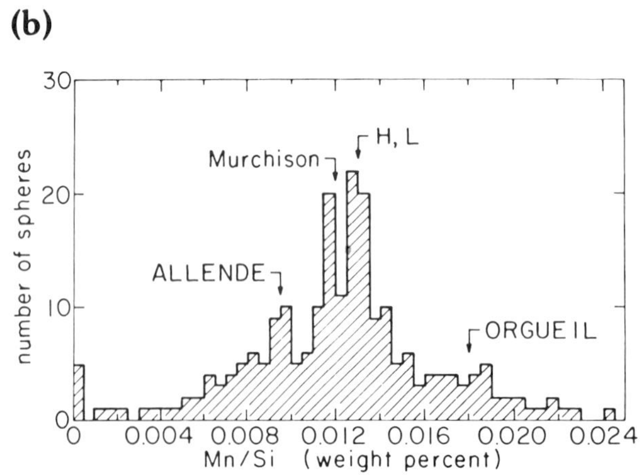

Fig. 3b Histogram of Mn/Si (wt.%) values. The peak, near Murchison (CM) and ordinary chondrite values, is superimposed on a broad background.

(c)

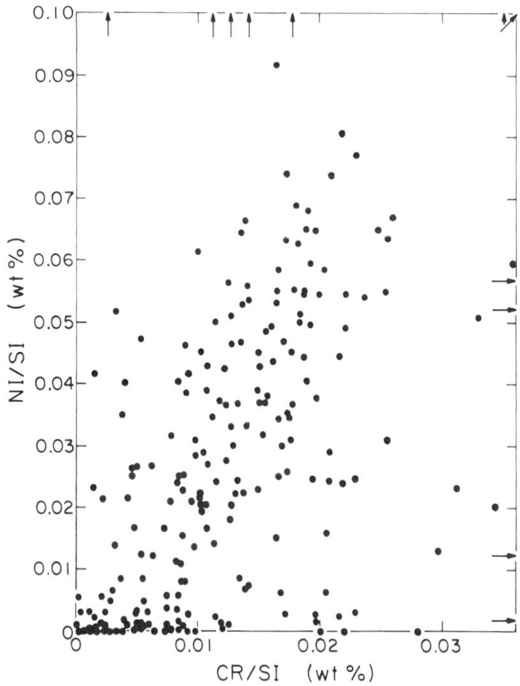

Fig. 3c Plot of Ni/Si (wt.%) and Cr/Si (wt.%) values. Both elements are depleted in spheres compared to chondritic values, yet are correlated. This indicates loss of a Ni,Cr-rich metal phase, probably during a period of intense reduction of precursor material during entry into the atmosphere.

(d)

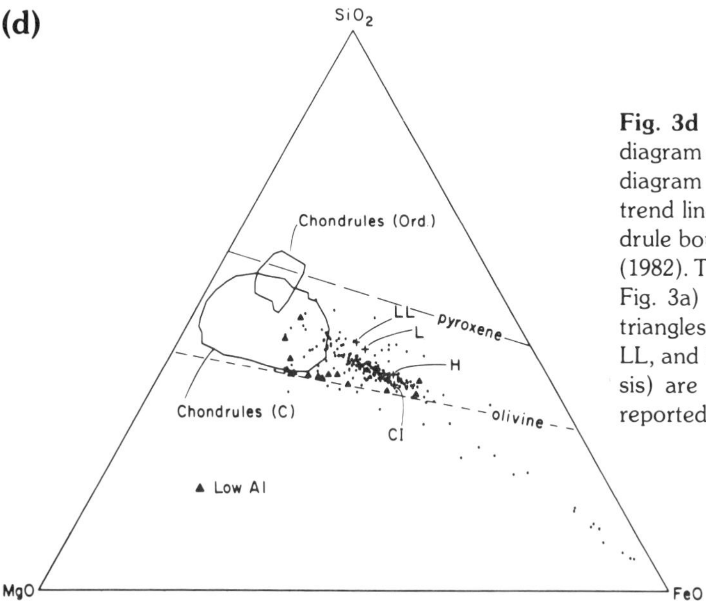

Fig. 3d SiO_2-FeO-MgO ternary diagram of spheres. Shown on the diagram are the olivine-pyroxene trend lines and contours of chondrule boundaries from Scott *et al.* (1982). The low Al/Si particles (see Fig. 3a) are represented as solid triangles. The locations of CI, L, LL, and H chondrites (bulk analysis) are also indicated. All Fe is reported as FeO.

this fine-grained material can efficiently reduce a significant fraction of the iron, Ni, and Cr from the silicate. We envision the sphere precursors to be a fine-grained, highly oxidized CI material in which rapid and intense heating and reaction with carbonaceous material reduces siderophiles from original silicates and produces a particle composed of silicate melt and a small metal spherule. After all the carbon is utilized or lost by vaporization, then the metal would begin to oxidize and reenter the silicate melt. Metal formation in Murchison samples heated in the laboratory has been reported by Matza and Lipschutz (1978) and Hashimoto *et al.* (1979). To test whether this can happen on short time scales we pulse-heated 10 μm particles of Murchison and chondritic micrometeorites with an electron beam in a 10^{-6} Torr vacuum. Within seconds a silicate sphere formed that was covered with Fe-Ni alloy mounds. Similar particles, called Mickey Mouse or metal mound silicate spheres (MMS), have been collected in the stratosphere (Fig. 4) (Brownlee *et al.*, 1982). The metal content of the unmelted stratospheric micrometeorites is very small and the existence of the MMS indicates that metal formation during atmospheric entry does in fact take place. As long as the reducing conditions are strong, then Cr can also be put into the metal phase. If the Cr/Ni correlation in Fig. 4 was not due to *in situ* reduction but due to loss of a pre-atmospheric metal phase, then the metal phase would have to contain a very high Cr content.

We believe that the observed Cr and Ni depletion in type S ablation spheres is due to formation and loss of metal by rapid carbothermic reduction in a fine-grained chondritic composition material. Although many metal beads are seen on the small sphere in Fig. 4, for the rare large spheres that still contain metal beads, the number of beads is limited to one and very rarely to two.

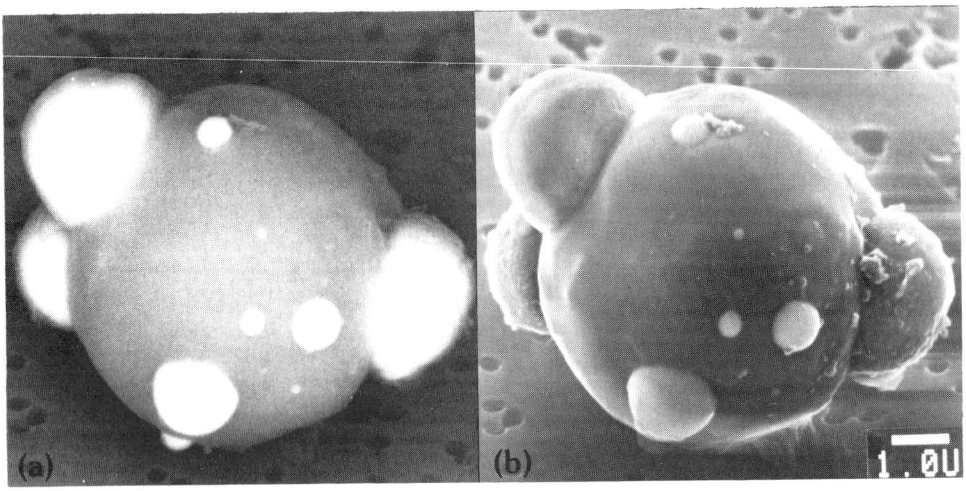

Fig. 4 Conventional SEM and backscatter images of a metal mound silicate sphere (MMS or Mickey Mouse Sphere) collected in the stratosphere. The Fe-Ni metal mounds (bright in backscatter) coat a Ni-free silicate sphere. These particles often occur with carbon and are believed to be products of *in situ* reduction from a carbon-rich chondritic precursor. Scale bar = 1 μm.

In these cases the spheres are distorted, suggesting that inertial effects (centrifugal or linear) were playing a strong role in separating the silicate and metal phases, which differ in density by a factor of 2. When two beads are seen they are on opposite ends of the distorted sphere like oranges glued to opposite points of a football. Inertial forces apparently rapidly coagulate what must have originally been many small metal beads. The strong influence of inertial forces on the distribution and assumed loss of metal from ablation spheres contrasts with the case for chondrules where the casual distribution of metal implies minimal centrifugal or other inertial effects. If reduction from silicates is the major source of metal spheres, then one would expect that the number of metal spheres hitting the Earth's surface would approximately equal the number of silicate spheres. This is consistent with what is observed for 200–600 μm deep-sea spheres. If siderophile depletion in chondrules occurred by an analogous process, then one would also expect that for every silicate chondrule there would be a pure Fe-Ni alloy sphere, but these are not commonly observed in meteorites.

Monomineralic precursors

For spheres produced by simple melting of irregular precursors it should be possible to determine the composition of any monomineralic precursors. To our knowledge no chondrule group has been identified as the melt product of a specific mineral precursor. In contrast, there is evidence that several percent of the ablation spheres were nearly pure olivine grains before atmospheric entry. These spheres are seen as the low Al group in the Al/Si histogram in Fig. 3a. In the Mg-Si-Fe ternary (Fig. 3d), these low Al spheres are plotted as solid triangles and they are seen to have unusual non-chondritic major element compositions that closely match the olivine stoichiometry line. These results indicate that several percent of 300 μm ablation spheres were produced from nearly pure olivine grains that had an appreciable Fe content. Unfortunately, the distribution of points along the olivine line does not faithfully relate the iron distribution in precursor olivine grains because highly forsteritic compositions would not yield magnetite and could not be magnetically collected. The lack of points along the pyroxene line implies that pyroxene precursors in this size range are very rare.

TEXTURE AND MINERALOGY

SEM photos of polished sections of type S deep-sea spherules are shown in Fig. 5 and ultra-thin sections viewed by crossed polars are shown in Fig. 6. All of the observed textures formed at cooling rates of $100° - 1000°$ C s^{-1}; there is no conceivable energy source that could slow the cooling of small meteor ablation spherules. The stony spheres that have been examined are composed of olivine, glass, and magnetite (Blanchard *et al.*, 1980). More than 85% of the spheres have barred olivine textures and most of the remainder have porphyritic textures. The high abundance of barred textures strongly contrasts with chondrites where barred chondrules are an order of magnitude less abundant (McSween, 1977). The bars are composed of interlocked

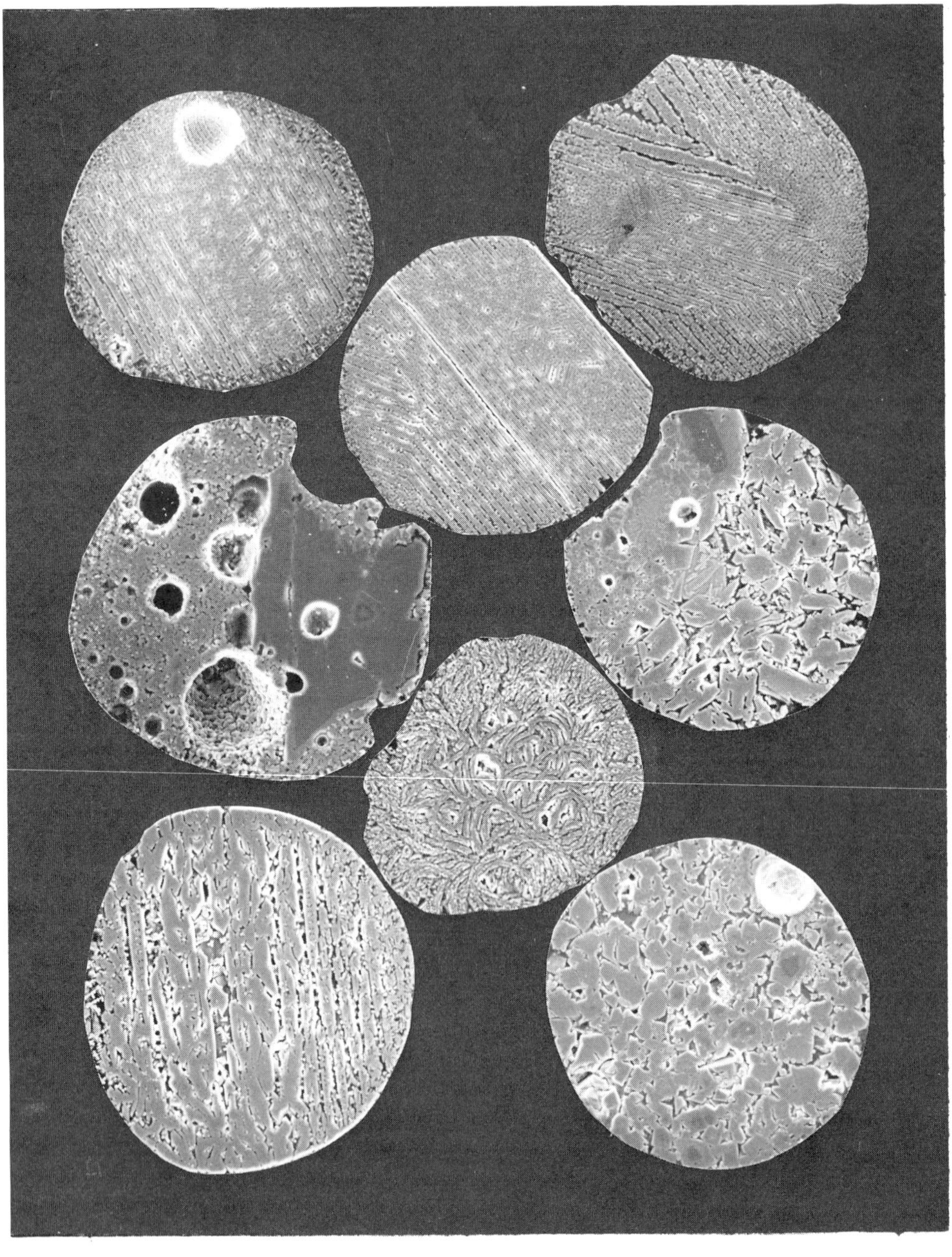

Fig. 5 SEM mosaic illustrating the variety of textures seen in polished sections of type S deep-sea spherules. The two center spheres and one of the corner spheres contain relic forsterite grains.

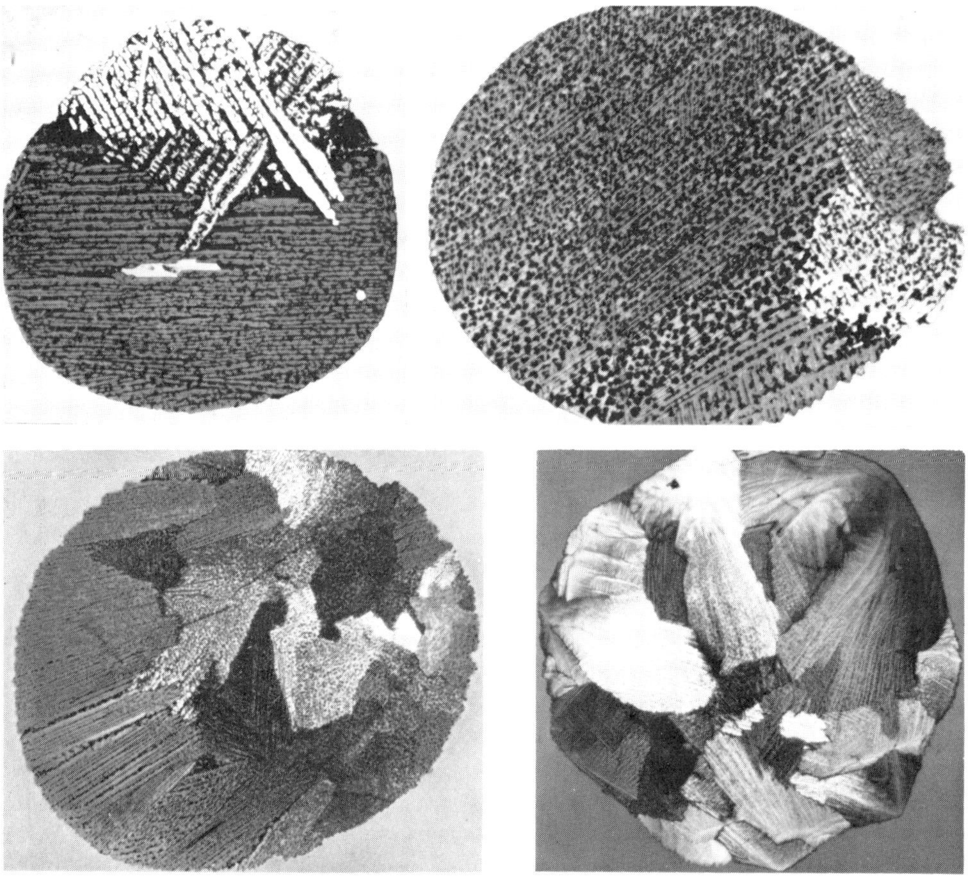

Fig. 6 Ultra-thin sections of type S deep-sea spherules viewed in crossed polars, showing optical continuity in barred olivine textures. The largest sphere has a smooth cavity similar to cavities in related spheres that contain an Fe-Ni bead. The fan-textured grain is primarily olivine and glass as are the other spheres. The opaque phase is magnetite.

micron and submicron olivine crystals (Fig. 7), which have optical continuity over large areas but which are only rarely optically continuous across an entire spherule. With rare exceptions, spheres do not have the circumferential olivine rims common around barred meteoritic chondrules. The barred spheres show a wide range in grain size even for spheres with identical compositions. Because the heating times cannot vary greatly for ablation spheres, textural variations may be the result of nucleation difficulties and supercooling as discussed by Lofgren (1982). The fan texture seen in Fig. 6 is very rare and appears to be a case where nucleation began at the sphere surface. The growth centers for the fans protrude from the sphere surface as faceted mounds giving the sphere surface a turtleback morphology.

Fig. 7 A crossed-polar view of an ultra-thin section of a type S deep-sea spherule. The width of the image is 100 μm.

Porphyritic ablation spheres have been seen with euhedral olivine crystals as large as 50 μm. An important aspect of the porphyritic spherules is that they often contain relic grains while barred spheres never contain relic materials. This implies that barred spheres were heated to higher temperatures. The formation of one texture rather than the other may be related to the peak temperature and the survival of nucleation centers (Lofgren, 1982). For ablation spheres, the preference of barred over porphyritic texture cannot be determined by cooling times because the cooling times of all spheres are short.

A quite remarkable property of the spheres is that with the exception of rare relic grains, olivine is the only crystalline silicate phase. In a detailed SEM study of polished sections of 300 stony deep-sea spherules, Zeitner (personal communication, 1983) found that olivine was the only silicate phase. The fact that the sphere compositions are olivine normative and that the spheres cooled quickly favors olivine formation, but it is extraordinary that olivine forms exclusively. This mineralogy is strikingly distinct from chondrules. The fan-textured sphere in Fig. 6 resembles a radiating pyroxene chondrule, but the fans are in fact composed of olivine.

CONCLUSIONS

Spherule production by aerodynamic melting is an ongoing process in the present solar system and it certainly played a role in the early solar system. While it is possible that this process could have even been the major mechanism for the formation of spherules in the solar nebula, it is clear that aerodynamic melting of the type that occurs in planetary atmospheres did not play a significant role in producing chondrules found in chondrites. Aerodynamic formation of chondrules from pre-solar particles falling into solar nebula gas has been discussed by Wood (1983). Atmospheric meteor ablation spherules are certainly chondrule-like objects, but when examined in detail they significantly differ from chondrules. Some of these differences are:

Ablation Spheres	*Chondrules*
Wide range of thermal effects.	Narrow range of thermal effects.
Large depletion of the alkalies and sulfur.	Only minor depletion of alkalies and sulfur.
Barred texture dominates.	Porphyritic texture dominates.
Inertial loss of metal.	Observed distribution of metal not influenced by inertial forces.
Pyroxene not formed during sphere formation.	Pyroxene common.
Some spheres had monomineralic precursors.	No monomineralic precursors identified.
Non-siderophile and non-volatile element abundances close to chondritic.	Wide range of compositions.

The results of the meteor ablation spherule study do not support chondrule formation models that include flash heating and rapid cooling of fine-grained carbonaceous chondrite precursor materials. The precursors of chondrules must have been coarse-grained and they probably were heated for relatively long periods of time. The apparent lack of highly vesicular chondrules implies either that the precursors did not contain volatiles or that there was sufficient time for complete outgassing and collapse of vesicles. The fact that olivine is the only silicate formed by quenching of ablation spheres may be an indication that other silicates require longer cooling times for nucleation and that chondrules cooled at a much slower rate. The common loss of alkalies, the occasional partial volatilization of Fe and Si, and the occasional survival of only mildly heated materials indicate an extreme range of thermal effects for meteor spheres in comparison with chondrules. The apparent buffering of chondrule peak temperatures argues against violent uncontrolled chondrule formation processes such as aerodynamic heating, shock, or electrical discharge. In the rare stony ablation spheres that contain metal, inertial forces coagulate the metal into a single bead or sometimes two beads located at the end or ends of the long axis of the inertially disturbed prolate spheroid. The lack of inertial effects on the distribution of iron in chondrules implies that chondrules were not rapidly spinning or significantly accelerating at their time of solidification.

Acknowledgment. This work was supported by NASA grant NSG 9052.

REFERENCES

Blanchard M. B., Brownlee D. E., Bunch T. E., Hodge P. W., and Kyte F. T. (1980) Meteor ablation debris in deep sea sediments. *Earth Planet. Sci. Lett.* **46**, 178–190.

Brownlee D. E. (1978) Microparticle studies by sampling techniques. In *Cosmic Dust* (J. A. M. McDonnell, ed.), pp. 295–336. Wiley, New York.

Brownlee D. E. (1981) Extraterrestrial components. In *The Sea*, Vol. 7 (C. Emiliani, ed.), pp. 733–762. Wiley, New York.

Brownlee D. E., Olszewski E., and Wheelock M. (1982) A working taxonomy for micrometeorites (abstract). In *Lunar and Planetary Science XIII*, pp. 71–72. Lunar and Planetary Institute, Houston.

Brownlee D. E., Pilachowski L. B., and Hodge P. W. (1978) Meteorite mining on the ocean floor (abstract). In *Lunar and Planetary Science X*, pp. 157–158. Lunar and Planetary Institute, Houston.

Brownlee D. E., Tomandl D. A., and Olszewski E. (1977) Interplanetary dust; A new source of extraterrestrial material for laboratory studies. *Proc. Lunar Sci. Conf. 8th*, pp. 149–160.

Bruun A. F., Langer E., and Pauly H. (1955) Magnetic particles found by raking the deep sea bottom. *Deep Sea Res.* **2**, 230–246.

Esat T. M., Brownlee D. E., Papanastassiou D. A., and Wasserburg G. J. (1979) Magnesium isotopic composition of interplanetary dust particles. *Science* **206**, 190–197.

Gooding J. L., Keil K., Fukuoka T., and Schmidt R. A. (1980) Elemental abundances in chondrules from unequilibrated chondrites: evidence for chondrule origin by melting of pre-existing materials. *Earth Planet. Sci. Lett.* **50**, 171–180.

Grossman J. N. and Wasson J. T. (1982) Evidence for primitive nebular components in chondrules from the Chainpur chondrite. *Geochim. Cosmochim. Acta* **46**, 1081–1099.

Grossman J. N., Kracher A., and Wasson J. T. (1979) Volatiles in Chainpur chondrules. *Geophys. Res. Lett.* **6**, 597–600.

Hashimoto A., Kumazawa M., and Onuma M. (1979) Evaporation metamorphism of primitive dust material in the early solar nebula. *Earth Planet. Sci. Lett.* **43**, 13–21.

Hertogen J. (1982) Arguments in favor of a volatilization-recondensation model for chondrule formation (abstract). In *Papers Presented to the Conference on Chondrules and their Origins*, p. 25. Lunar and Planetary Institute, Houston.

Lofgren G. E. (1982) The importance of heterogeneous nucleation for the formation of microporphyritic chondrules (abstract). In *Papers Presented to the Conference on Chondrules and their Origin*, p. 41. Lunar and Planetary Institute, Houston.

Lux G., Keil K., and Taylor G. J. (1980) Metamorphism of the H-group chondrites: implications from compositional and textural trends in chondrules. *Geochim. Cosmochim. Acta* **44**, 841–856.

Matza D. S. and Lipschutz M. E. (1978) Thermal metamorphism of primitive meteorites–VII. *Geochim. Cosmochim. Acta* **42**, 1655–1667.

Marvin U. B. and Einaudi M. T. (1967) Black magnetic spheres from Pleistocene beach sands. *Geochim. Cosmochim. Acta* **31**, 1871–1889.

McSween H. Y. (1977) Chemical and petrographic constraints on the origin of chondrules and inclusions in carbonaceous chondrites. *Geochim. Cosmochim. Acta* **41**, 1843–1860.

Millard H. T. and Finkelman R. B. (1970) Chemical and mineralogical compositions of cosmic and terrestrial spherules from a marine sediment. *J. Geophys. Res.* **75**, 2125–2133.

Murray S. and Renard A. F. (1891) *Rept. Sci. Results Voyage H.M.S. Challenger*, Vol. 3, Neill and Co., Edinburg.

Öpik E. J. (1958) *Physics of Meteor Flight in the Atmosphere*. Interscience Tracts on Physics and Astronomy, No. 6. Wiley, New York. 174 pp.

Papanastassiou D. A., Wasserburg G. J., and Brownlee D. E. (1983) Chemical and isotopic study of extraterrestrial particles from the ocean floor. *Earth Planet. Sci. Lett.* In press.

Pettersson H. and Fredriksson K. (1958) Magnetic spherules in deep sea deposits. *Pacific Sci.* **12**, 71–81.

Rambaldi E. R. and Wasson J. T. (1982) Fine, nickel poor Fe-Ni grains in the olivine of unequilibrated ordinary chondrites. *Geochim. Cosmochim. Acta.* **46**, 929–939.

Scott E. R. D., Taylor G. J., and Keil K. (1982) Origins of ordinary and type 3 carbonaceous chondrites and their components (abstract). In *Lunar and Planetary Science XIII*, pp. 704–705. Lunar and Planetary Institute, Houston.

Tsuchiyama A., Nagahara H., and Kushiro I. (1981) Volatilization of sodium from silicate melt spheres and its application to the formation of chondrules. *Geochim. Cosmochim. Acta* **45**, 1357–1367.

Wood J. A. (1983) Formation of chondrules and CAI's from interstellar grains accreting into the solar nebula. *Proc. 8th Symposium on Antarctic Meteorites.* Mem. Natl. Inst. of Polar Res., Spec. Issue. In press.

Chemical State of Pre-solar Matter

Donald D. Clayton
Department of Space Physics and Astronomy, Rice University, Houston, Texas 77001

We consider several features of the chemical state of interstellar matter that may be relevant to the observed properties of chondrules if they are formed from aggregates of pre-existing particles. First, a brief summary of the numbers and types of interstellar particles as indicated by the extinction of starlight is given. Second, a model of the free decay interval required by the low solar abundance of ^{129}I is advanced. This model can support variations in the $^{129}I/^{127}I$ ratio that are larger than typical isotopic anomalies. Third, a new steady-state model of the chemical state of the interstellar medium is presented as a first step toward a quantitative distribution of the abundances among different chemical forms. Last, the advantages of fusing chondrules from aggregates of pre-solar dust by the action of exothermic chemistry are enumerated.

I. INTRODUCTION

Although the chemical state of matter in a cold molecular cloud is very hard to specify with confidence, we can be fairly confident that it must be relevant to the understanding of some questions relating to isotopic composition and chemical disequilibration in early solar system aggregates (Clayton, 1978). This paper describes only a few specific considerations relevant to the types of dust structures that were probably present at the time of solar collapse. Although it is known that refractory elements do not reside in the gaseous state within molecular clouds, one can only speculate on the kinds of dust that must have contained them and in what relative proportions. Refractory dust will exist by virtue of near-equilibrium condensation sequences during mass loss from stars, but by no means will all refractory elements have been in this form. Some atoms will not have condensed at those times, some will have been sputtered away by the interstellar history of supernova shock waves, and some will have been evaporated by recycling through the outer envelopes of stars, followed by partial recondensation upon ejection. Thermal sputtering in the hot interstellar medium (ISM) will accentuate the refractory nature of dust therein and will impose some isotopic fractionation (Clayton, 1981a).

Whether any dust aggregates can occur before cloud collapse remains uncertain. Refractory gaseous atoms will reaccrete onto dust in all ISM phases to some degree, but, in the molecular cloud prior to collapse, virtually all remaining atoms and molecules will reaccrete in cold mantles also bearing volatiles and the possibility of later exothermic chemical energy. Millimeter-sized aggregates of these multicomponent particles can be expected to exist in the turbulent cloud during its collapse to a solar disk, whereafter the thermal and transport properties of the disk will moderate the growth of disk particles. Heating of the first aggregates, even of chondrule size, either chemically or environmentally, will be accompanied by chemical alteration as the aggregate seeks a more mineralized structure. Because water-ice is contained in the chemically reactive grain mantles, some hydrothermal flow will be expected when it suddenly melts. Chemical features resulting from the loss of this water from the pre-chondrule aggregate may still be visible in the chondrule itself.

II. INTERSTELLAR DUST

Many good reviews of interstellar dust exist. Greenberg (1978) presented a particularly good discussion of its many aspects, and the reader interested in the problem of the chemical distribution of dust may profit from his account. Huffman (1977) reviews data on interaction between dust and light. However, it is much less clear how the chemical state of interstellar dust may impact the origin of chondrules. If the initial chemical state had been erased by thermal vaporization, considerations of it would of course not be needed. But if cold-cloud matter remains cold ($T < 100°K$) while it rains down onto a turbulent pre-solar accretion disk, wherein it is turbulently diffused and aggregated (Morfill, 1983), a variety of chemical memories will be distributed through that disk, locked up in the chemical microstructure of the early aggregates. Clayton (1982) has argued that such cosmic chemical memory should be recoverable and constitutes a new field of astronomy, and the reader is referred to that work for the general arguments.

Detailed knowledge of interstellar dust is sparse. As far as the chemical structure of solids is concerned, clear indications are seen in infrared spectra only of water-ice via the 3.07 μm stretching of the O–H bond and of ill-defined silicates via their 9.7 μm transitions. These two features are often seen together, in emission or in absorption, from a variety of astronomical objects, extrasolar and cometary. The ice feature probably arises from mantles acquired by the large refractory particles while they reside in cold interstellar clouds. The relative masses of these two components are hard to estimate reliably, but appear to be of the same order of magnitude. However, because silicate dust cannot comprise 1–2% of the total mass of the ISM, as needed for observed levels of extinction, it seems that the H-, C-, and O-bearing mantles must be somewhat more massive in total.

The extinction of starlight in the visible and ultraviolet (UV) has a characteristic shape that is commonly analyzed in terms of three components, as shown in Fig. 1. Relatively large (0.1–0.2 μm) particles comprised of icy mantles on silicate cores seem to be needed for the infrared (IR) features mentioned above and for the visual extinction feature (part a in Fig. 1) bearing negative second derivative with frequency. A class of small particles, perhaps graphite (Gilra, 1972; Huffman, 1977) or perhaps organic (Wickramasinghe *et al.*, 1977), produces an extinction peak near $\lambda = 2200$ Å. Another class of small particles ($\lesssim 0.01 \mu$m), probably bare and refractory, account for the rising second derivative into the far ultraviolet (part c in Fig. 1). These small particles outnumber the layer core-mantle particles by about 10^3 to 1. The small λ2200-producing particles are somehow physically separable from those producing UV extinction, because the relative amounts of 2200 Å features and far UV extinction differ from one line of sight to another. This observation attests to the common presence of some type of variable sorting in the interstellar

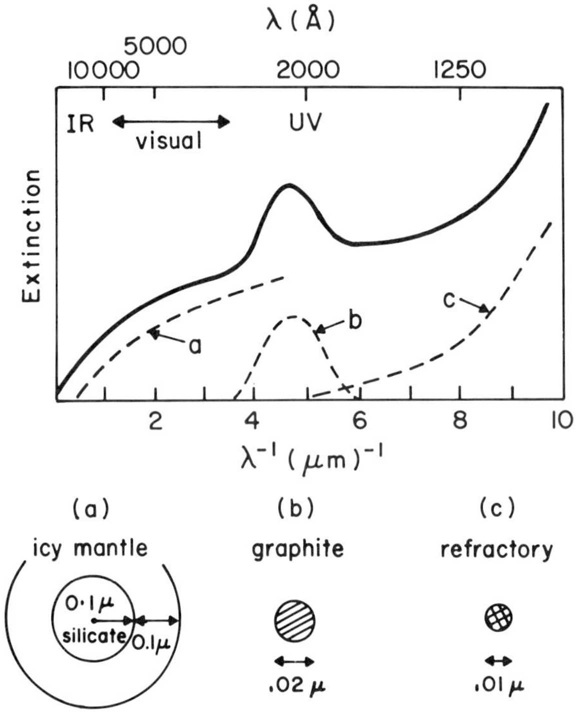

Fig. 1 The variable extinction of starlight along different lines of sight in the ISM has a form like the heavy curve. It is usually interpreted (e.g., Greenberg, 1978) in terms of three types of particles, indicated by a, b, and c. The larger composite particles (a) produce a negative second derivative, whereas the tiny bare refractories (c), at least 10^3 times more numerous, produce a positive second derivative. The λ2200 peak is due to independent small particles, shown here as graphite, but perhaps also carbonaceous polymer.

medium, which suggests that one must not discount grain sorting as a means of chemical fractionation in the solar system itself.

III. INTERSTELLAR MEDIUM PHASES AND EXTINCT RADIOACTIVITIES

The interstellar medium consists primarily of three phases: molecular clouds (perhaps 45% of the interstellar mass), neutral HI clouds (perhaps 45%), and warm media, where $T > 5000°K$ (perhaps 10%). Matter is not fixed in the respective phases, but transmutes from one phase to another. At the same time, the chemical microstructure changes. In the molecular clouds, heavy gas atoms are expected to be non-existent because they strike (and presumably stick to) grains within 10^4 to 10^6 yr. But in HI clouds and warm media, free gaseous atoms of refractory elements exist and are even liberated by sputtering. Similarly, grains seem to be larger in molecular clouds, probably because the icy mantles grow in those cold ($T{\sim}20°K$) and dense ($n_H{\sim}10^4$ cm^{-3}) environs.

These phases change into one another in times probably of order 10^8 yr. This will not allow much difference in the average chemical and isotopic abundances between separate phases *except* for the extinct radioactivities. Their *average* concentrations in the separate phases differ because of time delays. We envision two characteristic time scales. Over times $T_1 \approx 10^8$ yr, the large molecular complexes are grown by amalgamation of small molecular clouds (Kwan, 1979) and, on the same time scale, coordinated star formation breaks them up again into many smaller clouds, which are permeated by HI clouds that surround individual molecular clouds. This cyclic process is naturally accompanied by a slow transfer of material between molecular clouds and the surrounding HI clouds. On another time scale, T_2, the associated large HI clouds around molecular clouds are both condensed from and evaporated to the warmer atomic H medium. Average isolated HI clouds are chemically mixed with warm media owing to evaporation by individual supernova shock waves on time scales of order 10^6 yr. Clearly then, HI clouds must also be continuously formed from the atomic H medium by condensation on the same short time scale. Thus, for purposes of radioactive concentrations, the mass of small to average HI clouds (say, $M_{cloud} < 10^2 M_\odot$) may be regarded as part of the warm ISM, whereas the clouds too large for disruption by a passing supernova shock have a longer lifetime and a different radioactive concentration.

Clayton (1983) has analyzed this type of ISM model in search of an explanation of the small ratio $^{129}I/^{127}I = 10^{-4}$ found in meteorites. If the mixing times $T_1 = T_2 = 10^8$ yr, then $^{129}I/^{127}I$ in molecular clouds is only 7% of its average value over the entire ISM. The HI clouds and the warm media have progressively larger 129/127 ratios. This argument assumes, with supportive reasoning, that the new nucleosynthesis products (viz., ^{129}I) appear first in the hot ISM and then in the warm ISM, as expected if ^{129}I is ejected into supernovae cavities or large stellar wind bubbles around massive star remnants. The point for early chondrule-sized aggregates in a collapsing molecular cloud (e.g., the sun) is this: If there is any spatially inhomogeneous admixture from surrounding or permeating HI clouds or from the warm media, large $^{129}I/^{127}I$ isotopic anomalies can result. We are speaking here of live ^{129}I, not fossil anomalies. For example, the ratio in massive HI clouds is five times greater than in molecular clouds, so that 10% variations in mixing would yield 50% variations in initial iodine. Figure 2 shows a schematic diagram of this cycling process. Chondrules of different "iodine age" in the same meteorite may perhaps be explained in this way, but only if most of the iodine remains in solids during pre-chondrule dust aggregation. Should that be the case, some aggregates should also be found appearing to have excess ^{129}Xe in the trapped component. That is, the "unsupported ^{129}Xe" residing in the grains from pre-solar decay (Clayton, 1975a) should probably be expected to augment the trapped Xe component when that grain aggregate is heated. Since all of these isotopic features are seen in Bjurböle chondrules (Caffee *et al.*, 1982), the relevance of the pre-solar distribution of iodine to the problem must be taken seriously. This same model predicts, by the way, that $^{244}Pu/^{238}U$ should have a molecular-cloud value only about one-third as great as the global average over the entire interstellar medium. This mixing model is advanced in place of the more arbitrary 10^8 yr "waiting period," or "free-decay interval," usually discussed with extinct radioactivities.

Although this isotopic fractionation between ISM phases is hereby expected to be fascinatingly large for the extinct radioactivities, no such fractionation is anticipated among stable isotopes. This is because the overall time scales for nucleosynthesis are much greater than the 10^8 yr mixing times between the ISM phases. Local inhomogeneities owing to local nucleosynthesis may exist for a short time, of course, and many workers choose to interpret stable element anomalies in this way (e.g., Reeves, 1978). I have preferred to interpret oxygen anomalies, say, in terms of differing

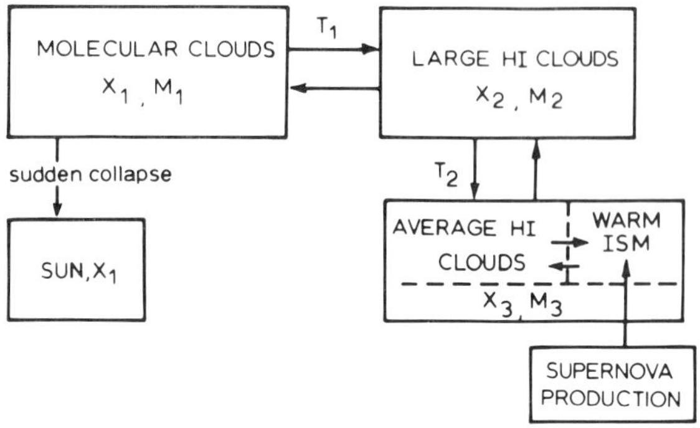

Fig. 2 Schematic diagram for the transport of fresh radioactivities made in supernova explosions into the collapsing solar system. The concentration X_1 in molecular clouds is determined by mass exchange with surrounding large HI clouds on a time scale T_1. The concentration X_2 in the large HI clouds is governed by the time scale T_2 with which they exchange matter with average HI clouds. Because supernova shocks evaporate average HI clouds, their mass is included with that of the hot media in the initial reservoir for fresh radioactivities.

isotopic composition in different types of dust that are distributed homogeneously (Clayton, 1977a, 1978) in space. Although supernovae do occur in molecular clouds, eventually disrupting them, I believe that the nucleosynthesis products remain in a hot phase and are not mixed with the remaining dense cold parts of the fragmenting cloud. To begin to assemble physical expectations for stable elements, it should be more relevant to construct a model of the global steady-state chemical condition of the ISM.

IV. STEADY-STATE INTERSTELLAR MEDIUM

For this discussion, let us temporarily suppress the differences between ISM phases and seek a global chemical balance between production and destruction of chemical forms. What is then relevant is the global rate of production and destruction of chemical forms through physical processes operative in establishing the average chemical state of the ISM. Let us first define time scales for some processes of known importance, along with rough estimates of their values from the astronomical literature. For each element (Z) we must distinguish between at least three chemical forms:

$G(Z)$ = gas mass of Z;

$D_o(Z)$ = dust mass of Z in initial *su*per *no*va *con*densates (SUNOCON's);

$D_m(Z)$ = dust mass of Z in mixed (processed) form;

$M(Z)$ = $G + D_o + D_m$ = total interstellar mass of Z.

In fact, each heavy element Z will at any time exist in a bewildering variety of chemical forms, comprising the total chemical memory of the ISM (Clayton, 1978, 1982). The SUNOCON component (Clayton, 1975a, b, 1978, 1979) must be explicitly included for at least three reasons: (1) it contains special isotopic anomalies characteristic of nucleosynthesis sources (Clayton, 1975a, b, 1978, 1979); (2) it associates the abundance of one element with isotopic anomalies in another element [for example, aluminum with ^{16}O (Clayton, 1977a, b, 1978, 1981a, b)]; and (3) it resides at the centers of refractory cores of composite dust particles where it may be protected from sputtering and moderate thermal effects. The remainder of the dust $D_m(Z)$ is of mixed type, including isotopic mixing during thermal condensation in stellar atmospheres that have not been greatly enriched in fresh nucleosynthesis products (e.g., Si in red giant atmospheres), and chemical mixing owing to accretion of gaseous atoms of Z. The gas G(Z) is of course the mediator between vaporization, either by sputtering or by incorporation into stars followed by reejection, and condensation, either thermally in stellar mass loss or non-thermally in mantle accretion in the ISM.

To build even the simplest model of these components requires time scales for these physical processes:

τ_{SN} = time for supernovae to replenish $M(Z) \approx 4 \times 10^9$ yr;
τ_{ej} = time for stellar mass loss to replenish $M(Z) \approx 4 \times 10^9$ yr;
α_1 = efficiency for Z to condense into SUNOCON's in SN;
α_2 = efficiency for Z to condense into STARDUST in mass loss;
τ_* = time for astration (incorporation of ISM into stars) $\approx 2 \times 10^9$ yr;
τ_s = time for sputtering from $D_m(Z) \approx 1 \times 10^9$ yr;
ϵ = inhibition factor for sputtering $D_o(Z)$ owing to core shielding ≈ 0.1(?); and
τ_G = time for gas G(z) to stick to grains $\approx 10^9 \mathrm{yr}/n_H(\mathrm{cm}^{-3})$.

Several of these quantities depend upon the phase of the ISM being considered. Be reminded that for a preliminary modeling we take these quantities here as weighted averages over the ISM phases. For example, $G(Z) \approx 0$ in molecular clouds because all atoms have adhered to grain mantles, but $G(Z) > 0$ in hot regions or in sputtered regions. Similarly, the time for a gas atom to stick to a grain (τ_G) must be averaged over the gas atoms in the ISM, although its value is about 10^9 yr/n_H in any one medium (Watson, 1975). This effectively means $\tau_G \approx 10^7 - 10^8$ yr, because 10^7 yr is roughly the time required in an HI cloud wherein $n_H \sim 10 - 10^2$. Furthermore, $10^7 - 10^8$ yr is probably also the time required to convert an HI cloud into a molecular cloud, wherein the sticking is rapid; however, much astrophysical modeling remains to be done in order to treat this aspect correctly.

The condensation efficiencies during stellar mass ejection are very uncertain. We have argued (e.g., Clayton, 1982) that they must be almost unity for very refractory elements in order to account for large depletion (e.g., $10^2 - 10^3$ for Al and Ca) from interstellar gas; the following analysis will partially confirm that conclusion but will also call some of our earlier conclusions into question. Because thermal condensation is *probably* the major source of the *refractory components* of interstellar grains, it is of clear potential relevance to the origin of chondrules.

With these definitions, the creation and destruction of the three defined components can be written for each element Z (not explicitly noted) as:

$$\frac{dG}{dt} = -\frac{G}{\tau_*} + (1-\alpha_2)\frac{M}{\tau_{ej}} - \frac{G}{\tau_G} + \frac{D_m}{\tau_s} + \epsilon\frac{D_o}{\tau_s} + (1-\alpha_1)\frac{M}{\tau_{SN}}, \quad (1)$$

$$\frac{dD_o}{dt} = -\frac{D_o}{\tau_*} - \epsilon\frac{D_o}{\tau_s} + \alpha_1\frac{M}{\tau_{SN}}, \text{ and} \quad (2)$$

$$\frac{dD_m}{dt} = -\frac{D_m}{\tau_*} - \frac{D_m}{\tau_s} + \frac{G}{\tau_G} + \alpha_2 \frac{M}{\tau_{ej}} . \tag{3}$$

These equations have a simple sum

$$\frac{d}{dt}(G + D_o + D_m) = -\frac{G + D_o + D_m}{\tau_*} + \frac{M}{\tau_{ej}} + \frac{M}{\tau_{SN}} \tag{4}$$

or

$$\frac{dM}{dt} = -\frac{M}{\tau_*} + \frac{M}{\tau_{ej}} + \frac{M}{\tau_{SN}} , \tag{5}$$

which relates the timescale for consumption by astration to the time scales for replenishment by mass ejection. If the mass M(Z) of element Z in the ISM is either decreasing (owing to the reduction of total ISM mass) or increasing (owing to heavy element nucleosynthesis), we might generally equate $dM/dt \equiv -M/\tau_{ISM}$ to define a timescale for that reduction. Then for self-consistency we must have the time scales related according to:

$$\frac{1}{\tau_*} = -\frac{1}{\tau_{ISM}} + \frac{1}{\tau_{ej}} + \frac{1}{\tau_{SN}} . \tag{6}$$

In this work, however, we examine only steady-state solutions, imagining all time derivatives as vanishing, so that each component is in equilibrium between creation and destruction. Then $\tau_{ISM} \to \infty$ expresses $dM/dt = 0$, so that $1/\tau_* = 1/\tau_{ej} + 1/\tau_{SN}$. The numbers roughly estimated from various literature sources (e.g., Salpeter, 1977) can be reasonably chosen to satisfy this requirement by $\tau_* = 2 \times 10^9$yr and $\tau_{ej} = \tau_{SN} = 4 \times 10^9$ yr. The latter requires special comment, because the galactic rate of mass loss from red giants, about 2 $M_\odot$/yr, is much greater than the rate from supernovae, about 5 $M_\odot$, say, each 40 yr. But because the mass fraction of those elements synthesized in supernovae ($Z \geq 6$) averages some 15 times more in SN ejecta than in unenriched stellar ejecta, the time scales for replenishment of $Z \geq 6$ are comparable in magnitude and are chosen to be 4×10^9 yr for numerical simplicity in the example.

The SUNOCON component is easiest to evalute. The steady state eq. (2) yields:

$$\frac{D_o}{M} = \frac{\alpha_1/\tau_{SN}}{(1/\tau_* + \epsilon/\tau_s)} = \frac{\alpha_1}{2 + 4\epsilon} , \tag{7}$$

where the last equality depends upon the explicit choices made for the time scales listed previously. The fraction of element Z residing in SUNOCON cores ranges from $\alpha_1/2$ if the cores are totally shielded from sputtering ($\epsilon = 0$) to $\alpha_1/6$ if the cores are sputtered as easily as the mixed dust D_m is ($\epsilon = 1$). If the supernovae do not condense that element ($\alpha_1 = 0$) there are of course no

SUNOCON's containing that element. The SUNOCON component also fractionates Al from Mg, as an example, as we have argued in the past (Clayton, 1977b, 1978, 1981a, 1982), owing to two effects. I expect α_1(Al)> α_1 (Mg) because Al condenses first and ϵ (Al) < ϵ (Mg) because Al is more shielded from sputtering than is Mg (for example, see Fig. 3 of Clayton, 1981a). Thus if $\alpha_1(\mathrm{Al}) = 1$ and $\alpha_1(\mathrm{Mg}) = 0.5$, and $\epsilon(\mathrm{Al}) = 0$, whereas $\epsilon(\mathrm{Mg}) = 0.5$, we have a SUNOCON fraction that is 50% for Al but only 12.5% for Mg. Analogous differences for all elements exist; however, there is not nearly enough known to numerically evaluate them. The SUNOCON fractions are also important if one associates ^{16}O with Al, for example.

The steady state for the mixed dust $D_m(Z)$ comes from eq. (3):

$$D_m\left(\frac{1}{\tau_*}+\frac{1}{\tau_s}\right) = \frac{G}{\tau_G} + \alpha_2\frac{M}{\tau_{ej}}, \tag{8}$$

which is not immediately evaluated without considerations of the G/τ_G term. To that end it is more useful to turn to the steady state for the gas from eq. (1), eliminating D_o and D_m by substitution of eqs. (7) and (8). This procedure yields for G(Z):

$$G\left(\frac{1}{\tau_*} + \frac{\tau_s/\tau_*}{\tau_G(1+\tau_s/\tau_*)}\right) = \frac{M}{\tau_{ej}}\left(1-\alpha_2\frac{\tau_s/\tau_*}{1+\tau_s/\tau_*}\right) + \frac{M}{\tau_{SN}}\left(1-\alpha_1\frac{\tau_s/\tau_*}{1+\tau_s/\tau_*}\right). \tag{9}$$

The ratio G(Z)/M(Z) is observed by astronomical absorption in the interstellar medium, and its reciprocal is commonly discussed as "depletion factors" from the interstellar gas (e.g., Salpeter, 1977). Taking $\tau_G = 3\times10^7$ yr for the average gas atom to strike a grain, along with other times mentioned earlier, yields an interesting dependence on condensation efficiencies $\alpha_{1,2}$ and sputtering shielding ϵ:

$$\frac{G(Z)}{M(Z)} \approx 0.02\left(2 - \frac{\alpha_2}{3} - \frac{\alpha_1}{1+2\epsilon}\right), \tag{10}$$

which is of order a few percent for all elements that stick efficiently to cold grains, almost irrespective of the values of α and ϵ. It is at first surprising that rather large variations over the range $0<\alpha<1$ do not more strongly influence the gas concentration, but the answer is evident from the astration of dust ($\tau_* \approx 2\times10^9$ yr) and the assumed rapid sputtering ($\tau_s \approx 1\times10^9$ yr) of the mixed dust condensed during stellar mass loss. Even a protection from sputtering ($\epsilon \rightarrow 0$) in the SUNOCON component makes only a small (say, ~30%) variation in G(Z)/M(Z). This provides little support for our argument that the preferential shielding of Al in SUNOCON cores, in contrast with Mg, accounts for the roughly ten times smaller value of G/M associated with Al *unless* the Al also condenses in protected cores in STARDUST as well (which is quite believable) or *unless* the time scales are seriously misevaluated. A difference in τ_G owing to different sticking coefficients would appear to be at least as effective. Likewise, differences in condensation efficiently are muted. The difference between $\alpha = 0$ and $\alpha = 1$ makes only about a 30% difference. Apparently the complicated differences in depletion have no *single* simple cause. Of course, in this simple model G/M is an average over ISM phases, between which exist large variations, so that it cannot be compared directly to the solar neighborhood observations.

The purpose of this section has been to show the nature of the theoretical models that one must construct in order to interpret the average chemical condition of interstellar matter. From it the differences in the average chemical condition between the separate phases can be computed. And finally, special effects owing to spatial inhomogeneity can be added *if required by chemical data.* I personally remain skeptical of the relevance of spatial inhomogeneity to solar system chemistry, except for that spatial inhomogeneity that is established in the solar disk (Morfill, 1983).

V. CHEMICAL ENERGY

Greenberg (1978) has explained reasons to expect the ultraviolet flux in the interstellar medium to maintain free radicals in the icy grain mantles. Recent experiments (d'Hendecourt *et al.*, 1982) confirm the explosive reaction of grain mantles as they are heated. The icy mantles luminesce and explosively emit gas when they have been first exposed to more than 0.1 UV photons per mantle molecule and then heated from 10°K to 27°K. It is not yet known why T = 27°K triggers the explosive disruption, but it suggests that the icy mantles on interstellar grains may be destroyed by the shock transition associated with the cold infall toward a primitive solar disk, which does heat from 10°K to about 50°K. Nonetheless, these reactive mantles may: (1) create the large interstellar molecules in their explosion in the interstellar medium (d'Hendecourt *et al.*, 1982); (2) create carbonaceous polymer when aggregates of mantled grains are heated in a solar disk; or (3) trigger subsequent exothermic chemistry that may complete the transformation of a millimeter-sized aggregate of primitive grains into a chondrule (Clayton, 1980). The latter paper argued that an amorphous flake of frozen monoxide molecules could in principle liberate 340 kcal/mole by the reaction and crystallization:

$$(MgO + SiO + FeO)_{amorphous} \rightarrow MgSiO_3 + Fe + heat.$$

Since this is roughly ten times the heat input required to melt crystals, even a 10% admixture of such energy sources in a grain aggregate could perhaps produce the appearance of resolidification from a melt if that energy can be liberated rapidly when triggered.

It may be important that many kinds of chemical energy may be available from atoms accreted in the ISM. Oxidation of Si atoms or SiO molecules, as above, should be a major source. Exothermic reactions with molecules like MgH, TiN, or SiC may be important in the brief quasi-melt. Although I am very far from a credible scenario, at least a few obvious advantages of chemical energy for the origin of chondrules should be listed:

(1) Chemical energy should be widely available, because millimeter-sized aggregates should

occur while the ambient temperature is still below 100°K.

(2) The subsequent cooling back to ambient temperature will be fast, leaving many glassy and unequilibrated features.

(3) Oxygen isotopic composition of the parent aggregate will be largely inherited by the chondrule.

(4) Chondrule formation could be fast enough to trap anomalous ^{129}Xe and ^{40}Ar if they are in excess in the dust aggregates.

(5) Fused chondrule pairs or interpenetrating chondrules could in principle occur by collision of the cold aggregates *before* the major exothermic chemistry transforms them, rather than after in a plastic state.

(6) Explosions of balls much larger than chondrules, if such cold aggregates formed, could produce a spray of rapidly cooling drops, exploded by the internal pressure of volatiles.

(7) The bifurcation into reduced (E-type) and oxidized (FeO-bearing) chondrules can be sought in different zones of the aggregation process, depending on the coaggregation of either an oxidizing or a reducing agent in one of the zones.

The real question remains that of whether chondrules can in fact be formed by an exothermic chemical transformation. This writer believes that serious chemical experiments and exothermic reactants will be needed to ascertain this.

Acknowledgments. Many conversations with B. Wieneke have contributed to the ideas of this paper. The research was supported in part by the Robert A. Welch Foundation and in part by NASA grant NSG-7361.

REFERENCES

Caffee M. W., Hohenberg C. M., and Swindle T. D. (1982) Isotopic measurements in individual chondrules (abstract). In *Papers Presented to the Conference on Chondrules and their Origins,* p. 6. Lunar and Planetary Institute, Houston.

Clayton D. D. (1975a) Extinct radioactivities: trapped residuals of presolar grains. *Astrophys. J.* **199,** 765–769.

Clayton D. D. (1975b) ^{22}Na, Ne-E, extinct radioactive anomalies and unsupported ^{40}Ar. *Nature* **257**, 36–37.

Clayton D. D. (1977a) Solar system isotopic anomalies: Supernova neighbor or presolar carriers. *Icarus* **32**, 255–269.

Clayton D. D. (1977b) Cosmoradiogenic ghosts and the origin of Ca-Al-rich inclusions. *Earth Planet. Sci. Lett.* **35**, 398–410.

Clayton D. D. (1978) Precondensed matter: key to the early solar system. *Moon and Planets* **19**, 109–137.

Clayton D. D. (1979) Supernovae and the origin of the solar system. *Space Sci. Rev.* **24**, 147–226.

Clayton D. D. (1980) Chemical energy in cold-cloud aggregates: The origin of meteoritic chondrules. *Astrophys. J. Lett.* **239**, L37–L41.

Clayton D. D. (1981a) Origin of Ca-Al-rich inclusions II. Sputtering and collisions in a three-phase interstellar medium. *Astrophys. J.* **251**, 374–386.

Clayton D. D (1981b) Some key issues in isotopic anomalies: Astrophysical history and aggregation. *Proc. Lunar Planet. Sci. 12B,* pp. 1781–1802.

Clayton D. D (1982) Cosmic chemical memory: A new astronomy. *Quart. J. Roy. Astron. Soc.* **23**, 174–212.

Clayton D. D (1983) Extinct radioactivities: A three-phase mixing model. *Astrophys. J.* **268**, 381–384.

Gilra D. P. (1972) Collective excitations and dust particles in space. In *The Scientific Results from the Orbiting Astrophysical Observatory OAO-2* (A. D. Code, ed.), pp. 259–319. NASA SP-310.

Greenberg J. M. (1978) Interstellar dust. In *Cosmic Dust* (J. A. M. McDonnell, ed.), pp. 187–294. Wiley, New York.

d'Hendecourt L. B., Allamandola L. J., Baas F., and Greenberg J. M. (1982) Interstellar grain explosions: Molecule cycling between gas and dust. *Astron. Astrophys.* **109**, L12–L14.

Huffman D. R. (1977) Interstellar grains: The interaction of light with a small-particle system. *Adv. Phys.* **26,** 129–230.

Kwan J. (1979) The mass spectrum of interstellar clouds. *Astrophys. J.* **229,** 567–577.

Morfill G. (1983) Some cosmochemical consequences of a turbulent protoplanetary cloud. *Icarus* **53,** 41–54.

Reeves H. (1978) The bing-bang theory of the origin of the solar system. In *Protostars and Planets* (T. Gehrels, ed.), pp. 399–425. Univ. of Arizona Press, Tucson.

Salpeter E. E. (1977) Formation and destruction of dust grains. *Ann. Rev. Astron. Astrophys.* **15,** 267–294.

Watson W. D. (1975) Interaction of gas and dust in the interstellar medium. In *The Dusty Universe* (G. B. Field and A. G. W. Cameron, eds.), pp. 113–129. Neale Watson Academic, New York.

Wickramasinghe N.C., Hoyle F., and Nandy K. (1977) Organic molecules in interstellar dust: A possible signature at λ2200 Å, *Astrophys. Space Sci.* **47**, L9–L13.

Oxygen Isotopic Compositions of Chondrules in Allende and Ordinary Chondrites

Robert N. Clayton[1], Naoki Onuma[2], Yukio Ikeda[2], Toshiko K. Mayeda[3], Ian D. Hutcheon[3], Edward J. Olsen[4], and Carol Molini-Velsko[5]

[1]*Enrico Fermi Institute, Department of Chemistry and Department of Geophysical Sciences, University of Chicago, Chicago, Illinois 60637*
[2]*Department of Earth Sciences, Ibaraki University, Mito, Japan*
[3]*Enrico Fermi Institute, University of Chicago, Chicago, Illinois 60637*
[4]*Field Museum of Natural History, Chicago, Illinois 60605*
[5]*Department of Chemistry, University of Chicago, Chicago, Illinois 60637*

The ferromagnesian chondrules in Allende follow a trend in the oxygen three-isotope plot that diverges significantly from the ^{16}O-mixing line defined by light and dark inclusions and the matrix of the meteorite. The trend probably results from isotopic exchange with an external gaseous reservoir during the process of chondrule formation sometime after the establishment of the isotopic compositions of the inclusions and matrix. The Allende chondrules approach, but do not reach, the isotopic compositions of chondrules in unequilibrated ordinary chondrites, implying exchange with a similar ambient gas, but isotopically different solid precursors for the two types of meteorite.

INTRODUCTION

Variations of oxygen isotopic abundances among solar system objects appear to be due primarily to two types of processes: (1) mixing of reservoirs with different degrees of ^{16}O-enrichment, and (2) mass-dependent fractionation. Reservoir mixing is especially evident in meteorites when the scale of sampling is less than a centimeter. Thus the largest ^{16}O variations in Allende are revealed in mineral separates from calcium-aluminum rich inclusions (CAI's), and the largest variations in ordinary chondrites are found in analyses of individual chondrules and clasts of the unequilibrated ordinary chondrites (UOC's). In both cases, the isotopic heterogeneity may be due to incomplete isotopic exchange between a solid (or liquid) condensed phase and a surrounding gas phase. The event during which this exchange took place required high temperatures (to permit exchange) and short times (to prevent complete equilibration), and thus may be the chondrule-forming event itself. Detailed oxygen isotope studies of chondrules may give information on the nature of precursor materials and on the kinetics of the melting and freezing processes.

Anhydrous phases from the C2 and C3 carbonaceous chondrites form an ^{16}O-mixing line with an experimentally determined slope of 0.94 and a range in $\delta^{18}O$ and $\delta^{17}O$ of about 45‰ (Clayton *et al.*, 1977). Chondrules from UOC's form a second ^{16}O-mixing line, displaced toward lower ^{16}O, with a slope of 1.00. This latter line is determined best from clasts of the LL3 meteorite ALHA 76004 (Mayeda *et al.*, 1980), which cover a range of 10‰, and also is the best fit to the data from 33 chondrules from 10 different UOC's, which cover a range of about 3‰ (Clayton *et al.*, 1981). These two ^{16}O-mixing lines intersect at a value of $\delta^{18}O$ near −40‰, the apparent isotopic composition of the precursors of Allende CAI's. It has therefore been suggested that the carbonaceous chondrite minerals and the UOC chondrules might have had a common ^{16}O-rich source of pre-melting solids that subsequently exchanged with gases of different compositions (Clayton, 1981).

Dodd (1981) has maintained that chondrules in carbonaceous chondrites have a different origin from chondrules in ordinary chondrites. Most of the chondrules of carbonaceous chondrites are thought to have formed from melt droplets, which were produced either by condensation or by melting of solids. He believes that chondrules from UOC's, on the other hand, have had a more complicated history, usually involving fragmentation of larger solid bodies, with or without a final remelting. An oxygen isotope study comparing chondrules of these two meteorite groups may reveal the similarities and differences in their origins.

SAMPLES AND ANALYTICAL METHODS

Two suites of samples from Allende were analyzed, one selected and characterized by N. Onuma and Y. Ikeda (identified by the symbol NO in the data tables), and one selected and characterized by I. D. Hutcheon (identified by IH in the data tables). Chondrules were split in two, one half mounted in epoxy and polished for petrographic and chemical characterization, and one half used for isotopic analyses. The NO samples were studied in thin section by petrographic microscope and were analyzed chemically by electron microprobe, using a 50 μm defocused beam for whole chondrule analysis. The IH samples were studied in polished section in the scanning electron microscope (SEM), using back-scattered electron images and energy-dispersive X-ray analysis using a focused beam rastered over a 50 × 50 μm area. Bulk silicate compositions were calculated from metal-free areas. Most samples were round olivine- or pyroxene-rich chondrules, but several lithic inclusions and dark inclusions were also analyzed. Oxygen isotopic data on chondrules from UOC's have been discussed previously (Clayton *et al.*, 1981), and are not tabulated here. Textural and chemical characterization of a selected subset of these chondrules was done by E. Olsen, using SEM and electron microprobe.

Oxygen isotopic analyses for $^{18}O/^{16}O$ and $^{17}O/^{16}O$ ratios were carried out on O_2 samples prepared by the BrF_5 reaction on crushed fragments of individual chondrules. Due to limitation of sample size, replicate analyses were not possible. Silicon isotopic analyses for $^{30}Si/^{28}Si$ and $^{29}Si/^{28}Si$ ratios were carried out on SiF_4 samples also produced by the BrF_5 reaction.

RESULTS AND DISCUSSION

Isotopic data for the Allende samples are given in Table 1 and are plotted, along with the data on chondrules from UOC's, in Fig. 1. The line labelled "Allende mixing line" in Fig. 1 was determined from earlier analyses of separated phases from CAI's (Clayton *et al.*, 1977), and not by the data in this paper. The new Allende samples fall into two groups: the dark inclusions, which fall on the extrapolation of the CAI mixing line, and the chondrules, which fall above the CAI mixing line. The chondrule trend terminates in a cluster of six samples, with indistinguishable isotopic compositions, at $\delta^{18}O = 3.7‰$, $\delta^{17}O = +1.3‰$. All of these, and two additional chondrules only slightly lower in $\delta^{18}O$ and $\delta^{17}O$, are of the barred olivine or radial pyroxene texture types. Microporphyritic and granular chondrules all have significantly lower $\delta^{18}O$ and $\delta^{17}O$.

Table 1. Oxygen and silicon isotopic compositions of Allende chondrules and lithic inclusions.

Sample No.	$\delta^{18}O‰$ *	$\delta^{17}O‰$	$\delta^{30}Si‰$ *	$\delta^{29}Si‰$	Wt. % FeO	Texture type †
			Chondrules			
IH-11	3.82	1.32	−0.52	−0.25	9.3	B
IH-25	3.63	1.23	−0.21	−0.12	13.5	B
IH-8	3.82	1.28	−0.65	−0.31	5.0	B
IH-7	3.67	1.25	−0.33	−0.17	13.8	B
NO-11	3.77	1.45	−0.73	−0.35	13.0	B
NO-13	3.68	1.50	−1.06	−0.54	13.7	B
IH-24	3.20	0.42	−0.92	−0.46	4.0	B
NO-6	3.13	0.14	−1.14	−0.58	–	B
IH-10	2.14	−1.15	−0.54	−0.26	4	I
IH-1	2.12	−1.32	−0.85	−0.47	5	I
IH-5	1.70	−2.13	−0.93	−0.51	5	II
IH-22	1.23	−2.25	−0.46	−0.32	4	I
NO-1	0.84	−2.87	−0.17	−0.04	6.5	II
NO-2	1.20	−2.42	−0.50	−0.23	6.1	II
NO-4	1.47	−2.12	–	–	5.4	II
NO-12	0.11	−3.32	−0.94	−0.47	8.1	II
AL6-S-4‡	0.29	−3.48	−0.19	−0.09	–	B
IH-4	−0.92	−4.45	−0.04	−0.05	5.6	I?
IH-15	−0.97	−4.56	–	–	5.9	?
L-211‡	−1.09	−4.67	−0.52	−0.26	5.1	–
NO-15 core	−1.31	−4.77	–	–	–	L
NO-15 mantle	0.74	−2.42	–	–	–	L
HN-GC‡	−2.65	−6.51	−0.36	−0.18	–	II?
		Lithic and Dark Inclusions and Matrix				
Al-2‡	3.61	−0.97	–	–	–	D
Al-3‡	4.62	−0.21	−0.69	−0.32	–	D
Al-5‡	2.48	−1.91	−0.50	−0.24	–	D
NO-8B	1.40	−2.42	−1.40	−0.72	16.4	L or D
NO-9	3.72	−0.89	−0.89	−0.43	32.3	M
NO-10	5.24	0.81	−1.23	−0.63	32.4	D
NO-14	0.99	−2.47	–	–	–	D
NO-7	0.22	−3.54	−0.50	−0.10	–	L or D

* $\delta^{18}O$, $\delta^{17}O$ relative to SMOW; $\delta^{30}Si$, $\delta^{29}Si$ relative to NBS 28 quartz.

† B = barred olivine and radial pyroxene; L = lithic; D = dark inclusion; I, II classes of McSween (1977); M = matrix.

‡ From Clayton *et al.* (1977).

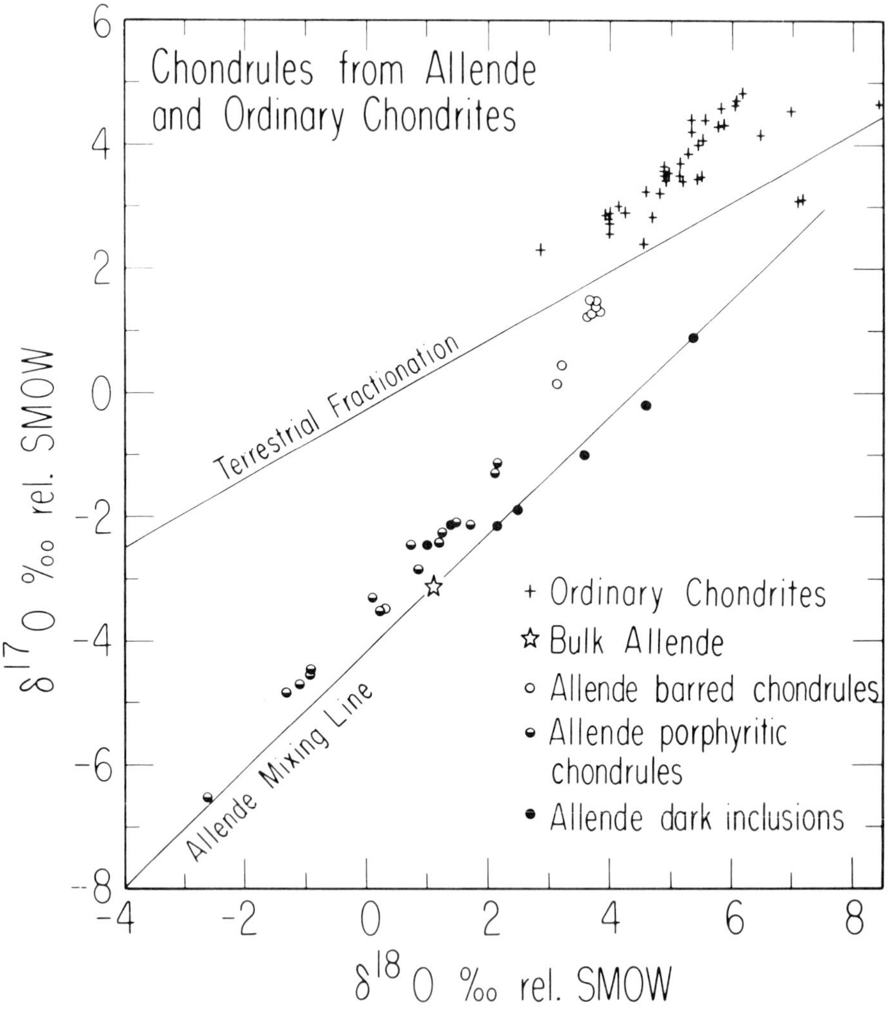

Fig. 1 Oxygen isotopic compositions of chondrules from Allende and several ordinary chondrites (Allegan, Bishunpur, Bjurböle, Borkut, Cape Girardeau, Chainpur, Chervattez, Dhajala, Hallingeberg, Krymka, Semarkona, Tieschitz). The "Allende mixing line" is determined by the compositions of both calcium-aluminum-rich inclusions (mostly far off-scale to the lower left) and dark inclusions, shown as solid circles. Allende olivine-pyroxene chondrules lie above this line and trend towards the compositions of chondrules from ordinary chondrites. Chondrules with barred or radiating textures cluster at the upper end of the trend, whereas porphyritic chondrules lie closer to the mixing line. Two isotope exchange or mixing events are implied for the Allende sample, the first producing the "Allende mixing line," and the second producing the chondrules from the products of the first. The mass-dependent fractionation line for terrestrial samples is shown for reference.

The oxygen isotopic compositions of CAI's are heterogeneous on a scale of micrometers, and vary systematically with mineralogy. The mixing lines representing these heterogeneities have been interpreted as resulting from post-crystallization exchange of oxygen isotopes between the solids and an ambient nebular gas (Clayton and Mayeda, 1977). The solid is presumed to have been ^{16}O-rich, and the gas ^{16}O-depleted. If the same mechanism operated for the Allende chondrules, then those with the highest $\delta^{18}O$ and $\delta^{17}O$ values have undergone the greatest extent of exchange with the gas. The correlation with texture could then reflect the degree of melting of the various chondrules during chondrule formation: complete melting and extensive exchange for the barred and radial chondrules, and incomplete melting and incomplete exchange for the granular and microporphyritic chondrules. One exceptionally large, 8-mm-diameter chondrule with radial texture, AL6-S-4, was found to have $\delta^{18}O = +0.29‰$, suggestive of incomplete isotopic exchange. Another very large, olivine-rich chondrule, HN-GC, has the lowest $\delta^{18}O$ and $\delta^{17}O$ of the group. Further evidence for the direction of the exchange is provided by chondrule NO-15, which has a concentric structure with a microporphyritic core ($\delta^{18}O = -1.3‰$) surrounded by a fine-grained mantle ($\delta^{18}O = +0.7‰$). The core appears to have undergone less isotopic exchange than the mantle.

If the Allende chondrule mixing line and the CAI (plus dark inclusions) mixing line both represent gas-solid exchange in some high-temperature event, the fact that there are two distinct trends requires two distinct events, involving gas reservoirs of different isotopic compositions.

Although there is no overlap in isotopic composition between the Allende chondrules and chondrules from UOC's, the two groups approach one another closely and might form a single trend. If the ^{16}O-mixing array for the UOC chondrules is also the result of a gas-solid or gas-liquid isotopic exchange, it must be determined which reservoir was ^{16}O-rich and which ^{16}O-depleted, i.e., in which direction the chondrules were displaced from their initial compositions. It was argued previously (Clayton, 1981) that the solid precursors of UOC chondrules were ^{16}O-rich, and might have had the same isotopic composition as the most ^{16}O-rich minerals in CAI's in Allende. However, no firm evidence is available to determine the direction of exchange of the UOC chondrules. The apparent convergence of UOC chondrules and Allende chondrules could result if both sets of chondrules had exchanged with a common external reservoir that had a composition lying in the narrow gap between them. The existence of a reservoir with such an isotopic composition would not be surprising, since that gap contains the isotopic composition of almost all of the achondrites, as well as the Earth and the moon.

In Allende, textural evidence for complete melting is characteristic of the chondrules with the greatest extent of isotopic exchange. We do not have similar textural information on all of the UOC chondrules to determine whether texture type is related to oxygen isotopic composition, so that such information could be used to identify the direction of exchange. However, for the UOC chondrules at the low $\delta^{18}O$ and $\delta^{17}O$ end of the range ($\delta^{17}O < 3.5‰$), nine of the eleven with known textures have either barred or radial textures, suggestive of complete melting. Thus, there is fairly strong evidence that the UOC chondrules became *depleted* in heavy isotopes rather than enriched. The observed convergence in isotopic compositions of UOC chondrules and Allende chondrules could then reflect their exchange with the same external reservoir, but derivation from quite different solid precursors.

The only correlation between chemical composition and oxygen isotopic composition of Allende chondrules is the tendency of the isotopically heavy group to be iron-rich, with FeO $\gtrsim 9\%$. The correlation is not unique, as one barred chondrule with high $\delta^{18}O$, IH-8, has only ~5% FeO. The correlation may simply reflect the effect of iron in decreasing melting temperatures, thus facilitating isotopic exchange.

Silicon isotopic compositions for most of the Allende chondrules are given in Table 1. On a three-isotope graph, the data fall on a mass-dependent fractionation line of slope 0.505. The range of $\delta^{30}Si$ is from 0 to −1.4‰, and averages −0.66‰, compared to the value of −0.80‰ for bulk Allende (all relative to NBS 28 quartz standard). The range of variation in silicon isotopic composition is much smaller than that of CAI's in Allende (Molini-Velsko *et al.*, 1983). There is no correlation between $\delta^{30}Si$ and $\delta^{18}O$ ($r = -0.10$) and no evident correlation between silicon isotopic composition and chemical or textural properties of the chondrules. This fact is consistent with our interpretation that the oxygen isotope variations result from a two-reservoir exchange process in which the gaseous reservoir contains oxygen compounds (CO and H_2O) but no silicon compounds. Thus the silicon isotopic variations result from fractionation effects in the chondrule precursors.

CONCLUSIONS

Subsamples of Allende form two different oxygen isotope trends: (1) a mixing line involving the CAI's, dark inclusions, and the matrix, and (2) a line involving the olivine- and pyroxene-rich chondrules. The chondrule line appears to be a branch from the inclusion line, intersecting near the composition of the bulk meteorite. Thus the "Allende mixing line" must have been formed first by exchange of ^{16}O-rich solids with an ^{16}O-poor gas, and then in a later event, some of the material was converted to chondrules and underwent exchange with a gas of different isotopic composition. This second gas also provided an exchange reservoir for the chondrules of the UOC's. The second gas reservoir was very similar in isotopic composition to the one that controlled the isotopic compositions of the achondrites, the moon, and Earth. The existence of gaseous reservoirs with different oxygen isotopic compositions does not require inhomogeneity in the gas phase, but could result from modification of the gas composition with time due to the various processes of gas-solid exchange. However, this model is more complicated than that previously proposed for chondrule precursors (Clayton, 1981), in that two isotopically different solid reservoirs are required: an ^{16}O-enriched material as the source of carbonaceous chondrites, and an ^{16}O-depleted material as the source of the ordinary chondrites.

It is essential that more isotopic and petrographic data be obtained for chondrules from UOC's to determine the direction of their isotopic displacement and the isotopic compositions of their precursors. Data on other carbonaceous chondrites are also needed. It should then be possible to establish the sequence of chondrule- and chondrite-forming processes for the different groups of chondrites.

Acknowledgments. This research was supported in part by NASA grant NGL 14-001-169 and NSF grant EAR 78-23680. Discussions with L. Grossman, G. J. MacPherson, and A. M. Davis were helpful.

REFERENCES

Clayton R. N. (1981) Isotopic variations in primitive meteorites. *Phil. Trans. Roy. Soc. London* **A303**, 339–349.

Clayton R. N. and Mayeda T. K. (1977) Correlated oxygen and magnesium isotope anomalies in Allende inclusions, I: oxygen. *Geophys. Res. Lett.* **4**, 295–298.

Clayton R. N., Mayeda T. K., Gooding J. L., Keil K., and Olsen E. J. (1981) Redox processes in chondrules and chondrites (abstract). In *Lunar and Planetary Science XII*, pp. 154–156. Lunar and Planetary Institute, Houston.

Clayton R. N., Onuma N., Grossman L., and Mayeda T. K. (1977) Distribution of the pre-solar component in Allende and other carbonaceous chondrites. *Earth Planet. Sci. Lett.* **34**, 209–224.

Dodd R. T. (1981) *Meteorites, a Petrologic-Chemical Synthesis*. Cambridge University Press, Cambridge. 368 pp.

Mayeda T. K., Clayton R. N., and Olsen E. J. (1980) Oxygen isotope anomalies in an ordinary chondrite (abstract). *Meteoritics* **15**, 330–331.

McSween H. Y., Jr. (1977) Chemical and petrographic constraints on the origin of chondrules and inclusions in carbonaceous chondrites. *Geochim. Cosmochim. Acta* **41**, 1843–1860.

Molini-Velsko C., Mayeda T. K., and Clayton R. N. (1983) Silicon isotopes in components of the Allende meteorite (abstract). In *Lunar and Planetary Science XIV*, pp. 509–510. Lunar and Planetary Institute, Houston.

Crystallinity, Recrystallization, Equilibration, and Metamorphism in Chondrites

Kurt Fredriksson

Department of Mineral Sciences, Smithsonian Institution, Washington, DC 20560

"Indeed the confused structure shown by many meteoric stones has given rise to an equal confusion on the part of students as to their causes."

—Merrill, 1929

A partial history of the controversial concept of thermal equilibration in chondritic meteorites is reviewed. Four typical H-group chondrites were selected to illustrate the difficulties with assigning chondrites to petrographic types. Textural studies of ultrathin sections of Allegan (H5) along with data on its ultrafine matrix show that even this apparently "homogeneous" rock did not undergo post-accretionary, secondary recrystallization or equilibration; rather, it is unequilibrated and certainly not a metamorphic rock. Analyses of individual chondrules from Allegan (H5), Ochansk (H4), Tieschitz (H3), and the complex breccia Weston (H3,4,5,6,7?) exemplify the confusion in applying the terms "equilibrated" and "unequilibrated" to chondrites. In the first two samples Fe/Mg is constant (by definition), while Al, Ca, and Ti are notably variable as are their *ratios* (by a factor of >10); the reverse occurs in the latter two. From these differences, especially between Allegan and the "primitive" Tieschitz, it is inferred that few, if any, chondrites are parts of metamorphic rock sequences. Because most chondrites are breccias, or "mixtures," containing fragments with varying degrees of crystallinity and recrystallization, the concept of thermal metamorphism and the petrographic type classification are still controversial; more importantly, they are possible deterrents to the understanding of the origin of chondrites.

INTRODUCTION AND SCOPE

Most early students of meteorites, mainly petrologists, recognized the fragmental and brecciated nature of chondrites; some even considered extraterrestrial impacts as the cause of brecciation. The varying degrees of crystallinity and induration, both among different stones and within single chondrites, were also commonly noted. Furthermore, it was considered a fact that chondrules, distinguished from lithic fragments, once had been individual droplets that crystallized and subsequently accreted with comminuted chondrules, other fragments, and dust to form chondrites.

Merrill (1929), in his last work, reviewed much of these earlier studies and wrote about chondrites, "Occasionally meteorites of this tuffaceous and chondritic type show signs of changes subsequent to their consolidation, which are comparable *to a form* of metamorphism produced in terrestrial rocks by heat." However, in his "Handbook" (Merrill, 1916) he was already patently aware that conventional thermal metamorphism could not be implied, "since a heat sufficient to render crystalline the pisolites [~corresponding to fragments in chondrites] in a tuff, as argued by some, would certainly produce a more marked degree of metamorphism in the surrounding matrix." Earlier, in his description of San Emigdio (Merrill, 1888), he stated, "There are no evidences to indicate that after the first period of solidification and crystallization was brought to a close by cooling, there was a second rise in temperature sufficient to allow certain of the silicate constituents to take on more perfect forms." These perceptive arguments are still viable.

In the early 1950's Wahl (1952) reemphasized the brecciated structure of many chondrites. However, being unaware of *in situ* high pressure shock transformations, he exaggerated the abundance of polymict breccias over monomict, which are now described as genomict types (Wasson, 1974, p. 76). The importance of the breccia concept has recently been reviewed in great detail by Keil (1982, 171 references).

The renewed interest in chondrites led to the discovery of the remarkably similar bulk composition of chondrites (Urey and Craig, 1953), exclusive of total and oxidized iron. Later, through the work of Mason (1963) and others (e.g., Keil and Fredriksson, 1964), the paradoxical constancy in composition of olivines and pyroxenes in most individual chondrites was established. Also, during this period many considered the possibility of a common chondrite parent material, and Wood (e.g., 1962, 1963) advanced a radical metamorphic model implying a direct relation between chondrites of different crystallinity with exchange and equilibration of Fe and Mg between Fe-rich, "oxidized" matrix and the chondrules. This concept, even as moderated by Dodd (1969), was questioned not only by petrologists but also by some chemists and rare gas specialists (Suess and Wänke, 1967). However, the "metamorphic" concept provided the basis for a petrologic classification scheme for chondrites within the different chemical (iron) groups (Van Schmus and Wood, 1967). Because of its simplicity [a letter for chemical group and a number 1–6(7) for petrologic type], this classification rapidly gained wide usage. Indeed it is useful for curators in quickly describing and comparing different specimens and in searching for like samples, particularly the now popular type 3 ("unequilibrated") chondrites. In spite of later warnings by the authors, the 1 through 6 grades (petrographic vs. petrologic; Van Schmus, 1969) were sometimes improperly used as a quantitative, if not linear, scale. Thus various correlations such as higher rare gases and volatile elements in the lower types were searched for and sometimes found to exist (for references, see Wasson, 1974). However, Tandon and Wasson (1968) reported data showing that, contrary to expectations, In and Ar, geochemically quite different, experienced similar fractionation with the In/Ar ratio remaining constant throughout the petrologic sequence. Although the volatilization is difficult to rationalize, they concluded this could have happened via a metal-silicate fractionation during the accretion of the parent body, which apparently also necessitated a closed system. Still, the linear relationship of In and Ar contradicts the metamorphic sequence concept and remains unexplained.

Other researchers who have also observed correlations between volatiles and some of the petrographic types occasionally differ in their analytical interpretations (for reviews and references see, e.g, Lipschutz and Ikramuddin, 1978; Lipschutz *et al.*, 1982; Alaerts and Anders, 1979). Some of the results require an open system (at least for a few volatiles), others apparently a closed system, while a few need both. These different requirements may be explained in part by the fact that modern research, carried out with sophisticated techniques, often deals with small samples (fragments, individual chondrules, and matrix grains), whereas in Merrill's time researchers considered whole rocks.

Studies regarding evidence for metamorphism, including heat sources, have been summarized by Wasson (1974, pp. 190–193) and more recently by Dodd (1981, pp. 89–103). However, chondrites are breccias (or at least "mixtures") that may contain fragments recrystallized to different degrees, chondrules, and even "primitive" (perhaps "primordial") matrix materials that, although extremely fine-grained, exhibit no recrystallization and are in complete disequilibrium with other constituents (Rambaldi *et al.*, 1981). These observations alone, reviewed briefly below, should make it clear that chondrites are not metamorphic rocks.

It is not within the scope of this paper to review all of the numerous arguments for or against thermal equilibration in various chondritic meteorites; references included should be sufficient for those interested. Rather, the purpose here is to present evidence indicating that at least one large, "simple," apparently homogeneous, "equilibrated" chondrite is in fact chemically and structurally *unequilibrated*, did not become recrystallized or thermally equilibrated after the final accretion, and never existed in a state similar to material classified as grade 3.

CHONDRULES VS. CHONDRITES

The rather large (~35 kg), "homogeneous" Allegan meteorite fell in 1899. Extensive (~1/10 m^2), fresh, broken surfaces of different pieces in the Smithsonian collection show minimal brecciation or shock effects, only millimeter-sized chondrules and fragments in an evenly grey groundmass. Mason and Maynes (1967) have published a petrographic and chemical description. Several analyses of small samples indicate that the meteorite is chemically homogeneous on the <0.5 g scale, and olivines and pyroxenes (in grains large enough for probe analyses; Keil and Fredriksson, 1964; and others) are constant in composition. Figure 1 shows the structure in a polished ultrathin (~4 μm) section. Note the considerable variation in grain size of olivine and pyroxene (<1 μm to >0.5 mm), the skeletal pyroxenes, and the cryptocrystalline matrix in the chondrules—all of which refute the possibility of extensive recrystallization. Such a process should affect smaller grains by increasing their average size, which involves a chemical equilibration between them; this has not occurred.

Rambaldi *et al.* (1981) included Allegan in an analysis of the ultrafine matrix, "Holy Smoke," from six different chondrites. They found volatile (e.g., Na, Zn, Sb) and refractory (e.g., W, Ir, Au) elements strongly enriched relative to the bulk, even to C1 chondrites. Also, some approximately micron-sized olivine or pyroxene-like grains were found to have variable Fe/Mg ratios in contrast to the constant composition of the larger olivines and pyroxenes. This finding contradicts a final chemical equilibration, and indicates that Allegan is an accretionary rock primarily composed of pre-existing chondrules and lithic and mineral fragments "equilibrated" with regard to Fe/Mg, but which also contains small amounts of "unequilibrated dust" perhaps similar to "cosmic" dust or some "Brownlee (1978) particles." Thus the accretionary and

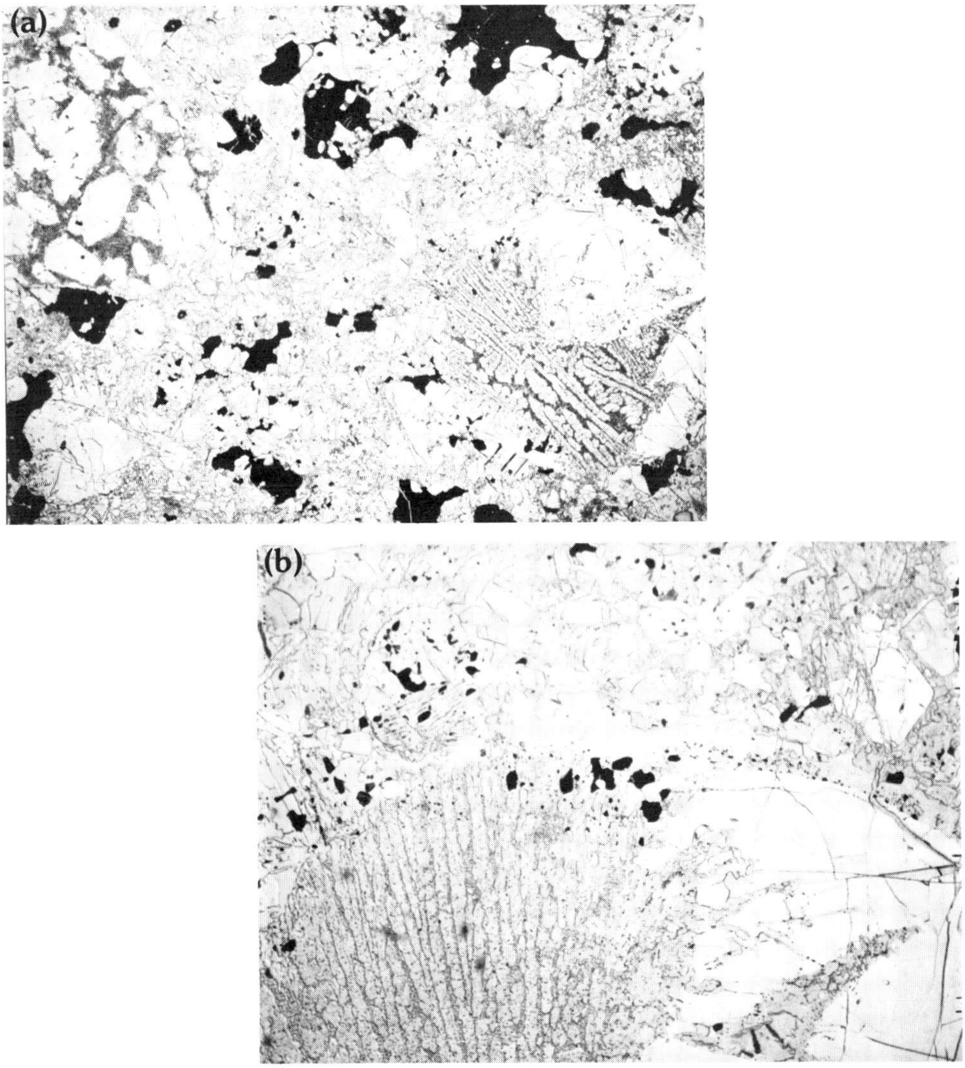

Fig. 1 Photomicrographs of an ultrathin (~4 μm) section of the H5 chondrite Allegan. (Section courtesy of R. Beauchamp, Battelle, N.W.). **(a)** Porphyritic olivine chondrule (upper left) with interstitial fine-grained to cryptocrystalline groundmass of pyroxene and plagioclase. At the lower right is a compound chondrule of porphyritic olivine, radiating skeletal pyroxene, and a dark matrix similar to that in the porphyritic chondrule. Length of section: 3.6 mm. **(b)** Detail of the compound chondrule in **(a)**. Olivine (lower right) shows partial corrosion. All olivines and pyroxenes, including those in the groundmass, have constant (i.e., "equilibrated") composition (~17% and ~10.8% FeO, respectively). Clearly, this section does not illustrate a conventional metamorphic structure or fabric, nor is post-accretionary recrystallization indicated (see text). Length of section: 1.1 mm.

induration processes did not drastically change the chemistry of the Allegan meteorite constituents. The cause, however, of the constant Fe/Mg ratios of most olivines and pyroxenes in the ordinary chondrites—the alleged evidence of equilibration—has been addressed by Fredriksson (1963), Kurat (1968), and Reid and Fredriksson (1967). These authors, to counter the necessity of thermal metamorphism, argue that the Fe/Mg ratios are inherited from processes involving quenching from a molten state.

Fredriksson (1982) has made electron probe analyses of pressed pellets of over 300 separated and pulverized individual chondrules (metal removed) from various chondrites. Figures 2 and 3 show partial results of the bulk analyses of chondrules from four H-group chondrites: Allegan

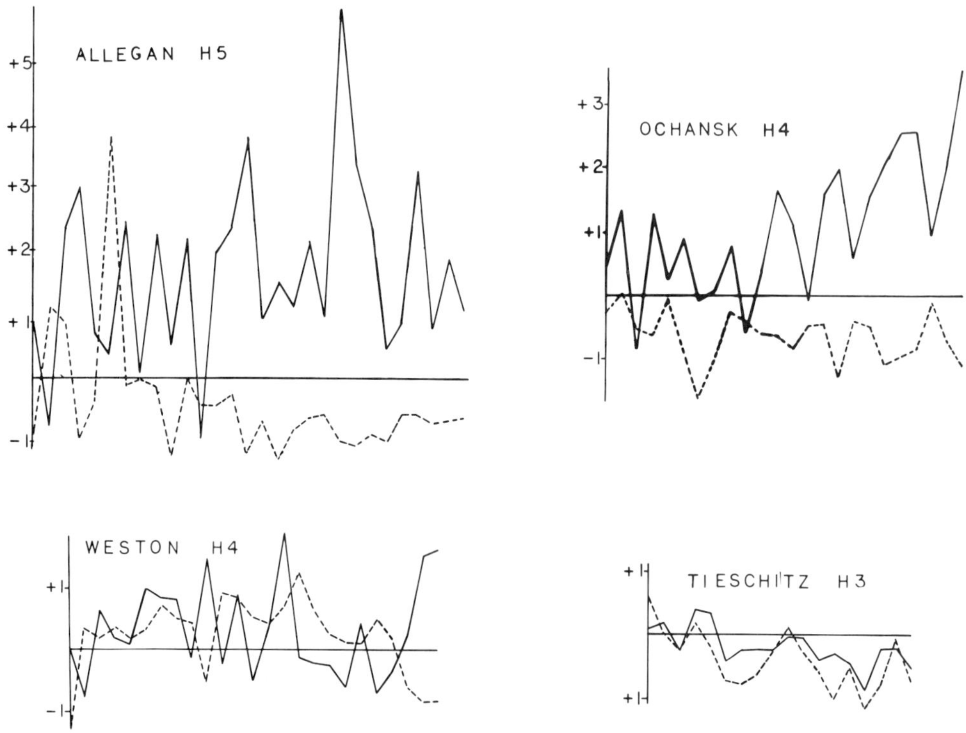

Fig. 2 CaO (solid lines) and Al_2O_3 (dashed lines) contents (wt.%) of individual chondrules from four H-chondrites, *expressed as deviation from bulk compositions* represented by horizontal lines. CaO: 2.0, 2.1, 1.7, and 1.8; Al_2O_3: 2.4, 2.4, 1.7, and 1.8 for Allegan, Ochansk, Tieschitz, and Weston, respectively. Thus the highest CaO content of an Allegan chondrule is $2.0+5.8=7.8$; the lowest Al_2O_3, $2.4-1.3=1.1$. The X-axis is non-dimensional, although the analyses were sorted such that the SiO_2 content *generally* increases from left (~42%) to right (~57%). Al and Ca are *negatively* correlated for grades 4 and 5, but *positively* correlated for grade 3; this is also the case for all other H as well as L and LL chondrites studied. Weston is a complex breccia, which is also indicated in this diagram by the varying Al-Ca correlation.

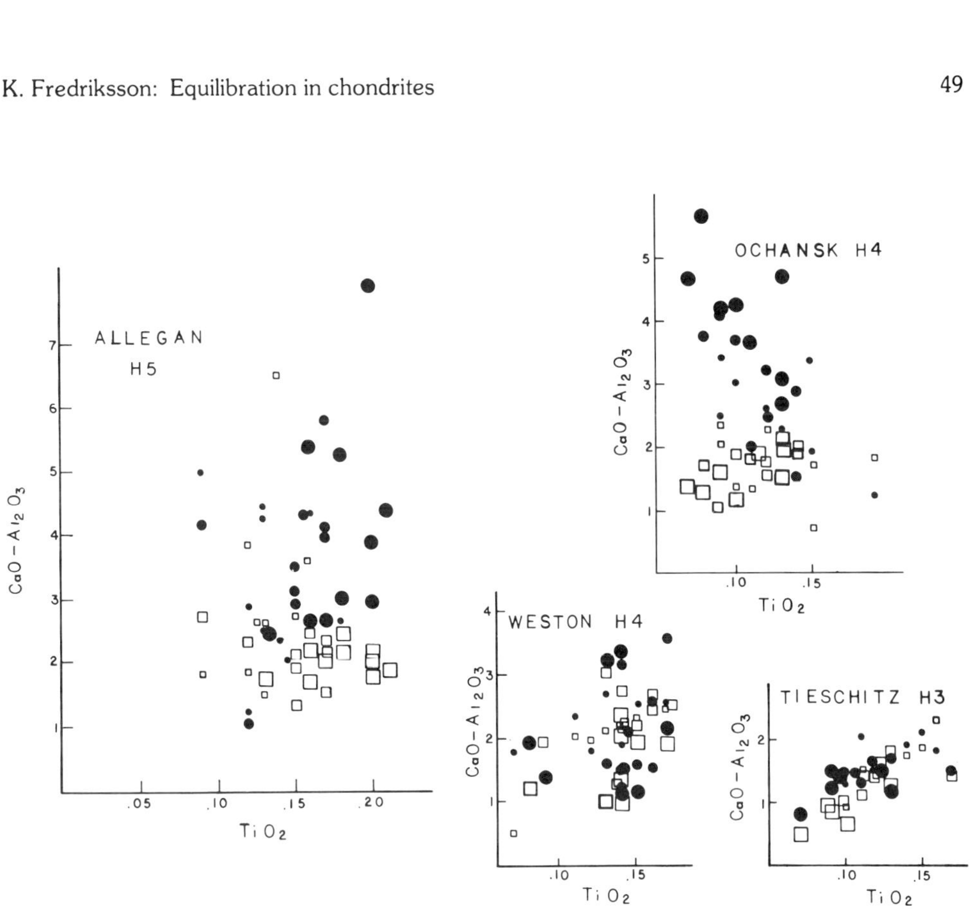

Fig. 3 CaO (circles) and Al_2O_3 (squares) vs. TiO_2 in individual chondrules from the same chondrites as in Fig. 2. Increasing size of symbols denotes increasing SiO_2, centered at 50 to 52%. The precision for TiO_2 is ~±10%. Note the strong *positive* correlation for the "unequilibrated" H3 Tieschitz, contrasted to *no correlation* in type 5 and weak (?) correlation in 4. Again, the complex brecciated nature of Weston is reflected in the varying degrees of correlation. Similar trends are observed in L and LL chondrites.

(H5), the rather similar Ochansk (H4), the complex breccia Weston (H3,4,5,6,7?; Noonan and Nelen, 1976), and the "unequilibrated" Tieschitz (H3). Noteworthy is that Ca and Al are *negatively* correlated (Fig. 2) in the "equilibrated" Allegan and Ochansk chondrules (Fe and Mg are positively correlated by definition). Also, the Ca/Al ratio for these chondrules generally increases with increasing SiO_2; the average CaO content is *30 to 50% higher* and the Al_2O_3 content somewhat lower than in the bulk samples (horizontal lines). The ratios of Al_2O_3 and CaO to TiO_2 (Fig. 3) scatter from ~5 to ≳50, while the bulk sample has ratios near the average (which is practically constant) for all chondrites, ~16. In the Allegan bulk sample Al_2O_3/TiO_2 is 16, and CaO/TiO_2 is 13 (Mason and Maynes, 1967). In strong contrast, chondrules from the type 3 Tieschitz have average Al_2O_3 and CaO contents near those of the bulk samples or somewhat lower (Fig. 2), decreasing with increasing SiO_2. Furthermore, Al and Ca are *positively* correlated. The Al_2O_3/TiO_2 and CaO/TiO_2 ratios are essentially constant, ~14 and ~17 respectively, and close to those of the bulk samples and the average for chondrites (Fig. 3). The Weston breccia, as expected, shows mixed trends, i.e., partially correlated Al, Ca, and Ti (Figs. 2 and 3). Similar extreme compositional differences between type 3 unequilibrated chondrites and the "equilibrated" chondrites of higher petrographic types occur in the L-group (Fredriksson, 1982) and LL-group stones. These results negate the supposition that type 3 chondritic material is the precursor of the "higher" chondritic types.

DISCUSSION AND CONCLUSION

These data and observations provide evidence that the Allegan (H5) and Ochansk (H4) chondrites are chemically unequilibrated, structurally "cataclastic," and consequently should not be classified as metamorphic rocks; Keil (1982) concludes this for other chondritic breccias. Wlotzka and Fredriksson (1980) have described another "unequilibrated" H5 chondrite, Moro do Roccio. In addition to a few "deviating olivines," this meteorite contains free SiO_2, indicative that it has not been thermally metamorphosed. Apparently such mineral associations in ordinary chondrites are neither uncommon nor "anomalous" as previously believed. Rather, they are another manifestation of the unequilibrated nature of these rocks.

Currently, there seems to be no rational explanation for the "equilibrated" Fe-Mg of the olivines and pyroxenes but grossly variable Al, Ca, and Ti in bulk chondrules of type 4 and higher vs. the reverse trends in the type 3 ("unequilibrated") chondrites. This problem possibly can be resolved by ion or X-ray microprobe (or INAA?) analysis of trace elements in the 336 samples of homogenized individual chondrules we have available.

If one supposes, however, that Allegan had a progenitor similar to petrographic type 3 material (e.g., Tieschitz) and was thermally metamorphosed to its present composition and structure, such a process would necessarily have to accomplish: (1) equilibration (by diffusion) of olivine and pyroxene; (2) addition of Ca to most chondrules; (3) separation of Ca from Al (and Ti) and subsequently, (4) mixing of the variable chondrules and different matrix components (recrystallized or not) to achieve the common bulk compositions for H, L, and LL chondrites as well as their constant Al/Ti and Ca/Ti ratios; and (5) retention of the "primitive(?)" composition of the ultrafine matrix components. This apparently would require a well-closed system during or after chondrule formation, continuing throughout the accretion and final break-up. The solution to many of these problems may lie in multiple, repetitious processes such as impacts, which would produce the gross textural and structural differences observed in chondrites, while preserving the similar major element compositions of the chondrite silicate. Differences in other elements, including rare gases, would then be due to mixing ratios of minor exotic components present in the fine-grained matrix.

Acknowledgments. I am grateful to Drs. B. Mason and S. R. Taylor for reading the original manuscript. A rather helpful review by Dr. E. Olsen is also acknowledged. Ms. P. Brenner produced the diagrams, and the final editing was done by Ms. B. J. Fredriksson. Mr. S. Dillon Ripley provided financial support through the S.I. Fluid Research Fund and the Universities Space Research Association provided funds through the Lunar and Planetary Institute.

REFERENCES

Alaerts L. and Anders E. (1979) On the kinetics of volatile loss from chondrites. *Geochim. Cosmochim. Acta* **43**, 547–553.

Brownlee D. E. (1978) Interplanetary dust: Possible implications for comets and pre-solar interstellar grains. In *Protostars and Planets* (T. Gehrels, ed.), pp. 134–150. Univ. of Arizona Press, Tucson.

Dodd R. T. (1969) Metamorphism of the ordinary chondrites: A review. *Geochim. Cosmochim. Acta* **33**, 161–203.

Dodd R. T. (1981) *Meteorites.* Cambridge University Press, N.Y. 368 pp.

Fredriksson K. (1963) Chondrules and the meteorite parent bodies. *Trans. N.Y. Acad. Sci.*, ser. 2, **25**, 756-769.

Fredriksson K. (1982) Chondrule compositions in different type chondrites (abstract). In *Lunar and Planetary Science XIII,* pp. 233–234. Lunar and Planetary Institute, Houston.

Keil K. (1982) Composition and origin of chondritic breccias. In *Workshop on Lunar Breccias and Soils and Their Meteoritic Analogs* (G. J. Taylor and L. L. Wilkening, eds.), pp. 65–83. LPI Tech. Rpt. 82-02. Lunar and Planetary Institute, Houston.

Keil K. and Fredriksson K. (1964) The iron, magnesium, and calcium distribution in olivines and rhombic pyroxenes of chondrites. *J. Geophys. Res.* **69**, 3487–3515.

Kurat G. (1968) The formation of chondrules and chondrites and some observations on chondrules from the Tieschitz meteorite. In *Meteorite Research* (P. M. Millman, ed.), pp. 185–190. Reidel, Dordrecht.

Lipschutz M. E. and Ikramuddin M. (1978) Condensation and/or metamorphism: Genesis of E- and L-group chondrites from studies on artificially heated primitive congeners. In *Comets-Asteroids-Meteorites* (A. H. Delsemme, ed.), pp. 375–383. Univ. Toledo, Ohio.

Lipschutz M. E., Biswas S., and McSween H. Y. (1982) Chemical-petrologic characteristics and origin of gas-rich H-chondrites (abstract). In *Lunar and Planetary Science XIII*, pp. 439–440. Lunar and Planetary Institute, Houston.

Mason B. (1963) Olivine composition in chondrites. *Geochim. Cosmochim. Acta* **27**, 1011–1024.

Mason B. and Maynes A. D. (1967) The composition of the Allegan, Bur-Gheluai and Cynthiana meteorites. *Proc. U.S. Nat. Mus.* **124**, 1–12.

Merrill G. P. (1888) On the San Emigdio meteorite. *Proc. U.S. Nat. Mus.*, XI, 161–167.

Merrill G. P. (1916) Handbook and descriptive catalogue of the meteorite collections in the United States National Museum. *U.S. Nat. Mus. Bull. 94.* 207 pp.

Merrill G. P. (1929) The story of meteorites. *Smithsonian Scientific Ser.* **3**, 1–115.

Noonan A. F. and Nelen J. A. (1976) A petrographic and mineral chemistry study of the Weston, Connecticut, chondrite. *Meteoritics* **11**, 111–130.

Rambaldi E. R., Fredriksson B. J., and Fredriksson K. (1981) Primitive ultrafine matrix in ordinary chondrites. *Earth Planet. Sci. Lett.* **56**, 107–126.

Reid A. M. and Fredriksson K. (1967) Chondrules and chondrites. In *Researches in Geochemistry, 2* (P. H. Abelson, ed.), pp. 170–203. Wiley, N.Y.

Suess H. E. and Wänke H. (1967) Metamorphosis and equilibrium in chondrites. *J. Geophys. Res.* **72**, 3609–3612.

Tandon S. N. and Wasson J. T. (1968) Gallium, germanium, indium and iridium variations in a suite of L-group chondrites. *Geochim. Cosmochim. Acta* **32**, 1087–1109.

Urey H. C. and Craig H. (1953) The composition of the stone meteorites and the origin of the meteorites. *Geochim. Cosmochim. Acta* **4**, 36–82.
Van Schmus W. R. (1969) The mineralogy and petrology of chondritic meteorites. *Earth Sci. Rev.* **5**, 145–184.
Van Schmus W. R. and Wood J. A. (1967) A chemical-petrological classification for the chondritic meteorites. *Geochim. Cosmochim. Acta* **31**, 747–765.
Wahl W. (1952) The brecciated stony meteorites and meteorites containing foreign fragments. *Geochim. Cosmochim. Acta* **2**, 91–117.
Wasson J. T. (1974) *Meteorites*. Springer-Verlag, N.Y. 316 pp.
Wlotzka F. and Fredriksson K. (1980) Morro do Roccio, an unequilibrated H-5 chondrite (abstract). *Meteoritics* **15**, 387–388.
Wood J. A. (1962) Metamorphism in chondrites. *Geochim. Cosmochim. Acta* **26**, 739–749.
Wood J. A. (1963) On the origin of chondrules and chondrites. *Icarus* **2**, 152–180.

Constraints on the Heating and Cooling Processes of Chondrule Formation

Naoyuki Fujii
Department of Earth Sciences, Kobe University, Nada-ku, Kobe 657, Japan

Masamichi Miyamoto
Department of Pure and Applied Sciences, College of General Education, The University of Tokyo, Komaba, Meguro-ku, Tokyo 153, Japan

Constraints on the temperature history of chondrules are examined on the basis of petrological observations. The chondrule formation processes are divided into two parts, i.e., the heating of pre-chondrule particles by external heat sources, and the cooling of chondrules surrounded by solar nebula. Relations between the duration of the surface temperature and the maximum attainable temperature inside chondrules are obtained from thermal diffusion calculations for externally heated pre-chondrule particles. It is concluded that the duration of each heating event should be less than 0.01 sec for chondrules that have a typical radius of 0.3 mm and include relict minerals. The duration of each heating event could not vary widely because of the limited size distribution of chondrules and the common occurrence of relict minerals in chondrules. By assuming radiative conduction cooling of the gas-dust mixture from near the liquidus of olivine to the ambient temperature of solar nebula, the cooling rate becomes 1 to 0.1 K/sec for a radius of about 600 m, provided that the chondrules and the solar nebula surrounding them cool as a whole. The number density of pre-chondrule particles would be so large that mutual collisions of melted or partially melted chondrules would occur frequently with low relative velocity in the central part of the chondrule-forming region. Chondrules thus produced would be reheated at least several times and would have complex thermal histories because of the wide variety of cooling processes. These conditions of the temperature history of chondrule evolution seem to be concordant with the "lightning" discharge model, although other mechanisms cannot be excluded.

INTRODUCTION

Models of chondrule formation processes generally involve some violent and quick heating to melt silicate minerals followed by a rapid cooling (e.g., Smith, 1982). Quantitative loss of volatile elements (e.g., Larimer and Anders, 1967) suggests that chondrules were formed in the primordial solar nebula with an ambient temperature of some 550 K and pressure of 10^{-2} to 10^{-6} atm (e.g., Whipple, 1972). However, mechanisms of chondrule formation have been the subject of controversy because a variety of possible events could take place; for example, (1) fusion of dust or aggregate grains by the lightning discharge (Whipple, 1966; Cameron, 1966; Rasmussen and Wasson, 1982) or acceleration of electrons by the reconnection of interplanetary magnetic field (Sonett, 1979); (2) shock melting by high velocity collisions (Wasson, 1972; Whipple, 1972; Kieffer, 1975; Grossman *et al.*, 1979; King, 1982); and (3) direct condensation from the solar nebula or reheated nebula with infalling dust particles by shear friction or shock waves

(Wood, 1963, 1983; Wood and McSween, 1977; Podolak and Cameron, 1974; Cameron, 1982).

Among various mechanisms for the chondrule formation, melting of pre-existing solids would be favored from the petrological investigations including textures (e.g., Dodd, 1978; Ikeda, 1980; Christophe Michel-Lévy, 1981; Gooding and Keil, 1981) and chemical studies (Gooding *et al.*, 1980). The recent recognition of relict minerals in chondrules (Nagahara, 1981; Rambaldi, 1981; Rambaldi and Wasson, 1981, 1982) and a wide variety of cooling rates (a few to 10^{-4} K/sec) in dynamic crystallization experiments that reproduce chondrule-like texture (Blander *et al.*, 1976; Tsuchiyama *et al.*, 1980; Hewins *et al.*, 1981) contribute to restrict conditions of evolutional processes of chondrules. These observations appear to preclude direct condensation from the solar nebula and high velocity impact as mechanisms of chondrule formation (Lux *et al.*, 1981; Hewins, 1982) and throw light on the heating and cooling mechanisms of chondrule formation. This paper presents some constraints on the heating duration and the maximum temperatures experienced for the chondrule precursors and physical conditions from the requirements of observed cooling rates of melted or partially melted chondrules.

HEATING DURATION AND THE MAXIMUM TEMPERATURE

For a chondrule including a relict olivine, the melting of pre-chondrule materials occurs only partially from the surface. It is noteworthy that relict olivines exist mostly inside the chondrules and very few exist near the rims. The maximum attainable temperature should be lower than the liquidus of olivines for such a chondrule heated externally. The duration of heating events should

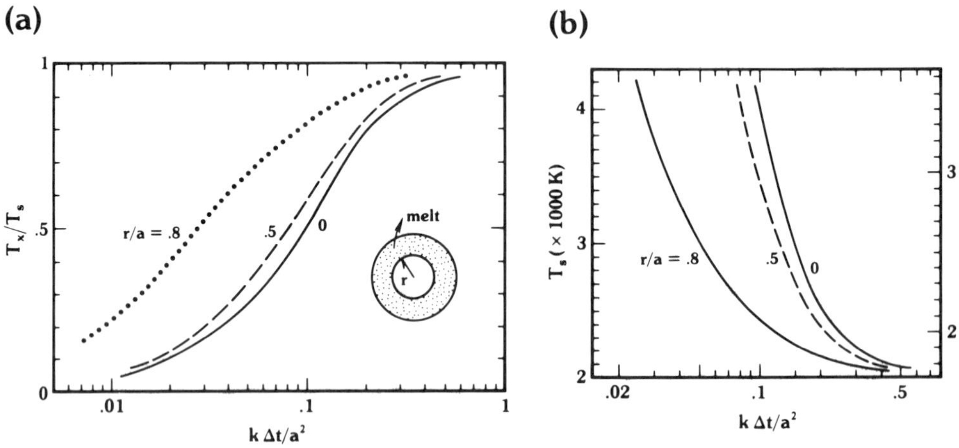

Fig. 1 **(a)** Normalized maximum attainable temperature (Tx/Ts) vs. the heating duration ($k\,\Delta t/a^2$) at radial distances (r/a) of 0.8, 0.5, and 0 for an externally heated chondrule. **(b)** The surface temperature (Ts) vs. the heating duration of Ts for liquidi of 2000 K (left-hand scale) and of 1700 K (right-hand scale).

have been limited if the surface temperature would exceed their liquidus. By assuming a constant radius (a) and thermal diffusivity (k) for externally heated spherical particles, a temperature distribution can be calculated (Carslaw and Jaeger, 1959). The relation between the duration of external heating (Δt) and the maximum attainable temperature (Tx) at a radial distance (r/a) has previously calculated with a radiation boundary condition (Miyamoto and Fujii, 1980; Fujii *et al.*, 1982). In this analysis, the surface temperature Ts is maintained constant during the time interval Δt, and is measured from the initial uniform temperature (i.e., ambient solar nebula). Figure 1a shows the maximum attainable temperature (Tx/Ts) vs. a non-dimensional heating duration of Ts ($k\, \Delta t/a^2$) for radial distances r/a of 0.8, 0.5, and 0. If the surface temperature (Ts) is 3000 K and the liquidus of olivine is 2000 K, the heating duration of Ts should be shorter than about 0.1 as illustrated in Fig. 1b, in which the left-hand scale for Ts corresponds to a liquidus of 2000 K and the right-hand to 1700 K, respectively. For a radius of 0.3 mm and a thermal diffusivity of 0.01 cm^2/sec, the actual duration becomes about 0.01 sec. It is, however, noted that this duration tends to infinity as Ts approaches to liquidus or between solidus and liquidus.

COOLING RATES OF CHONDRULES

From experimentally simulated textures of chondrules, the cooling rate ranges from a few to 10^{-4} K/sec at the temperatures between 1600 and 1200 C (Tsuchiyama *et al.*, 1980; Hewins *et al.*, 1981; Tsuchiyama and Nagahara, 1981; Planner and Keil, 1982). However, the cooling rate of isolated chondrules with a radius of about 0.3 mm is 1000 to 10 K/sec, when the ambient temperature is around 550 K. These are much faster than those observed in the simulation experiments. As a consequence, a "dense" gas is required to explain this discrepancy (e.g., Nagahara, 1982; Tsuchiyama *et al.*, 1982). We consider that the initial stage of chondrule formation could be responsible for the requirement of the observed cooling rates.

The radiative heat transfer depends highly on the absolute temperature and density of the gas surrounding chondrules. It may be likely that the radiation energy from the melted or partially melted chondrules is absorbed by the gas. In addition, by absorbing the radiation energy from external heat sources dust particles smaller than millimeter-size may have evaporated and have subsequently condensed within a shorter time scale than the cooling time of the surrounding gas-dust mixture. This process could have contributed to raising the temperature of the gas around the chondrules as well. Thus, cooling rates of chondrules appear to be controlled by the cooling of the gas-dust (chondrule) mixture as a whole. In the following, we consider cooling of the gas-dust mixture by radiative conduction with the approximation that the optical length of the gas-dust mixture is small in this medium, which has cylindrical symmetry for convenience. The high number density of dust particles and chondrules in the chondrule-forming region may also contribute to shortening the optical length.

The initial temperature (To) may be as high as the temperature of melted silicates at the inside of the cylindrical medium and as low as that of the solar nebula at the outside. By assuming a radiation cooling boundary condition (Carslaw and Jaeger, 1959), the temperature variations

(T/To) at the center (C) and the average one (A) are shown as a function of normalized elapsed time ($\hat{k}t/R^2$) in Fig. 2a, in which $\hat{k}$ and R are thermal diffusivity of the gas-dust mixture and radius of the medium, respectively. By differentiating temperature with respect to time, the variation of cooling rates ($\dot{T} \cdot R^2/To \cdot \hat{k}$) with elapsed time ($\hat{k}t/R^2$) is shown in Fig. 2b, in which $\dot{T}$ denotes the time derivative of temperature. The thermal diffusivity of the gas-dust mixture can be approximated by that of molecular hydrogen, and is about 60 m^2/sec for 1500 ~ 2000 K and 10^{-4} atm. The cooling rate at T ~ 1600 K (for To ~ 2000 K) is 1 to 0.1 K/sec when the radius R is about 600 m and the corresponding time is about 10 min.

DISCUSSIONS AND CONCLUSIONS

From the observations of the presence of relict minerals in chondrules with the most common texture (Nagahara, 1981; Rambaldi, 1981; Rambaldi and Wasson, 1981, 1982) and of volatile elements such as sodium (Tsuchiyama *et al.,* 1981), the maximum temperature of chondrules could not be very high for long duration. The duration of 0.01 sec for heating the surface above the

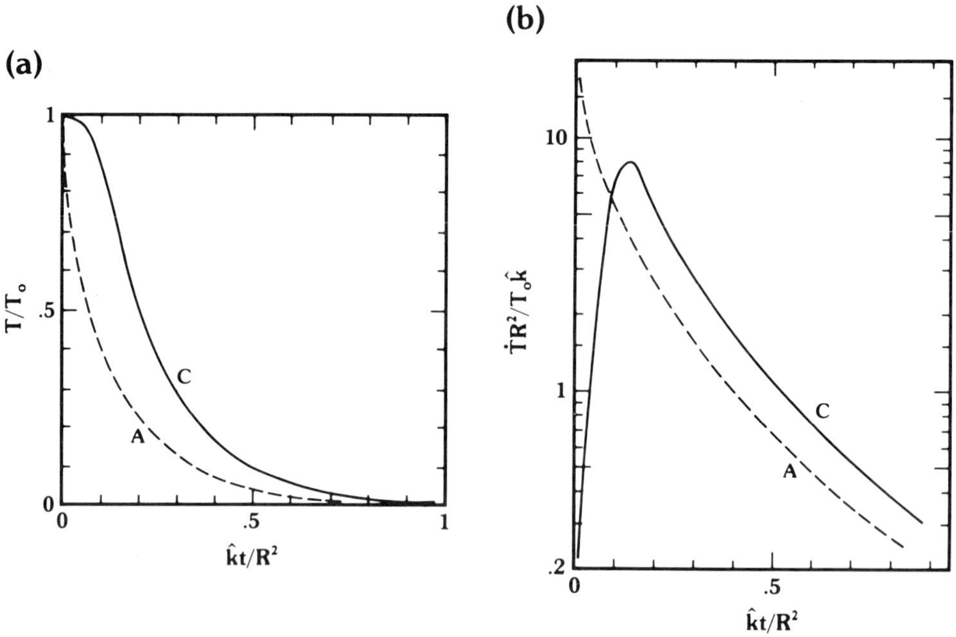

Fig. 2 **(a)** Temperature variations (T/To) of the cylindrical medium as a function of elapsed time ($\hat{k}t/R^2$), where To is the initial constant temperature and $\hat{k}$ and R are thermal diffusivity and radius of the medium, respectively. Labels A and C indicate the average temperature and the temperature at the center, respectively. **(b)** Cooling rates ($\dot{T} \cdot R^2/To \cdot \hat{k}$) of the medium. Other symbols are the same as in **(a)**.

liquidus temperature is a crude limit (Fujii *et al.*, 1982). Because only thermal diffusion was considered in this study, this limit may change by some amount. If the maximum temperature were between liquidus and solidus, there would be no limitation for its duration. It seems, however, unlikely to consider such a mechanism that can raise the temperature of pre-chondrule materials just between liquidus and solidus. Although the evaporation experiments may suggest longer heating duration, evaporation rates of sodium in chondrules are strongly influenced by the oxygen fugacity and the amount of superheat. As the fugacity is lower and temperature is higher, sodium is more easily evaporated from silicate melt and the required duration of the maximum temperature is far shorter under actual chondrule formation conditions than in experiments (Tsuchiyama *et al.*, 1981). Further investigations could make quantitative constraints on the heating duration from the evaporation rate of volatiles for chondrules without relict silicate minerals.

The obtained conditions could not specify the actual mechanism of heat sources to produce melt droplets, but could restrict the heating and cooling history by following a particular heat source. As an example of such a heat source, we examine the case of the "lightning" discharge model (Whipple, 1966). During the formation process of planetesimals with a mass of $10^{18} \sim 10^{20}$ g by gravitational instability of the dust layer (Goldreich and Ward, 1973), the lightning discharge could occur frequently throughout this layer. At this stage, dust particles have grown to about 0.1 mm in average size through sedimentation and accretion processes (Safronov, 1969; Kusaka *et al.*, 1970). These pre-chondrule dust particles would be so fluffy that the radiation energy emitted by lightning discharge could be effectively absorbed and could produce submillimeter-sized melted silicate particles within some distance from the lightning paths (Whipple, 1966). The heating duration of about 0.01 sec is about three orders longer than the duration of lightning in the Earth's atmosphere, but is concordant with the mechanism proposed by Sonett (1979). Because the pre-chondrule particle is assumed to be a solid spherule in this estimation, the limit of 0.01 sec for the heating duration may be an overestimate. The radiation energy would be more effectively absorbed for fluffy or sponge-like aggregates (Whipple, 1966). The temperature of such particles may increase more instantaneously than demonstrated in Fig. 1. If the lightning would occur in a turbulent solar nebula (Rasmussen and Wasson, 1982), radiated energy would be transported by the turbulent convection. However, it would be probable that pre-chondrules could be heated by radiation and part of the gas-dust mixture could cool by radiation conduction when the eddies in the turbulent nebula were far larger than a few kilometers.

If the primordial gas-dust mixture were not so dense that radiative heat transfer would be dominant, the cooling rate estimated by radiative conduction alone may be too slow. In the course of planetesimal formation by gravitational instability, dust particles would become concentrated in the central part and would be surrounded by a dense primordial nebula. We expect at some stage of this process that the radiative conduction would become dominant, although we could not quantitatively postulate actual values of the gas pressure and number density of pre-chondrule materials at this moment. Because of the probable occurrence of frequent heating events, the chondrules would have complex thermal histories, or would be reheated several times. Data for cooling history in the lower temperature range are desirable to constrain the cooling process more clearly. It may be probable that the number density of chondrules would be so large that chondrules would collide with each other with slow relative velocities and would easily become compound (Ikeda, 1980; Lux *et al.*, 1981; Gooding and Keil, 1981). Most parts of the condensed dust layer (the central part of the chondrule-forming region) would be affected by such heating processes. There would remain, however, unirradiated material that would later become matrices in ordinary chondrites through the mixing with various parts of the dust layer and planetesimals. It appears reasonable under these circumstances that chondrules occupy more than half and sometimes 70% of the chondrites' volume (Wasson, 1972, 1974; Hutchison *et al.*, 1980). After the events of chondrule formation, the primordial gas would escape and solid material would accumulate to form kilometer-sized planetesimals that became consolidated with the aid of interchondrule matrix materials (Fujii *et al.*, 1981).

It is concluded that the duration of each heating event should be less than 0.01 sec for relict minerals to exist commonly in chondrules. The heating events would be closely spaced so that the cooling of chondrules was governed by the surrounding gas-dust mixture a few kilometers wide. The gas and smaller dust would also be heated by the radiation from lightning and melted chondrules. A wide variety of cooling rates can be interpreted as the cooling time scale of this gas-dust mixture in the evolving planetesimals. As far as these constraints are satisfied, other mechanisms such as a drag heating of infalling dust particles through the nebula (Wood, 1983) could not be precluded. The constraints seem concordant with the "lightning" model, although detailed investigations of other heating mechanisms are obviously needed.

Acknowledgments. We would like to thank Drs. K. Ito and Y. Kobayashi, who gave us constructive suggestions and encouragement. We also thank Dr. H. Takeda for his helpful comments. We appreciate very much the suggestions and comments of Drs. E. Levy and R. Hewins, which greatly improved the manuscript. This work was supported in part by the Grant in Aid for Scientific Research from the Ministry of Education, Science and Culture, Japan. We are grateful to Ms. J. Asakura for typing the manuscript.

REFERENCES

Blander M., Planner H. N., Keil K., Nelson L. S., and Richardson N. L. (1976) The origin of chondrules: Experimental investigation of metastable liquids in the system Mg_2SiO_4-SiO_2. *Geochim. Cosmochim. Acta* **40**, 889–896.

Carslaw H. S. and Jaeger J. C. (1959) *Conduction of Heat in Solids*, 2nd ed. Oxford University Press, N. Y. 510 pp.

Cameron A. G. W. (1966) The accumulation of chondritic material. *Earth Planet. Sci. Lett.* **1**, 93–96.

Cameron A. G. W. (1982) Chondrule-related processes in the primitive solar nebula (abstract). In *Papers Presented to the Conference on Chondrules and their Origins*, pp. 9–11. Lunar and Planetary Institute, Houston.

Christophe Michel-Lévy M. (1981) Some clues to the history of the H-group chondrites. *Earth Planet. Sci. Lett.* **54**, 67–80.

Dodd R. T. (1978) Compositions of droplet chondrules in the Manych (L-3) chondrite and the origin of chondrules. *Earth Planet. Sci. Lett.* **40**, 71–82.

Fujii N., Miyamoto M., Ito K., and Kobayashi Y. (1981) Effects of minor components on the consolidation of planetesimals and chondrites. *Mem. Natl. Inst. Polar Res., Spec. Issue 20* (T. Nagata, ed.), pp. 372–383. National Institute of Polar Research, Tokyo.

Fujii N., Miyamoto M., Kobayashi Y., and Ito K. (1982) Some constraints on the formation of chondrules: The case of lightning discharge. *Proc. 15th Lunar Planet. Symp.* ., pp. 268–277. INstitute for space and Astronautical Sciences, Tokyo.

Goldreich P. and Ward W. R. (1973) The formation of planetesimals. *Astrophys. J.* **183**, 1051–1061.

Gooding J. L. and Keil K. (1981) Relative abundances of chondrule primary textural types in ordinary chondrites and their bearing on conditions of chondrule formation. *Meteoritics* **16**, 17–43.

Gooding J. L., Keil K., Fukuoka T., and Schmitt R. A. (1980) The origin of chondrules as secondary objects. (abstract). In *Lunar and Planetary Science XI,* pp. 345–347. Lunar and Planetary Institute, Houston.

Grossman J. N., Kracher A., and Wasson J. T. (1979) Volatiles in Chainpur chondrules. *Geophys. Res. Lett* **6**, 597–600.

Hewins R. H. (1982) Dynamic crystallization experiments as constraints on chondrules (abstract). In *Papers Presented to the Conference on Chondrules and their Origins*, p. 26. Lunar and Planetary Institute, Houston.

Hewins R. H., Klein L. C., and Fasano B. V. (1981) Conditions of formation of pyroxene excentroradial chondrules. *Proc. Lunar Planet. Sci. Conf. 12B*, pp. 1123–1133.

Hutchison R., Bevan A. W. R., Agrell S. O., and Ashworth J. R. (1980) Thermal history of the H-group of chondritic meteorites. *Nature* **287**, 787–790.

Ikeda Y. (1980) Petrology of Allan Hills-764 chondrite (LL3). *Mem. Natl. Inst. Polar Res., Spec. Issue 17*, (T. Nagata, ed.), pp. 50–82. National Institute of Polar Research, Tokyo.

Kieffer S. W. (1975) Droplet chondrules. *Science* **189**, 333-340.

King E. A. (1982) Refractory residues, condensates and chondrules from solar furnace experiments (abstract). In *Lunar and Planetary Science XIII*, pp. 389–390. Lunar and Planetary Institute, Houston.

Kusaka T., Nakano T., and Hayashi C. (1970) Growth of solid particles in the primordial solar nebula. *Prog. Theor. Phys.* **44**, 1580–1595.

Larimer J. W. and Anders E. (1967) Chemical fractionations in meteorites—II. Abundance patterns and their interpretation. *Geochim. Cosmochim. Acta* **31**, 1239–1270.

Lux G., Keil K., and Taylor G. J. (1981) Chondrules in H3 chondrites: textures, compositions, and origins. *Geochim. Cosmochim. Acta* **45**, 675–685.

Miyamoto M. and Fujii N. (1980) A model of the ordinary chondrite parent body: An external heating model. *Mem. Natl. Inst. Polar Res., Spec. Issue 17*, (T. Nagata, ed.), pp. 291–298. National Institute of Polar Research, Tokyo.

Nagahara H. (1981) Evidence for secondary origin of chondrules. *Nature* **292**, 135–136.

Nagahara H. (1982) Effect of heating temperature on the texture of chondrules with special reference to the porphyritic chondrules (abstract). In *Papers Presented to the Conference on Chondrules and their Origins*, p. 47. Lunar and Planetary Institute, Houston.

Planner H. N. and Keil K. (1982) Evidence for the three-stage cooling history of olivine-porphyritic fluid droplet chondrules. *Geochim. Cosmochim. Acta* **46**, 317–330.

Podolak M. and Cameron A. G. W. (1974) Possible formation of meteoritic chondrules and inclusions in the precollapse of Jovian protoplanetary atmosphere. *Icarus* **23**, 326–333.

Rambaldi E. R. (1981) Relict grains in chondrules. *Nature* **293**, 558–561.

Rambaldi E. R. and Wasson J. T. (1981) Metal and associated phases in Bishunpur, a highly unequilibrated ordinary chondrite. *Geochim. Cosmochim. Acta* **45**, 1001–1015.

Rambaldi E. R. and Wasson J. T. (1982) Fine, nickel-poor Fe-Ni grains in the olivine of unequilibrated ordinary chondrites. *Geochim. Cosmochim. Acta* **46**, 929–939.

Rasmussen K. L. and Wasson J. T. (1982) A new lightning model for chondrule formation (abstract). In *Papers Presented to the Conference on Chondrules and their Origins*, p. 53. Lunar and Planetary Institute, Houston.

Safronov V. S. (1969) *Evolution of the Protoplanetary Cloud and Formation of the Earth and Planets*. NASA TTF-677. 211 pp.

Smith J. V. (1982) Heterogeneous growth of meteorites and planets, especially earth and moon. *J. Geology* **90**, 1–125.

Sonett C. P. (1979) On the origin of chondrules. *Geophys. Res. Lett.* **6**, 677–680.

Tsuchiyama A., Nagahara H., and Kushiro I. (1980) Experimental reproduction of textures of chondrules. *Earth Planet. Sci. Lett.* **48**, 155–165.

Tsuchiyama A., Nagahara H., and Kushiro I. (1981) Volatilization of sodium from silicate melt spheres and its application to the formation of chondrules. *Geochim. Cosmochim. Acta* **45**, 1357–1367.

Tsuchiyama A. and Nagahara H. (1981) Effects of precooling thermal history and cooling rate on the texture of chondrules: A preliminary report. *Mem. Natl. Inst. Polar Res., Spec. Issue 20* (T. Nagata, ed.), pp. 175–192. National Institute of Polar Research, Tokyo.

Tsuchiyama A., Nagahara H., and Kushiro I. (1982) Conditions of chondrules formation: Experimental reproduction of texture and volatilization of sodium from chondrules (abstract). In *Papers Presented to the Conference on Chondrules and their Origins*, p. 59. Lunar and Planetary Institute, Houston.

Wasson J. T. (1972) Formation of ordinary chondrites. *Rev. Geophys. Space Phys.* **10**, 711–749.

Wasson J. T. (1974) *Meteorites*. Springer-Verlag, N. Y. 316 pp.

Whipple F. L. (1966) Chondrules: suggestion concerning the origin. *Science* **153**, 54–56.

Whipple F. L. (1972) On certain aerodynamic processes for asteroids and comets. In *From Plasma to Planet* (A. Elvius, ed.), pp. 211–232. Wiley, N. Y.

Wood J. A. (1963) On the origin of chondrules and chondrites. *Icarus* **2**, 152–180.

Wood J. A. (1983) Formation of chondrules and CAI's from interstellar grains accreting to the solar nebula (abstract). In *Papers Presented to the 8th Symposium on Antarctic Meteorites*, pp. 1–3, 15. National Institute of Polar Research, Tokyo.

Wood J. A. and McSween H. Y., Jr. (1977) Chondrules as condensation products. In *Comets, Asteroids, Meteorites: Interrelations, Evolution, and Origins* (A. H. Delsemme, ed.), pp. 365–373. University of Toledo, Ohio.

Survey of Chondrule Average Properties in H-, L-, and LL-Group Chondrites: Are Chondrules the Same in All Unequilibrated Ordinary Chondrites?

James L. Gooding

Planetary Materials Branch, NASA/Johnson Space Center, Houston, Texas 77058

As a guide for future chondrule studies, especially statistical analyses of large data bases, petrogenetic properties of chondrules in different unequilibrated ordinary chondrites (UOC's) are compared and contrasted using chondrule-suite average values computed from analyses made on individual whole chondrules from Tieschitz (H3), Dhajala (H3,4), Weston (H3,4), Ochansk (H4), Hallingeberg (L3), Saratov (L4), Tennasilm (L4), Semarkona (LL3), Chainpur (LL3), Hamlet (LL3,4), and Soko-Banja (LL4). Mean (and standard deviation of the mean) end-member compositions of olivine (mol.% Fa) and pyroxene (mol.% Fs) are used as indicators of the relative degree of "equilibration" of each chondrule suite. To a first approximation, respective bulk chondrule geometric-mean abundances of Na, Mg, Al, Ca, Sc, V, Cr, Mn, Fe, Co, Ni, Zn, La, Sm, Eu, Yb, Lu, Ir, and Au are the same from one UOC to another and show no major systematic trends related to the H-, L-, or LL-group parentage of the host chondrites. However, subtle but systematic interchondrite variations may occur and, in fact, may be exemplified by covariation of chondrule-average bulk Zn abundance with chondrule-average Fa content of olivine. That trend cannot be excluded as a possible chondrule primary formational feature but may be at least partly a "metamorphic" effect. Rare-earth element abundance patterns of some chondrules also may have been changed by "metamorphism." Despite the near constancy of chondrule-average bulk Co abundance among UOC's, Co enrichment in Ni-Fe metal may constitute a primary, systematic variation that distinguishes chondrules in LL-group UOC's from those in H- and L-group UOC's. Furthermore, chondrule-average sphericity and density seem to be lower whereas size is greater in LL-group UOC's, relative to H- and L-group UOC's, although the density/size inverse covariation may be, at least in part, a "metamorphic" effect. Although chondrule-average oxygen isotopic compositions may not vary systematically among UOC's, absence of essential data currently prevents a similar conclusion from being made for the isotopic composition of any other chemical element in chondrules. None of these findings conflict with previous conclusions that UOC chondrules formed by melting of pre-existing, heterogeneous solids. Instead, they require that future studies of UOC chondrules sample as many different UOC's as possible so that interchondrite systematic variations will be recognized and accommodated in interpretive models. Multivariate statistical analyses of "pooled" UOC chondrule data appear justified for chondrule bulk compositions, providing that resultant statistical factors based on elements sensitive to "metamorphic" redistribution are not misinterpreted as primary petrogenetic features of chondrules. Unfortunately, similar statistical analyses that attempt to simultaneously pool bulk compositions with phase compositions or physical properties of chondrules (the latter two of which may vary systematically among UOC's) may be less justifiable and more susceptible to error.

INTRODUCTION

Can any given ordinary chondrite be as useful as any other ordinary chondrite in studies on the origins of chondrules? Based on current wisdom, meteorite petrologists would undoubtedly agree that the answer is emphatically "no." Chondrules in equilibrated ordinary chondrites [EOC's; members of the H4-6, L4-6, and LL4-6 groups in the classification scheme of Van Schmus and Wood (1967)] apparently experienced significant post-formational changes that have obscured their primary, formational properties. Consequently, investigations of the origins of chondrules have been restricted largely to chondrules in unequilibrated ordinary chondrites [UOC's; members of the H3, L3, and LL3 groups of Van Schmus and Wood (1967)], although the question of possible sytematic differences among respective chondrule populations in various UOC's has seldom been raised and never thoroughly investigated.

The chronic absence of a thorough, comparative study of chondrules in various UOC's can be blamed on the lack of adequate samples. Although many studies based on thin sections have described the properties of chondrules in UOC's such as Bremervörde, Sharps, Tieschitz, Bishunpur, Hallingeberg, Krymka, Manych, Chainpur, and Semarkona, few UOC's have been sufficiently available as whole-rock specimens to support the studies of physical properties, trace element compositions, and isotopic compositions of whole chondrules that are needed to complete the desired data base. Consequently, it has not been demonstrated that chondrules in any given UOC can or should be considered equivalent to those in any other UOC for the purposes of deciphering chondrule origins. Ideally, if one or a few of the more readily available UOC's could be certified as representative samples of the total UOC chondrule population, future studies could comfortably concentrate on those samples and progress toward solution of the chondrule problem would be enhanced. However, it is more realistic to expect that no "ideal" samples will emerge so that analysis of chondrules from a variety of UOC's will continue as a necessary pursuit. In any event, it is important to thoroughly understand how chondrule properties vary from one UOC to another in order to protect interpretations from being led astray by inadequate sampling.

This paper attempts to outline some similarities and differences among chondrule populations in various UOC's in an attempt to address the question posed in the title. Only a survey-level treatment is presented, and emphasis is placed on identifying features that deserve further study. However, a few speculations about chondrule petrogenesis are also offered.

Previously unpublished bulk compositional data for chondrules from seven mildly to moderately unequilibrated chondrites (petrologic types 4 and transitional 3/4) are summarized and compared with those for the four highly unequilibrated (type 3) chondrites reported by Gooding *et al.* (1980a). As a supplement to the report by Gooding and Keil (1981), all eleven sample suites are also intercompared in terms of their average physical and major mineralogical properties. It is found that, when allowances are made for compositional differences among chondrule textural subgroups and differential sensitivities of various elements to "metamorphic" redistribution, average bulk elemental compositions of chondrules are remarkably constant from one UOC to another. However, average kamacite and taenite compositions and average whole-chondrule size, shape, and density may be examples of properties that significantly and systematically vary from one UOC to another. Consequently, future studies seeking the origins of chondrules must carefully guard against overinterpretation of results for chondrules from any single UOC.

DATA AND METHODS

Gaps in literature data

Meteorite literature is replete with chemical compositional data for UOC and EOC chondrules, although they are not always supported by other types of data. Whole chondrules have been fused into glasses with subsequent compositional determination by electron microprobe analysis (EMPA) (Walter, 1969; Walter and Dodd, 1972; deGasparis *et al.*, 1975; Fredriksson *et al.*, 1978; Evensen *et al.*, 1979) and also analyzed intact by non-destructive instrumental neutron activation analysis (INAA) (Schmitt *et al.*, 1965, 1967; Osborn *et al.*, 1973; Grossman *et al.*, 1979; Grossman and Wasson, 1982). Compositions of chondrules in whole-chondrite polished sections have been determined by broad-beam ("defocused-beam") EMPA (McSween, 1977; Lux *et al.*, 1980, 1981; Dodd, 1978b; Ikeda and Takeda, 1979; Kimura and Yagi, 1980; Fujimaki *et al.*, 1981; Nagahara, 1981) and by combined point-beam ("focused-beam")EMPA and modal phase analyses (Dodd, 1978a; Kimura *et al.*, 1979; Kimura and Yagi, 1980). Published analyses of chondrule minerals (and mesostasis fractions) are included in several of the papers cited above and also in other papers too numerous to review here.

Physical properties (size, shape, mass, density) of chondrules have been less often and less comprehensively reported in meteorite literature than have chemical-petrological properties. Apparent sizes of chondrules in thin sections have been reported as companion data in several of the studies cited above (e.g., Lux *et al.*, 1981; Ikeda and Takeda, 1979; Nagahara, 1981), although studies directed specifically toward size measurements of chondrules in UOC thin sections have been rare (Hughes, 1978b; King and King, 1979). Published size measurements of whole chondrules are likewise not very abundant (Lang *et al.*, 1975; Martin and Mills, 1976, 1978; Hughes, 1978b).

Quantitative data on chondrule shapes are even less abundant than data on sizes. Although chondrule apparent sizes measured in thin section must be interpreted cautiously, chondrule shapes are even more difficult to correctly deduce from the random slices made available for individual chondrules in whole-chondrite thin sections. However, whole-chondrule measurements reported by Martin and Mills (1976) did include numerical representations of chondrule three-dimensional shapes.

Measurements of chondrules masses can only be made using whole chondrules. Although Hughes (1977, 1978a, b) derived approximate chondrule mass-distribution curves by applying an assumed average density to measured size-distribution curves, the only measured mass-distribution results seem to be those of Lang *et al.* (1975) and Das Gupta *et al.* (1978). A related property, the chondrule density distribution, has remained largely unstudied, judging from the paucity of published measurements. However, by placing the entire suite in a single pycnometer bottle, Hughes (1978b) was able to determine an average bulk density for 955 Bjurböle (L4) chondrules. Later, Hughes (1980) combined size and mass measurements of individual Bjurböle chondrules to find their densities. Unfortunately, he did not publish the actual mass distribution, which was an incidental product of that study.

Measurements of stable and radiogenic isotopes in chondrules have not kept pace with other types of measurements. However, limited data can be found in the literature regarding isotopic compositions of noble gases (Alexander and Manuel, 1969; Gopalan *et al.*, 1977; Caffee *et al.*, 1982), oxygen (Onuma *et al.*, 1972; Robert *et al.*, 1979), lithium (Dews, 1966), and cadmium (Rosman and De Laeter, 1978) in chondrules. In addition, Rb-Sr ages have been reported for individual chondrules (Evensen *et al.*, 1979; Hamilton *et al.*, 1979; Minster and Allègre, 1981).

Despite the wide variety of measurements that have been reported for chondrules, the accumulated data base remains inadequate in the search for possible systematic trends among chondrules from one UOC to another. The large number of published analyses has been performed on an almost equally large number of different samples. Few attempts have been made to comprehensively study a single suite of UOC chondrules by a variety of techniques. Work by Evensen *et al.* (1979) was a step in the right direction but, unfortunately, was restricted to chondrules from Richardton (H5), an EOC. Perhaps because of its friability and relative availability in whole-rock form, another relatively equilibrated chondrite, Bjurböle (L4), has been a favorite candidate in other chemical, petrological, and isotopic studies of whole chondrules. Likewise, we find our cumulative knowledge of chondrule physical properties hinged principally on results for one H5 chondrite (Allegan) and one L4 chondrite (Bjurböle). Chainpur (LL3) is the only genuine UOC for which both chondrule physical properties and integrated chemical-petrological properties have been reported. Consequently, the comprehensive data base for UOC chondrules required to address the question posed in this paper does not yet exist in the literature. However, the new data described below should contribute toward that end.

New data

During the time that much of the previously cited recent work was being performed or published (papers dated 1977 and later), additional data on the physical, chemical, and petrological properties of UOC chondrules were being collected by the author. Suites of individual whole chondrules from the chondrites Tieschitz, Ochansk, Weston, Saratov, Tennasilm, Chainpur, and Soko-Banja were selected from among a larger set that had been collected and carefully preserved by Professor Roman A. Schmitt and non-destructively studied by INAA by T. W. Osborn (Osborn *et al.*, 1973). In addition, the author mechanically separated fresh batches of whole chondrules from the chondrites Dhajala, Hallingeberg, Semarkona, and Hamlet. All eleven chondrites were observed falls.

In surveying both "chondrule" inventories, separated objects, which appeared upon careful binocular stereomicroscopic examination to be irregular clasts or inclusions rather than self-contained spheroids, were rejected from further study. Despite the lack of a consensus on the definition of "chondrule," the screening criteria that were applied reflect the author's belief that the term "chondrule" should be applied only to naturally occurring extraterrestrial objects that display evidence for prior existence as molten or partially molten, independent droplets of generally submillimeter to centimeter size. [Petrographic criteria for recognizing such objects were briefly reviewed by Gooding and Keil (1981).] The resultant 219 acceptable "chondrules" were studied for their physical properties and a subgroup of 141 was selected for detailed chemical-petrological study.

Complete details of collection, certification, and preliminary interpretation of the data can be found in Gooding (1979) and, in summary forms, in Gooding *et al.* (1980a) and Gooding and Keil (1981). Briefly stated, each whole chondrule was analyzed physically (surface characterization and three-dimensional sizing by optical stereomicroscopy; weighing), then subjected to non-destructive bulk elemental analysis by INAA. Selected chondrules were then studied by scanning electron microscopy and eventually all chondrules were individually sliced and prepared as polished sections for petrographic study and phase analysis by EMPA. Scanning electron microscopy was also performed on some of the polished sections. Data from EMPA reported here were collected with a computer-automated ARL EMX-SM instrument operated at 15 kV. Mineral standards and Bence-Albee corrections were used for olivine and pyroxene analyses, whereas pure metal standards and Z-A-F corrections were used for Ni-Fe metal analyses. The procedure used to analyze metal was not significantly affected by overlap of Fe and Co X-ray peaks. Replicate analyses made periodically on pure Fe metal and Ni-Fe metal alloys showed, first, that "apparent Co" was always effectively at detection limit (determined to be 0.10% Co) and, second, that "apparent Co" did not vary systematically with Fe concentration. Accepted analyses of even the most Co-poor chondrule metal typically gave Co concentrations at least three times the detection limit. Data from EMPA for other phases were also obtained but are more appropriate as parts of a separate report.

Ten relatively large but petrologically representative chondrules were selected from the chemical-petrological subgroup and, after INAA, each was precisely split with half subjected to oxygen isotopic analyses by R. N. Clayton. Those informative results require separate discussion but are summarized here in a later section.

The chondrule chemical-petrological study group is summarized in Table 1. The group totals 138, rather than 141, because one chondrule from Tennasilm and two from Semarkona, which were composed dominantly of Ni-Fe metal, have been excluded. Metallic chondrules [M chondrules described by Gooding and Keil (1981)] comprise <1% of all chondrules in UOC's, yet if included in the small numbers of chondrules considered here, they could drastically and erroneously perturb the "average" properties being sought.

The classification assumed for each chondrite in Table 1 is that given by Van Schmus and Wood (1967), except for Dhajala, which appears to be transitional between H3 and H4 (Bhandari *et al.*, 1976; Noonan *et al.*, 1976), and Weston, which appears to be a breccia composed of H3 clasts in an H4 host rock (Noonan and Nelen, 1976).

The author readily admits that the data base presented here contains imperfections common to most, if not all, previous studies of whole chondrules. Relative to several studies based on whole-chondrite thin sections, fewer chondrules were analyzed and, despite the best of intentions,

Table 1. Textural characteristics of chondrules used in computation of "average" chondrule properties for various chondrites.

Chondrule	No. Chondrules Per Textural Type*							
	POP	PO	PP	BO	RP	GOP	C	Total
Tieschitz (H3)	7	2	2	0	3	1	2	17
Dhajala (H3,4)	7	2	1	0	3	1	1	15
Weston (H3,4)	6	2	1	1	1	0	0	11
Ochansk (H4)	2	0	1	0	2	2	1	8
Hallingeberg (L3)	12	2	4	2	3	1	0	24
Saratov (L4)	2	1	1	1	7	1	0	13
Tennasilm (L4)	4	0	0	0	2	0	0	6
Semarkona (LL3)	4	0	2	3	3	3	0	15
Chainpur (LL3)	6	1	0	0	6	1	0	14
Hamlet (LL3,4)	4	1	1	1	1	0	0	8
Soko-Banja (LL4)	5	0	1	0	0	1	0	7

* POP = porphyritic olivine-pyroxene; PO = porphyritic olivine; PP = porphyritic pyroxene; BO = barred olivine; RP = radial pyroxene; GOP = granular olivine-pyroxene; C = cryptocrystalline [see Gooding and Keil (1981) for descriptions of chondrule textural types].

sample selection was probably biased to some degree toward large (>1 mm) chondrules (an advantage in trace-element analysis) and also toward non-friable chondrules (an advantage in all analyses). Furthermore, as is common to INAA-oriented studies, bulk concentrations of Si were not directly measured for the chondrules (although the available EMPA and petrographic data permit estimates of the bulk SiO_2 content of each chondrule). Still, the author is unaware of any other single data base that is currently more capable of supporting the investigation attempted here.

"Average" results as petrogenetic guides

Compositional heterogeneity among individual UOC chondrules is known to be large so that investigation of compositional trends among individual chondrules remains an important task. However, dispersion among measured values of chondrule parameters is often so great, especially when the number of samples is large, that the search for order among the chaos is sometimes discouraging. Progress may be achieved by reducing the number of variables through computation of summary parameters, which might more readily reveal trends among the data. Toward that end, multivariate analysis has been applied to various sets of chondrule bulk compositional data (Lux *et al.*, 1980, 1981; Grossman and Wasson, 1982). However, such analyses are not attempted here and, in fact, should more logically be applied after completion of the present analysis. *Major* petrogenetic differences between chondrules from one chondrite to another (e.g., similar in magnitude to differences observed between chondrite groups) should become visible upon comparison of best-estimate central tendencies of the parameters under study. Resolvable differences among "average" values would clearly imply that chondrules from different chondrites could not be considered members of a single grand population. Multivariate analyses might then be constrained to treat chondrules from only one chondrite at a time in order to avoid confusing results. In contrast, absence of major sytematic differences among chondrule suite "average" properties would imply that consanguinity does exist among the overall UOC chondrule population and that either intensive study of chondrules from a single well-chosen UOC, or "data pooling" for chondrules from different UOC's, are justifiable approaches to understanding the origins of chondrules. Briefly stated, this paper offers a direct intercomparison of chondrules from various UOC's as a necessary first step toward certifying and guiding the application of future, more elegant statistical analyses. Given that objective and the utility of "average" results as compact descriptors, an acceptable index of "average" behavior remains to be defined.

The arithmetic mean is the parameter most widely used to represent "average" results. However, it accurately reflects the best estimate of the central tendency only for a normal distribution (Meyer, 1975), whereas elemental concentrations in rocks commonly exhibit highly skewed frequency distributions, which approach or match log-normal functions (Ahrens, 1965). Bulk concentrations of some elements in chondrules also appear to be nearly log-normal in their frequency distributions (Osborn *et al.*, 1973; Gooding, 1979). In addition, both the mass and size distributions of chondrules are approximately log-normal (Martin and Mills, 1976, 1978; Hughes, 1977, 1978a; Gooding, 1979). For a log-normal frequency distribution, a good measure of central tendency is provided by the geometric mean (Meyer, 1975). Computationally, the geometric mean is less perturbed than is the arithmetic mean by extreme values in a data set containing wide dispersion. Furthermore, the standard deviation computed for a geometric-mean value consists of two unequal positive and negative elements (unlike the single value defined for the arithmetic mean) that, by their relative magnitudes, readily show the direction of skewness in the data set. However, it can be shown that as the frequency distribution of a data set approaches an unskewed normal condition, computed values of its arithmetic and geometric means (and their respective standard deviations) converge.

In this paper, geometric-mean values are adopted as the best-estimate "average" values for chondrule variates that are known or suspected to be log-normally distributed (e.g., bulk elemental concentrations, physical properties). For other normally distributed parameters (e.g., mineral compositions), arithmetic-mean values are used because they do not differ significantly from their geometric-mean counterparts and are more easily compared with existing data in the literature.

COMPARISON AND CONTRAST

Bulk elemental and isotopic compositions

For reasons discussed above, the geometric mean of individual chondrule analyses is the parameter used to represent the "average" chondrule bulk composition compiled for each of the chondrites listed in Table 2. Each geometric-mean entry in Table 2 is accompanied by the value of the standard deviation of the geometric mean (SDGM) representing dispersion toward higher (+)

Table 2. "Average" bulk elemental compositions of chondrule suites determined by INAA.*

Chondrite:	Dhajala (H3,4)		Weston (H3,4)		Ochansk (H4)	
No. Chondrules†	(15)		(11)		(8)	
	GM	+/- SDGM	GM	+/- SDGM	GM	+/- SDGM
% Al_2O_3	2.6	1.0/0.7	3.0	1.2/0.8	2.5	1.3/0.8
% Cr_2O_3	0.59	0.21/0.16	0.57	0.22/0.16	0.38	0.15/0.11
% Fe (total)	12.5	4.9/3.5	9.88	2.53/2.01	9.65	1.70/1.45
% MnO	0.40	0.14/0.10	0.35	0.15/0.10	0.38	0.03/0.03
% MgO	27.4	3.2/2.9	28.9	2.6/2.4	27.3	4.2/3.6
% CaO	1.8	0.6/0.4	2.7	2.1/1.2	2.7	2.6/1.3
% Na_2O	1.09	0.87/0.48	1.25	0.60/0.40	1.12	0.61/0.39
ppm Sc	8.51	3.48/2.47	14.5	7.0/4.7	14.8	7.6/5.0
ppm V	86	27/20	98	34/26	82	17/14
ppm Zn	40	19/13	47	31/19	24	22/11
ppm La	0.39	0.11/0.08	0.29	0.35/0.16	0.16	0.17/0.08
ppm Sm	0.26	0.09/0.07	0.23	0.17/0.10	0.14	0.11/0.06
ppm Eu	0.076	0.040/0.026	0.078	0.040/0.027	0.064	0.064/0.032
ppm Yb	0.27	0.09/0.07	0.28	0.18/0.11	0.22	0.18/0.10
ppm Lu	0.042	0.014/0.011	0.045	0.017/0.012	0.028	0.024/0.013
ppm Hf	0.15	(10) 0.20/0.09	0.18	(9) 0.16/0.09	0.21	0.18/0.10
ppm Ta	—	—	0.096	(3) 0.063/0.038	—	—
ppm Co	101	169/63	41	88/28	18	34/12
ppm Ni	2030	5400/1480	1040	2090/697	964	(7) 1940/643
ppb Ir	96	188/63	58	96/36	45	36/20
ppb Au	28	62/19	17	(10) 26/10	13	(7) 30/9

* Geometric mean (GM) ± one standard deviation of the geometric mean (+/- SDGM).

† For elements where fewer chondrules were successfully analyzed, no. chondrules given in parentheses.

and lower (−) values. The general relationship between the geometric-mean entries in Table 2 and their arithmetic-mean counterparts is the same as displayed previously in Table 1 of Gooding *et al.* (1980a) for chondrules from four other UOC's. Namely, for cases where +SDGM/−SDGM >1, the geometric mean is systematically lower than the arithmetic mean. [Note: an apparent exception to this rule erroneously occurs for Co in Tieschitz chondrules, for which the mean value of 136 ppm was misprinted as "36 ppm" in Table 1 of Gooding *et al.* (1980a)]. That generality applies to all elements analyzed, but is especially evident in results for siderophile elements (Co, Ni, Ir, Au and, to a lesser extent, Fe), which by virtue of their tendency toward values of +SDGM/−SDGM >>1 appear to be clearly skewed toward log-normal frequency distributions. As a further demonstration of the relationship between geometric-mean and arithmetic-mean values, reference can be made to Table 2 of Gooding *et al.* (1980a), which shows that, for replicate analyses of the Orgueil (C1) carbonaceous chondrite, +SDGM/−SDGM ≃ 1 and the geometric and arithmetic

Table 2. *continued*

Saratov (L4) (13)		Tennasilm (L4) (6)		Hamlet (LL3,4) (8)		Soko-Banja (LL4) (7)	
GM	+/- SDGM	GM	+/- SDGM	GM	+/- SDGM	GM	+/- SDGM
2.1	0.6/0.4	2.6	0.3/0.3	2.8	0.9/0.7	2.7	0.9/0.7
0.59	0.12/0.10	0.66	0.08/0.07	0.56	0.13/0.11	0.63	0.22/0.16
12.3	3.3/2.6	13.9	6.6/4.5	11.6	2.1/1.8	12.9	3.1/2.5
0.42	0.06/0.05	0.39	0.09/0.07	0.32	0.09/0.07	0.36	0.08/0.06
26.4	3.2/2.8	25.5	3.1/2.8	27.3	5.3/4.4	26.8	1.4/1.3
1.8	0.6/0.5	1.7	0.9/0.6	2.6	1.2/0.8	2.3	1.7/1.0
1.04	0.27/0.22	1.24	0.20/0.18	1.03	0.44/0.31	1.04	0.26/0.21
9.62	2.20/1.79	10.5	1.8/1.6	10.4	3.1/2.4	10.3	4.0/2.9
78	11/9	86	10/9	92	15/13	84	13/11
49	18/13	50	42/23	38	7/6	41	16/11
0.30	0.12/0.09	0.32	0.10/0.08	0.38	0.23/0.14	0.30	0.15/0.10
0.19	0.03/0.03	0.17	0.07/0.05	0.29	0.11/0.08	0.22	0.07/0.05
0.055 (12)	0.034/0.021	0.084	0.093/0.044	0.12	0.07/0.04	0.056	(6) 0.061/0.029
0.26	0.08/0.06	0.22	(5) 0.05/0.04	0.34	0.16/0.11	0.23	0.09/0.06
0.042	0.017/0.012	0.034	0.015/0.010	0.045	0.013/0.010	0.029	0.014/0.010
0.10	(5) 0.06/0.04	0.12	(2) 0.01/0.01	0.23	(6) 0.08/0.06	0.11	(2) 0.03/0.02
0.07	(2) 0.10/0.04	—	—	—	—	—	—
40	105/29	79	324/63	93	136/55	58	82/34
2300	5080/1580	1770	5740/1350	1860	5060/1360	1420	2440/898
76	105/44	64	191/48	109	219/73	54	156/40
29	67/20	40	96/28	30	53/19	21	42/14

means are virtually identical for each element because the dispersion is attributable to normally distributed random analytical errors.

Bulk compositions of the various chondrule suites are conveniently compared using elemental abundance diagrams as shown in Figs. 1–3. Each diagram was compiled for each chondrule suite by computing an abundance of $A = (X_i)_{GMC} \cdot (Fe)_{GMO} / (X_i)_{GMO} \cdot (Fe)_{GMC}$, where $(X_i)_{GMC}$ is the geometric-mean concentration of element i in the chondrules, $(X_i)_{GMO}$ is the geometric-mean concentration of i in the Orgueil (C1) carbonaceous chondrite, and $(Fe)_{GMC}$ and $(Fe)_{GMO}$ are the geometric-mean concentrations of total iron in chondrules and Orgueil, respectively. Figures 1–3 are improved versions of Fig. 1 in Gooding *et al.* (1980a), which was constructed by computing $A = (X_i)_{GMC}/(X_i)_{GMO}$. The original diagram was constructed to summarize the relative abundances of all analyzed elements, whereas the diagrams presented here as Figs. 1–3 sacrifice the abundance data points for one element, Fe, in order to normalize the others to a common basis that is independent of concentration effects related to the high (and variable) hydration state of Orgueil.

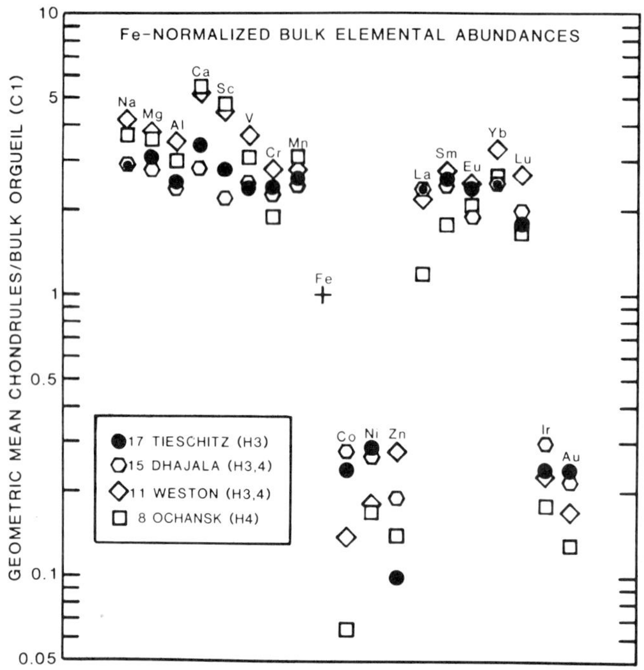

Fig. 1 "Average" elemental abundances in whole chondrules separated from mildly to highly unequilibrated H-group chondrites. This plot was constructed by normalizing geometric-mean data sets in Table 2 (and Table 1 of Gooding *et al.*, 1980a) to total Fe concentration and then, for each element, dividing the result by the Fe-normalized abundance in the bulk Orgueil (C1) chondrite as derived from the geometric-mean data reported for Orgueil in Table 2 of Gooding *et al.* (1980a). Elements are listed, from left to right, in order of increasing atomic number.

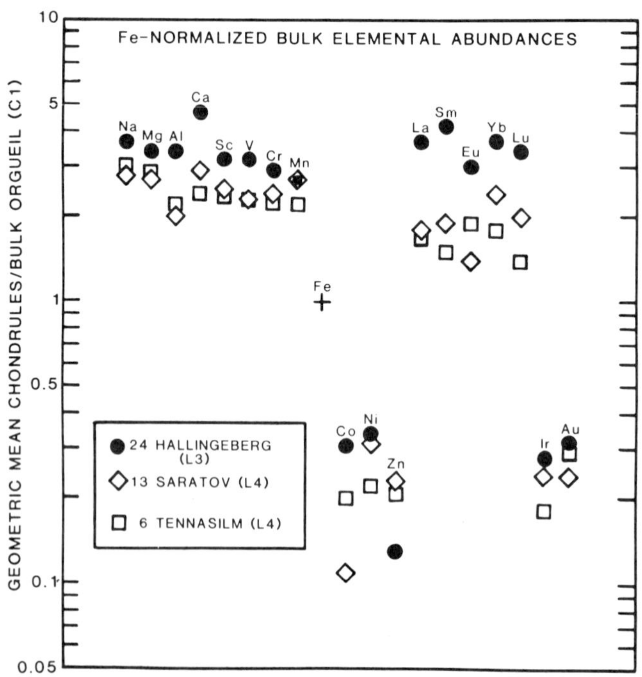

Fig. 2 "Average" elemental abundances in whole chondrules separated from mildly to highly unequilibrated L-group chondrites. Details are the same as given for Fig. 1.

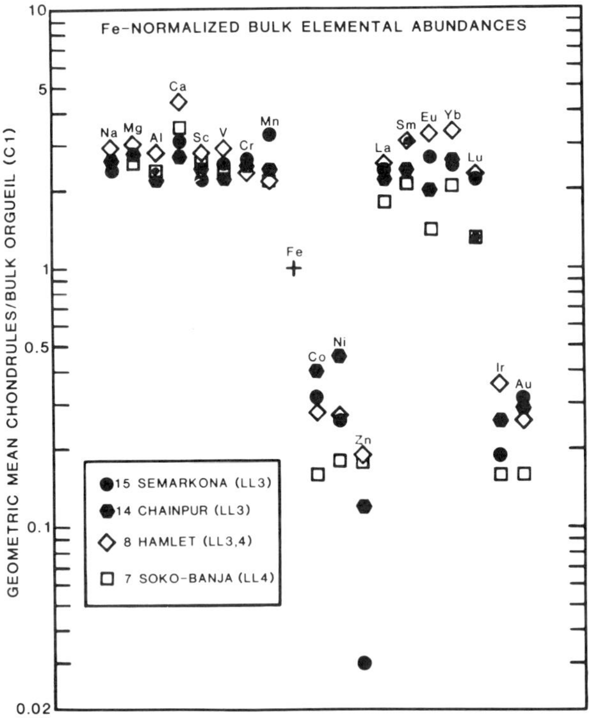

Fig. 3 "Average" elemental abundances in whole chondrules separated from mildly to highly unequilibrated LL-group chondrites. Details are the same as given for Fig. 1.

In both the original diagram and Figs. 1–3, the Orgueil data were taken from Table 2 of Gooding *et al.* (1980a) and represent analyses made in parallel with those of chondrules, which constitute the sample suites discussed here as well as in the earlier paper.

The choice of Fe (vs. Si, Mg, etc.) as the normalizing element is mostly arbitrary but does offer two advantages. First, Fe is one of the most precisely and accurately determined major elements in both the chondrule and Orgueil analyses such that its use as the normalization divisor should introduce only minimal errors. Second, Fe is the only major element in bulk chondrules and chondrites that is diversely distributed as a major element among silicates, oxides, sulfides, and Ni-Fe metal. Because the silicate/metal abundance ratio varies widely among chondrules, normalization of bulk elemental abundances to total Fe concentration helps facilitate intercomparison of metal-poor and metal-rich chondrules. Final normalization to the bulk Orgueil composition yields elemental abundances relative to those estimated for the condensible bulk solar system.

The principal reason one might expect differences among chondrule-suite average compositions is that significant differences among chondrite whole-rock bulk compositions are known to exist. Among both UOC's and EOC's, total Fe, Ni-Fe metal abundance, and Fe/Ni ratio in metal all systematically decrease from H- to L- to LL-chondrites (e.g., Gomes and Keil, 1980, p. 31). In fact, bulk compositional parameters can be used to separate the whole-rock chondrites involved in the present report into H-, L-, and LL-groups (e.g., Fig. 3 of Gooding *et al.*, 1980b), yielding individual iron-group assignments that agree with those given in Table 1. Only Tieschitz is unusual in that its bulk composition (Dodd *et al.*, 1967) reflects the total Fe concentration of H-chondrites but the Ni-Fe metal abundance and Fe/Ni ratio expected for L-chondrites. Nonetheless, there is no reason to expect *a priori* that chondrule compositions should be the same in H-, L-, and LL-chondrites. Thus, it is somewhat surprising that the chondrule data in Table 2 and Figs. 1–3 show no major systematic trends related to iron groups of the whole-rock chondrite parents. In fact, chondrule compositional dispersions within any single iron group appear as great as dispersions between different iron groups and can be easily attributed to non-uniform representation of different chondrule types in the various chondrule analytical suites. Although the relative abundances of chondrule textural types are approximately the same in various UOC's (Gooding and Keil, 1981), bulk compositions of different chondrule textural types may vary significantly (Gooding *et al.*, 1980a; Lux *et al.*, 1981) so that differences should be expected among average compositions computed for suites involving small numbers of randomly selected chondrules. Therefore, although some systematic differences based on chondrule type or degree of "equilibration" (see following sections) may exist, the data do not indicate major systematic differences among chondrule-suite bulk compositions that are correlatable with the iron groups of the host chondrites.

The foregoing conclusion was reached by Gooding *et al.* (1980a) for four UOC's (Tieschitz, Hallingeberg, Semarkona, and Chainpur), but was criticized by Wlotzka (1981). Wlotzka did not actually dispute the apparent mutual indistinguishability of the four chondrule-suite compositions, but argued against assignment of the UOC hosts to the H-, L-, and LL-groups. However, Wlotzka's arguments relied heavily on the characteristics of Tieschitz (an atypical case) and on the use of total Fe concentration as a classification criterion (only one of several important taxonomic parameters). In that regard, it is significant that the average bulk composition of Dhajala chondrules appears to be virtually the same as the average bulk composition of Tieschitz chondrules (Fig. 1). The Dhajala and Tieschitz analyzed chondrule suites contain nearly identical proportions of the various chondrule textural types (Table 1) but with significantly different degrees of "equilibration" (see following sections). However, by several different criteria (including, but not limited to, total Fe content), Dhajala is clearly an H-chondrite (Noonan *et al.*, 1976). A similar inference can be made for the LL-chondrites based on intercomparison of the Semarkona and Hamlet analyzed chondrule suites. Therefore, the absence of major differences in average chondrule bulk compositions among H-, L-, and LL-chondrites is not simply an artifact or defect of chondrite classification. Many UOC's display whole-rock attributes consistent with H-, L-, or LL-group identities, yet their respective chondrule populations remain relatively constant in average bulk elemental composition.

Although few data are currently available, oxygen isotopic compositions of chondrules also seem unrelated to the iron groups of whole-rock chondrites. Oxygen isotopic compositions of ten UOC chondrules (two Dhajala, four Hallingeberg, four Semarkona) from the set described in this paper show no systematic relationships to those of their host chondrites or to those of previously determined EOC groups (Gooding *et al.*, 1982). Oxygen isotopic analyses of additional UOC and EOC chondrules (Clayton *et al.*, 1982) support that conclusion.

Data for isotopic compositions of elements other than oxygen are rather meager for chondrules and, unfortunately, are more abundant for EOC chondrules than for UOC chondrules. For Xe isotopes, Alexander and Manuel (1969) found that radiogenic Xe-129 is enriched in chondrules relative to matrix for both the Bruderheim (L6) and Bjurböle (L4) chondrites. Detailed analyses of individual Bjurböle chondrules by Caffee *et al.* (1982) showed that a variety of I-Xe apparent ages could be derived. The data of Gopalan *et al.* (1977) for Dhajala appear to be the only noble-gas data available for UOC chondrules.

Dews (1966) found no difference in the isotopic compositions of Li in Bruderheim chondrules relative to Bjurböle chondrules. Once again, though, both host rocks were L-group EOC's and no data for UOC's are available for comparison.

In their survey of Cd isotopic compositions of meteorites, Rosman and De Laeter (1978) found that two H3 chondrites, Brownfield and Tieschitz, presented the only anomalous results among the samples studied. At least for Tieschitz, it was determined that the isotopically anomalous Cd was located mostly in chondrules. However, no such anomaly was found for the UOC Chainpur (LL3) nor for chondrules from the EOC Richardton (H5).

Hamilton *et al.* (1979) obtained a Rb-Sr model age of 4.53 Gy for Parnallee (LL3) chondrules. Minster and Allègre (1981) were unable to obtain a Rb-Sr isochron for Chainpur (LL3) chondrules because of scatter in the data, but found a model age of 4.45 Gy for Soko-Banja (LL4) chondrules. Unfortunately, no chondrule Rb-Sr data are available for UOC's of iron groups other than LL, although chondrules from the EOC Richardton (H5) gave a Rb-Sr age of 4.39 Gy (Evensen *et al.*, 1979).

Even though more analyses are needed, oxygen is the only element for which sufficient data currently exist to support at least a provisional conclusion that chondrule isotopic parameters show no systematic change from one UOC to another. The case for a common oxygen reservoir among all UOC chondrules is at least sufficiently clear to be further tested. For isotopes of other elements, though, the ground is much less firm and it would be wrong to conclude at this time that, in general, chondrules from one UOC are isotopically equivalent to chondrules from any other UOC. Without question, many more isotopic analyses of key cosmochemical elements in chondrules (e.g., O, Mg, and Si) are needed.

Phase compositions and "metamorphic" effects

Thanks to the pioneering work of Keil and Fredriksson (1964), it is well known that the compositions of olivine and pyroxene in EOC's define coherent groups that are diagnostic for the H-, L-, and LL-groups. However, olivine and pyroxene compositions scatter widely both within and between individual UOC's and, therefore, are of little use in classification systems for their host chondrites. Among the chondrule suites described here, all appear to be unequilibrated

Table 3. Average olivine and pyroxene compositions in chondrules.*

Chondrite	Olivine		Pyroxene		
	No. Chondrules (No. Analyses)	Mol. % Fa ± 1 SDM	No. Chondrules (No. Analyses)	Mol. % ± 1 SDM Fs	Wo
Tieschitz (H3)	11 (96)	9.8 ± 5.4	14 (97)	12.8 ± 11.2	2.0 ± 1.6
Dhajala (H3,4)	10 (48)	19.9 ± 2.6	13 (69)	15.6 ± 8.2	3.0 ± 3.4
Weston (H3,4)	10 (52)	15.0 ± 4.8	9 (44)	13.4 ± 3.6	3.2 ± 4.0
Ochansk (H4)	4 (18)	18.7 ± 0.4	8 (43)	16.3 ± 0.7	2.0 ± 2.0
Hallingeberg (L3)	21† (129)	15.8 ± 9.7	21 (124)	11.6 ± 7.8	3.0 ± 8.3
Saratov (L4)	4 (30)	23.0 ± 0.3	11 (83)	18.7 ± 2.3	1.4 ± 0.9
Tennasilm (L4)	5 (27)	22.2 ± 0.3	7 (41)	19.3 ± 1.0	1.3 ± 1.0
Semarkona (LL3)	8 (39)	10.7 ± 7.7	13 (73)	16.1 ± 12.2	1.9 ± 1.6
Chainpur (LL3)	7 (43)	12.2 ± 6.5	13 (105)	13.5 ± 9.4	1.7 ± 1.4
Hamlet (LL3,4)	6 (34)	20.0 ± 8.2	6 (35)	18.1 ± 7.8	1.6 ± 1.3
Soko-Banja (LL4)	7 (34)	27.0 ± 1.4	7 (36)	22.0 ± 3.6	2.0 ± 0.8

*Mean ± one standard deviation of the mean (SDM), as determined by electron microprobe analyses.
† Includes two different compositional types of olivine in chondrule HG-7.

to various degrees, based on their olivine and pyroxene average compositions (Table 3, Fig. 4). However, four (Ochansk, Saratov, Tennasilm, Soko-Banja) are noticeably less unequilibrated than the others and possess chondrule average olivine and pyroxene compositions that are comparable to those expected for the iron groups of their respective EOC hosts (Fig. 4). Consequently, it is possible to make a first-order assessment of the importance of "equilibration" in controlling observable chemical-petrological properties of chondrules.

Referring again to Figs. 1–3, it is apparent that the most highly "fractionated" (i.e., non-flat) trace-element patterns displayed by the various chondrule suites belong to those suites that are composed of nearly "equilibrated" chondrules (i.e., chondrules with nearly constant mol.% Fa in olivine and mol.% Fs in pyroxene). Rare-earth elements (REE) serve as a useful case in point.

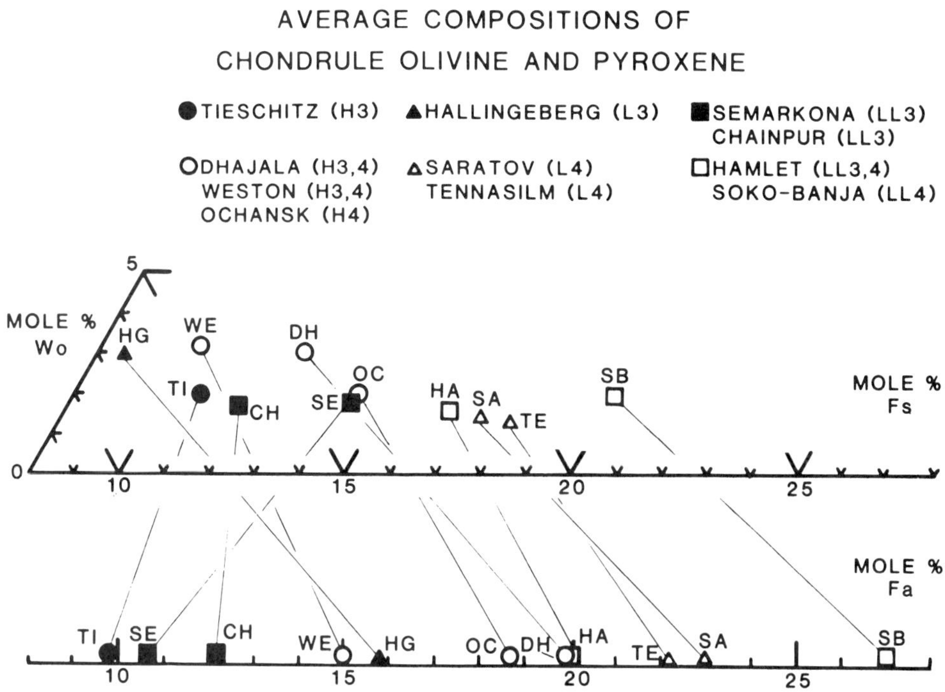

Fig. 4 Average end-member compositions of olivine and pyroxene in whole chondrules separated from mildly to highly unequilibrated H-, L-, and LL-group chondrites. Chondrules depicted here are the same as those in Figs. 1–3. Data are from Table 3. Tie lines join olivine and pyroxene compositions for individual chondrule suites. Average mol.% Fa/Fs values expected in olivine/pyroxene of "equilibrated" chondrules are 18.8/17.2 (H), 24.6/21.3 (L), and 28.5/24.1 (LL) (Gomes and Keil, 1980, p. 31). It is apparent that only the respective chondrule suites of OC, SA, TE, and SB approach "equilibrated" states. Abbreviations: CH = Chainpur; DH = Dhajala; HA = Hamlet; HG = Hallingeberg; OC = Ochansk; SA = Saratov; SB = Soko-Banja; SE = Semarkona; TE = Tennasilm; TI = Tieschitz; WE = Weston.

Because the Hallingeberg (L3) chondrule analytical suite is composed of highly unequilibrated chondrules that, in proportions of the various chondrule primary textural types (Table 1), approximate the chondrule population observable in most individual UOC's (Gooding and Keil, 1981), the Hallingeberg data (Fig. 2) can be taken as an average "unequilibrated" chondrule reference pattern. Figures 1–3 show that most of the other chondrule suites give REE patterns that closely resemble that of the Hallingeberg suite. However, the Ochansk and Tennasilm chondrule suites (each composed of approximately "equilibrated" chondrules) give REE patterns that noticeably deviate from the Hallingeberg-type pattern. Two other examples of nearly "equilibrated" chondrule suites, Saratov and Soko-Banja, display REE patterns approximating that of Hallingeberg, whereas a significantly more unequilibrated chondrule suite from Hamlet displays a REE pattern distinctively unlike that of Hallingeberg. Unequal sampling of various chondrule textural types is not an adequate explanation of REE pattern diversity. Although the average Eu/Sm ratio is greater for porphyritic than for non-porphyritic chondrules, no major heavy/light REE fractionation between the two groups is apparent (Gooding *et al.*, 1980a). The basis of the possible correlation between REE pattern and degree of chondrule equilibration remains unclear, but the data in Figs. 1–3 suggest that heavy/light REE fractionation may be associated with "metamorphic" equilibration and that the normalized Yb/La ratio, for example, may be a useful index of post-formational trace-element changes in chondrules.

Despite the ostensible constancy of siderophile element relative abundances, which seems to apply to chondrule-suite "average" compositions, there is reason to suspect that "average" chondrule Ni-Fe phase compositions are not constant from one UOC to the next. The clearest systematic trend seems to be the Co enrichment in taenite that distinguishes chondrules in LL-chondrites from those in H- and L-chondrites (Table 4, Fig. 5). As shown in Fig. 5, three of the

Table 4. Average Ni-Fe metal compositions (wt. %) in chondrules.*

Chondrite	Kamacite			Taenite		
	No. Chondrules (No. Analyses)	%Ni	%Co	No. Chondrules (No. Analyses)	%Ni	%Co
Tieschitz (H3)	12 (49)	5.6 ± 1.5	0.90 ± 0.18	12 (45)	44.4 ± 3.3	0.29 ± 0.04
Dhajala (H3,4)	14 (60)	5.1 ± 1.8	0.67 ± 0.13	5 (16)	36.5 ± 1.2	0.38 ± 0.12
Weston (H3,4)	8 (36)	6.6 ± 1.5	0.72 ± 0.19	3 (9)	42.2 ± 2.1	0.25 ± 0.02
Ochansk (H4)	6 (19)	5.7 ± 1.5	0.69 ± 0.15	3 (12)	34.7 ± 5.1	0.33 ± 0.06
Hallingeberg (L3)	19 (58)	4.3 ± 0.5	0.96 ± 0.18	7 (24)	42.5 ± 3.5	0.28 ± 0.04
Saratov (L4)	4 (16)	5.5 ± 0.5	0.84 ± 0.07	10 (50)	41.2 ± 3.1	0.33 ± 0.08
Tennasilm (L4)	6 (21)	6.0 ± 0.6	0.90 ± 0.03	2 (4)	38.2 ± 5.1	0.28 ± 0.04
Semarkona (LL3)	8 (30)	6.5 ± 1.1	0.68 ± 0.18	7 (26)	47.4 ± 9.8	2.0 ± 0.8
Chainpur (LL3)	13 (56)	4.6 ± 0.5	1.2 ± 0.2	12 (41)	47.5 ± 3.4	0.42 ± 0.36
Hamlet (LL3,4)	4 (10)	5.4 ± 1.0	1.4 ± 0.3	5 (21)	30.5 ± 2.3	0.75 ± 0.11
Soko-Banja (LL4)	5 (15)	4.3 ± 0.2	1.6 ± 0.4	6 (31)	34.1 ± 4.7	1.0 ± 0.2

*Mean ± one standard deviation of the mean (SDM), as determined by electron microprobe analyses.

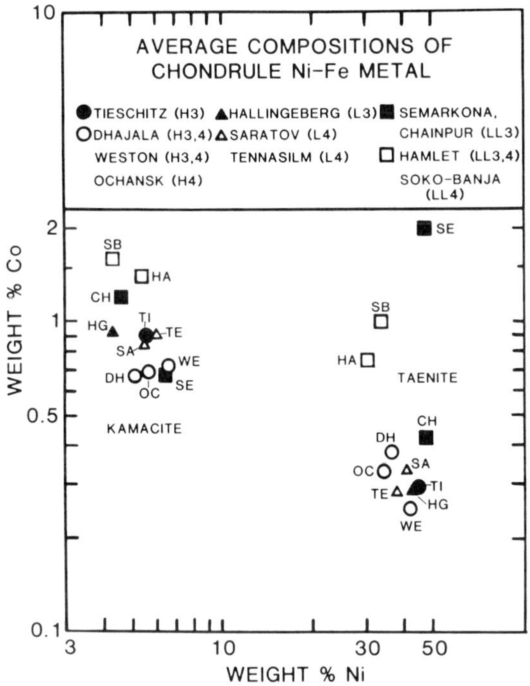

Fig. 5 Average Co and Ni concentrations in Ni-Fe metal phases of chondrules separated from mildly to highly unequilibrated H-, L-, and LL-group chondrites. Chondrules depicted here are the same as those in Figs. 1–4. Data are from Table 4 and abbreviations are the same as given for Fig. 4.

four LL-chondrite chondrule suites display average taenite compositions that are unmistakably richer in Co than the average taenites of H- and L-chondrite chondrules. Only Chainpur chondrules present a possible exception to that trend, although the relatively large standard deviation of the mean (Table 4) reflects the fact that some Chainpur chondrules did contain Co-rich taenite. Although they did not study chondrules specifically, Sears and Axon (1976) also found Co-rich (1–2 wt.% Co) taenite in several of the LL-group EOC's. However, the Co-enrichment trend among chondrule taenites cannot be attributed solely to "metamorphic" effects because the enrichment is strongly evident among the highly unequilibrated Semarkona chondrules. Systematic enrichment of Co in chondrule kamacite among LL-chondrites is also suggested by the data in Fig. 5. However, Semarkona chondrule kamacite appears to be unexpectedly poor in Co relative to the other three LL-chondrite chondrule suites and relative to some of the H- and L-chondrite chondrule suites. Thus, the basis of the Co-indexed compositional variations among average compositions of chondrule metal phases remains unexplained. However, it is clear that systematic trends may exist and that such trends may be primary, formational features of chondrules rather than effects of "metamorphic" equilibration. Additional detailed studies of metal phases in chondrules are needed to answer this important question.

The covariation of bulk Zn abundance with silicate FeO concentration offers one example of a possible systematic correlation between the average bulk and phase compositions of chondrules. As shown in Fig. 6, chondrule bulk Zn abundance appears to vary sympathetically with average

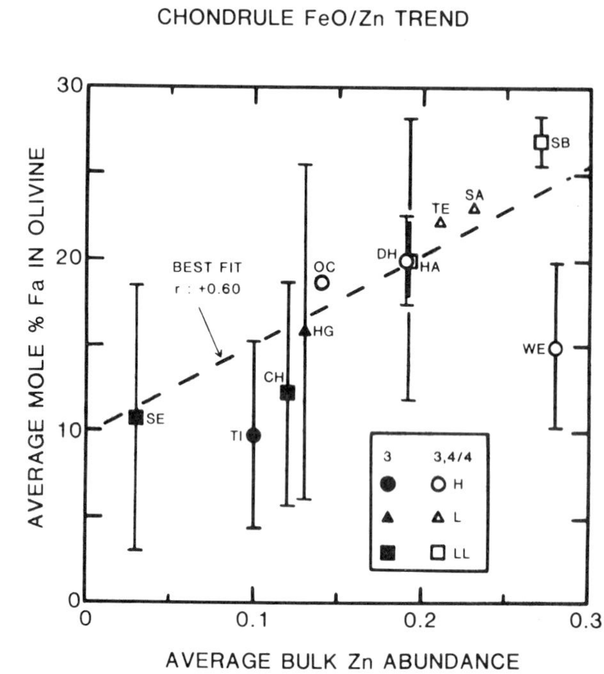

Fig. 6 Covariation between average bulk Zn abundance (Fe- and Orgueil-normalized, as described in text) and average olivine composition in whole chondrules separated from mildly to highly unequilibrated H-, L-, and LL-group chondrites. Chondrules depicted here are the same as in Figs. 1– 5. Data are from Tables 2 and 3 and abbreviations are the same as given for Fig. 4. Note that for all chondrules, bulk Zn abundance tends to increase with FeO content of olivine. In addition, Zn abundance tends to increase as degree of "equilibration" increases among chondrules of the individual H-, L-, and LL-chondrite groups. "Equilibration" is indicated by small standard deviations for olivine compositions that, for OC, TE, and SA chondrules, lie within the symbols. The linear least-squares best-fit line and correlation coefficient (r) are shown for reference.

mol.% Fa in chondrule olivine. The observed FeO/Zn trend seems to encompass all the chondrule suites (with the possible exception of Weston), regardless of their respective degrees of equilibration. However, it further seems that the most highly unequilibrated chondrules seem to be systematically lower both in bulk Zn abundance and in olivine Fa content relative to unequilibrated chondrules. Thus, the FeO/Zn trend may be partly a primary chondrule formational feature and partly a "metamorphic" feature. Evidence that the trend has a significant "metamorphic" component follows from the fact that it becomes much weaker when mol.% Fs in pyroxene is used as the chondrule FeO index (correlation coefficient, $r = +0.28$). It is well known (and can be seen in Table 3) that, for a given transitional UOC/EOC, olivine homogeneity is more complete than pyroxene homogeneity. Consequently, relative degrees of equilibration may be less obvious when pyroxene compositions, instead of olivine compositions, are used as the "equilibration" index.

Although the data in Fig. 6 do not suggest that the FeO/Zn trend partitions chondrules according to the iron groups of their host chondrites, it is clear that systematic variations between chondrule bulk and phase compositional parameters may exist and that such coupled variations may complicate the use of chondrules from any single UOC as the basis for models intended to represent chondrules in all UOC's.

Physical properties

In most of the current literature on chondrule physical properties, the assumption seems to have been implicitly made that a single set of distribution curves can be found to describe, respectively, the sizes nd masses of chondrules in all UOC's. However, review of both the data from existing literature and new data presented here suggest that interchondrite systematic trends may actually exist for sizes and related properties of chondrules. Specifically, there is reason to suspect that in LL-chondrites, relative to H- and L-chondrites, average chondrule size is greater, whereas average chondrule sphericity and density are both smaller. The evidence of this is summarized in Table 5 and in Figs. 7 and 8.

Figure 7 shows that degree of chondrule sphericity may vary antisympathetically with chondrule size. As discussed in a preceding section, chondrule selection in the new data presented here was probably biased slightly toward large chondrules. However, there is no reason to believe that the bias was greater for LL-chondrites than for H- and L-chondrites or that non-spherical chondrules were preferentially extracted from LL-chondrites. Consequently, the apparent trend in Fig. 7 must be considered as possibly being a real phenomenon. As summarized by Gooding and Keil (1981, p. 34), the limited data available in the literature are also consistent with that interpretation. Accordingly, "data pooling" of chondrule size and shape measurements should beware of mixing data for LL-chondrites with data for H- or L-chondrites. Possibly, statistical analyses of chondrule physical properties may need to treat each UOC iron group separately.

Figure 8 supports the proposition by Hughes (1980) that chondrule density varies antisympathetically with chondrule size. However, Hughes' data were limited to Bjurböle (L4) chondrules, whereas the data in Fig. 8 suggest that chondrules from H-, L-, and LL-chondrites may comprise a similar grand trend. Obviously, if it is real, that trend must reflect a size-dependent compositional variation among chondrules, although neither Hughes (1980) nor the present paper attempts to explain the phenomenon. However, the effect might be more strongly developed among "equilibrated" chondrules than among highly unequilibrated chondrules, as suggested by the fact that the value of the correlation coefficient for the presumed linear trend is much greater for a selected subgroup of four "equilibrated" chondrule suites than for the entire group of eleven chondrule suites. Specifically, the best-fit line through the Group 2 data points (defined in Table 5 and Fig. 7) for Ochansk, Saratov, Tennasilm, and Soko-Banja gives $r = -0.97$, whereas a value of only $r = -0.66$ applies to a similar regression line computed for Group 2 data points of all analyzed chondrule suites. Consequently, Hughes' (1980) discovery of the inverse density/size trend may have been facilitated by his use of "equilibrated" chondrules from Bjurböle (L4). It might then be argued that the physical basis of the density/size trend, though as yet unidentified, is somehow related to "metamorphic" equilibration. In any case, multivariate analyses of chondrule properties may need to include chondrule size as a variable. However, because chondrule density also appears to vary systematically, the use of chondrule *mass* as an input parameter may not yield the same results as would the use of actual chondrule *size*. Clearly, our current understanding of possible relationships between physical and other properties of chondrules is not adequate.

Table 5. "Average" values of chondrule physical properties.*

Chondrite	Maximum Dimension (mm)			Minimum/Maximum Dimension ratio			Mass (mg)			Density (g/cm^3)		
	No.	GM	+/−SDGM	No.	GM	+/−SDGM	No.	GM	+/−SDGM	No.	GM	+/−SDGM
Tieschitz (H3)	17	1.09	0.34/0.26	14	0.84	0.06/0.06	17	1.60	1.73/0.83	17	3.00	0.42/0.37
	26	1.04	0.41/0.30	20	0.85	0.10/0.09	26	1.40	1.79/0.78	26	3.10	0.54/0.46
Dhajala (H3,4)	13	1.00	0.38/0.28	10	0.87	0.09/0.08	15	1.23	1.86/0.74	13	3.10	0.37/0.33
	14	1.00	0.36/0.27	11	0.86	0.09/0.08	27	0.62	1.45/0.44	14	3.13	0.37/0.33
Weston (H3,4)	11	0.92	0.34/0.25	11	0.90	0.05/0.05	11	1.14	1.40/0.63	11	3.28	0.31/0.28
	16	0.86	0.35/0.25	16	0.89	0.08/0.08	16	0.93	1.34/0.55	16	3.33	0.36/0.32
Ochansk (H4)	8	0.95	0.54/0.34	8	0.90	0.05/0.05	8	1.24	3.35/0.91	8	3.28	0.34/0.31
	9	0.93	0.49/0.32	9	0.89	0.06/0.05	9	1.15	2.84/0.82	9	3.34	0.36/0.33
Hallingeberg (L3)	22	0.89	0.31/0.23	16	0.86	0.11/0.10	24	0.98	1.41/0.58	22	3.12	0.47/0.41
	—			—			32	0.85	1.17/0.49	—		
Saratov (L4)	13	1.16	0.51/0.36	12	0.87	0.11/0.10	13	2.07	4.22/1.39	13	3.24	0.44/0.39
	20	1.08	0.53/0.36	16	0.88	0.10/0.09	20	1.70	3.61/1.16	20	3.21	0.43/0.38
Tennasilm (L4)	6	0.90	0.40/0.27	4	0.83	0.09/0.08	6	0.98	1.74/0.62	6	3.36	0.24/0.22
	12	0.92	0.47/0.31	6	0.83	0.09/0.08	12	1.03	2.07/0.69	12	3.38	0.25/0.23
Semarkona (LL3)	15	1.39	0.89/0.54	3	0.81	0.07/0.06	15	2.27	6.91/1.71	15	2.96	0.52/0.44
	17	1.28	0.89/0.53	5	0.77	0.11/0.09	20	1.42	5.20/1.12	18	2.96	0.51/0.43
Chainpur (LL3)	14	1.59	0.24/0.21	10	0.79	0.12/0.10	14	4.65	3.18/1.89	14	3.11	0.67/0.55
	20	1.39	0.38/0.30	13	0.78	0.11/0.09	20	2.78	4.27/1.68	20	3.08	0.58/0.48
Hamlet (LL3,4)	8	0.94	0.29/0.22	5	0.81	0.07/0.06	8	0.90	0.60/0.36	8	3.03	0.30/0.28
	10	0.94	0.26/0.20	6	0.80	0.06/0.06	10	0.93	0.54/0.34	10	3.05	0.32/0.29
Soko-Banja (LL4)	7	1.53	0.25/0.21	4	0.82	0.15/0.13	7	4.08	2.74/1.64	7	2.99	0.34/0.30
	15	1.32	0.46/0.34	8	0.83	0.11/0.10	15	2.54	3.19/1.42	15	3.11	0.38/0.34

* Geometric mean (GM) ± one standard deviation of the geometric mean (SDGM); for each chondrite, the first line represents results for a petrologically characterized group of chondrules (Group 1), whereas the second line comprises results (Group 2) obtained by combining data for Group 1 with data for chondrules that were not petrologically characterized.

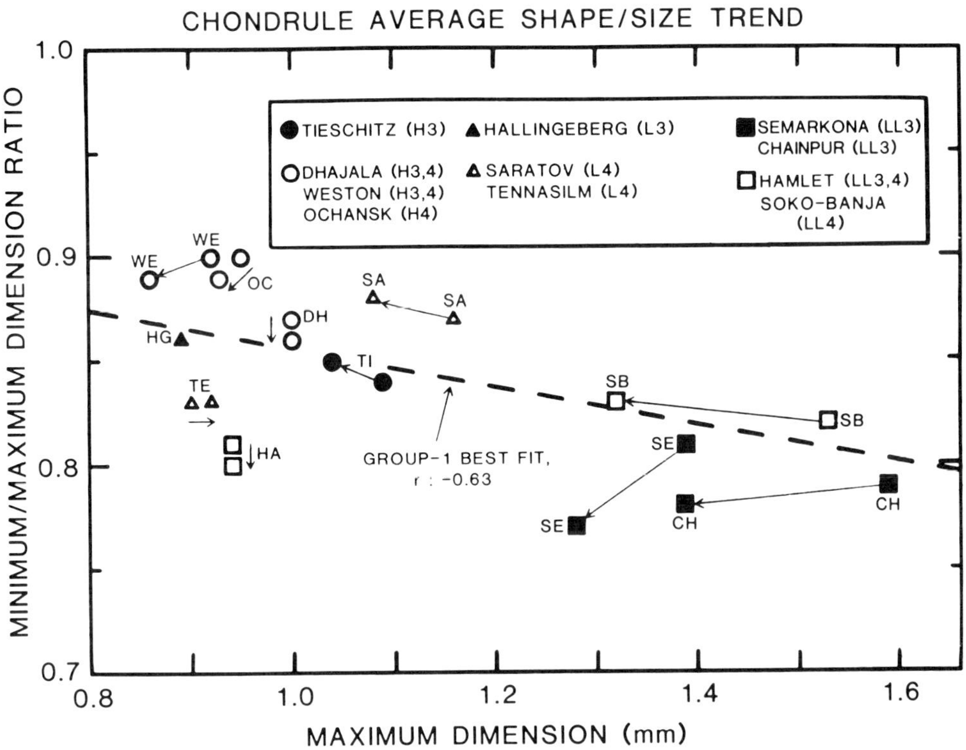

Fig. 7 Inverse covariation between "average" shapes and sizes of whole chondrules separated from mildly to highly unequilibrated H-, L-, and LL-chondrites. Data are from Table 5 and abbreviations are the same as given for Fig. 4. For each chondrite, except HG, two points are shown. Group 1 points (points at tails of arrows) represent groups of chondrules characterized chemically and petrologically (the same chondrules depicted in Figs. 1–6), whereas the Group 2 points (points at heads of arrows) represent movement of Group 1 points upon addition of other chondrules for which no chemical-petrological data are available. The linear least-squares best-fit line and correlation coefficient (r) for Group 1 points are shown for reference.

IMPLICATIONS FOR CHONDRULE ORIGINS AND EVOLUTION

The preceding sections have concentrated on outlining the similarities and differences that may exist among different suites of chondrules. The message resulting from that descriptive treatment should now be clear: Some petrogenetic properties of chondrules may vary systematically from one UOC to another so that chondrule formational models based on observational results for chondrules from single UOC's should not be considered complete. Given that message as the principal intention of this paper, it seems worthwhile to temporarily set aside the descriptive approach and to briefly point out some of the petrogenetic implications that might follow if the data trends discussed above are assumed to be real.

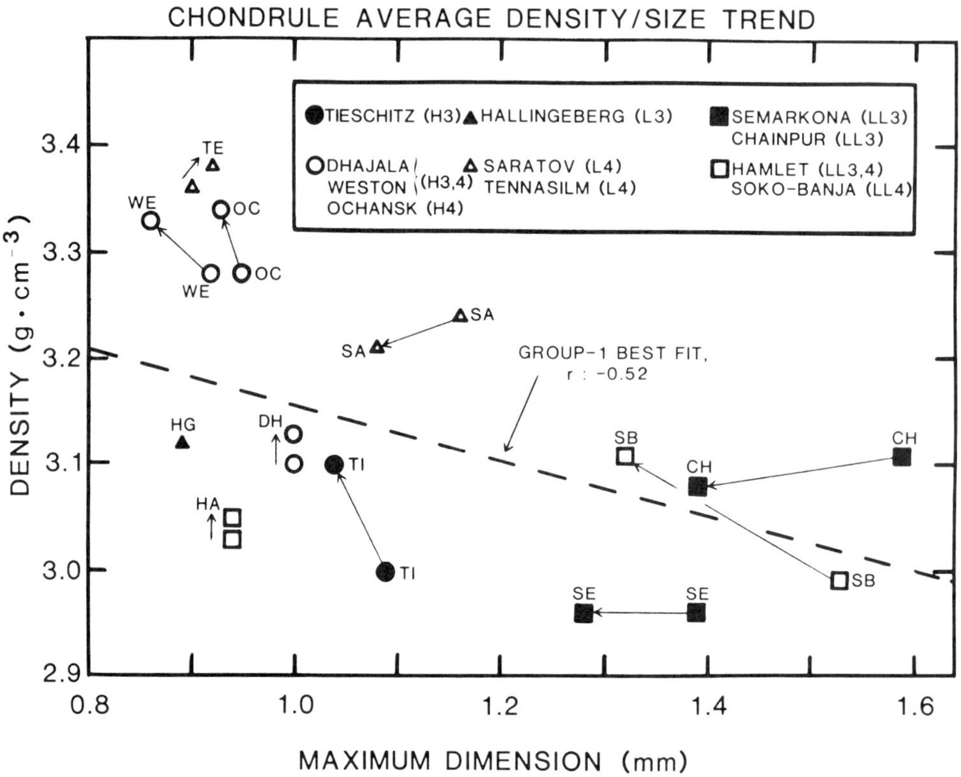

Fig. 8 Inverse covariation between "average" densities and sizes of whole chondrules separated from mildly to highly unequilibrated H-, L-, and LL-chondrites. Data are from Table 5 and details are the same as given for Fig. 7.

Despite minor or subtle systematic trends that may await discovery, the approximate constancy of chondrule bulk elemental compositions from one UOC to another is both striking and significant. Clearly, that virtual constancy implies that chondrules in all UOC's must have been derived from a common material reservoir. The reservoir must have been significantly heterogeneous at the scale of chondrule sizes (millimeter range) but effectively "homogeneous" with respect to the chondrule inventories, which it supplied to the parent bodies of the H-, L-, and LL-chondrites. As discussed by Gooding *et al.* (1980a), the elemental abundance patterns of chondrules are not consistent with fractional nebular condensation as the dominant chondrule-forming process but can be understood as the result of melting of pre-existing materials that had already experienced some degree of geochemical fractionation (i.e., separation of siderophiles from lithophiles). Oxygen isotopic compositions of chondrules also conflict with condensation models but can be explained by melting models (Gooding *et al.*, 1982).

If chondrules did indeed form by melting of pre-existing solids, then the covariation of chondrule olivine FeO content with chondrule bulk Zn content might reflect the nature of the processes involved. Loss of Fe from silicates by chemical reduction to form metal might be accompanied by the extraction of Zn into a companion sulfide phase, presuming that Zn exhibited dominantly chalchophilic behavior. Therefore, the FeO/Zn trend could be interpreted as evidence for the systematic production and loss of sulfide (and metal) from evolving chondrule droplets. Unfortunately, related trends that should be expected to follow are not evident. As shown by Gooding *et al.* (1980b), there appears to be no systematic trend between metal abundances and olivine Fa contents of UOC chondrules, contrary to what would be expected if progressive reduction and loss of Fe as metal was a dominant process during chondrule formation. However, Grossman and Wasson (1982) noted that a subgroup of their Chainpur (LL3) chondrules exhibited a negative covariation of bulk chondrule Ir/Au ratio with chondrule olivine Fa content. Grossman and Wasson interpreted that trend as evidence that the Ir-rich metal, which they postulated as a phase in the precursor materials that were melted to form chondrules, was associated with "refractory" (i.e., Fe-poor or Mg-rich) silicates. Accordingly, the FeO/Zn trend might be viewed as a "volatility" trend that was established by mixing of an Fe- and volatile-rich (i.e., Zn-rich) component with an Fe- and volatile-poor (i.e., Zn-poor) component prior to chondrule melt formation. However, an alternative interpretation of the FeO/Zn trend might be that it is principally a "metamorphic" effect, rather than a primary chondrule formational effect. If so, then the increase of bulk Zn with silicate FeO would imply that Zn closely followed Fe during chondrule "metamorphism," thereby requiring that Zn behaved dominantly as a lithophile rather than as a chalcophile element.

Systematic differences in the compositions of chondrule metal phases among UOC's would imply that either metal phases are highly sensitive to "metamorphic" disturbance or that systematic differences in the nature or behavior of metal phases occurred as part of the chondrule formational process. Although "metamorphic" diffusion of Ni has long been the basis of cooling rate estimates for taenite particles in chondrites (e.g., Wood, 1967), the Co-enrichment trend discussed in this paper may apply both to kamacite and taenite and is not well explained as a "metamorphic" effect. Consequently, one is forced to postulate that either complex, unidentified oxidation/reduction processes selectively created Co-rich metal in some chondrules, or the Co-rich metal reflects heterogeneity that existed among the metal phases in the chondrule precursors before melting. In either case, one would need to explain why the Co-rich metal occurs predominantly in chondrules from the LL-chondrites.

Possible systematic trends involving chondrule physical properties are equally important but, unfortunately, less likely to attract the attention of most meteorite petrologists. Still, the simple existence of features such as the inverse shape/size trend would imply either that chondrule formation was not exactly the same for the chondrule populations in all UOC's, or that sorting by shape or size occurred sometime after chondrule formation but before incorporation of chondrules into the UOC parent bodies. Alternatively, the existence of smaller, more perfectly spherical chondrules in H- and L-chondrites, relative to LL-chondrites, might be interpreted as evidence that H- and L-chondrite parent bodies sampled chondrules that were more fully molten (or possibly more vigorously dispersed from a larger body of melt) than their counterparts in LL-group UOC's, or that the large chondrules of LL-chondrites are less perfectly spherical because they were molten (or partially molten) for longer time intervals so that they were more susceptible to dynamic distortions. However, the author is not aware of any evidence that supports either of these highly speculative notions. They are mentioned only to illustrate the fact that useful information remains to be gleaned from the physical properties of chondrules.

The foregoing speculations are presented as stimuli for further research. Those speculations are not suggested as necessarily the only or best explanations of the trends in question but point out the need for additional, detailed work in order to resolve key difficulties. Briefly stated, despite recent progress in understanding chondrules, the complexity of the chondrule problem should not be underestimated. Efforts to create "pooled" data bases for analysis by powerful statistical methods must recognize that it may be counterproductive to pool all types of chondrule measurements without first screening the data for interchondrite variations that might introduce unnecessary confusion.

CONCLUSIONS

(1) To a first approximation, the average bulk elemental compositions of chondrules are constant from one UOC to another for many major, minor, and trace elements. Subtle, systematic trends related to either *primary* (formational) or *secondary* ("metamorphic" equilibration) processes may exist although no major, systematic bulk compositional trends related to iron-group (H, L, or LL) parentage of the chondrule host rocks are obvious. Therefore, "pooling" bulk compositional analyses for chondrules from different UOC's to support multivariate statistical analyses seems justifiable, within limits. However, derivative statistical factors based on elements sensitive to "metamorphic" redistribution (including but not limited to Fe, Zn, and rare-earths) must be cautiously interpreted with respect to their bearing on chondrule *primary* properties.

(2) Average concentrations of Co in Ni-Fe metal phases may be systematically greater in chondrules from LL-chondrites than in chondrules from H- or L-chondrites. However, it is doubtful that "metamorphism" can adequately explain the effect so that the trend should be further explored as a feature possibly established during chondrule formation. Multivariate statistical analyses of "pooled" chondrule data might be perturbed by uninformed use of metal compositions as input variables.

(3) Average sizes of chondrules are apparently greater in LL-chondrites, whereas average sphericities and densities of chondrules may be lower, relative to H- and L-chondrites. Data for all three chondrite groups suggest a systematic decrease in both chondrule average sphericity and density with increasing chondrule average size. The inverse density/size trend may be partly a consequence of "metamorphism" although the inverse shape/size trend is probably a chondrule

formational feature. Use of chondrule physical properties data as input variables may complicate the multivariate statistical analyses of "pooled" chondrule data bases.

Acknowledgments. The original data reported in this paper were collected as part of the author's work done in fulfillment of requirements for the degree of Doctor of Philosophy in Geology at the University of New Mexico. Professor Klaus Keil suggested the research topic and provided vital and unfailing guidance and support throughout all aspects of the work. The author will be forever grateful for that encouragement, help, and advice. Neutron activation analyses were performed at Oregon State University where the author greatly benefited from the generous cooperation of Professor R. A. Schmitt and the invaluable laboratory assistance of Dr. T. Fukuoka. R. A. Schmitt and R. S. Clarke, Jr. (U.S. National Museum of Natural History) provided samples. Data collection was financially supported mostly by NASA grant NGL 32-004-064 (K. Keil, Principal Investigator) and partly by NASA grant NGL 32-002-039 (R. A. Schmitt, Principal Investigator). Manuscript preparation was supported by the NASA/Johnson Space Center.

REFERENCES

Ahrens L. H. (1965) *Distribution of the Elements in Our Planet.* McGraw-Hill, New York. 110 pp.

Alexander E. C., Jr. and Manuel O. K. (1969) Xenon in chondrules and matrix material of chondrites. *Geochim. Cosmochim. Acta* **33**, 298–301.

Bhandari N., Lal D., Trivedi J. R., and Bhatnagar A. (1976) The Dhajala meteorite shower. *Meteoritics* **11**, 137–147.

Caffee M. W., Hohenberg C. M., Swindle T. D., and Hudson B. (1982) I-Xe ages of individual Bjurböle chondrules. *Proc. Lunar Planet. Sci. Conf. 13th*, in *J. Geophys. Res.* **87**, pp. A303–A317.

Clayton R. N., Mayeda T. K., and Olsen E. J. (1982) Oxygen isotopes in ordinary chondrites: what is the origin of the iron groups? (abstract). *Meteoritics* **17**, 198–199.

Das Gupta S. P., Chakrabarti C., Sen Gupta P. R., and Dube A. (1978) A statistical analysis of chondrules in Chainpur meteorite (abstract). *Meteoritics* **13**, 434–435.

deGasparis A. A., Fredriksson K., and Brenner P. (1975) Composition of individual chondrules in ordinary chondrites (abstract). *Meteoritics* **10**, 390–392.

Dews J. R. (1966) The isotopic composition of lithium in chondrules. *J. Geophys. Res.* **71**, 4011–4020.

Dodd R. T. (1978a) The composition and origin of large microporphyritic chondrules in the Manych (L3) chondrite. *Earth Planet. Sci. Lett.* **39**, 52–66.

Dodd R. T. (1978b) Compositions of droplet chondrules in the Manych (L3) chondrite and the origin of chondrules. *Earth Planet. Sci. Lett.* **40**, 71–82.

Dodd R. T., Jr., Van Schmus W. R., and Koffman D. M. (1967) A survey of the unequilibrated ordinary chondrites. *Geochim Cosmochim. Acta* **31**, 921–951.

Evensen N. M., Carter S. R., Hamilton P. J., O'Nions R. K., and Ridley W. I. (1979) A combined chemical-petrological study of separated chondrules from the Richardton meteorite. *Earth Planet. Sci. Lett.* **42**, 223–236.

Fredriksson K., O'Keefe J., and Brenner P. (1978) The bulk composition of individual chondrules (abstract). *Meteoritics* **13**, 464.

Fujimaki H., Matsu-ura M., Sunagawa I., and Aoki K. (1981) Chemical compositions of chondrules and matrices in the ALH-77015 chondrite (L3). *Mem. Natl. Inst. Polar Res., Spec. Issue No. 20* (T. Nagata, ed.), pp. 161–174. National Institute of Polar Research, Tokyo.

Gomes C. B. and Keil K. (1980) *Brazilian Stone Meteorites.* Univ. of New Mexico Press, Albuquerque. 159 pp.

Gooding J. L. (1979) Petrogenetic properties of chondrules in unequilibrated H-, L-, and LL-group chondritic meteorites. Ph.D. thesis, Univ. of New Mexico, Albuquerque. 392 pp.

Gooding J. L. and Keil K. (1981) Relative abundances of chondrule primary textural types in ordinary chondrites and their bearing on conditions of chondrule formation. *Meteoritics* **16**, 17–43.

Gooding J. L., Keil K., Fukuoka T., and Schmitt R. A. (1980a) Elemental abundances in chondrules from unequilibrated chondrites: evidence for chondrule origin by melting of pre-existing materials. *Earth Planet. Sci. Lett.* **50**, 171–180.

Gooding J. L., Keil K., Fukuoka T., and Schmitt R. A. (1980b) The origins of chondrules as secondary objects: evidence from chemical-petrological heterogeneities (abstract). In *Lunar and Planetary Science XI*, pp. 345–347. Lunar and Planetary Institute, Houston.

Gooding J. L., Mayeda T. K., Clayton R. N., Keil K., Fukuoka T., and Schmitt R. A. (1982) Oxygen isotopic compositions of chondrules in unequilibrated chondrites: further petrological interpretations (abstract). In *Lunar and Planetary Science XIII*, pp. 271–272. Lunar and Planetary Institute, Houston.

Gopalan K., Rao M. N., Suthar K. M., and Venkatesan T. R. (1977) Cosmogenic and radiogenic noble gases in the Dhajala chondrite. *Earth Planet. Sci. Lett.* **36**, 341–346.

Grossman J. N. and Wasson J. T. (1982) Evidence for primitive components in chondrules from the Chainpur chondrite. *Geochim. Cosmochim. Acta* **46**, 1081–1099.

Grossman J. N., Kracher A., and Wasson J. T. (1979) Volatiles in Chainpur chondrules. *Geophys. Res. Lett.* **6**, 597–600.

Hamilton P. J., Evensen N. M., and O'Nions R. K. (1979) Chronology and chemistry of Parnallee (LL3) chondrules (abstract). In *Lunar and Planetary Science X*, pp. 494–496. Lunar and Planetary Institute, Houston.

Hughes D. W. (1977) A comparison between the mass distribution indices of chondrules and cometary meteoroids. *Earth Planet. Sci. Lett.* **33**, 428–436.

Hughes D. W. (1978a) Chondrule mass distribution and the Rosin and Weibull statistical functions. *Earth. Planet. Sci. Lett.* **39**, 371–376.

Hughes D. W. (1978b) A disaggregation and thin section analysis of the size and mass distribution of the chondrules in the Bjurböle and Chainpur meteorites. *Earth Planet. Sci. Lett.* **38**, 391–400.

Hughes D. W. (1980) The dependence of chondrule density on chondrule size. *Earth Planet. Sci. Lett.* **51**, 26–28.

Ikeda Y. and Takeda H. (1979) Petrology of the Yamato-74191 chondrite. *Mem. Natl. Inst. Polar Res., Spec. Issue No. 12* (T. Nagata, ed.), pp. 38–58. National Institute of Polar Research, Tokyo.

Keil K. and Fredriksson K. (1964) The iron, magnesium, and calcium distribution in coexisting olivines and rhombic pyroxenes of chondrites. *J. Geophys. Res.* **69**, 3487–3515.

Kimura M. and Yagi K. (1980) Crystallization of chondrules in ordinary chondrites. *Geochim. Cosmochim. Acta* **44**, 589–602.

Kimura M., Yagi K., and Onuma K. (1979) Petrological studies on chondrules in Yamato-74 meteorites. *Mem. Natl. Inst. Polar Res., Spec. Issue No. 12* (T. Nagata, ed.), pp. 114–133. National Institute of Polar Research, Tokyo.

King T. V. V. and King E. A. (1979) Size frequency distributions of fluid drop chondrules in ordinary chondrites. *Meteoritics* **14**, 91–96.

Lang B., Stakheav Y., Baryshnikova G. V., and Lavrukhina A. K. (1975) Mass/size frequency distribution of chondrules from four L-chondrites (abstract). *Meteoritics* **10**, 437–439.

Lux G., Keil K., and Taylor G. J. (1980) Metamorphism of the H-group chondrites: implications from compositional and textural trends in chondrules. *Geochim. Cosmochim. Acta* **44**, 841–855.

Lux G., Keil K., and Taylor G. J. (1981) Chondrules in H3 chondrites: textures, compositions, and origins. *Geochim. Cosmochim. Acta* **45**, 675–685.

Martin P. M. and Mills A. A. (1976) Size and shape of chondrules in the Bjurböle and Chainpur meteorites. *Earth Planet. Sci. Lett.* **30**, 281–291.

Martin P. M. and Mills A. A. (1978) Size and shape of near-spherical Allegan chondrules. *Earth Planet. Sci. Lett.* **38**, 385–390.

McSween H. Y., Jr. (1977) Chemical and petrographic constraints on the origin of chondrules and inclusions in carbonaceous chondrites. *Geochim. Cosmochim. Acta* **41**, 1843–1860.

Meyer S. L. (1975) *Data Analysis for Scientists and Engineers*. Wiley, New York. 513 pp.

Minster J. F. and Allègre C. J. (1981) Rb-87/Sr-87 dating of LL chondrites. *Earth Planet. Sci. Lett.* **56**, 89–106.

Nagahara H. (1981) Petrology of chondrules in ALH-77015 (L3) chondrite. *Mem. Natl. Inst. Polar Research, Special Issue No. 20* (T. Nagata, ed.), pp 145–160. National Institute of Polar Research, Tokyo.

Noonan A. F. and Nelen J. A. (1976) A petrographic and mineral chemistry study of the Weston, Connecticut, chondrite. *Meteoritics* **11**, 111–130.

Noonan A. F., Fredriksson K., Jarosewich E., and Brenner P. (1976) Mineralogy and bulk, chondrule, size-fraction chemistry of the Dhajala, India, chondrite (abstract). *Meteoritics* **11**, 340–343.

Onuma N., Clayton R. N., and Mayeda T. K. (1972) Oxygen isotope cosmothermometer. *Geochim. Cosmochim. Acta* **36**, 169–188.

Onuma N., Nishida N., Ohtsuka Y., Kimura M., and Yanai K. (1979) Origin and evolution of chondrules based on Na/Al-dispersion and convergence in Yamato-74 ordinary chondrites. *Mem. Natl. Inst. Polar Res., Special Issue No. 15* (T. Nagata, ed.), pp. 206–214. National Institute of Polar Research, Tokyo.

Osborn T. W., Smith R. H., and Schmitt R. A. (1973) Elemental composition of individual chondrules from ordinary chondrites. *Geochim. Cosmochim. Acta* **37**, 1909–1942.

Robert F., Merlivat L., and Javoy M. (1979) Deuterium concentration in the early solar system: hydrogen and oxygen isotope study. *Nature* **282**, 785–789.

Rosman K. J. R. and De Laeter J. R. (1978) Cadmium isotopic anomalies in stony meteorites (abstract). *Meteoritics* **13**, 619–621.

Schmitt R. A., Smith R. H., and Goles G. G. (1965) Abundances of Na, Sc, Cr, Mn, Fe, Co, and Cu in 218 individual chondrules via activation analysis, 1. *J. Geophys. Res.* **70**, 2419–2444.

Schmitt R. A., Smith R. H., Ehmann W. D., and McKown D. (1967) Silicon abundances in meteoritic chondrules. *Geochim. Cosmochim Acta* **31**, 1975–1985.

Sears D. W. and Axon H. J. (1976) Metal of high Co content in LL chondrites. *Meteoritics* **11**, 97–101.

Van Schmus W. R. and Wood J. A. (1967) A chemical-petrological classification for the chondritic meteorites. *Geochim. Cosmochim. Acta* **31**, 747–765.

Walter L. S. (1969) The major-element composition of individual chondrules of the Bjurböle meteorite. *Meteorite Research* (P. M. Millman, ed.), pp. 191–205. D. Reidel, Dordrecht.

Walter L. S. and Dodd R. T. (1972) Evidence for vapor fractionation in the origin of chondrules. *Meteoritics* **7**, 341–352.

Wlotzka F. (1981) Classification of unequilibrated ordinary chondrites (abstract). *Meteoritics* **16**, 403–404.

Wood J. A. (1967) Chondrites: their metallic minerals, thermal histories, and parent planets. *Icarus* **6**, 1–49.

The Compositions of Chondrules in Unequilibrated Chondrites: An Evaluation of Models for the Formation of Chondrules and their Precursor Materials

Jeffrey N. Grossman* and John T. Wasson*†

Institute of Geophysics and Planetary Physics, University of California, Los Angeles, California 90024

The compositions of chondrules from unequilibrated ordinary and carbonaceous chondrites can be used to constrain the nature of chondrule melting, the materials from which chondrules formed, and the relationship between chondrules and other chondritic components. Published chemical/petrographic studies of individual chondrules form a large data base of several hundred analyses, and are sufficient for evaluating many problems concerning chondrule origin. Chondrules are objects that show evidence both for melting and for incorporation of low-temperature materials. Many once-molten but refractory-rich objects in carbonaceous chondrites are probably not chondrules; few such objects exist in ordinary chondrites. Chondrules with outlying compositions cannot be used as constraints on models for chondrule formation since they may have formed under unusual conditions or from unusual precursors. There are differences between the compositions of different chondrule textural types in ordinary chondrites, in agreement with previous work; such differences also exist in carbonaceous chondrite chondrules, but they show a pattern distinct from that in ordinary chondrites. These variations demonstrate the necessity for producing both chemical and petrographic data while studying chondrules. The grand mean compositions of ordinary and carbonaceous chondrite chondrules reflect the compositions of the host chondrites for lithophile elements; siderophiles are uniformly depleted, Fe is intermediate. Opaque matrices show different compositions. Chondrules must have formed from the same components that were involved in the chemical fractionations seen between chondrite groups. That mean chondrules are not significantly depleted in volatiles compared to whole rocks precludes a condensation origin for chondrule melts. Chondrule compositions show wide variations in all elements. This heterogeneity is inconsistent with direct condensation models. Chondrules formed by the melting of pre-existing solids. Volatile elements are mainly in chondrule interiors rather than on surfaces. Their patterns of variation demonstrate that chondrule melts generally behaved as closed systems. Fractional vaporization was not an important mechanism for creating chondrule diversity. The melting and subsequent cooling of chondrules must have been brief. As a result, chondrules preserve chemical variations that existed among their precursors. Published models involve different sets of precursor solids. The covariations among the non-volatile lithophile elements in chondrules are much more consistent with models involving heterogeneous nebular precursors than with processed (igneous or metamorphic) or homogeneous solid precursors.

INTRODUCTION

The "solar" compositions and unequilibrated mineral assemblages of many chondrites indicate formation in the solar nebula with at most minor alteration by planetary igneous or

*Also at Department of Earth and Space Sciences
†Also at Department of Chemistry and Biochemistry

metamorphic processes. Chondrites are generally divided into coarse-grained "chondrules," fine-grained "matrix," and minor amounts of other petrographic assemblages. The chondrules are an extremely diverse array of objects accounting for an estimated 80% of the volume of some ordinary chondrites (Dodd, 1981), 30–50% of the CV and CO group carbonaceous chondrites (McSween, 1977a), and about 5% of the CM chondrites. No chondrules are found in CI chondrites.

The bulk of the mass of the chondrules resides in objects in the millimeter-size range. We define "matrix" to include the fine-grained, opaque materials described by Huss *et al.* (1981), which sometimes appears to be recrystallized, isolated metal and silicate grains, and the fine-grained "rims" around chondrules (Allen *et al.*, 1979, 1980). Other components found in some chondrite classes include aggregates and various types of refractory inclusions. The latter objects are much more abundant in carbonaceous chondrites than in other groups, occupying as much as 10–20% of the volume in some CO and CV chondrites (McSween, 1977b,c). Refractory chondrules and inclusions have also been reported in ordinary chondrites (Noonan, 1975; Nagahara and Kushiro, 1982; Bischoff and Keil, 1983). These tend to be small, and probably occupy far less than 1% of the total volume.

Although there is no consensus among researchers as to which objects should be classified as true chondrules, most would agree that chondrules were once molten or partially molten, and that they existed at least briefly as independent entities. We discuss the detailed definition of chondrules in a later section.

Major questions regarding chondrules are: (1) What was the mechanism responsible for the production of chondrule melts? (2) What was the nature of the chondrule precursor material? (3) What is the relationship between chondrules and the other chondritic components? (4) To what degree and how were the chondrules and the other chondritic components fractionated from each other during chondrite agglomeration?

The high abundance of chondrules in ordinary chondrites suggests that either the chondrule-melting mechanism was very efficient, converting most of the material in the ordinary chondrite formation location into chondrules, or that chondrite components have been mechanically separated, and are not representative of the total solid material in the nebula at that formation location. The scarcity of chondrule-poor chondrites implies that the formation mechanism was efficient. However, there is also evidence that chondrules have been size-sorted (Dodd, 1976; King and King, 1978), as well as evidence that chondrite elemental fractionations are due to the differential agglomeration of coarse and fine components (Grossman and Wasson, 1983). It seems probable that chondrites preserve a record of all types of material present in the solar system at the time of chondrule formation, despite the fact that agglomeration efficiency may have been different for the various components.

If it can be shown that chondrules behaved as closed systems since the time that they became molten, then their compositions can be used directly to deduce the compositions of the chondrule precursor materials. If chondrules were open systems, experiencing mass transport across their surfaces either when they were liquid or after solidification, then the estimation of precursor compositions is more difficult. Determination of the compositions of chondrule precursors would provide important constraints on the mechanism(s) responsible for chondrule melting, and information on physical conditions and processes early in the history of the solar system. A first-order question is whether the precursor was a gas or a solid. If it was solid, were the precursors nebular condensates, rocks on asteroidal surfaces, interstellar grains, etc.? In this case, the melting mechanism(s) must be able to operate under the conditions in which the precursors are expected to exist.

In this paper we review the progress toward answering some of these questions made through the study of chondrule compositions. Special emphasis is placed on studies in which both chemical and textural information was gathered on the same chondrule set. In light of the historical controversy over which objects are true chondrules, the value of a data set is greatly increased when petrographic descriptions accompany the analyses. Because of the compositional differences between textural types of chondrules (Gooding *et al.*, 1980), it is necessary to have petrographic data in order to correct for sampling biases.

It is also essential that the chondrule data be obtained from unequilibrated chondrites (in general, type 3 chondrites). Systematic changes in chondrule composition with petrologic type have been noted by Schmitt *et al.* (1968), Osborn (1972), Gooding (1979), and Lux *et al.* (1981). The unequilibrated-chondrite chondrules best preserve the record of processes affecting the precursors.

DEFINITION OF CHONDRULE

Every researcher has his own working definition or set of criteria designating which objects to call chondrules. The common element in almost all definitions is the requirement that the objects have been molten to an extensive degree. Other criteria have commonly included chondrule outline, size, mineralogy, and texture. Each researcher tends to select a different subset of the population of chondritic objects for analysis. The result of these differences is that meaningful comparisons of data from different studies can only be done when petrographic data are also present.

It is probable that there is more than one genetic class of chondrule-like objects. In carbonaceous chondrites, many refractory inclusions show clear evidence of melting, but are likely to be unrelated to chondrules. These objects are quite different from the ferromagnesian silicate-rich chondrules in composition and in size frequency distribution. The volatile elements in refractory inclusions are commonly contained in materials associated with late-stage (secondary) alteration (Grossman, 1980). By comparison, the interiors of silicate-rich chondrules almost always contain volatile elements such as Na, K, and S; apparently, these elements were primary features of the melts, and not associated with post-melting alteration. Many coarse-grained inclusions are quite large, reaching centimeter-size in CV chondrites. Silicate-rich chondrules, which are more abundant, only rarely reach this size.

Many refractory inclusions must have ceased equilibrating with nebular gas at very high temperature; it seems probable that many are distillation residues, indicating orders-of-magnitude more massive starting materials (Kurat, 1970; Chou *et al.*, 1976; Hashimoto *et al.*, 1979; Kornacki, 1981). In contrast, the presence of primary volatiles in silicate-rich chondrules indicates some equilibration of the precursor material with nebular gases at much lower temperatures. If distillation processes were responsible for the differences between refractory inclusions and silicate-rich chondrules, then, paradoxically, the latter objects seem to have experienced less volatile loss despite their smaller sizes. Thus, there is evidence in the carbonaceous chondrites for at least two populations of igneous objects differing in origin. In light of these differences, it is unwise to combine the data on these two types of materials. It is conceivable that they share some aspects of their formation histories (such as the melting mechanism), but it seems equally likely that they do not. To avoid unnecessary constraints on our attempts to formulate models for chondrule formation, it is preferable to treat them as separate populations. We recommend that these refractory objects continue to be designated "inclusions" rather than "chondrules."

In ordinary chondrites, there are also refractory-rich objects that were once molten. These commonly contain volatiles such as Na and K and tend to be quite small, $<300\ \mu m$ (Bischoff and Keil, 1983). Presently available evidence does not indicate that these objects form a distinct population; it seems reasonable to classify them as chondrules. They are rare enough that their inclusion will probably not perturb a mass-weighted mean. If it later appears that some properties of these refractory-rich objects are inconsistent with the formational processes required for the mafic chondrules, it may be desirable to reassign them to a separate population, as we recommend for the refractory inclusions in carbonaceous chondrites.

Several "metallic" chondrules appeared in studies of ordinary chondrites by Osborn (1972) and Gooding and Keil (1981). Our own observations show that these commonly contain mafic minerals, and show textural evidence for the melting of low-temperature solids (occurrence of FeS and, in Semarkona, magnetite). These objects also seem best classified as chondrules; they only differ from the rest of the population in their content of minor components. In this sense they are similar to refractory-rich chondrules.

We suggest the following criteria: (1) Chondrules must show evidence for melting. Evidence for melting has been summarized by King and King (1978) and Gooding and Keil (1981), and includes the presence of igneous glass, quench texture, or circular to partially circular outline along the original droplet surface. (2) Based on the preceding discussion, chondrules must show evidence of the incorporation of materials formed at low nebular temperatures (<700 K). The best evidence is the presence of sulfides and oxidized Fe in the chondrule interior in non-negligible amounts [S/Si and FeO/(FeO + MgO) atom ratios >0.01]. Chondrules with extreme compositions, such as refractory-rich and metal-rich ones, should not be used to constrain models for chondrule formation because they may record information about different or uncommon nebular processes and materials. While this information is important, it should be evaluated separately from information derived from a study of the common, mafic-rich chondrules. In most ordinary chondrites, the large majority of the objects having sizes >0.3 mm seem to be chondrules.

STUDIES OF CHONDRULE COMPOSITIONS

Neutron activation analysis (NAA)

Before the development of neutron activation techniques around 1960 there were very few attempts to measure the compositions of chondrules. Nearly all of the research was directed toward the petrographic study of chondrules in meteorite thin sections. Schmitt *et al.* (1965) were able to determine Na, Sc, Cr, Mn, Fe, Co, and Cu by instrumental neutron activation analysis (INAA) in several hundred chondrules separated from ordinary and carbonaceous chondrites. This data set was later extended to include Si (Schmitt *et al.*, 1967) and, by radiochemical neutron activation analysis (RNAA), the rare earth elements (REE) (Schmitt *et al.*, 1968). Osborn *et al.* (1973, 1974) used improved INAA techniques to obtain more precise data on Fe, Al, Na, Mn, Cr, Sc, Co, Si, Ni, Ca, and V in similar chondrule suites. The complete data from the Schmitt and Osborn work are listed in Osborn (1972), where the compositions of over 600 chondrules are tabulated.

The shortcomings of the early INAA studies of chondrules were the lack of petrographic data accompanying the chemical results and the fact that most of the chondrules were from equilibrated chondrites. Many coarse-grained objects that can be separated from chondrites, and especially carbonaceous chondrites, fall into categories that may not share a common origin with chondrules as defined above. As we have stated, the working definition of "chondrule" should always be defined and must include textural/mineralogical parameters as well as physical ones in order to prevent ambiguity during the interpretation of the data. Because INAA is a non-destructive technique, recent researchers were able to prepare polished thin sections of chondrules for which chemical data were obtained. For this reason, and due to vast improvements since the mid-1970's in gamma-ray detector technology, recent NAA studies are providing more useful data.

Gooding *et al.* (1980) and Gooding and Keil (1981) report analyses of ~140 chondrules from a suite of ordinary chondrites. More than half of the chondrules were from type 3 chondrites. They obtained extensive petrological data as well as nearly complete NAA bulk chemical data for Al, V, Cr, Fe, Mn, Mg, Ca, Na, Sc, Zn, La, Sm, Eu, Yb, Lu, Hf, Co, Ni, Ir, and Au. These data are tabulated in Gooding (1979). Grossman has similarly obtained data for 11 Bjurböle (L4) chondrules, 36 Chainpur (LL3.4) chondrules, and 34 Semarkona (LL3.0) chondrules (Grossman *et al.*, 1978, 1979; Grossman, 1982b; Grossman and Wasson, 1982, 1983). These data include the same element set as Gooding's work, as well as K, Ga, Ge, As, Se, Cd, Ce, Nd, Tb, Ho, Ru, and Os. Combined textural and NAA data have also been obtained for 10 Chainpur chondrules by Kurat *et al.* (1982).

Electron probe microanalysis (EPMA)

The electron probe is another tool that has been widely used to determine chondrule bulk compositions. It differs from NAA in that no trace elements are determined; the elements Mg, Al,

Si, Ca, and Fe are determined with high precision, and Na, K, Ti, Cr, Mn, P, S, and Ni with somewhat lower precision. Several variations of this method may be found in the literature.

Walter (1969) separated 82 chondrules from Bjurböle on the basis of shape and surface texture, fused them to produce a uniform glass, and analyzed them with a focused electron beam. This work was extended to include 61 Chainpur chondrules and 49 Allende (CV3) chondrules (Dodd and Walter, 1972; Walter and Dodd, 1972). DeGasparis *et al.* (1975) and Fredriksson (1982) separated ~340 ordinary chondrite (OC) chondrules, crushed them, and pressed them into pellets before analysis with a defocused beam. As in the early NAA studies, there were no petrographic controls in these studies; however, they have an advantage over many other EPMA studies in that sample heterogeneity is not a major problem.

Evensen *et al.* (1979) and Hamilton *et al.* (1979) were able to characterize their samples petrographically by crushing only a part of each chondrule prior to analysis and reserving the remaining piece for the preparation of a thin section. They measured both chemical and isotopic parameters on the powdered material using EPMA and isotope-dilution techniques. Bulk chemical data for major elements, Rb, Sr, and REE were thus obtained for 12 Richardton (H5) and 4 Parnallee (H3.6) chondrules.

EPMA of chondrules in chondrite thin sections using a defocused beam also has been reported. In this technique, several analyses covering the surface of a chondrule are averaged to give the bulk composition. This has been done by McSween (1977a; unpublished report, 1977) for ~120 carbonaceous chondrite (CC) chondrules and ~60 chondrules from Bishunpur (L3.1), Tieschitz (H3.6), and Chainpur; Snellenburg (1978) for ~50 Krymka (L3.0) and Semarkona chondrules; Lux *et al.* (1980, 1981) for ~370 chondrules from H-chondrites (including 120 from type 3 H-chondrites); Fujimaki *et al.* (1981) for ~60 and Nagahara (1981) for ~100 Allan Hills A77015 (L3.5) chondrules.

Several researchers have used the above method only for fine-grained samples; for coarse samples they have determined modal mineral abundances and reconstructed the bulk composition using these and focused beam analyses of the individual minerals [Dodd (1978a,b) for 27 chondrules from Manych (L3.4); Kimura and Yagi (1980) for ~20 Yamato 74191 (L3) chondrules; Simon and Haggerty (1980) for ~30 Allende chondrules]. This method tends to be less precise for elements such as P, S, Cr, and Ni, which are mainly sited in minor and trace phases.

The disadvantage of studies of chondrules in thin sections is the possibility of non-representative sectioning of inhomogeneous chondrules. Major sampling problems are possible if the dimensions of the major phases are comparable to those of the host chondrule, as commonly occurs in porphyritic chondrules. As a result, bulk data may be less precise than those from whole or fused chondrule studies. While chondrule mean compositions will tend to be similar between the two types of studies, chemical trends can be different. This will be evident later in the discussion.

Table 1 summarizes the work that has been done on the chemical/petrographical characterization of chondrules. The carbonaceous chondrites have only been subjected to EPMA studies, mainly by McSween; therefore, almost nothing is known about the trace elements in their

Table 1. Chemical/petrographic studies of chondrules.

Sample	Type	No.	Ref.	Method*	Sample	Type	No.	Ref.	Method*
Carbonaceous Chondrites:									
Allende	CV3	15	k	EPMA	Kainsaz	CO3	48	k	EPMA
		~30	m	EPMA	Kakangari	anom	19	k	EPMA
Arch	CV3	8	k	EPMA	Karoonda	anom	10	k	EPMA
Bali	CV3	17	k	EPMA	Lance	CO3	43	k	EPMA
Coolidge	CV3	21	k	EPMA	Leoville	CV3	13	k	EPMA
Efremovka	CV3	17	k	EPMA	Mokoia	CV3	9	k	EPMA
Felix	CO3	32	k	EPMA	Ornans	CO3	30	k	EPMA
Grosnaja	CV3	16	k	EPMA	Renazzo	anom	11	k	EPMA
Isna	CO3	26	k	EPMA	Vigarano	CV3	15	k	EPMA
Kaba	CV3	15	k	EPMA	Warrenton	CO3	39	k	EPMA
CM chondrites: none									
Unequilibrated Ordinary Chondrites:									
A77015	L3.5	~100	l	EPMA	Manych	L3.4	32	a	EPMA
Bishunpur	L3.1	19	k	EPMA	Parnallee	H3.6	4	f	EPMA,ID
Bremervörde	H3.9	30	i	EPMA	Semarkona	LL3.0	18	c	NAA
Chainpur	LL3.4	14	c	NAA			33	e	NAA
		36	e	NAA			39	n	EPMA
		10	h	NAA	Sharps	H3.4	30	i	EPMA
		17	k	EPMA	Tieschitz	H3.6	17	c	NAA
Dhajala	H3.8	15	c	NAA			30	i	EPMA
		30	i	EPMA			26	k	EPMA
Hallingeberg	L3	24	c	NAA	Y74191	L3	~20	g	EPMA
Krymka	L3.0	8	n	EPMA					
Equilibrated Ordinary Chondrites:									
Bjurböle	LL4	11	d	NAA	Saratov	L4	13	c	NAA
Hamlet	LL4	8	c	NAA	Soko-Banja	LL4	7	c	NAA
H-group	H4-6	~250	j	EPMA	Tennasilm	L4	7	c	NAA
Ochansk	H4	8	c	NAA	Weston	H4	11	c	NAA
Richardton	H5	10	b	EPMA,ID					
Enstatite Chondrites: none									

* Analytical methods are EPMA: electron probe microanalysis, NAA: neutron activation analysis, ID: isotope dilution.

References:

a) Dodd (1978a,b)
b) Evensen *et al.* (1979)
c) Gooding *et al.* (1980)
d) Grossman *et al.* (1978)
e) Grossman and Wasson (1982, 1983)
f) Hamilton *et al.* (1979)
g) Kimura and Yagi (1980)
h) Kurat *et al.* (1982)
i) Lux *et al.* (1980)
j) Lux *et al.* (1981)
k) McSween (1977a)
l) Nagahara (1981)
m) Simon and Haggerty (1980)
n) Snellenburg (1978)

chondrule populations (Osborn, 1972, had no petrographic controls). There are *no* published chemical/petrographical studies of CM or EH (enstatite) chondrites, although the latter group is currently being studied by Smith *et al.* (1983). In effect, the bulk of our knowledge of chondrules comes from studies of unequilibrated ordinary chondrites, especially Chainpur, Semarkona, and Tieschitz.

THE BULK COMPOSITION OF CHONDRULES

Ordinary chondrite chondrules

In order to produce the best estimate of the mean composition of all chondrules, we have combined all of the chemical/petrographic data sets in the literature having more than 10 analyzed samples. For unequilibrated ordinary chondrite chondrules, data from eight sources (McSween, 1977a; Dodd, 1978a,b; Snellenburg, 1978; Gooding, 1979; Kimura and Yagi, 1980; Lux *et al.*, 1981; Nagahara, 1981; Grossman and Wasson, 1982, 1983) were utilized.

Gooding and Keil (1981) divided silicate-rich chondrules by texture and mineralogy into several categories, including porphyritic olivine (PO), porphyritic pyroxene (PP), porphyritic olivine-pyroxene (POP), radial pyroxene (RP), cryptocrystalline (C), granular (G), and barred olivine (BO). Petrographic descriptions found in the eight data sets allow the chondrules to be assigned into three general textural types. We have combined data for those chondrules with (1) porphyritic texture (PO + PP + POP); (2) non-porphyritic or "droplet" texture (RP + C + G); and (3) barred olivine texture (BO). First, an average composition was calculated for each category. Then the three compositions were averaged together, using the modal data for ordinary chondrites from Gooding and Keil (1981) for weighting, to produce a grand mean composition. The three groups (H, L, LL) were combined since no large differences in chondrule modes or chondrule mean compositions are found (Gooding *et al.*, 1980). The compositions of each textural group and the grand mean chondrule, as well as the modal abundance data, are listed in Table 2.

Clear differences are seen between the mean compositions of the different chondrule textural categories, in agreement with previous observations by Gooding *et al.* (1980), Lux *et al.* (1981), Nagahara (1981), and Grossman and Wasson (1982). These are shown in Fig. 1, in which the mean compositions of the two less abundant types (non-porphyritic and BO) are normalized to that of the porphyritic type. The data are arranged in approximate order of decreasing condensation temperature and grouped by geochemical affinity. As expected, Mg/Si ratios reflect the mineralogy: they are the lowest in the pyroxene-dominated non-porphyritic chondrules, highest in the olivine-dominated BO chondrules, and intermediate in the porphyritic chondrules. Refractory element/Si ratios are correlated with Mg/Si, being high in BO and low in non-porphyritic chondrules. Porphyritic chondrules are also rich in siderophile elements compared to non-porphyritic chondrules (Gooding *et al.*, 1980); the statistics are poor for BO chondrule siderophiles, but they seem to be depleted in metal. Our interpretation of the origin of these differences will be discussed later. The existence of compositional differences between textural types, and

Table 2. The mean concentrations of elements in ordinary chondrites and their components.*

Sample	Vol.	Lithophile Elements												
		Al	Ca	Ti	Sc	La	Sm	V	Mg	Si	Cr	Mn	Na	K
	%	mg/g	mg/g	mg/g	μg/g	ng/g	ng/g	μg/g	mg/g	mg/g	mg/g	mg/g	mg/g	mg/g
Porph.[a]	65	15.7	15.2	0.91	10.6	400	275	98	192	218	3.9	2.9	8.5	1.01
Non-porph.	12	13.3	14.1	0.72	7.7	320	233	78	162	247	4.3	3.8	8.1	1.48
Barred Ol.	3	23.8	17.5	1.32	12.6	570	380	97	205	199	3.5	2.2	9.7	1.01
Mean cdl.	80	15.7	15.1	0.90	10.3	400	273	95	189	222	3.9	3.0	8.5	1.08
H, L, LL[b]		12.0	12.4	0.71	7.8	315	196	77	147	183	3.6	2.4	6.3	0.84
Opaq. Mx.[c]	12	13.7	7.5	0.48	—	—	—	—	121	170	3.0	3.1	11.7	2.23
CI		8.6	9.3	0.49	5.9	240	150	56	96	105	2.6	1.9	5.0	0.55

[a] Chondrule compositions compiled from sources a, c, e, g, i, k, m, and n from Table 1.
[b] Chondrite bulk analyses are from Mason (1971) and from unpublished analyses done at UCLA.
[c] Matrix data compiled from unequilibrated ordinary chondrite analyses by Huss *et al.* (1980) and by Fujimaki *et al.* (1980).
* Also shown are data for CI chondrites used for normalization.

non-trivial differences in the distribution of textural types in literature data sets, point out the advantages gained in chemical/petrographic studies. They are also the reason why the largest set of compositional data (Osborn, 1972) cannot be used here.

Carbonaceous chondrite chondrules

Only two significant chemical/petrographic data sets for carbonaceous chondrite chondrules exist: McSween (1977a) and Simon and Haggerty (1980). Each calculated both modes and

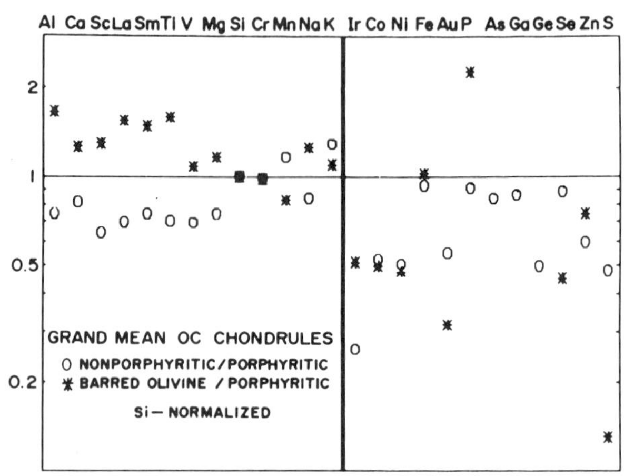

Fig. 1 Variations in composition between different textural types of chondrules in unequilibrated ordinary chondrites. The data are normalized to the composition of the most abundant type of chondrule (porphyritic) and to Si. The non-porphyritic category comprises chondrules described as having RP, C, and G textures (Gooding and Keil, 1981); the barred olivine category comprises only BO chondrules; the porphyritic chondrules used for normalization comprise the PO, POP, and PP textures. The barred chondrules have high and the non-porphyritic chondrules have low refractory lithophile/Si ratios compared to porphyritics. Both groups are comparatively low in siderophiles. The data were compiled from sources a, c, e, g, i, k, l, and n from Table 1.

Table 2. *continued*

Siderophile and Other Elements									
Ir ng/g	Co μg/g	Ni mg/g	Fe mg/g	Au ng/g	As μg/g	Se μg/g	Zn μg/g	P mg/g	S mg/g
290	238	5.4	130	69	1.00	7.9	26	0.34	4.2
85	141	3.1	135	43	0.95	8.0	17	0.35	2.3
134	108	2.4	120	20	—	3.3	18	0.70	0.5
251	218	5.2	131	63	0.98	7.7	24	0.35	3.7
510	600	13.3	231	177	1.67	9.1	56	0.97	22.1
—	—	6.1	246	—	—	—	—	1.90	3.3
450	505	10.6	182	142	1.70	20.0	330	0.84	59.1

mean compositions for his own petrographic classification scheme. These were combined to produce the CC chondrule grand means. The differences between textural types in the McSween set alone are shown in Fig. 2. The sample data basically are grouped in the same way as in Fig. 1; however, in McSween's classification system the BO chondrule data were combined with some petrographically related porphyritic chondrules (his type II). A different trend is observed here than was seen in OC chondrules. While Mg/Si ratios still reflect the chondrule mineralogy, the non-porphyritic (RP) chondrules have high refractory element/Si ratios and the type II chondrules have low ones. Data from the same analyst (McSween, 1977a) on OC chondrules are quite consistent with the rest of the data in Fig. 1, showing that this effect is probably not due to analytical error. This is the first indication of a fundamental compositional difference between CC and OC chondrules.

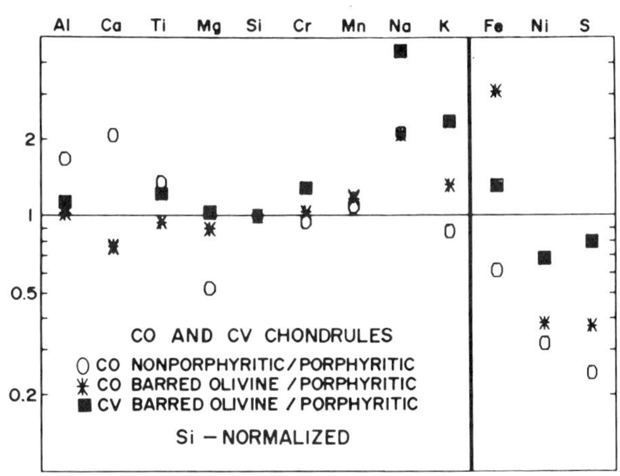

Fig. 2 Variations in composition between different textural types of chondrules in CO and CV chondrites from McSween (1977a,b). Plotted are type II and type III categories normalized to type I in each chondrite group. Type III are mainly RP chondrules; type II contains both BO and some related porphyritics. Type I comprises the abundant porphyritic category. Type III chondrules are absent from CV. In contrast with OC chondrules, non-porphyritic chondrules in CO chondrites have high refractory lithophile/Si ratios. Type II chondrules in both CO and CV groups have comparable refractory contents to type I; this cannot be directly compared to the OC barred chondrule data, since type II contains some porphyritic chondrules.

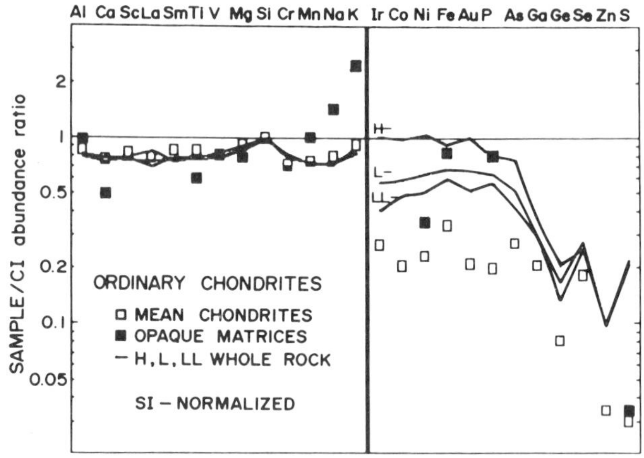

Fig. 3 The grand mean composition of ordinary chondrite chondrules is plotted together with the compositions of fine-grained matrix (Huss *et al.*, 1981; Fujimaki *et al.*, 1981) and the three groups H, L, and LL. The data are normalized to CI chondrite composition (Table 2) and to Si. Chondrule and whole rock abundance ratios for the lithophile elements are nearly identical. Opaque matrix shows a different pattern. Chondrules are depleted in all siderophile elements.

In all, 375 analyses of OC chondrules, 135 of which included trace elements by NAA, and 390 analyses of CC chondrules, none of which included trace elements, were used in the calculations.

Comparison of the compositions of chondrules, chondrites, and matrices

The grand mean OC chondrule composition is presented graphically in Fig. 3, normalized to CI chondrites and to Si. Refractory element abundances are barely resolved from those in the host chondrites, but may be a few percent higher. Although the *concentrations* of these elements tend to be higher in the chondrules than in the chondrites (Gooding *et al.*, 1980; Table 2), there appear to be only small net differences in *Si-normalized abundances*. The mean Mg/Si ratios in the chondrules also are close to those in the bulk chondrites, but perhaps 4–6% higher in chondrules than in chondrites.

That OC chondrules do not appear to differ greatly in refractory elements relative to whole rock is hardly surprising because the rocks may be 80% chondrules. Mass balance shows that ratios in the matrix 2× lower would only manifest themselves as 10% enrichments in the chondrules. The averaged data on OC matrices by Huss *et al.* (1981) and Fujimaki *et al.* (1981) (also shown in Fig. 3 and Table 2) show depletions in Ca, Ti, and Mg, and a modest enrichment in Al. Thus, opaque matrix is complementary to the chondrules for Ca, Ti, and Mg if the small enrichments of those elements in chondrule/whole rock ratios are real. The case with Al is not so simple. As discussed in some detail by Grossman and Wasson (1983), the Al/Ca ratio in the opaque matrix from the three most unequilibrated ordinary chondrites is about 2× higher than the CI ratio; in contrast, the bulk Al/Ca ratio in chondrules and chondrites is the same as the CI ratio. Grossman and Wasson (1983) discussed various possible cosmic explanations for this, and found none that seemed plausible.

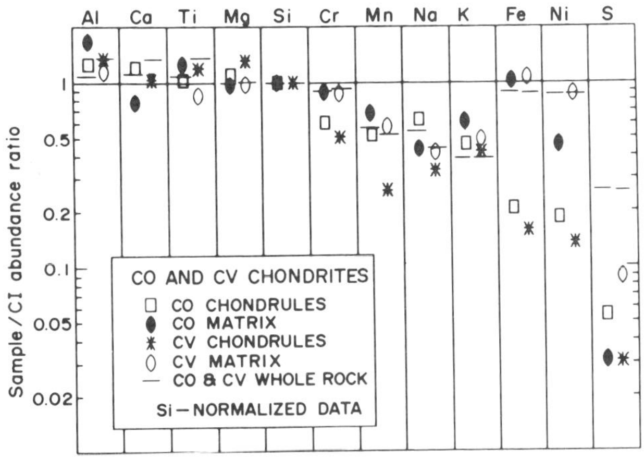

Fig. 4 The grand mean compositions of CO chondrules and CV chondrules (McSween, 1977a,b; Simon and Haggerty, 1980) are plotted together with the compositions of coexisting fine-grained matrix (McSween and Richardson, 1977) and whole-rock CO and CV chondrites. The data are normalized to CI chondrites and to Si. Chondrule lithophile/Si ratios are similar to those in whole rock except for Mg/Si, which is higher, and Cr/Si and Mn/Si, which are lower than in whole rock. Matrices show variations similar to those in ordinary chondrites.

They suggested that the analyzed matrices may not have been representative of the bulk non-chondrule matter in these meteorites.

The CO and CV chondrites contain a far smaller fraction of chondrules than ordinary chondrites, and therefore might be more likely to show differences in refractory abundances between chondrules and whole rock. The abundance of chondrules in CO chondrites may not be well-defined because of petrographic difficulties in distinguishing between chondrules (and their fragments) and non-chondrule material; we use the observations of McSween (1977a,b), which show a chondrule fraction in these meteorites of about half that in ordinary chondrites. In addition, meteorites in these groups contain refractory inclusions in amounts of 1–9%. Therefore it is somewhat surprising that CO and CV chondrules also have refractory abundances that are nearly the same as the bulk meteorites (Fig. 4, Table 3). The mean Mg/Si abundance ratios in these chondrules are considerably higher than those in the bulk meteorites (Ahrens *et al.*, 1973), and are also higher than the Ca/Si and Ti/Si abundance ratios; it is possible that this may reflect high sampling uncertainties because there are too few components with low Mg/Si ratios to provide a mass-balance. Data from McSween and Richardson (1977) generally yield low CC matrix to chondrule ratios for Ca, Ti, and Mg, but not for Al, similar to the patterns found in ordinary chondrites.

The OC chondrule/bulk ratios are near one for Na, K, Cr, and Mn (Fig. 3); matrix/bulk ratios are higher for all but Cr. CC chondrules show quite a different pattern (Fig. 4); chondrule/bulk ratios are less than one for Cr and Mn, but close to one for Na and K. Matrix/bulk ratios for these elements in CO and CV chondrites are unity to within the precision of the data.

All measured siderophile and chalcophile elements, including Fe, are significantly depleted in OC and CC chondrules. This has been shown previously by Gooding *et al.* (1980) and Grossman

Table 3. The mean concentrations of elements in carbonaceous chondrites and their components.

Sample	Vol. %	Al mg/g	Ca mg/g	Ti mg/g	Mg mg/g	Si mg/g	Cr mg/g	Mn mg/g	Na mg/g	K mg/g	Ni mg/g	Fe mg/g	S mg/g
CO Porph.[a]	34	21.4	21.9	1.00	214	209	3.12	1.92	5.9	0.51	4.1	70	6.7
CO Non-porph.	2	41.0	52.0	1.55	128	238	3.35	2.39	14.2	0.50	1.47	49	1.87
CO Barred Ol.	1	18.5	14.1	0.80	162	177	2.74	1.89	10.4	0.56	1.32	184	2.13
Mean CO cdl.	37	21.6	22.1	1.00	210	208	3.11	1.93	6.3	0.51	3.9	75	6.4
CO group[b]	—	14.3	15.9	0.75	145	160	3.56	1.65	4.1	0.33	14.0	246	21.2
CO Opaq. Mx.[c]	39	17.3	8.7	0.74	113	125	2.76	1.54	2.6	0.40	5.8	225	2.21
CV Porph.	43	22.3	18.6	1.12	243	202	2.52	0.94	3.0	0.44	2.77	55	3.5
CV Non-porph.	—	—	—	—	—	—	—	—	—	—	—	—	—
CV Barred Ol.	<1	20.1	11.1	1.09	199	160	2.58	0.89	10.6	0.82	1.50	57	2.20
Mean CV cdl.	43	22.2	18.5	1.12	242	201	2.52	0.94	3.2	0.45	2.74	55	3.48
CV bulk	—	17.5	18.6	1.05	145	156	3.61	1.46	3.3	0.32	13.6	237	22.9
CV Opaq. Mx.	39	11.9	12.2	0.50	112	125	2.74	1.30	2.5	0.32	11.1	233	6.2

[a] All chondrule analyses are by McSween (1977b). The three types of chondrules are actually McSween's Type I (porphyritic), Type III (radial pyroxene), and Type II (barred plus some porphyritic).
[b] CO and CV analyses are from Kallemeyn and Wasson (1981, 1983).
[c] Matrix analyses are from McSween and Richardson (1977).

and Wasson (1982). For the most part, the depletions relative to whole rock seem to be unrelated to volatility, although Semarkona chondrules have consistently low Ge/Ni ratios relative to whole rock. The S values on each plot are surprisingly low (in chondrules and in matrix). Only EPMA data are available for S, and we suspect that large sulfide grains may have been avoided. Iron is less depleted than the other elements in the mean composition and in many individual chondrules, reflecting the fact that it may have been partially oxidized at the time when some or most chondrules formed. We see no evidence for other elements being in both metal and silicate phases in OC chondrules at the time of chondrule metal-silicate fractionation. However, the depletion of Cr and Mn in CC chondrules is a possible indication that those elements were partially reduced in CO and CV. The OC matrix measurements show low Ni and S abundances; we attribute this to the exclusion of coarse metal and sulfide particles in those microprobe measurements. Iron abundances in the matrix are high.

The fact that chondrules have nearly the same abundances of non-volatile lithophiles as the host chondrite has significance in terms of the formation of the chondrites. From the fact that chondrules and bulk ordinary chondrites show similar fractionation effects (with respect to CI chondrites), Grossman and Wasson (1983) concluded that OC chondrules formed from the same chondritic components that were involved in the nebular fractionations seen between chondrite groups. Although similar analyses have never been done for CC chondrules, we assume that this is also true for them. The only known processes that can produce these fractionations involve separating nebular solids from each other and from gas (Suess, 1965; Larimer and Anders, 1970; Baedecker and Wasson, 1975; Larimer, 1979; Kerridge, 1979). Therefore, it follows that chondrules had to form *after* the condensation of solids, and so cannot have formed by the condensation of stable or metastable liquids directly from the gas.

The lack of severe volatile depletions also is evidence that chondrules formed from previously condensed solids. Equilibrium condensation calculations indicate that the alkalis condense below 1000 K under equilibrium conditions (see summary in Wasson, 1983). This is below the temperatures at which silicate liquids can exist even with 100–200 degrees of undercooling. Equilibrium condensation (Wood, 1962; Suess, 1963; Podolak and Cameron, 1974; Wood and McSween, 1976; Herndon and Suess, 1977) and constrained equilibrium condensation of chondrule droplets (Blander and Katz, 1967; Blander and Abdel-Gawad, 1969) all require the entry of alkalis and other volatiles into previously solidified chondrules. Such theories thus predict a dramatic enrichment of moderately volatile lithophiles in fine-grained material, such as matrix. However, by mass balance we find that most of the alkalis in unequilibrated chondrites are inside chondrules. The presence of primary chondrule sulfides offers a similar argument against condensation of chondrules (Larimer and Anders, 1970; Dodd, 1978b; Grossman *et al.*, 1979), although it is conceivable that FeS could condense at relatively high temperatures if Fe-Ni metal could not condense from the gas (Blander, 1971).

DIVERSITY OF CHONDRULE COMPOSITIONS

It has long been recognized that chondrules represent a wide variety of textures and mineralogies (Tschermak, 1883). They also show an equally wide diversity of chemical compositions. In studies involving a large suite of elements (Gooding *et al.*, 1980; Grossman and Wasson, 1982) chondrules are seen to show a minimum of a factor of two range in the concentration of every element; siderophiles and chalcophiles show enormous variation (less than three orders of magnitude), refractory and volatile lithophiles show less variation (about one order of magnitude), and the "common" lithophile elements (Mg and Si) show the least (about a factor of two). These variations add constraints to the chondrule formation process and provide direct information about the compositions of chondrule precursor materials.

These authors have interpreted the siderophile data to indicate either non-representative sampling of pre-existing solids in which metal/silicate ratios were variable or the loss of metal/sulfide assemblages as immiscible liquids during melting. Evidence that the latter process did occur in some chondrules was presented by Grossman (1982a). The relative importance of the processes is not known. These processes left some chondrules nearly free of metal and some with chondritic proportions of metal; the typical chondrule generally is depleted in siderophiles by a factor of four relative to Si. A very few chondrules managed to sample material with siderophile element/Si and siderophile element/Fe ratios greater than the same ratios in chondrites. In order to resolve true variations among chondrule lithophile elements a correction must be made for the variable dilution effect of metal and sulfide components. This may be accomplished by normalizing the lithophile

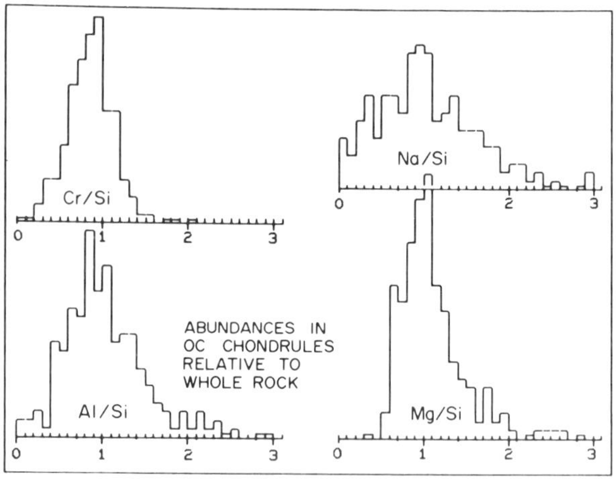

Fig. 5 Histograms of ordinary chondrite chondrule compositions. These data are normalized to mean ordinary chondrite values: Al/Si = 0.081, Mg/Si = 0.88, Cr/Si = 0.12, and Na/Si = 0.043 g/g. Four hundred individual chondrule analyses were used, 100 of which had no Cr determination. The widest ranges occur for Al and Na, with volatility far from that of Si, the normalizing element; the smallest ranges are seen in Cr and Mg, which are similar in volatility to Si.

data to a reference element such as Si, or by making dilution corrections similar to those in Grossman and Wasson (1983).

Data for several elements in OC chondrules normalized to Si and to bulk ordinary chondrites are plotted as histograms in Fig. 5. Even after normalization, refractory Al, non-volatile Mg and Cr, and moderately volatile Na all show wide ranges, reflecting real differences in silicate compositions among the chondrules. The widest ranges are obtained for elements that have large differences in volatility compared to the normalizing element. Thus, in Fig. 5, Al and Na show much wider ranges than Mg and Cr. Were the normalizing element chosen to be a refractory element such as Ca, then Al would show a small range, and the others large ranges.

Large ranges of compositions are inconsistent with the formation of chondrules directly from a gas. In equilibrium condensation processes, one would expect that all liquid droplets that formed at one nebular location would acquire similar bulk compositions because the mole fraction of each constituent in the melt is set by equilibrium with the same vapor phase. Variations in refractories may be possible if chondrule droplets nucleated over the range of temperatures during which refractories and common lithophiles were condensing. However, it is doubtful that this could happen in a nebula containing abundant unevaporated (refractory) dust grains available to serve as condensation nuclei. In addition, if there were a large range of nucleation temperatures, then one should find many droplets that formed after the condensation of most of the refractory elements (such a temperature range would be needed to produce a wide range of Mg/Si ratios). Chondrules having very low refractory lithophile/Si ratios are only rarely reported (Brigham *et al.*, 1982).

Blander and Katz (1967; Blander and Abdel-Gawad, 1969) imposed kinetic constraints on the condensation of Fe-Ni to explain the presence of FeO in chondrules, since equilibrium condensation (e.g., Grossman, 1972) only yields FeO/(FeO + MgO) ratios >0.1 at temperatures <700 K. Because all metastable liquid droplets should crystallize in the same temperature range, they should all have similar FeO/(FeO + MgO) ratios. By this model, the metal that is found inside chondrules that bear FeO would either have formed during later reduction or have been introduced after solidification. The metal found in OC chondrules shows textural evidence against subsolidus introduction; droplets of metal are common, as are crystals enclosing metal. The metal also commonly bears moderately volatile to volatile siderophiles such as Au, As, Ga, and Ge (Grossman *et al.*, 1979; Grossman and Wasson, 1982). The general lack of fractionations between siderophile elements in chondrules and the presence of siderophile element abundance patterns that are a smooth function of volatility show that the metal almost certainly formed by nebular condensation rather than by later reduction of mafic silicates.

Together with the previous arguments concerning volatiles, the wide range of chondrule compositions indicates that chondrules could not have formed by condensation of liquids directly from the gas phase. The simplest and seemingly best explanation for producing these properties is that chondrules formed through the melting of pre-existing solids. If some of these solids formed at low nebular (or interstellar) temperatures, then the precursor would contain volatile elements and FeO. Diversity of chondrule compositions could then be produced either by melting heterogeneous precursors, by fractionating homogeneous precursors during open-system melting and volatilization, or both.

VOLATILE ELEMENTS

Volatile elements can provide constraints on the nature of chondrule precursor materials and on the nature of the chondrule melting event. In this section, we shall review the evidence for the presence of volatiles inside chondrules, and evaluate to what extent elements of different volatility may have been lost from chondrules during melting. The abundances of volatiles give an indication of the nebular temperatures when chondrule melting occurred. The degree of volatile loss probably is related to the intensity and duration of chondrule heating and to the physical properties of the precursor solids.

The siting of volatile elements

Several studies have proven that the largest fractions of the volatile elements are located in chondrule interiors, and are not merely present in surficial coatings. Petrographic work has shown that sulfide particles were primary constituents of chondrule melts. Rambaldi and Wasson (1981) reported Ca- and Cr-bearing FeS inside OC chondrules; Grossman and Wasson (1982) observed FeS interstitial to radiating pyroxene crystals in droplet chondrules, and FeS-bearing spherules intergrown with olivine crystals in porphyritic chondrules. Na- and K-rich mesostases have long been reported in chondrules. These characteristics are not likely to be the result of secondary or subsolidus effects on the chondrules from highly unequilibrated chondrites.

Grossman *et al.* (1979) and Grossman and Wasson (1982) chemically etched Chainpur chondrules to remove the surfaces. They found enhancements of siderophile and chalcophile elements in the etch solutions, consistent with petrographic observations of metal and sulfide-rich rims (e.g., Allen *et al.*, 1980). The moderately volatile lithophile, Na, was also somewhat enriched (relative to non-volatile Cr); however, because the rim mass is but a tiny fraction of the chondrule mass, most Na was in the chondrule interior. There remained the possibility that the chemical etches were mineralogically selective, and thus not representative of the rim material. Boynton *et al.* (1982) performed similar experiments on Chainpur chondrules by abrasively removing chondrule surfaces prior to analysis. Their results were consistent with the earlier studies, confirming that most chondrule volatiles are in the interiors. The enrichments of siderophile and chalcophile elements on chondrule surfaces handicap these studies; because of the high activities produced by these elements, the NAA technique used by both research groups cannot easily detect any lithophile elements more volatile than (moderately volatile) K. Thus, there is no information on the siting of highly volatile lithophile elements, such as Cs and Rb. It is possible to determine several highly volatile non-lithophiles (Zn, Ge, Ga, Cd, Se, Br); their abundances tend to be proportional to those of siderophiles such as Ni.

The presence of volatile elements and FeO in chondrule interiors indicates that some chondrule precursor components equilibrated with the nebula at temperatures at or below 600 K. At such low temperatures, very little matter less volatile than H_2O remains in the nebular gas.

Evidence for fractionation as a result of vaporization

Compositional studies can help to answer the question of whether variations in volatile lithophiles arose from loss of vapor during chondrule melting or from precursor heterogeneity. The 100 kPa (1 atm) melting temperature of chondritic silicates is ~1500 K. This is nearly 200 K above the condensation temperatures of the common elements Si, Mg, and Fe in a nebula having an H_2 pressure of 1–10 Pa. Thus, during the moment of melting, and even through 100 K of supercooling prior to crystallization, all common and volatile elements were evaporating from the chondrule surface. If material is transported from the interior to the surface by convection, then each species will evaporate in proportion to its vapor pressure (fractional vaporization). If transport is limited (as in the case of diffusion), the surface will gradually evaporate; however, little compositional change will occur in the interior (residual) materials.

Individual OC chondrules that have properties consistent with severe heating and fractional evaporation (low Si, alkalis, reduced Fe) have been identified (Dodd and Walter, 1972; Dodd, 1974). If severe volatile loss occurred frequently, then it seems probable that small chondrules, due to their high surface/volume ratios, would be more likely to be lower in volatiles than large ones. If volatilization occurred to the extent that major elements such as Si were lost, then refractory elements might anticorrelate with volatiles.

Walter (1969) and Walter and Dodd (1972) found no relationship between Si or Al content with the apparent chondrule diameter in thin section. However, the apparent diameter is not necessarily a true measurement of chondrule size. Separated chondrules studies avoid the problem of off-center sectioning, and provide direct information on chondrule masses. The combined OC data of Osborn (1972), Gooding (1979), and Grossman and Wasson (1982, 1983) were used in

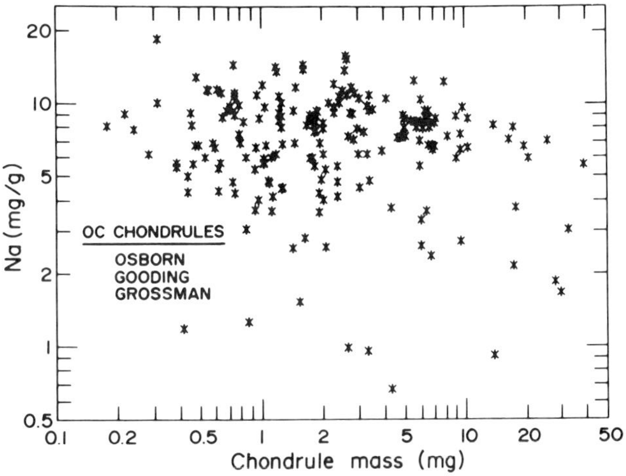

Fig. 6 Sodium concentration is plotted against chondrule mass for chondrules separated from unequilibrated ordinary chondrites by Osborn (1972), Gooding (1979), and Grossman and Wasson (1982, 1983). The loss of volatiles upon extended heating should be most complete in the smaller chondrules. However, no systematic depletion of sodium with decreasing mass is observed.

Fig. 6, a plot of Na concentration vs. chondrule mass, and Fig. 7, Na/Al vs. mass. Dilution effects of metal/sulfide were neglected. As in the work on thin sections, the predicted correlations are not observed.

It has been proposed that many porphyritic chondrules are fragments of pre-existing rock (Dodd, 1978a) and were never present as individual molten droplets. If this is true, then these would not be expected to show volatile-mass relationships. In Fig. 8, only non-porphyritic and barred olivine chondrules from the latter two data sets are plotted. These chondrules, which almost certainly were once discrete molten droplets, show no evidence for ubiquitous evaporative loss of Na.

The "microchondrules" recently recognized by Rubin *et al.* (1982) were not included in these NAA studies. These tiny spherules occupy a miniscule fraction of the total chondrule mass.

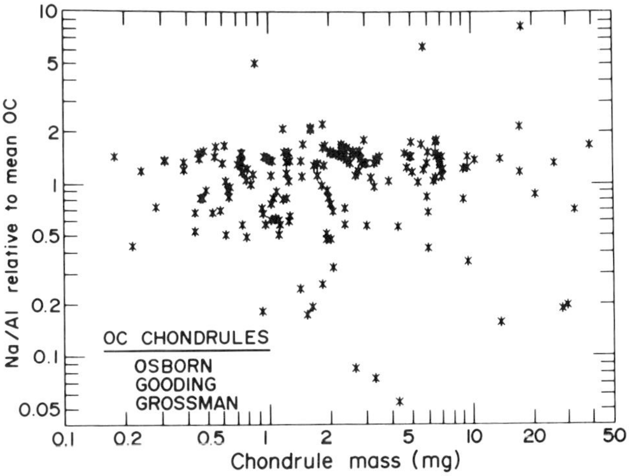

Fig. 7 Sodium/Al ratios are plotted against chondrule mass for the same samples as Fig. 6. Had major elements such as Si been partially lost during chondrule heating, then the ratio of a volatile element such as Na to a refractory element such as Al would become very low. Low Na/Al ratios should be seen most frequently in small chondrules if this process occurred. As in Fig. 6, no correlation is observed.

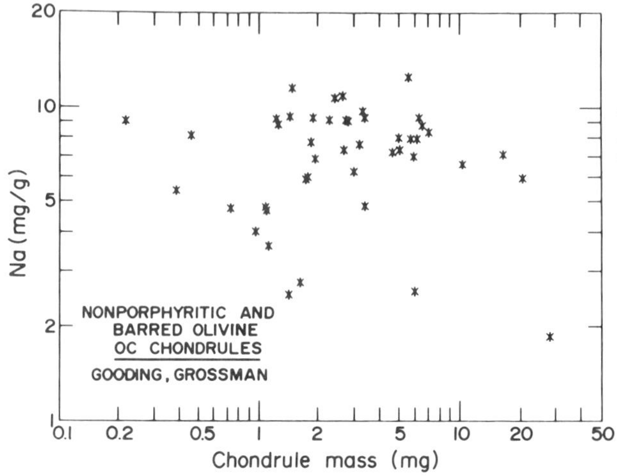

Fig. 8 Sodium concentration vs. chondrule mass for only those chondrules that clearly were fully molten droplets. The non-porphyritic chondrules from Gooding (1979) and from Grossman and Wasson (1982, 1983) are shown. Even for this subset, no correlation of volatile content with mass is observed.

Their relationship to the main population of chondrules has not been adequately investigated, and little is known about their compositions, including whether they show vaporization effects.

Thus, fractional evaporation was either independent of chondrule size or it was not an important process for producing chondrule variability. Relationships among the volatile elements can be used to discriminate between these possibilities. Fractional vaporization should affect both the lithophile and siderophile elements in chondrule melts. In Fig. 9, volatile/refractory element ratios for lithophiles and siderophiles are compared by plotting Na/Al vs. Au/Ir for unequilibrated OC chondrules from Gooding (1979) and Grossman and Wasson (1982, 1983). A weak trend in the direction predicted by fractionational vaporization (a positive correlation) is evident among the highly scattered data. No correlation at all is seen when the volatiles are normalized to common non-volatile elements in a similar plot of Na/Mg vs. Au/Ni, implying that the weak positive trend in Fig. 9 is due to correlations among refractory Ir and Al, and not to the common evaporative loss of volatile Au and Na.

Grossman (1982b) showed that in Semarkona chondrules, there was a trend toward increasing Na/K ratio with increasing Na content (Fig. 10). This was interpreted to result during vaporization, in which the volatile-poor chondrules had lost K more efficiently than Na (in accordance with the higher vapor pressure of K). Some volatile-rich chondrules seemed to have scavenged alkalis from a K-enriched vapor, but these were few in number. Because this trend was not correlated with mass or any petrographic characteristics, it was assumed that the fractionation occurred during the sudden melting of the chondrule precursor solids. Because less volatile elements such as Si, Cr, and Mn were not correlated with the alkalis, it was concluded that the duration of the heating event was brief: only those elements in the most labile sites were lost. An equally plausible hypothesis is that the Na-K fractionation is not due to evaporation, but occurred during the condensation of alkalis. Sodium/K ratios may indicate that vaporization fractionated the alkalis in some chondrules. In light of alternate hypotheses for producing the Na-K trend and

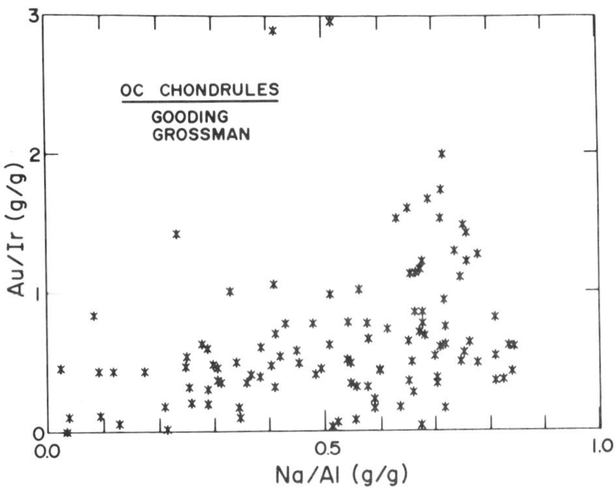

Fig. 9 Au/Ir is plotted against Na/Al for separated OC chondrules. If fractional evaporation processes affected chondrule compositions independent of chondrule mass, then both metal and silicates should show evidence for it. The volatile content of the metal is weakly related to that of the silicates. However, this effect may be due to primary differences in precursor compositions.

other evidence against this mechanism, we conclude that vaporization was not an important pathway toward producing chondrule diversity.

Interestingly, K showed much more scatter in chondrules from Chainpur (Grossman and Wasson, 1982) such that no trend similar to that in Fig. 10 could be resolved. This suggests that even the mild differences in metamorphism between Semarkona and Chainpur (Sears *et al.*, 1980) might be sufficient to obscure what may be subtle nebular effects, but existing data are inadequate to test this idea.

To summarize, several important conclusions can be reached from the study of the more volatile elements in chondrules: (1) The presence of a major fraction of the volatile elements in chondrule interiors is indicative of low nebular temperatures at the time of chondrule formation.

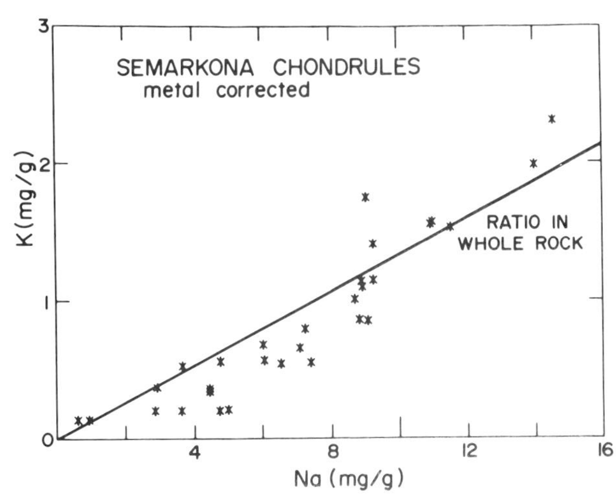

Fig. 10 Potassium vs. Na for Semarkona chondrules. The data have been corrected for the diluting effects of metal and sulfide components (see Grossman and Wasson, 1983). These elements correlate strongly with a trend toward increasing K/Na ratio with increasing alkali content. This may result either from the fractional evaporation of the most volatile elements in some chondrules or from constraints on the condensation of Na and K.

(2) Chondrules do not show the trends expected from partial evaporation of the major elements. Most compositional trends were not the result of vaporization, and the duration of chondrule heating must have been brief. (3) In a few chondrules, appreciable fractions of the more volatile elements may have been vaporized, but this may be more related to the properties of the precursor material than to those of the chondrule melt. Data are lacking for highly volatile elements but the very low abundances of these elements disallows the possibility that their evaporation could be responsible for variations seen among major and minor elements.

THE NATURE OF CHONDRULE PRECURSOR COMPONENTS

Possible precursor materials

We have deduced in previous sections that chondrules formed by melting pre-existing solids after nebular temperatures were $\leq$600 K. Further, vaporization did not play a dominant role, suggesting that information concerning the compositions of the precursor solids can still be obtained from chondrule compositions. For most lithophile elements, the chondrules did behave as closed systems during the melting event. We shall examine the chemistry of chondrules in light of several models for the origin of the precursors. It is important to note here that these models are not necessarily consistent both with current ideas about astrophysical processes and with the large body of meteoritic data. We chiefly attempt to assess the relative merits of each model in light of what is known about chondrule chemistry.

Possible chondrule precursors can be placed into three categories: homogeneous nebular (or pre-solar), heterogeneous nebular (or pre-solar), and heterogeneous igneous or metamorphic. Melting of the first type of material would produce uniform chondrules, unless melting occurred in an open system. Melting of the latter two types of material produces chondrules of variable composition; the expected patterns of variation among chondrules should be different for these two precursors, especially if the material contained some grains having dimensions comparable to those of the resulting chondrules.

Examples of homogeneous precursor models are those proposed by Whipple (1966), Cameron (1966), and Larimer and Anders (1967, 1970). In these, chondrules are presumed to be produced by the melting of fine-grained condensates. The authors offer no discussion of a mechanism for introducing chemical variations (the models pre-date most of the compositional work on chondrules). One candidate for such a mechanism would be fractionation via evaporative loss, which, as we have shown, was probably unimportant. Other mechanisms such as crystal/liquid differentiation and subsolidus alteration seem equally unlikely. Dodd (1978a,b) showed that chondrule variation could not have resulted from the differentiation of olivine (the most common primary liquidus phase for chondrule compositions). It seems very doubtful that the chondrule precursor material was homogeneous.

Chondrule formation models involving heterogeneous, igneous precursors generally call for the impact melting of rocks on the surfaces of asteroids; the models of Fredriksson (1963) and

Dodd (1971) involve chondritic surface compositions, but probably highly recrystallized as a result of impact heating. Dodd (1978a,b) has suggested that the precursors were igneous rocks of chondritic composition less metal and sulfide. He suggested that most porphyritic chondrules represent fragments of the original rock, while non-porphyritic chondrules are remelted equivalents. In this case, the chemical trends among the chondrules reflect non-representative sampling of different mineral phases in the precursor rocks. The main prediction of this model is that variations among chondrules can be related directly to the equilibrium partitioning of elements among the minerals of the target rock.

Several models calling for heterogeneous, nebular chondrule precursors have been recently proposed. One involving the melting of coexisting coarse refractory and fine low-temperature nebular condensates was first stated by Wasson (1972) and later developed by Grossman and Wasson (1982, 1983). Such a model requires either incomplete attainment of equilibrium during condensation processes or heterogeneity produced by mixing solids formed at widely different nebular locations. The important prediction of this model is that elements in chondrules should vary in a fashion consistent with that expected from their cosmochemical partitioning in a cooling nebula.

A second example of a model involving heterogeneous, nebular material is that of Scott *et al.* (1982), who suggested that chondrite matrix material was the chondrule precursor. They cite evidence for unmelted lumps of (precursor) matrix enclosed within chondrules. For this model to work, it is necessary that similar heterogeneity and covariation of elements exist among chondrite matrices as is found in the chondrules themselves. Furthermore, because all matrix materials seem to contain large amounts of oxidized iron, it is necessary to have reduction and metal loss operating on a wide scale in order to form the (relatively) FeO-poor chondrules. This model does not so much make predictions about the composition of chondrules as it does about that of matrix. As we showed earlier, published data show resolvable differences between mean chondrule and matrix compositions, inconsistent with matrix being the immediate chondrule precursor. If new data on matrices should confirm in detail its similarity to chondrules, then the problem of chondrule precursors may be solved, but an entirely new problem arises: What is the origin of heterogeneous fine-grained matrix?

We also anticipate problems with the matrix-precursor model in its application to carbonaceous chondrite chondrules. In those groups, the scenario is complicated by the need to reoxidize metal lost from chondrules. Ordinary chondrites contain abundant extra-chondrule metal, some of which may have originated inside chondrule melts. Despite the relatively low FeO contents of CC chondrules (McSween *et al.*, 1983), carbonaceous chondrites do not contain much free metal. Any metal produced by chondrule formation has either been lost from the chondrites or reoxidized. Bulk carbonaceous chondrites do not show evidence for a net loss of siderophiles; they show little fractionation of Fe, Ni, and Co from Si relative to CI (Kallemeyn and Wasson, 1981, 1983), implying that little metal has been lost from the system. Thus the matrix-melting model requires reduction and metal loss from chondrules, and finally, reoxidation in carbonaceous chondrites. This seems less likely than simply including metal grains among chondrule precursors.

The direct melting of interstellar grains has also been proposed for the formation of chondrules (Clayton, 1980; Rambaldi and Wasson,1981); interstellar grains were also proposed as a possible source for Scott *et al.* (1982) matrix. It is difficult to make compositional predictions from these models because so little is known about the nature of interstellar grains. Proponents of such models favor them because current star-formation calculations indicate that the nebula was never very hot. We note that the commonly accepted picture of interstellar grains is that they are tiny (<1 μm diameter) and highly oxidized. If this is correct, then one would expect the chondrules produced by melting these grains to have uniform compositions. Because so little concrete knowledge about interstellar grains exists, it is possible to account for chondrule heterogeneity by proposing that coarser grains also were present; however, such a proposal would be *ad hoc*.

Tests of the models for chondrule precursors

The interstellar precursor model offers few testable predictions. The matrix-precursor model predicts similarities between chondrule and matrix compositions, the latter of which are poorly constrained at the present time. Therefore, we will focus our attention on examining whether chondrule compositions are better explained by mineralogical or cosmochemical processes.

One approach to this question, done by Grossman and Wasson (1982) for Chainpur chondrules, is to use the multivariate technique of factor analysis. This tool allows the grouping of covarying elements into a small number of "factors." These elemental groupings can be used to deduce precursor components and assess the validity of precursor models. The Chainpur study revealed several independent groups of intercorrelated elements. The groups were: (1) siderophiles and 1 chalcophiles (Ir, Ni, Co, Au, As, Se); (2) refractory lithophiles (Ca, Al, Sc, REE's, some V and Mg); (3) common lithophiles (some Mg, Cr and Mn, and, by assumption, Si); and (4) moderately volatile lithophiles (Na, K, some Ga and Mn). These were interpreted to represent chondrule precursor components, namely: (1) metal/sulfide; (2) reduced silicates formed at high temperature; (3) silicates that formed when the "common" lithophiles were condensing; and (4) either low-temperature silicate condensates or material lost by partial evaporation. The metal-rich component was further divided into a refractory-rich (Ir-bearing) component and a "common" metal component (with composition similar to ordinary chondrite metal).

Figure 11 shows the most important factor loading plot from a factor analysis of our Semarkona chondrule data. This is very similar to the analogous loading plot for Chainpur chondrules in Grossman and Wasson (1982). The siderophile elements and Se are strongly controlled by one factor (factor 2), and are intercorrelated. Strong positive correlations are seen among Mg and the refractory lithophile elements; anticorrelations exist between those elements and FeO. Together with modal olivine abundance, which also correlates with refractory lithophiles, these elements load on factor 1 in Fig. 11. Using these same data, Grossman and Wasson (1983) further characterized chondrule components 2 and 3 in Semarkona as olivine-rich, refractory-bearing, low-FeO silicates that essentially ceased equilibrating with nebular gas follow-

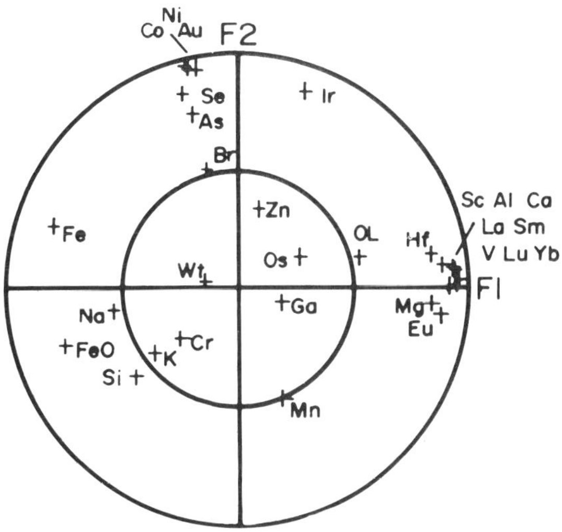

Fig. 11 A factor loading diagram from a factor analysis of Semarkona chondrules. Plotted are the two most important factors extracted, accounting for 39% of the total variance in F1 and 19% in F2. Evidence for several important chondrule precursor components can be seen. F1 is dominated by refractory lithophile elements and Mg (loading positively) and iron (loading negatively). Thus, the refractory lithophiles intercorrelate, and they anticorrelate with both total and oxidized Fe. The positive loading of modal olivine on F1 shows the tendency of olivine-rich chondrules to be refractory-rich. Si loads negatively on F1. Two chondrule precursor components are probably represented by the elements loading positively and negatively on F1. F2 is dominated by siderophile and chalcophile elements (loading positively). These elements are intercorrelated, and are independent from the elements loading strongly on F1. Metal and sulfide comprised yet another chondrule precursor component.

ing the condensation of forsterite, and Si- and FeO-rich silicates formed at lower nebular temperatures.

These results are more consistent with the melting of nebular condensates than with the melting of igneous, pre-existing rock. The groupings of elements that were found are not consistent with equilibrium partitioning of elements between phases in coarse-grained rocks. In the context of the latter model, Ca and Al should be largely contained in feldspathic material. The maximum Ca/Al ratio possible would be 0.7 g/g in anorthite. As shown in Fig. 12d, a plot of Ca vs. Al for OC chondrules measured by INAA, the chondrules cluster tightly around the solar Ca/Al ratio of 1.1 g/g. As pointed out by McSween (1977a), this requires the existence of two Ca-bearing components in chondrules, always melted in nearly the same proportions (the second, high Ca/Al phase would probably be a clinopyroxene). The covariation of refractories with Mg is also not predicted in a model where Mg would be largely in olivine and/or pyroxene and Ca, Al, Sc, etc. would be in other phases. REE would probably be concentrated in minor phases, especially phosphates, in chondrule precursors, and should not necessarily correlate with other refractories as well as they do.

EPMA analyses of chondrules by Lux, McSween, Nagahara, Dodd, Kimura, and Snellenburg all confirm the strong covariation of Ca and Al in chondrules (Fig. 12c) and show that Ti, another refractory lithophile, belongs to this group (Fig. 12a,b). Titanium also is present in nearly solar proportions to other refractory lithophiles. It is very difficult to reconcile constant Ti/Al or

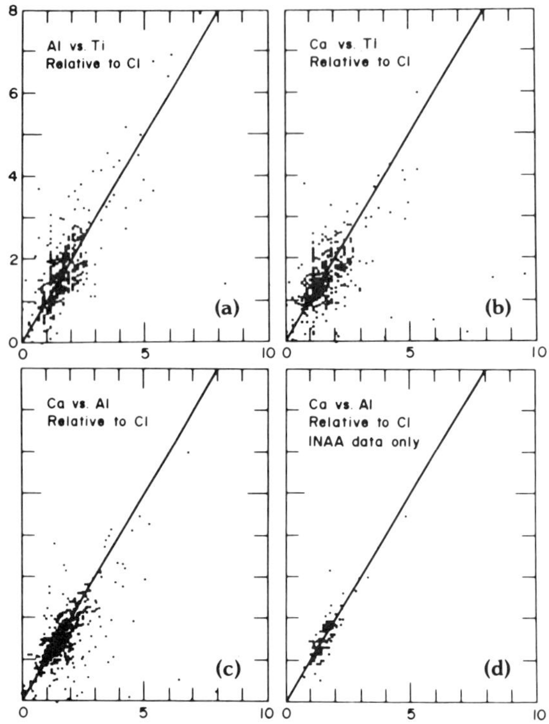

Fig. 12 Interelement plots between refractory lithophiles in ordinary chondrite chondrules: **(a)** Al-Ti, **(b)** Ca-Ti, **(c)** Ca-Al, and **(d)** Ca-Al from INAA studies alone. Calcium and aluminum correlate strongly in INAA studies **(d)**, and samples cluster near the CI-ratio (solid line). When EPMA studies are included **(c)**, Ca and Al show more scatter, but follow the same trend. Titanium can only be determined well in EPMA studies. When plotted against Al and Ca **(a, b)**, Ti shows good correlations that also cluster around the CI ratio. Chondrules probably acquired most refractories by sampling refractory-rich nebular grains, and not by sampling mineral grains from pre-existing rocks.

Ti/Ca ratios with the melting of pre-existing rocks, because Ti presumably would be concentrated in different phases (chromite and proxene) than the other elements.

It is noteworthy that EPMA analyses do not show the strong relationship between Mg and refractories observed in the NAA data. The OC chondrule data of both Gooding (1979) and Grossman and Wasson (1982, 1983) show strong positive correlations between Mg and Al, but the EPMA data show vitually no correlation. We suggest that this results from sampling problems in thin sections through coarse-grained (porphyritic) chondrules. Calcium, Al, and Ti are all excluded from the coarsest common phenocryst phase, olivine. Non-representative sampling of olivine would either preserve or enhance correlations among Ca, Al, and Ti, but would obscure relationships between Mg and elements in the phases other than olivine by superimposing negative correlations on any trends.

Grossman and Wasson (1982) pointed out that there was an anticorrelation between the Fe/(Fe + Mg) ratio in silicates and the content of refractory lithophiles in Chainpur chondrules. This effect is shown in Fig. 13, in which the Sc/Cr ratio is plotted against mol.% fayalite in olivine for OC chondrules from Grossman and Wasson (1982, 1983) and Gooding (1979). This relationship has been interpreted in terms of chondrule precursors that derive from coarse, high-temperature and fine, low-temperature nebular condensates. It is not consistent with the

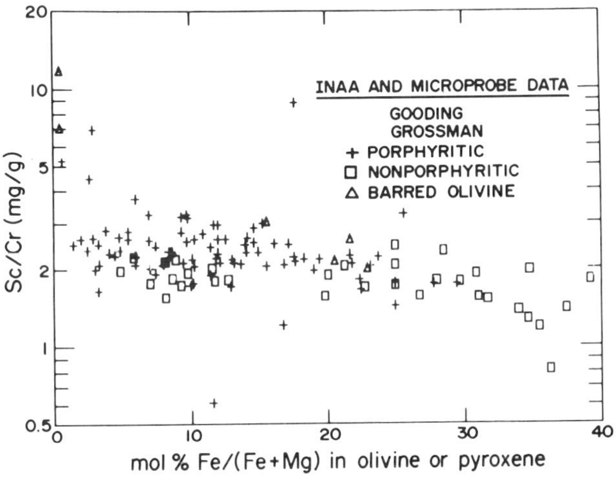

Fig. 13 Sc/Cr vs. mol.% Fe/(Fe + Mg) in ferromagnesian silicates of OC chondrules from Gooding (1979) and Grossman and Wasson (1982, 1983). The fayalite content of olivine was plotted if that mineral was present; otherwise, the ferrosilite content of low-Ca pyroxene was used. The anticorrelation between these parameters is probably caused by the mixing of olivine-rich, refractory-rich, FeO-poor and olivine-poor, refractory-poor, FeO-rich precursor components.

sampling of pre-existing coarse-grained rocks, because the Sc/Cr ratio should be independent of olivine composition.

One important elemental ratio that has received much attention in the literature is Na/Al. The relative proportions of Na and Al could be affected by the compositions of the precursors, as well as by evaporation and recondensation processes during chondrule melting. Figure 14 is a plot of Na vs. Al based on literature data for OC chondrules. EPMA data were included because both elements are excluded from olivine, and their ratio probably is not affected by sampling. The plot reveals: (1) there is not a good correlation between Na and Al; (2) there are data points both above and below the line of solar (= whole rock) ratio, with the greatest deviation occurring in Na-poor samples; and (3) there are few data points above the 1:1 molar ratio of Na/Al (INAA

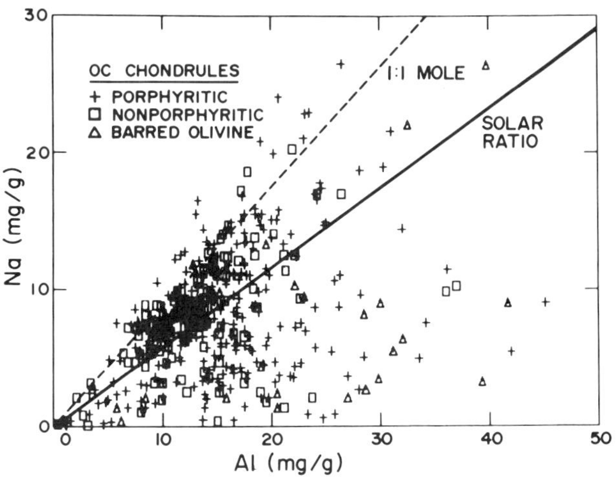

Fig. 14 Sodium vs. Al concentrations in unequilibrated OC chondrules. The solar ratio is shown as a solid line, and the 1:1 mol. ratio is shown as a dotted line. The mean chondrule composition is near the solar ratio; relatively few chondrules have Na/Al ratios above the 1:1 line, even at low Al contents. Recondensation of Na onto chondrule surfaces probably did not occur. The Na/Al ratio is probably controlled by condensation processes involving the precursor solids.

data alone show virtually none). In addition, there is no correlation between chondrule textural type and position on the Na-Al diagram.

The first conclusion one may draw from this plot is that recondensation of alkalis from the vapor following chondrule formation was extremely limited. If late-stage Na (re)condensation were important, then high Na/Al ratios should be found in some low-Al chondrules. We once again conclude that vaporization of Na during heating was minor.

At first glance, the properties of Fig. 14 seem easily understood if feldspathic material was a chondrule precursor. The 1:1 molar upper limit is coincident with the maximum ratio allowed in plagioclase (albite), while the low Na/Al ratios could be explained by volatilization. However, we rejected volatilization on the basis of the evidence cited above. These data are also consistent with the melting of nebular condensates. If Na condensation occurred only on grains with Al available at the surface then coarser grains, with low surface/volume ratios, would acquire low Na/Al ratios. Finer grains could acquire Na up to a 1:1 molar ratio of the elements. This type of condensation behavior presumably extends to K, and might help to explain the Na-K trend discussed above.

The REE also can provide constraints on the nature of precursor materials. If these elements partitioned between different phases in the precursors under equilibrium conditions, and if these phases were not uniformly sampled during chondrule formation, then fractionated REE patterns or Eu anomalies could result. Pre-existing solids that have undergone melting before chondrule formation have the potential to produce these effects. Nebular condensates might either preserve a record of very high temperature epochs as in some refractory inclusions (Boynton, 1975) or else they would have unfractionated patterns.

Gooding (1979) and Gooding and Fukuoka (1982) observed a tendency toward increasing degrees of REE fractionation and Eu-anomalies among chondrules with increasing chondrite metamorphism. They found that variability existed even in some unequilibrated ordinary chondrites among chondrule REE patterns, including Eu-anomalies and light-heavy fractionations. A careful reexamination of the data from Grossman and Wasson (1983) for highly unequilibrated Semarkona chondrules does not reveal any significant Eu-anomalies, and few significantly fractionated patterns among 33 chondrules. A slight (10%) enrichment in Yb and Lu over La and Sm was observed in the mean chondrule composition, in agreement with Gooding. Although we do not understand the differences between REE data from the various authors, the data are more consistent with unfractionated nebular precursors than with pre-existing igneous materials; further study is necessary to reach a firm conclusion.

Finally, it appears that the heterogeneity of chondrules is not limited to chemical composition, but extends to isotopic composition. Preliminary data in Gooding *et al.* (1982) indicate that OC chondrules define a mixing line with a slope nearly equal to one on a three-oxygen isotope plot, and that the relative enrichments in ^{18}O and ^{17}O correlate with several compositional parameters. In particular, refractory lithophiles seem to correlate and total Fe anticorrelates with the heavy oxygen isotopes. This strongly suggests a relationship between the chondrule precursor components suggested by Grossman and Wasson (1983) and the components responsible for the slope-1 oxygen isotope mixing line. It is difficult to conceive of a scenario involving "processed" chondrule precursors in which both the chemical and isotopic data could be reconciled.

Thus, the elemental covariations seen among OC chondrules favor the melting of nebular condensates rather than pre-existing coarse-grained rocks. Although the data are inconsistent with the melting of fine interstellar dust, the actual size frequency distribution and chemical variability among interstellar particles is unknown. The model of Grossman and Wasson (1982, 1983) involves the fractionation of refractory-rich, reduced, coarse and refractory-poor, oxidized, fine materials formed in the nebula. It is difficult to prove that material coarse enough to form the reduced chondrule precursors could have condensed directly from a gas; it may be necessary to call on *ad hoc* mechanisms (such as formation of some chondrules at very high temperatures, followed by fragmentation) to produce these coarse-grained particles. However, it is still more difficult to conceive of the alternative environment where fine-grained precursors such as matrix-like nebular or pre-solar particles could have preserved a wide range of refractory abundances and degrees of oxidation.

The systematic variations in chondrule compositions with textural type that were discussed earlier are probably the result of differences in the chemical and physical properties of the precursors. Fine-grained FeO-rich, refractory-poor precursor material, as proposed by Grossman and Wasson (1982, 1983) for one major chondrule component, would be more easily wholly melted, and, therefore, more likely to undercool and form RP, C, or G textures. Conversely, coarse-grained FeO-poor, refractory-rich material would be more likely to retain unmelted relicts that could serve as nucleation centers, preventing undercooling, and resulting in porphyritic textures. While most chondrules were dominated by these abundant precursor materials, occasionally a chondrule non-representatively sampled minor components. Metallic and refractory-rich chondrules may have formed in this way. According to this interpretation there is no need to associate chemical differences between the textural types with formation in different nebular locations, a model that requires mixing on a wide scale that is not achieved in simple fashion.

CONCLUSIONS

Chondrules from primitive chondritic meteorites seem to preserve a compositional record of the heterogeneous solid material that existed early in the history of the solar nebula. Recent advances in the techniques of electron probe microanalysis and neutron activation analysis have enabled researchers to compile a data base of several hundred compositions of texturally characterized chondrules. Using these data, it was possible to constrain the mechanism for generating chondrule melts and to characterize the material from which the melts formed.

The Si-normalized abundances of both volatile and non-volatile lithophile elements are nearly the same in chondrules and their host chondrites, independent of the proportions of chondrule and non-chondrule components in the meteorite. The refractory-rich compositions of CO and CV chondrites are not solely due to the incorporation of refractory inclusions into an otherwise CI-like rock. Chondrules from both carbonaceous and ordinary chondrites have sampled among their precursors the same solid components that must have been involved in the fractionation of refractory lithophile elements among the various chondrite groups. This is strong evidence that chondrules formed in a region where the right mixture of solids already existed, presumably the dusty midplane of the solar nebula (Rasmussen and Wasson, 1983).

Chondrules are somewhat depleted in Fe and greatly depleted in siderophile elements. Whether metal was fractionated from silicate before or during chondrule melting is unknown. However, some of the Fe was in an oxidized form at the time of this fractionation, indicating a low-temperature origin for some chondrule precursors. Siderophile abundances are not strongly related to volatility. The composition of the metal phase is more consistent with a condensation origin than with an origin by the reduction of completely oxidized interstellar grains or matrix-like material.

Chondrules show wide ranges of composition for all elements. Interelement lithophile ratios vary the least among elements of similar volatility. This diversity rules out the direct condensation of chondrule liquids from the nebula. It is more consistent with the melting of heterogeneous precursor solids than with homogeneous ones.

Volatile elements such as alkalis and S are in chondrule interiors as primary constituents; their siting is inconsistent with introduction through alteration or subsolidus diffusion processes. Again, this indicates that some chondrule precursor materials must have reacted with nebular gases at temperatures <600 K.

There is no correlation between volatile content and chondrule mass. Because the volatile lithophile elements, Na and K, in chondrites are largely inside the chondrules, devolatilization of chondrules must have been limited. The degree of volatile loss may have been a function of the properties of the precursor solids; precursor materials rich in volatile minor phases such as H_2O or hydrocarbons probably suffered the greatest volatile loss. In general, chondrules were nearly-closed systems during melting.

Sodium/Al molar ratios are highly variable, but rarely are greater than one. Sodium probably condensed onto surfaces of precursor grains where Al was available. The lack of high Na/Al ratios in Al-poor chondrules shows that the recondensation of evaporated volatiles onto chondrule surfaces did not usually occur.

The interelement variations among lithophiles in OC chondrules were used to show that the chondrule precursor solids could not have been pre-existing igneous or "processed" rocks. The melting of nebular condensates that equilibrated at high and low temperatures is more consistent with the observed trends. All refractory lithophiles intercorrelate, and generally are present in solar interelement abundances. Magnesium and V also correlate with refractories. These elements must have been present in an olivine-rich, FeO-poor precursor component formed at high temperatures. This and a low-temperature component rich in Si and FeO constitute the bulk of the chondrule precursor material.

Thus, we can understand chondrule precursor components in terms of models calling for condensation and reaction in the solar nebula. It seems unlikely that the same trends could be produced by remelting pre-solar solids, but our lack of knowledge about possible size and compositional distributions prevents us from ruling out this possibility.

Further work is necessary to better characterize the precursor components of chondrules at different nebular locations. The carbonaceous and enstatite chondrite data are too sparse to allow conclusions about their particular suites of precursor materials. Such information would prove to have great signficance in terms of understanding the origins of chondrite groups. Also, little is known about the relationship between oxygen isotopes and chemical compositions. Scanty data on OC chondrules by Gooding *et al.* (1982) suggest a correlation between refractory precursor components and deficiency in ^{16}O; heavy oxygen isotopes are enriched in refractory-rich chondrules. This type of data obviously has an important bearing on the possible origin of this component and the extent to which pre-solar grains may have been involved in chondrite formation.

Acknowledgments. We wish to thank Grace No for patient technical assistance. This research was mainly supported by NASA grant NGR 05-007-329.

REFERENCES

Ahrens L. H., Willis J. P., and Erlank A. J. (1973) The chemical composition of Kainsaz and Efremovka. *Meteoritics* **8**, 133–139.

Allen J. S., Nozette S., and Wilkening L. L. (1979) Chondrule rims: Composition and texture (abstract). In *Lunar and Planetary Science X*, pp. 27–29. Lunar and Planetary Institute, Houston.

Allen J. S., Nozette S., and Wilkening L. L. (1980) A study of chondrule rims and chondrule irradiation records in unequilibrated ordinary chondrites. *Geochim. Cosmochim. Acta* **44**, 1161–1175.

Baedecker P. A. and Wasson J. T. (1975) Elemental fractionations among enstatite chondrites. *Geochim. Cosmochim. Acta* **39**, 735–765.

Bischoff A. and Keil K. (1983) Ca-Al-rich chondrules and inclusions in ordinary chondrites. *Nature* **303**, 588–592.

Blander M. (1971) The constrained equilibrium theory: sulphide phases in meteorites. *Geochim. Cosmochim. Acta* **35**, 61–79.

Blander M. and Abdel-Gawad M. (1969) The origin of meteorites and the constrained equilibrium condensation theory. *Geochim. Cosmochim. Acta* **33**, 701–716.

Blander M. and Katz J. L. (1967) Condensation of primordial dust. *Geochim. Cosmochim. Acta* **31**, 1025–1034.

Boynton W. V. (1975) Fractionation in the solar nebula: condensation of yttrium and the rare earth elements. *Geochim. Cosmochim. Acta* **39**, 569–584.

Boynton W. V., Hill D. H., and Wilkening L. L. (1982) Trace elements in rim and interior of chondrules from Chainpur (abstract). In *Papers Presented to the Conference on Chondrules and their Origins*, p. 3. Lunar and Planetary Institute, Houston.

Brigham C., Murrell M. T., and Burnett D. S. (1982) SiO_2-rich chondrules in ordinary chondrites (abstract). *Meteoritics* **17**, 187–188.

Cameron A. G. W. (1966) The accumulation of chondritic material. *Earth Planet. Sci. Lett.* **1**, 93–96.

Chou C.-L, Baedecker P. A., and Wasson J. T. (1976) Allende inclusions: volatile element distribution and evidence for incomplete volatilization of presolar solids. *Geochim. Cosmochim. Acta* **40**, 85–94.

Clayton D. D. (1980) Chemical energy in cold-cloud aggregates: the origin of meteoritic chondrules. *Astrophys. J.* **239**, L37–L41.

DeGasparis A. A., Fredriksson K., and Brenner P. (1975) Composition of individual chondrules in ordinary chondrites (abstract). *Meteoritics* **10**, 390–392.

Dodd R. T. (1971) The petrology of chondrules in the Sharps meteorite. *Contrib. Mineral. Petrol.* **31**, 201–227.

Dodd R. T. (1974) The petrology of chondrules in the Hallingeberg meteorite. *Contrib. Mineral. Petrol.* **47**, 97–112.

Dodd R. T. (1976) Accretion of the ordinary chondrites. *Earth Planet. Sci. Lett.* **30**, 281–291.

Dodd R. T. (1978a) The composition and origin of large microporphyritic chondrules in the Manych (L-3) chondrite. *Earth Planet. Sci. Lett.* **39**, 52–66.

Dodd R. T. (1978b) Compositions of droplet chondrules in the Manych (L-3) chondrite and the origin of chondrules. *Earth Planet. Sci. Lett.* **40**, 71–82.

Dodd R. T. (1981) *Meteorites, a Petrologic-Chemical Synthesis*. Cambridge University Press, N. Y. 368 pp.

Dodd R. T. and Walter L. S. (1972) Chemical constraints on the origin of chondrules in ordinary chondrites. In *On the Origin of the Solar System* (H. Reeves, ed.), pp. 293–300. Centre National de Recherche Scientifique, Paris.

Evensen N. M., Carter S. R., Hamilton P. J., O'Nions R. K., and Ridley W. I. (1979) A combined chemical-petrological study of separated chondrules from the Richardton meteorite. *Earth Planet. Sci. Lett.* **42**, 223–236.

Fredriksson K. (1963) Chondrules and the meteorite parent bodies. *Trans. N. Y. Acad. Sci.* **25**, 756–769.

Fredriksson K. (1982) Chondrule compositions in different type chondrites (abstract). In *Lunar and Planetary Science XIII*, pp. 233–234. Lunar and Planetary Institute, Houston.

Fujimaki H., Matsu-ura M., Sunagawa I., and Aoki K. (1981) Chemical compositions of chondrules and matrices in the ALH-77015 chondrite (L3). In *Proceedings of the Sixth Symposium on Antarctic Meteorites*, pp. 161–174. National Institute of Polar Research, Tokyo.

Gooding J. L. (1979) Petrogenetic properties of chondrules in unequilibrated H-, L-, and LL-group chondritic meteorites. Ph.D. thesis, Univ. of New Mexico, Albuquerque. 392 pp.

Gooding J. L. and Fukuoka T. (1982) Rare-earth element abundances in chondrules from unequilibrated ordinary chondrites (abstract). *Meteoritics* **17**, 222–223.

Gooding J. L. and Keil K. (1981) Relative abundances of chondrule primary textural types in ordinary chondrites and their bearing on conditions of chondrule formation. *Meteoritics* **16**, 17–43.

Gooding J. L., Keil K., Fukuoka T., and Schmitt R. A. (1980) Elemental abundances in chondrules from unequilibrated chondrites: evidence for chondrule origin by melting of pre-existing materials. *Earth Planet. Sci. Lett.* **50**, 171–180.

Gooding J. L., Mayeda T. K., Clayton R. N., Keil K., Fukuoka T., and Schmitt R. A. (1982) Oxygen isotopic compositions of chondrules in unequilibrated chondrites: Further petrological interpretations (abstract). In *Lunar and Planetary Science XIII*, pp. 271–272. Lunar and Planetary Institute, Houston.

Grossman J. N. (1982a) Mechanisms for the depletion of metal in chondrules (abstract). In *Papers Presented to the Conference on Chondrules and their Origins*, p. 23. Lunar and Planetary Institute, Houston.

Grossman J. N. (1982b) The abundance and distribution of moderately volatile elements in Semarkona chondrules (abstract). In *Lunar and Planetary Science XIII*, pp. 289–290. Lunar and Planetary Institute, Houston.

Grossman J. N. and Wasson J. T. (1982) Evidence for primitive nebular components in chondrules from the Chainpur chondrite. *Geochim. Cosmochim. Acta* **46**, 1081–1099.

Grossman J. N. and Wasson J. T. (1983) Refractory precursor components of Semarkona chondrules and the fractionation of refractory elements among chondrites. *Geochim. Cosmochim. Acta* **47**, 759–771.

Grossman J. N., Kracher A., and Wasson J. T. (1979) Volatiles in Chainpur chondrules. *Geophys. Res. Lett.* **6**, 597–600.

Grossman J. N., Kracher A., Kallemeyn G. W., and Wasson J. T. (1978) Chemical-petrographic study of chondrules (abstract). In *Lunar and Planetary Science IX*, pp. 422–423. Lunar and Planetary Institute, Houston.

Grossman L. (1972) Condensation in the primitive solar nebula. *Geochim. Cosmochim. Acta* **36**, 597–619.

Grossman L. (1980) Refractory inclusions in the Allende meteorite. *Ann. Rev. Earth Planet. Sci.* **8**, 559–608.

Hamilton P. J., Evensen N. M., and O'Nions R. K. (1979) Chronology and chemistry of Parnallee (LL-3) chondrules (abstract). In *Lunar and Planetary Science X*, pp. 494–495. Lunar and Planetary Institute, Houston.

Hashimoto A., Kumazawa M., and Onuma N. (1979) Evaporation metamorphism of primitive dust material in the early solar nebula. *Earth Planet. Sci. Lett.* **43**, 13–21.

Herndon J. M. and Suess H. E. (1977) Can the ordinary chondrites have condensed from a gas phase? *Geochim. Cosmochim. Acta* **41**, 233–236.

Huss G. R., Keil K., and Taylor G. J. (1981) The matrices of unequilibrated ordinary chondrites: implications for the origin and history of chondrites. *Geochim. Cosmochim. Acta* **45**, 33–51.

Kallemeyn G. W. and Wasson J. T. (1981) The compositional classification of chondrites: I. The carbonaceous chondrite groups. *Geochim. Cosmochim. Acta* **45**, 1217–1230.

Kallemeyn G. W. and Wasson J. T. (1983) Mechanisms responsible for elemental fractionations among the carbonaceous chondrites. *Geochim. Cosmochim. Acta.* In press.

Kerridge J. F. (1979) Fractionation of refractory lithophile elements among chondritic meteorites. *Proc. Lunar Planet. Sci. Conf. 10th*, pp. 989–996.

Kimura M. and Yagi K. (1980) Crystallization of chondrules in ordinary chondrites. *Geochim. Cosmochim. Acta* **44**, 589–602.

King T. V. V. and King E. A. (1978) Grain size and petrography of C2 and C3 carbonaceous chondrites. *Meteoritics* **13**, 47–72.

Kornacki A. S. (1981) Are CAI's condensates or distillation residues? Evidence from a comprehensive survey of fine- to medium-grained inclusions in the Allende C3(V) carbonaceous chondrite (abstract). In *Lunar and Planetary Science XII*, pp. 562–564. Lunar and Planetary Institute, Houston.

Kurat G. (1970) Zur Genese der Ca-Al-reichen Einschlüssen im Chondriten von Lance. *Earth Planet. Sci. Lett.* **9**, 225–231.

Kurat G., Pernicka E., Herrwerth I., and El Goresy A. (1982) Geochemistry of Chainpur chondrules; evidence for evaporation and recondensation (abstract). In *Papers Presented to the Conference on Chondrules and their Origins*, p. 37. Lunar and Planetary Institute, Houston.

Larimer J. W. (1979) The condensation and fractionation of refractory lithophile elements. *Icarus* **40**, 446–454.

Larimer J. W. and Anders E. (1967) Chemical fractionations in meteorites—II. Abundance patterns and their interpretation. *Geochim. Cosmochim. Acta* **31**, 1239–1270.

Larimer J. W. and Anders E. (1970) Chemical fractionations in meteorites—III. Major element fractionations in chondrites. *Geochim. Cosmochim. Acta* **34**, 367–387.

Lux G., Keil K., and Taylor G. J. (1980) Metamorphism of the H-group chondrites: implications from compositional and textural trends in chondrules. *Geochim. Cosmochim. Acta* **44**, 841–855.

Lux G., Keil K., and Taylor G. J. (1981) Chondrules in H3 chondrites: textures, compositions and origins. *Geochim. Cosmochim. Acta* **45**, 675–685.

Mason B. (1971) *Handbook of Elemental Abundances in Meteorites*. Gordon and Breach, N. Y. 555 pp.

McSween H. Y. (1977a) Chemical and petrographic constraints on the origin of chondrules and inclusions in carbonaceous chondrites. *Geochim. Cosmochim. Acta* **41**, 1843–1860.

McSween H. Y. (1977b) Carbonaceous chondrites of the Ornans type: a metamorphic sequence. *Geochim. Cosmochim. Acta* **41**, 477–491.

McSween H. Y. (1977c) Petrographic variations among carbonaceous chondrites of the Vigarano type. *Geochim. Cosmochim. Acta* **41**, 1777–1790.

McSween H. Y. and Richardson S. M. (1977) The composition of carbonaceous chondrite matrix. *Geochim. Cosmochim. Acta* **41**, 1145–1161.

McSween H. Y., Fronabarger A. K., and Driese S. G. (1983) Ferromagnesian chondrules in carbonaceous chondrites. This volume.

Nagahara H. (1981) Petrology of chondrules in ALH-77015 (L3) chondrite. *Mem. Natl. Inst. Polar Res., Spec. Issue No. 20* (T. Nagata, ed.), pp. 145–160. National Institute of Polar Research, Tokyo.

Nagahara H. and Kushiro I. (1982) Calcium-aluminum-rich chondrules in the unequilibrated ordinary chondrites. *Meteoritics* **17**, 55–63.

Noonan A. F. (1975) The Clovis (No. 1), New Mexico, meteorite and Ca, Al and Ti-rich inclusions in ordinary chondrites. *Meteoritics* **10**, 51–59.

Osborn T. W. (1972) Elemental abundances in meteoritic chondrules. Ph.D. thesis, Oregon State University, Corvallis. 354 pp.

Osborn T. W., Smith R. H., and Schmitt R. A. (1973) Elemental composition of individual chondrules from ordinary chondrites. *Geochim. Cosmochim. Acta* **37**, 1909–1942.

Osborn T. W., Warren R. G., Smith R. H., Wakita H., Zellmer D. L., and Schmitt R. A. (1974) Elemental composition of individual chondrules from carbonaceous chondrites, including Allende. *Geochim. Cosmochim. Acta* **38**, 1359–1378.

Podolak M. and Cameron A. G. W. (1974) Possible formation of meteoritic chondrules and inclusions in the precollapse Jovian protoplanetary atmosphere. *Icarus* **23**, 326–333.

Rambaldi E. R. and Wasson J. T. (1981) Metal and associated phases in Bishunpur, a highly unequilibrated ordinary chondrite. *Geochim. Cosmochim. Acta* **45**, 1001–1015.

Rasmussen K. L. and Wasson J. T. (1983) A new lightning model for chondrule origin. *Icarus*. In press.

Rubin A. E., Scott E. R. D., and Keil K. (1982) Microchondrule-bearing clast in the Piancaldoli LL3 meteorite: a new kind of type 3 chondrite and its relevance to the history of chondrules. *Geochim. Cosmochim. Acta* **46**, 1763–1776.

Schmitt R. A., Smith R. H., and Goles G. G. (1965) Abundances of Na, Sc, Cr, Mn, Fe, Co, and Cu in 218 individual meteoritic chondrules via activation analysis, 1. *J. Geophys. Res.* **70**, 2419–2444.

Schmitt R. A., Smith R. H., and Olehy D. A. (1968) Rare earth abundances in meteoritic chondrules. In *Origin and Distribution of the Elements* (L. H. Ahrens, ed.), pp. 273–282. Pergamon, Oxford.

Schmitt R. A., Smith R. H., Ehmann W. D., and McKown D. (1967) Silicon abundances in meteoritic chondrules. *Geochim. Cosmochim. Acta* **31**, 1975–1985.

Scott E. R. D., Taylor G. J., Rubin A. E., Keil K., and Kracher A. (1982) Origin of chondrules by preaccretionary melting of silicate matrix material (abstract). In *Papers Presented to the Conference on Chondrules and their Origins*, p. 54. Lunar and Planetary Institute, Houston.

Sears D. W., Grossman J. N., Melcher C. L., Ross L. M., and Mills A. A. (1980) Measuring metamorphic history of unequilibrated ordinary chondrites. *Nature* **287**, 791–795,

Simon S. B. and Haggerty S. E. (1980) Bulk compositions of chondrules in the Allende meteorite. *Proc. Lunar Planet. Sci. Conf. 11th*, pp. 901–927.

Smith J. V., Smith M. R., and Schmitt R. A. (1983) Chemical composition of 24 chondrules in Indarch E4 chondrite (abstract). In *Lunar and Planetary Science XIV*, pp. 714–715. Lunar and Planetary Institute, Houston.

Snellenburg J. W. (1978) A chemical and petrographic study of the chondrules in the unequilibrated ordinary chondrites, Semarkona and Krymka. Ph.D. thesis, State University of New York, Stony Brook. 171 pp.

Suess H. E. (1963) Comments on the origin of chondrules. In *Origin of the Solar System* (R. Jastrow and A. G. W. Cameron, eds.), pp. 154–156. Academic, N. Y.

Suess H. E. (1965) Chemical evidence bearing on the origin of the solar system. *Ann. Rev. Astron. Astrophys.* **3**, 217–234.

Tschermak G. (1883) Beitrag zur Classification der Meteoriten. *Sitzber. Akad. Wiss. Wien. Math.-Naturw. Cl.* **85**, Abt. 1, 347–371.

Walter L. S. (1969) Major-element composition of individual chondrules of the Bjurböle meteorite. In *Meteorite Research* (P. M. Millman, ed.), pp. 191–205. Reidel, Dordrecht.

Walter L. S. and Dodd R. T. (1972) Evidence for vapor fractionation in the origin of chondrules. *Meteoritics* 7, 341–352.

Wasson J. T. (1972) Formation of ordinary chondrites. *Rev. Geophys. Space Phys.* **10**, 711–759.

Wasson J. T. (1983) *Meteorites: Their Record of Early Solar-System History*. In press.

Whipple F. L. (1966) Chondrules: Suggestions concerning their origin. *Science* **153**, 54–56.

Wood J. A. (1962) Chondrules and the origin of the terrestrial planets. *Nature* **197**, 127–130.

Wood J. A. and McSween H. Y. (1976) Chondrules as condensation products. In *Comets, Asteroids, Meteorites: Interrelations, Evolution, and Origins* (A. H. Delsemme, ed.), pp. 365–373. Univ. of Toledo, Ohio.

Dynamic Crystallization Experiments as Constraints on Chondrule Genesis

Roger H. Hewins
Department of Geological Sciences, Rutgers University, New Brunswick, New Jersey 08903

Silicate spheres with compositions and textures similar to those of chondrules have been produced in a number of dynamic crystallization studies. Pyroxene excentroradial chondrules have been simulated by cooling totally melted charges at low rates (pyroxene-olivine-normative composition) or at a large range of cooling rates (more Si-rich composition). Textures similar to those of barred olivine chondrules have been produced experimentally by cooling totally melted charges (pyroxene-olivine-normative and less Si-rich compositions) at moderate and high rates. Spherules comparable to olivine microporphyritic chondrules have been made in two ways: Cooling melts lacking obvious nucleii at moderate rates produced large hopper olivines, but melts rich in heterogeneous nucleii generated granular and granular-poikilitic textures with low or high cooling rates. This review of experimental studies shows that many chondrules have textures consistent with heating in the range 1400°–1600°C and cooling at rates from a few degrees to many thousands of degrees per hour. This constraint rules out mechanisms for chondrule genesis that require exclusively either nearly instantaneous or really slow cooling. Textural, chemical, and isotopic data suggest that chondrules were generated from primordial particles and several mechanisms might have operated to heat them as they settled towards the median plane of the nebula. Of these heating mechanisms, friction with nebular gas (Wood, 1983) and possibly heating by flares (Levy, personal communication, 1982) appear to have the potential, when fully worked out, of providing thermal histories for chondrules consistent with those required by dynamic crystallization experiments.

INTRODUCTION

In the last three years, studies of chondrules using a wide variety of techniques have very much reduced the number of possible interpretations of their origins. In short, the large fraction of chondrules that have textures resulting from the cooling of silicate melt can be explained by the rapid heating and cooling of independent primordial particles (Hewins *et al.*, 1981). Some of the more important evidence leading to this working hypothesis is discussed below.

Some chondrules include large olivine grains, often dusty and inclusion-rich but surrounded by clear olivine, which by virtue of their reversed zoning appear to be unmelted relict grains (Rambaldi, 1981; Rambaldi and Wasson, 1981, 1982; Nagahara, 1981). A second type of inclusion consists of very fine-grained silicate similar to some chondrite matrix (Scott *et al.*, 1982b). These occurrences require a solid (rather than, say, a vapor) precursor for these chondrules and the simplest assumption to make is that all igneous chondrules formed by melting pre-existing solids. If the chondrule-forming process was repeated, the fine-grained material might represent the original precursor material, whereas the large relict olivine grains might represent earlier chondrules or fragments. Some material in chondrites (matrix and accretionary chondrules) appears not to have been melted.

Analyses of separated chondrules show that chondrules from different kinds of unequilibrated chondrites have essentially the same average composition (Gooding *et al.*, 1980a; Scott *et al.*, 1982a). There is, of course, a range of chondrule compositions (e.g., Dodd, 1978b; Scott and Taylor, 1983; Bischoff and Keil, 1983), but in a general way chondrule compositions resemble bulk solar composition except for enrichments in lithophile elements and depletions in siderophile elements. Chondrules and chondrule fragments in ordinary and carbonaceous chondrites have been shown to be unequilibrated in oxygen isotopes (Mayeda *et al.*, 1980; Gooding *et al.*, 1980b; Onuma *et al.*, 1982). The chemical data on chondrules show that they (and their precursors) were never equilibrated in a planetary body.

The textures of droplet and microporphyritic chondrules have been reproduced in a number of experimental studies (e.g., Planner and Keil, 1982; Tsuchiyama and Nagahara, 1981; Hewins *et al.*, 1981). It has been shown previously that the thermal history data from these studies are compatible with the model for chondrule origin by melting of primordial particles (Hewins *et al.*, 1981). As this working hypothesis gains widespread support, the importance of the dynamic crystallization studies becomes clear. Such experiments offer our principal indication of the physical conditions obtaining at the time of chondrule genesis and thus directly constrain the nature of the heating mechanism. This paper, therefore, discusses the extent to which chondrule textures have been reproduced (for pyroxene-rich, olivine-rich, and refractory bulk compositions) by dynamic crystallization and the significance of the thermal histories involved.

REVIEW OF DYNAMIC CRYSTALLIZATION EXPERIMENTS

Lofgren (1980) has reviewed recent work on the dynamic crystallization of basaltic liquids and has shown that the interpretation of thermal histories of rocks is not at all simple. Chondrules are less well understood than basalts, and extrapolation from melt droplets produced on wire loops in vertical furnaces to the meteoritic chondrules is even more complex. A large number of assumptions must be made in applying experimental results to an individual chondrule. For example, to infer, say, the cooling rate of a chondrule by matching textures with those of experimental charges, the latter should have *identical* bulk composition, degree of superheat and/or number of nucleii, oxygen fugacities, etc., to the natural chondrule. It is unreasonable to expect exact duplication of every parameter pertaining to a single chondrule. However, when the textural variations within chondrule suites are compared to different groups of experimental charges, there is a better chance of having typical parameters duplicated within the experiments, and indeed general guides to chondrule thermal history have emerged. Furthermore, although these guides will be shown below to involve ranges of cooling rates, the differences seem relatively insignificant as the inferred cooling times are all very short on a cosmic scale.

Pyroxene excentroradial chondrules

Three sets of experiments have used highly pyroxene-normative bulk compositions (Planner and Keil, 1982, Hsil composition; Tsuchiyama and Nagahara, 1981; Hewins *et al.*, 1981). Hewins *et al.* (1981) produced exclusively pyroxene as a crystalline phase, whereas olivine was important in the other experiments. Tsuchiyama and Nagahara (1981) explained the difference as due to differences in chemical composition, since Hewins *et al.* (1981) used starting material with more SiO_2 (58%). Tsuchiyama and Nagahara produced barred-radial olivine with high and moderate cooling rates (down to 125°/hr) but barred-radial pyroxene at low cooling rates (5°–46°/hr). Planner and Keil (1982) did not use such low cooling rates and olivine is the only liquidus phase in their experiments. The less silica-rich compositions have a significant olivine field at the liquidus and presumably only olivine is found when kinetics are unfavorable for pyroxene nucleation. Planner and Keil made many runs in which the droplets were cooled to about 1460°C for up to 93 minutes but no pyroxene grew, even with this isothermal episode. Donaldson (1979) showed that even olivine can have its nucleation delayed for many hours, particularly when undercooling is not very great.

Both Tsuchiyama and Nagahara (1981) and Hewins *et al.* (1981) cooled some charges that were not totally melted. These charges contain stubby crystals, so that it is certain that pyroxene excentroradial chondrules cooled from a totally molten state. Hewins *et al.* (1981) found that cooling from the liquidus to 1000°C before quenching produced pyroxene-glass charges resembling excentroradial chondrules, whereas quenching from higher temperatures yielded more glass than is found in natural chondrules. It is inferred that the ambient temperature of the chondrule environment was at least as low as 1000°C.

Cooling rates for excentroradial chondrules may be estimated from the experimental run products, by comparing mean widths of the largest dendrites. Estimates derived from the Hewins experiments are ten times faster than from the Tsuchiyama experiments. This suggests that crystallization is sensitive to bulk compositions, with an increase in silicate content making pyroxene nucleation and/or growth easier. The experimental charges of Hewins *et al.* (1981) seem well suited to interpret chondrules in Manych (Dodd, 1978b) on which their composition was based. Pyroxene dendrite widths correlate with Fe-Al enrichment patterns in both the natural and synthetic droplets. In addition, Hewins *et al.* (1981) grew some very slender dendrites at high cooling rates, similar to those in some chondrules, whereas Tsuchiyama and Nagahara (1981) produced olivine at comparable rates using material poorer in SiO_2.

The total range of cooling rates from experiments that produced excentroradial pyroxene is 5°–3000°/hr. Since some natural chondrules contain pyroxene slightly coarser and finer than in the run products, they could have experienced a slightly wider range of cooling rates. The inferred rates are listed in Table 1 for easy comparison with the apparent cooling rates of other kinds of chondrules. Note that the cooling times are variable and not instantaneous, but all are short on a cosmic time scale.

Table 1. Summary of cooling rates used in experiments that produced textures comparable to those in chondrules.

Type	SiO_2	Nuclei	Rates (°C/hr)	Reference
Pyroxene excentroradial	58.2	—	50–3000	Hewins *et al.*, 1981
	53.5	—	5–46	Tsuchiyama and Nagahara, 1981
Barred olivine	53.5	—	125–500	Tsuchiyama and Nagahara, 1981
	53.2	—	330 (+ isothermal)	Planner, 1979
	46.1	—	3,000	Tsuchiyama *et al.*, 1980
Olivine microporphyry granular/poikilitic	~ 44.	+	5–800	Nagahara, 1983
granular	43.6	+	1200–4800	Tsuchiyama and Nagahara, 1981
granules and blades	44.9	+	1200–14400	Tsuchiyama *et al.*, 1981
granular	38.6	+	412–3678	Planner, 1979
hopper	38.6	—	223–421 (+ isothermal)	Planner, 1979
elongate hopper	38.6	—	3680	Planner, 1979
Ca-Al-rich	31.1	+	20	Paque and Stolper, 1982

Barred olivine chondrules

Spherules resembling barred olvine chondrules were produced by Tsuchiyama *et al.* (1980) with material containing 40% normative olivine. They showed that dendrite widths increased over a series of experiments as cooling rates decreased from about 8000°/hr to about 2000°/hr. They also showed dendrite widths increasing as initial degree of superheat was decreased (Tsuchiyama and Nagahara, 1981) in experiments on different bulk composition (see Table 1 and discussion below). Experimental charges cooled from below the liquidus contained small equant olivine grains, presumably partly relict unmelted material overgrown during cooling. It is therefore certain that barred olivine chondrules were totally melted. The charges most like natural barred olivine chondrules (although perhaps not as coarse and regular) were formed by cooling at about 3000°C/hr from close to the liquidus.

Planner (1979) illustrated in detail textures like those in barred olivine chondrules, which he produced in the experiments on highly pyroxene-normative material described above. The charges contain parallel groupings of highly elongate, somewhat skeletal olivine crystals. They might be described as varying from very close to natural barred olivine chondrules to transitional between microporphyritic and barred, although Planner termed the "transitional" examples porphyritic. The matrix to the olivine is unlike that in barred olivine chondrules because of differences in composition. The factors leading to these textures in silica-rich material seem to involve moderately high growth rate for olivine (initial cooling rates were 319°–382°/hr) and suppression of pyroxene nucleation. Final temperature and length of an isothermal episode do not seem to have major systematic effects on textures, possibly because much of the olivine grew at very high temperature.

Cooling rate estimates for barred olivine chondrules are summarized in Table 1. It is interesting that although the results of Planner (1979) and Tsuchiyama and Nagahara (1981) appear to agree closely, the latter authors produced rather fine olivine dendrites in many charges, which bear little resemblance to most barred olivine chondrules. Compositions differ, except for SiO_2, with 12% minor elements present in Tsuchiyama's charges. Furthermore, since silica-rich barred olivine chondrules are rare [e.g., Dodd (1978b) lists only one example for Manych], the best guide to the thermal history of barred olivine chondrules appears to be the work of Tsuchiyama *et al.* (1980). However, barred olivine chondrules are mainly more highly olivine-normative than their bulk composition (see Dodd, 1978b), and it would be valuable to extend dynamic crystallization experiments to these difficult, more refractory compositions.

Olivine microporphyry

The difficulties of applying controlled crystallization runs to chondrules are compounded in this category by semantic problems: the word "porphyritic" is used in very different senses by different petrographers and experimentalists. Dodd (1978a) used the term "microporphyry" for chondrules in a correct petrographic sense, in that the olivine crystal, with maximum size a few tenths of a millimeter, are microphenocrysts. Since chondrules are at most a few millimeters across, with no very big olivine grains, "porphyritic" has come to be used as synonymous with microporphyritic, which is regrettable, particularly in cases of very fine-grained olivine. McSween (1977) has appropriately termed olivine in some chondrules "granular," compatible with Donaldson (1976), and yet others have called such textures porphyritic or microporphyritic. Experimentalists have lumped a number of textures, involving relict granules, stubby blades, and elongate hopper crystals, into the general category of "porphyritic." The reader is therefore warned that the experiments described below may not closely constrain the textures most often associated with the term "porphyritic" [e.g., those illustrated by Dodd (1978a)].

Donaldson (1976) summarized the hierarchy of olivine morphologic types grown in *totally melted* basaltic material, in a sequence by crystal growth rate. Polyhedral olivine crystals, hopper crystals, elongate hoppers, chains, and more complex dendritic forms were found systematically, with the former in slower-cooled and the latter in faster-cooled experimental charges. Part of this hierarchy is represented in olivine microporphyritic chondrules, but barred olivine chondrules represent the higher growth rates. The microporphyritic chondures contain polyhedral, hopper or elongate hopper olivine grains (e.g., Dodd, 1978a; Klein *et al.*, 1980; Planner and Keil, 1982). In many cases, particularly those illustrated in the above papers, pyroxene dendrites nucleated in the chondrules containing polyhedral olivine but not in those with hopper olivine, consistent with the idea of more rapid cooling in the latter case.

Planner (1979) and Planner and Keil (1982) were able to match chondrule textures with hopper olivine in runs using initial cooling rates of 223°–421°/hr and with elongate hopper olivine in a single charge cooled at 3650°/hr. In Donaldson's work (1976), order of magnitude differences in cooling rate are found between polyhedral, hopper, and elongate crystals. Projecting similar differences between the experiments of Planner and Keil, an estimate of about 30°/hr for polyhedral olivine is reached. Since it is based on compositions closer to natural chondrules, this estimate supercedes the 1°/hr suggested by Klein *et al.* (1980) using Donaldson's data.

Most of Planner's (1979) experimental charges contain one or very few hopper crystals, whereas some microporphyritic chondrules contain more abundant crystals. However, he also performed three runs that produced abundant, fine, granular olivine (which he likened to "microporphyritic" chondrules). Planner showed that the abundant nucleation in these runs was due to the presence of trapped air bubbles. The dramatically different textures are due not to different cooling rates but to the presence of nucleii. Tsuchiyama and Nagahara (1981) also produced fine-grained olivine in runs cooled from below the liquidus at the same rates that produced barred olivine from initially totally melted charges. Many of these experimental charges, and also those of Tsuchiyama *et al.* (1980) with both relict granules and short olivine blades, do not closely resemble common microporphyritic chondrules. However, runs from just below the liquidus with larger crystals that are subhedral presumably because of overgrowth do look like microporphyry with polyhedral olivine. There appear to be (at least) two possible ways to simulate olivine microporphyry: cool from above the liquidus at low rates or cool from just below the liquidus at moderately high rates. The key question in chondrule thermal history therefore becomes one of the maximum temperature reached.

Many chondrules, particularly those with dendritic pyroxene or hopper olivine, look nothing like experimental charges cooled from below the liquidus, suggesting that a range of low cooling rates from above the liquidus applies. On the other hand, Nagahara (1981) and Rambaldi (1981) have identified certain olivine grains in chondrules as unmelted relicts, and more recently Nagahara (1982) has suggested that granular olivine in some poikilitic chondrules may also be relict. If this interpretation is correct, such chondrules might have experienced higher cooling rates comparable to those of barred olivine chondrules. However, Nagahara (1983) has also reported experiments simulating granular/poikilitic chondrules by cooling from below the liquidus as slowly as 5°/hr. This new result is consistent with Lofgren's (1980) finding that textures are not very sensitive to cooling rate when initial nuclei are present and underscores the fact that a wide range of thermal histories is possible for olivine-rich chondrules.

Planner and Keil (1982) argued that chondrules were quenched from high temperature, because continuous cooling of experimental charges to about 600°C failed to produce glass. The matrix to the larger olivine grains in their charges consists of olivine-silica and silica-magnetite intergrowths. Addition of elements found in chondrules, such as Al, to their simple bulk compositions would make glass formation much easier. New experiments on natural chondrule compositions are needed to see if glass is formed by continuous cooling to low temperatures. At this point, both quenching and continuous cooling seem possible histories for chondrules.

Planner and Keil (1982) developed a fractional crystallization model to relate olivine compositions and cooling rates. They concluded that olivine in natural chondrules did not have the appropriate composition for continuous cooling and that equilibrium (Fe-Mg exchange) with the liquid in an isothermal event was required. This kind of modeling is difficult: ideally, the effects of diffusion and composition gradients in the liquid should be incorporated (Onorato *et al.*, 1978). To check on the necessity of an isothermal event for olivine microporphyritic chondrules, which does not seem to be required for pyroxene excentroradial chondrules, it would be desirable to test the fractionation models with data from new experiments performed at the appropriate oxygen fugacities on compositions matching chondrules. Lofgren (1982) suggested that the composition range of olivine in chondrules would be greatly influenced by the presence of heterogeneous nuclei, possibly associated with unmelted relicts, which might explain the apparent partial equilibration in the chondrules studied by Planner and Keil. Additional petrographical, experimental, and analytical work should clarify how mineral compositions and textures vary for droplets with abundant and minimal nuclei.

Refractory chondrules

Chondrules and inclusions with refractory compositions (Ca-Al-rich) occur in many carbonaceous and ordinary chondrites (Reid *et al.*, 1974; McSween, 1977; Bischoff and Keil, 1983). Paque and Stolper (1982) conducted the only crystallization experiments on a composition that was at all close to that of the refractory chondrules, specifically representing Type B calcium-aluminum-rich inclusions. They achieved the closest approach to natural textures by cooling from below the liquidus (1550°C) at 20°/hr, and lower cooling rates are indicated to achieve the grain size of the coarse-grained inclusions. Some of the refractory chondrules contain glass and skeletal crystals and doubtless experienced more rapid cooling. In addition, they were probably totally melted, although the liquidus temperatures are unknown since the chondrules are more SiO_2-rich than the material studied by Paque and Stolper. The simplest models for the early solar system might suggest simple differences in temperatures or bulk compositions between melted, partly melted, and unmelted refractory materials in chondrites, but more work may very well reveal the absence of such simple patterns.

Volatilization data

An interesting by-product of dynamic crystallization studies involved melt-vapor equilibria and kinetics. Tsuchiyama *et al.* (1981) studied sodium volatilization as a function of temperature, oxygen fugacity, and melt-sphere size. Their sodium loss data suggest that any chondrules with moderate sodium contents have spent very little time (at most a few hours) at high temperature. Dodd's data (1978a,b) for chondrules show great variety in chondrule Na_2O contents. Barred olivine chondrules tend to have less Na_2O than olivine microporphyry (typically, 0.2–0.3 vs. 1–2% Na_2O). Since there is no theoretical reason to suppose a higher initial Na in the microporphyry and it most likely spent longer time at temperature (although peak temperatures are not known), the difference may be due to the greater size of the microporphyritic chondrules. The differences in Na_2O and size between barred and microporphyritic chondrules are of the order expected from Tsuchiyama *et al.*'s (1981) results, but it should be remembered that there are no simple correlations between size and Na content within each chondrule type (Dodd, 1978a,b). If barred olivine chondrules did initially have the maximum Na_2O content of microporphyry, they spent 1–10 hrs at liquidus temperatures according to the results of Tsuchiyama *et al.* (1981). Unless the lost Na was excluded from the chondrite aggregates, the initial Na could not have been much higher than this and, of course, it could have been lower (requiring a shorter time at temperature). The extraordinary variability of Na contents of chondrules (e.g., Dodd 1978a,b) makes it difficult to derive rigorous time-temperature information, but the more highly sodic chondrules must have spent minimal time at high temperature.

CONSTRAINTS ON CHONDRULE GENESIS

Given the presence of unmelted material in chondrules and oxygen isotope differences between chondrules, it seems reasonable to set aside condensation and regolith models for the origin of most chondrules. The above review indicates that dynamic crystallization studies are generally consistent with the working hypothesis of melting of primordial particles to generate chondrules. The challenge now is to extract from these studies information on the physical environment in which chondrules formed, so as to constrain the nature of the heating events or heating mechanism. Although it may not be possible to uniquely define the thermal history of every specific chondrule, because of individual fluctuations in bulk composition and maximum temperature, it should be clear that on a cosmic scale, most droplet chondrules had a similar history.

The cooling rates used in all experiments that produced spherules resembling chondrules are summarized in Table 1 and range from a few degrees to many thousand degrees per hour. The maximum and minimum liquidus temperatures of chondrule analogues used in these experiments are 1580°C and 1440°C [coincidentally, both are from Tsuchiyama and Nagahara (1981)]. Some chondrules undoubtedly reached temperatures somewhat above their liquidi (finely dendritic and nearly glassy chondrules); others were not totally melted (chondrules with relict grains and poikilitic chondrules). Accretionary chondrules and some inclusions were apparently not melted at all, while rare Al_2O_3-MgO droplets may have reached 2000°C (Macdougall, 1982). However, the success with which the textures of Fe-Mg chondrules have been duplicated suggests that many particles were heated to 1400°–1600°C. Also of concern are the final temperatures after chondrule cooling. Planner and Keil (1982) rarely produced glass in their Al-free runs and argued from their experiments that glass-bearing chondrules must have been quenched from very high temperatures. Hewins *et al.* (1981) produced abundant glass in charges continuously cooled to

1000°C, which suggests that a quenching episode may not be required, provided that the ambient temperature was less than the glass transition of the residual glass. The ambient temperature is thus undefined, except that it must be less than glass transition temperatures.

The thermal history of a very large number of particles in chondrites can thus be summarized as follows: Chondrules were heated to high temperatures, typically about 1500°C, spending at most only hours at these temperatures (based on Na contents) and were cooled to relatively low ambient temperatures in times ranging from a few seconds to a few days. The importance of this conclusion is that it may rule out a number of otherwise possible origins for chondrules (e.g., models requiring exclusively very rapid or very slow cooling). Condensation and impact models have already been set aside because of textural and isotopic constraints. Any types of "transient high-temperature events of unknown nature" (Wood, 1981) that demand instantaneous cooling may also be rejected, while any mechanism linked to the general cooling of the solar nebula seems totally inappropriate in time scale. The cooling rates of chondrules do require the presence of a blanketing medium, presumably an opaque or chondrule-rich portion of the nebula that prevented rapid radiative cooling (Hewins *et al.*, 1981). It seems that this portion of the nebula was relatively cool at the time of chondrule formation, although Tsuchiyama *et al.* (1980) entertained the idea that the nebula cooled rapidly along with the chondrules.

Several mechanisms for chondrule formation have recently been discussed that appear capable, when fully worked out, of yielding thermal histories compatible with those inferred from the chondrule crystallization experiments. Levy (personal communication, 1982), in an approach described briefly in Cameron and Mercer-Smith (1983), has proposed flares as the heating mechanism. A smooth transition is assumed from the protosun to the adjacent hot turbulent nebula with magnetic activity and therefore flares from the disk at considerable distances from the protosun. The flares would heat and melt dust particles settling towards the median plane of the nebula. What is not yet defined for this mechanism is whether the flares would subside gradually enough, or whether there would be a sufficient density of gas and dust far from the median plane, so as to provide for low enough cooling rates. This restriction would seem even more severe for the case of heating by lightning, and yet Fujii and Miyamoto (1983) have suggested that lightning may be effective shortly prior to planetesimal formation when the gas-dust mixture was adequately thick.

Wood (1983) has proposed that chondrules formed when interstellar dust particles, settling toward the median plane of the nebula, entered the nebula and were heated by friction. The analogy with particles entering the Earth's atmosphere makes this model intuitively very attractive. The heating is rapid and chondrules would have appropriate cooling rates if the median plane was particle-rich and at a moderate or low ambient temperature. Like heating by flares, this mechanism involves aspects of chance that could account for the variety of objects in chondrites. Particles falling in from short distances might not accelerate enough to be melted, while melt droplets could impinge on other particles to form complex objects. This scenario would also allow for partial evaporation of some melted material to produce Ca-Al-rich material (King, 1982) and should provide suitable cooling rates for any spatter chondrules formed as a by-product.

FUTURE WORK

It is desirable to extend dynamic crystallization studies of compositions closely matching chondrules under appropriate redox conditions, as listed below. One technique that might yield considerable insight into chondrule crystallization would be to mount the charge on the junction of a thermocouple (Paque, personal communication, 1983). Comparison of temperature measurements with the reference thermocouple would show when crystallization occurred and indicate the amount of undercooling before nucleation.

(1) Runs should be continuously cooled down to 600°C to check for glass or total crystallization. This will serve to constrain the ambient temperature and test whether a quenching event is necessary.

(2) Experiments are needed on a wider range of compositions, including those of highly olivine-rich and Ca-Al-rich chondrules.

(3) Most importantly, for a composition typical of olivine-bearing chondrules, systematic experiments using both low and high cooling rates are needed, starting both above and below the liquidus. This would show under what conditions textures most like chondrules occurred, and if similar textures can indeed be produced using different thermal histories. It will be important to test how closely mineral compositions, as well as textures, match those in natural chondrules.

(4) Once systematic results are obtained as in (3) above, it should be simple to test specific models of chondrule origins (e.g., by using specific bulk compositions or specific thermal or redox histories).

CONCLUSIONS

Dynamic crystallization studies suggest that many chondrules were heated to temperatures in the range 1400°–1600°C and were cooled at rates ranging from a few degrees to many thousand degrees per hour down to a relatively low ambient temperature. This constraint rules out any model for chondrule origins that requires exclusively very rapid or very slow cooling. Heating by infall of particles into nebular gas (Wood, 1983), and possibly heating by flares from the nebular disk (Levy, personal communication, 1982), appear capable of providing suitable thermal histories for chondrules. More detailed accounts of these mechanisms should be compared closely with results from new experiments that more accurately reproduce textures and phase compositions in chondrules.

Acknowledgments. Financial support for this work was provided by NASA grant NSG-7327, and the Lunar and Planetary Institute supported travel to the Conference on Chondrules and their Origins. Presentations and discussions at the conference contributed materially to this paper. Critical review by H. N. Planner, comments from G. E. Lofgren, and editorial suggestions from K. Keil are greatly appreciated.

REFERENCES

Bischoff A. and Keil K. (1983) Ca-Al-rich chondrules and inclusions in ordinary chondrites: Evidence for a related genesis of ordinary and carbonaceous chondrites (abstract). In *Lunar and Planetary Science XIV*, pp. 47–48. Lunar and Planetary Institute, Houston.

Cameron A. G. W. and Mercer-Smith J. A. (1983) Behavior of the sun during growth from the primitive solar nebula (abstract). In *Lunar and Planetary Science XIV*, pp. 88–89. Lunar and Planetary Institute, Houston.

Dodd R. T. (1978a) The composition and origin of large microporphyritic chondrules in the Manych (L-3) chondrite. *Earth Planet. Sci. Lett.* **39**, 52–66.

Dodd R. T. (1978b) Compositions of droplet chondrules in the Manych (L-3) chondrite and the origin of chondrules. *Earth Planet. Sci. Lett.* **40**, 71–82.

Donaldson C. H. (1976) An experimental investigation of olivine morphology. *Contrib. Mineral. Petrol.* **57**, 187–213.

Donaldson C. H. (1979) An experimental investigation of the delay in nucleation of olivine in mafic magmas. *Contrib. Mineral. Petrol.* **69**, 21–32.

Fujii N. and Miyamoto M. (1983) Constraints on the heating and cooling processes of chondrule formation. This volume.

Gooding J. L., Keil K., Fukuoka T., and Schmitt R. A. (1980a) Elemental abundances in chondrules from unequilibrated chondrites: evidence for chondrule origin by melting of pre-existing materials. *Earth Planet. Sci. Lett.* **50**, 171–180.

Gooding J. L., Keil K., Mayeda T. K., Clayton R. N., Fukuoka T., and Schmitt R. A. (1980b) Oxygen isotopic compositions of petrologically characterized chondrules from unequilibrated chondrites (abstract). *Meteoritics* **15**, 295.

Hewins R. H., Klein L. C., and Fasano B. V. (1981) Conditions of formation of pyroxene excentroradial chondrules. *Proc. Lunar Planet. Sci. 12B*, pp. 1123–1133.

King E. A. (1982) Refractory residues, condensates and chondrules from solar furnace experiments. *Proc. Lunar Planet Sci. Conf. 13th*, in *J. Geophys. Res.* **87**, pp. A429–A434.

Klein L. C., Fasano B. V., and Hewins R. H. (1980) Flow behavior of droplet chondrules in the Manych (L-3) chondrite. *Proc. Lunar Planet. Sci. Conf. 11th*, pp. 865–878.

Lofgren G. E. (1980) Experimental studies on the dynamic crystallization of silicate melts. In *Physics of Magmatic Processes* (R. B. Hargraves, ed.), pp. 487–551. Princeton University Press, Princeton.

Lofgren G. E. (1982) The importance of heterogeneous nucleation for the formation of microporphyritic chondrules (abstract). In *Papers Presented to the Conference on Chondrules and their Origins*, p. 41. Lunar and Planetary Institute, Houston.

Macdougall J. D. (1982) Refractory chondrules in the CM meteorites (abstract). In *Papers Presented to the Conference on Chondrules and their Origins*, p. 42. Lunar and Planetary Institute, Houston.

Mayeda T. K., Clayton R. N., and Olsen E. J. (1980) Oxygen isotopic anomalies in an ordinary chondrite (abstract). *Meteoritics* **15**, 330–331.

McSween H. Y., Jr. (1977) Petrographic variations among carbonaceous chondrites of the Vigarano type. *Geochim. Cosmochim. Acta* **41**, 1770–1790.

Nagahara H. (1981) Evidence for secondary origin of chondrules. *Nature* **292**, 135–136.

Nagahara H. (1982) Effect of heating temperature on the texture of chondrules with special reference to the porphyritic chondrules (abstract). In *Papers Presented to the Conference on Chondrules and their Origins*, p. 47. Lunar and Planetary Institute, Houston.

Nagahara H. (1983) Porphyritic chondrules: evidence for incomplete melting of chondrules (abstract). In *Papers Presented to the Eighth Symposium on Antarctic Meteorites*, p. 25. National Institute of Polar Research, Tokyo.

Onorato P. I. K., Uhlmann D. R., Taylor L. A., Coish R. A., and Gamble R. P. (1978) Olivine cooling speedometers. *Proc. Lunar Planet. Sci. Conf. 9th*, pp. 613–628.

Onuma N., Clayton R. N., Ikeda Y., Hutcheon I. D., Mayeda T. K., and Molini-Velsko C. (1982) Oxygen isotopes in chondrules of ordinary and carbonaceous chondrites (abstract). Paper given at the Conference on Chondrules and their Origins, November 15–17, 1982. Lunar and Planetary Institute, Houston.

Paque J. M. and Stolper E. (1982) Dynamic crystallization experiments on an average Ca-Al-rich inclusion composition (abstract). In *Papers Presented to the Conference on Chondrules and their Origins*, p. 49. Lunar and Planetary Institute, Houston.

Planner H. N. (1979) Chondrule thermal history implied from olivine compositional data. Ph.D. thesis, Univ. of New Mexico, Albuquerque. 150 pp.

Planner H. N. and Keil K. (1982) Evidence for the three-stage cooling history of olivine-porphyritic fluid droplet chondrules. *Geochim. Cosmochim. Acta* **46**, 317–330.

Rambaldi E. R. (1981) Relict grains in chondrules. *Nature* **293**, 558–561.

Rambaldi E. R. and Wasson J. T. (1981) Metal and associated phases in Bishunpur, a highly unequilibrated ordinary chondrite. *Geochim. Cosmochim. Acta* **45**, 1001–1015.

Rambaldi E. R. and Wasson J. T. (1982) Fine, nickel-poor Fe-Ni grains in the olivine of unequilibrated ordinary chondrites. *Geochim. Cosmochim. Acta* **46**, 929–939.

Reid A. M., Williams R. J., Gibson E. K., Jr., and Fredriksson K. (1974) A refractory glass chondrule in the Vigarano chondrite. *Meteoritics* **9**, 35–45.

Scott E. R. D. and Taylor G. J. (1983) Petrologic similarities among chondrules in H, LL, CO, CM and E chondrites (abstract). In *Lunar and Planetary Science XIV*, pp. 680–681. Lunar and Planetary Institute, Houston.

Scott E. R. D., Taylor G. J., and Keil K. (1982a) Origins of ordinary and carbonaceous type 3 chondrites and their components (abstract). In *Lunar and Planetary Science XIII*, pp. 704–705. Lunar and Planetary Institute, Houston.

Scott E. R. D., Taylor G. J., Rubin A. E., Keil K., and Kracher A. (1982b) Origin of chondrules by preaccretionary melting of silicate matrix material (abstract). In *Papers Presented to the Conference on Chondrules and their Origins*, p. 54. Lunar and Planetary Institute, Houston.

Tsuchiyama A. and Nagahara H. (1981) Effects of precooling thermal history and cooling rate on the texture of chondrules: a preliminary report. *Mem. Natl. Inst. Polar Res., Spec. Issue No. 20* (T. Nagata, ed.), pp. 175–192. National Institute of Polar Research, Tokyo.

Tsuchiyama A., Nagahara H., and Kushiro I. (1980) Experimental reproduction of textures of chondrules. *Earth Planet. Sci. Lett.* **48**, 155–165.

Tsuchiyama A., Nagahara H., and Kushiro I. (1981) Volatilization of sodium from silicate melt spheres and its application to the formation of chondrules. *Geochim. Cosmochim. Acta* **45**, 1357–1367.

Wood J. A. (1981) The interstellar dust as a precursor of Ca,Al-rich inclusions in carbonaceous chondrites. *Earth Planet. Sci. Lett.* **56**, 32–44.

Wood J. A. (1983) Formation of chondrules and CAI's from interstellar grains accreting to the solar nebula. *Mem. Natl. Inst. Polar Res., Spec. Issue.* National Institute of Polar Research, Tokyo. In press.

Conditions of Early Chemical Processing of Matter: Explosive Exhalations of Supernovae

D. Heymann

Department of Geology and Department of Space Physics and Astronomy, Rice University, Houston, Texas 77251

The overwhelming majority of atoms other than H and He in the solar system come from stellar nucleosynthesis. A very significant proportion of these come from supernovae. This paper reviews the chemical and isotopic stratifications of supernova exhalations, which are the first of all chemical environments of these atoms.

INTRODUCTION

Chemical and isotopic heterogeneity is a hallmark of the solar system and of interstellar media. Until one decade ago, most cosmochemists thought that the heterogeneities of the solar system were *endogenic*. When doubts arose, we suggested theories such as the occurrence of live superheavy elements in the nascent solar system (Anders and Heymann, 1969). In retrospect it is clear that the change in ideas began with the discovery of isotopically anomalous xenon in meteorites by Reynolds (1960). Further change was brought about by Black's (1972) interpretations of isotopic neon anomalies in the solar system, but the breakthrough to a wider audience than the inert-gas cosmochemists occurred with the discovery of oxygen anomalies in the Allende meteorite by Clayton *et al.* (1973). Today the study of *exogenic heterogeneities*, that is to say chemical and/or isotopic heterogeneities of the solar system that are "memories" from pre-solar environments, is a swiftly developing branch of cosmochemistry.

For several years after the discovery of the oxygen anomaly, it appeared that chondrules of ordinary chondrites could be studied without considerations about exogenic heterogeneities. This, too, is no longer possible, as the paper presented by R. N. Clayton at the Conference on Chondrules and their Origins has clearly demonstrated.

This paper is a brief overview of the chemical and isotopic stratifications of the exhalations from exploding stars (supernovae). These exhalations are not the sole sources for the replenishment of matter in interstellar media, hence of its heterogeneities, but they deserve close attention because they are thought to deliver roughly one-half of that matter.

SUPERNOVAE AND THEIR PROGENITOR STARS

This is a summary of a recent review by Arnould (1982). Supernovae are classified as Type I and Type II on the basis of their light curves. (The term supernova was originally proposed to represent the *visual* astronomical phenomenon associated with an exploding star. Over the years, it has come to stand for all aspects of the explosion.) Both have in common deduced explosive energies on the order of 10^{51} erg or greater and the emission of at least one solar mass ($1\ M_\odot = 1.9866 \times 10^{33}$ g) of matter into the immediately surrounding interstellar media with initial speeds of typically 10^3 km s^{-1} or faster. Model studies suggest that the progenitor stars of Type I supernovae are low-mass binaries of a comparatively old stellar population, whereas those of Type II supernovae are more massive stars from younger populations with the transition occurring near 8 $M_\odot$. However, there is no unanimity on this issue. The rate of supernova occurrence in our galaxy is estimated as 30 (+20/−10) y^{-1}. However, as Reeves (1978) has emphasized, the probability of supernova occurrence is much higher in regions of active star formation than elsewhere in the galaxy.

There is a general agreement that the cause of the explosions is "runaway-style" stellar instability, and that both the progenitor stars and their initial ($t = 0$) exhalations are chemically and isotopically stratified, even though there is very little direct observational evidence for this. Perhaps the most pertinent observations are those of Chevalier and Kirshner (1978, 1979), who found that the so-called quasi-stationary flocculi of the roughly 300-year-old Cassiopeia A nebula are strongly enriched in He and N (relative to "cosmic"), and that the "fast moving knots" of this nebula are enriched in O, S, Ar, and Ca, with substantial "knot-to-knot" variations of the S/O, Ar/O abundance ratios.

The detailed calculations described in the remainder of the paper are associated with Type II supernovae.

INITIAL CHEMICAL STRATIFICATIONS OF THE EXHALATIONS

Hence, what we "know" about the initial chemical stratifications of exhalations from supernovae comes primarily from astrophysical model studies of their progenitor stars. Such studies are complex because they must consider conditions of temperature, mass density, and pressure throughout the star. They must also consider the state of the stellar matter, nuclear reactions, energy budgets, convection, etc., all within the boundary conditions deduced from observations of the *surfaces* of stars. Most models assume spherical symmetricity of the stars of interest and absence of mass-loss during the hydrostatic legs of their evolutions. Investigators concerned with supernova exhalations assume that their initial chemical stratifications are identical to those of the progenitor stars after the cessation of explosive burning *sans* the innermost about 1.6 $M_\odot$ of matter that remains in the remnant entity (Clayton, 1977, 1978, 1979, 1981; Lattimer *et al.*, 1978). However, Clayton (1981) has suggested a model that assumes that the progenitor star expels, non-explosively, its C-, Ne-, and H-rich outer layers during hydrostatic evolution to form a dust-laden circumstellar "coccoon," and that the remnant, still-massive star explodes considerably later. The physics of the encounter of the explosive exhalation with a circumstellar coccoon has not been significantly studied, hence I do not consider this model here.

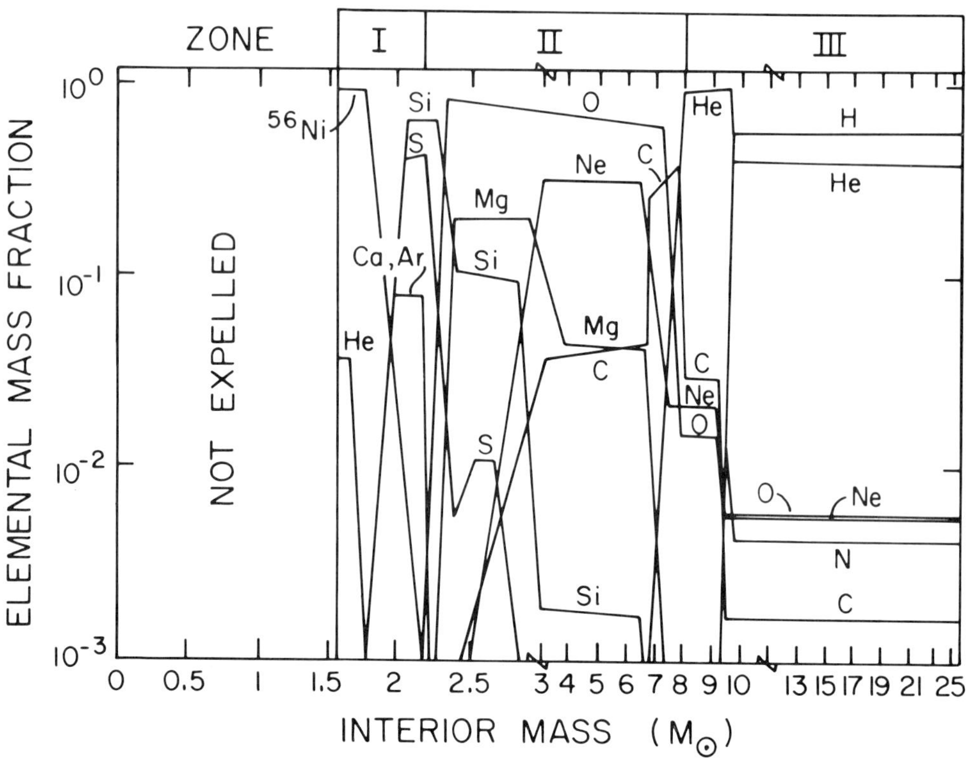

Fig. 1 Major element initial stratigraphy of the explosive exhalation of a 25 $M_\odot$ progenitor star. Vertical coordinate is mass fraction. Horizontal coordinate is "internal mass point" (see text). Constructed from data in Weaver *et al.* (1978) and Weaver and Woosley (1980).

Two major strategies have been employed for the model calculations of stellar evolution. In one, the star is thought to contain a number of large "single zones" of uniform conditions. In the other, a comparatively dense radial grid of points is used. I am using here results from Weaver and Woosley (1980), whose treatment employs the second of the two strategies. Figure 1 has been constructed by me from data in these papers. It shows the initial stratigraphy of elements with mass fractions greater than 10^{-3} in the explosive exhalation from the 25 $M_\odot$ progenitor star. (Mass fraction is a more useful parameter for grossly heterogeneous systems than abundance normalized to a single element.) The gross stratigraphy of this figure is pertinent for progenitor stars with masses of about 10 $M_\odot$ and larger. The variable on the horizontal axis is the "internal mass point" in units of 1 $M_\odot$, that is to say the radius of the sphere within which there is contained the mass given by the value of the variable. This variable is much more useful than linear distance from the star's center because of the swift expansion of the exhalation.

One recognizes three major chemical zones:

(1) Zone I. 1.6 $M_\odot$–2.2 $M_\odot$. Essentially oxygen-free, but "oxidizing" because of abundant S. Major elements are "Fe-group," Si, S, Ar, Ca, and He. ^{4}He at the base is from (γ, He) nuclear reactions during explosive burning. The zone contains abundant radioactive ^{56}Ni ($t_{1/2} = 6.1$ d), which transforms via ^{56}Co ($t_{1/2} = 78.5$ d) to stable ^{56}Fe. At $t = 0$, the decay of ^{56}Ni generates on the order of 10^{36} watts of power, and ^{56}Co generates on the order of 10^{35} watts one year later, but a significant fraction of the energy is lodged in neutrinos. Note that the zone does not contain enough Fe + Cr + Ca + Si to satisfy S chemically. Zone I is probably the most massive, albeit transitory, S-rich environment of interstellar media.

(2) Zone II. 2.2 $M_\odot$–8 $M_\odot$. Very rich in oxygen. Other major elements are Ne, Mg, Si, S, and C. ^{27}Al is not shown because its mass fraction is less than 10^{-3} everywhere, but Zone II contains the bulk of this element. Its stratigraphic trend follows that of Mg closely (Woosley and Weaver, 1980). Zone II also contains the bulk of ^{26}Al (Woosley and Weaver, 1980; Morgan, 1979). At $t = 0$, the specific activity of ^{26}Al for the bulk of the matter of Zone II is about one C per gram. Note that $O/C > 1$ and note the varying Mg/Si ratio. Note also that this zone does not contain enough Mg + Si + S + C to satisfy O chemically. One may distinguish three "subzones": (O,Mg), (O, Ne), and (O, C). In the case of the 15 $M_\odot$ progenitor, O/C is less than 1.0 in the (O, C) subzone (Weaver *et al.*, 1978). Zone II is probably the most massive, albeit transitory, O-rich environment of interstellar media.

(3) Zone III. 8.0 $M_\odot$–25 $M_\odot$. This zone is chemically reducing with C, He, and H the most abundant elements. It is compositionally the closest to "solar" of the three zones.

Since the trace elements are not shown in Fig. 1, two comments are in order: Zone I contains few, if any, nuclei of elements $Z > 30$ (Zn), and Zone II is the repository of the bulk of $Z > 30$ elements.

INITIAL ISOTOPIC STRATIFICATIONS OF THE EXHALATIONS

The initial isotopic stratifications of the exhalations are also thought to be replicas of the stratifications of their progenitor stars immediately after the explosion, but the isotopic stratigraphies have not yet been worked out with as much detail as the major element chemical stratigraphies. Therefore, I offer only a sketch for six groups of elements for the 25 $M_\odot$ progenitor.

(1) C, N, O. One must bear in mind that stars with $M_{STAR} > 10\ M_\odot$ are comparatively poor producers of these elements, hence interstellar heterogeneities of these elements are probably determined in the main from other sources. Zone I contains virtually no C, N, or O. In Zone II, C is near-monoisotopic ^{12}C and O is near-monoisotopic ^{16}O. The isotopic composition of what little N there is is uncertain here. In Zone III, C may be enriched in ^{13}C and O in ^{17}O + ^{18}O. N may be enriched here in ^{14}N, but the degree of enrichment may vary greatly from star to star.

(2) Ne, Mg, Si, S, Ar. Zone I contains virtually no Ne and Mg. Where its matter is dominated by explosive oxygen burning (near the top), Si is overwhelmingly ^{28}Si, but S and Ar

appear to be near-normal (Truran and Arnett, 1970; Woosley *et al.*, 1973). Zone II contains virtually no Ar. The zone's isotopic stratigraphy for the remaining elements is best "known" for two subregions, one about 2.2 $M_\odot$–2.8 $M_\odot$ (inner), the other (outer) 2.8 $M_\odot$–6 $M_\odot$ (Woosley and Weaver, 1980). The inner subregion contains near-monoisotopic ^{24}Mg, but ^{30}Si and ^{29}Si are enriched by factors of about 3.5 and 1.2, respectively. S is strongly enriched in ^{33}S and ^{34}S; the subregion is very poor in Ne. The outer subregion contains no ^{22}Ne; ^{21}Ne/^{20}Ne is about one-half normal, ^{26}Mg/^{24}Mg is near-normal, but ^{25}Mg is enriched by factors 1.5–2.0. ^{28}Si is depleted, but S is strongly enriched in ^{36}S. The only firm statement for the remainder of Zone II (6 $M_\odot$–8 $M_\odot$) is that it ought to be poor in ^{22}Ne. In Zone III, ^{21}Ne is almost certainly overabundant (see Clayton, 1981).

(3) Ti. Clayton (1981) has presented a gross stratigraphy for this element. The base of Zone I contains primarily ^{50}Ti; its top, ^{48}Ti + ^{49}Ti. Where matter of Zone II is dominated by explosive neon burning (near its base), ^{46}Ti is abundant. Where it is dominated by explosive carbon burning one finds primarily ^{47}Ti and some ^{49}Ti + ^{50}Ti, and where it is dominated by prior *s*-processing one finds primarily ^{50}Ti. At locations of "mini-*r*-processes" during explosive burning, one finds primarily ^{47}Ti and some ^{49}Ti + ^{50}Ti.

(4) The *s*-nuclei. $A > 70$. These nuclei occur with substantial overproduction factors in Zone II between about 3 $M_\odot$ and 8 $M_\odot$ (Lamb *et al.*, 1977). However, stars in the intermediate mass range of 2–6 $M_\odot$ are thought to be the dominant source of *s*-process matter of the cosmos. Detailed stratigraphies are unavailable.

(5) The *p*-nuclei. Owing to the study of Woosley and Howard (1978), their stratigraphy is fairly well established. These nuclei occur between about 2 $M_\odot$ and 3.5 $M_\odot$ where they were formed by the photoerosion of an *s*-seed during explosive burning. Figure 2 shows the stratigraphy of a number of selected *p*-nuclei. Maximum overproduction is seen to occur at deeper locations for "light" species such as ^{74}Se, ^{78}Kr, and ^{84}Sr, but at shallower locations for "heavy" species such as ^{184}Os. Note that *p*-nuclei find themselves in chemically different environments (compare Figs. 1 and 2).

(6) The *r*-nuclei. These are enigmatic. Their formation has been traditionally associated with supernovae. However, only "mini *r*-processes" have been identified by the model studies (Howard *et al.*, 1972; Truran *et al.*, 1978; Heymann and Dziczkaniec, 1979; Lee *et al.*, 1979). Mini *r*-processes generate interesting, "non-canonical" *r*-abundance curves with maxima near nuclei with closed neutron shells, but they do not form *r*-nuclei with $A > 200$ (including ^{232}Th, ^{235}U, ^{244}Pu) well, and their overproduction factors are no match for those of the *s*- and *p*-nuclei with $A < 200$. *The current understanding is that supernovae are comparatively poor producers of r-nuclei.* Cowan *et al.* (personal communication, 1981) have recently suggested that the "canonical" *r*-process may occur in "ordinary" low-mass stars that probably do not explode, but rather "exhale" *r*-nuclei non-explosively; however, the required great enhancement of ^{13}C in a degenerate helium region is still a difficulty of this model. This theory, if correct, means that $r/(s + p)$ abundance ratios in interstellar media could be much more variable than hitherto thought to be the case.

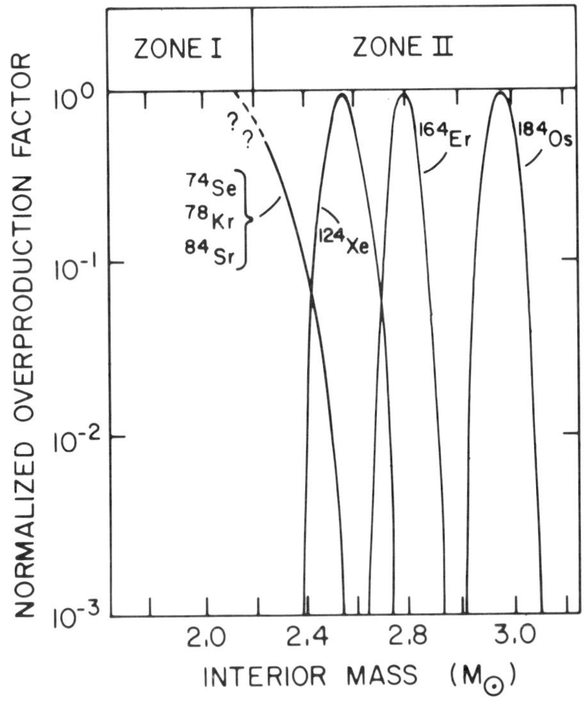

Fig. 2 Initial stratigraphy of *p*-nuclei in the explosive exhalation of a 25 $M_{\odot}$ progenitor star. Constructed from data in Woosley and Howard (1978). Only selected species are shown to avoid overcrowding of the figure. In every case, overproduction factors have been normalized to the maximum overproduction factor.

EVOLUTION OF EXHALATIONS

Exhalations allegedly start out as immediate post-supernova progenitor stars, that is to say very hot ($T \approx 10^{9}$°K), very dense ($\rho \approx 10^{6}$ g cm^{-3}), and fully ionized in their deep interiors. Radially outward, T and ρ decrease (for details see Weaver and Woosley, 1980). Interactions with the immediately surrounding interstellar medium have been studied. A shock develops at the leading edge, and it is there that matter from interstellar media is ingested. Observations of pertinent nebulae show that the spherical symmetricity is not maintained forever; the exhalations develop filamentary structures and "knots," but the time scale for this development is still poorly known.

The cosmochemist is more interested in the evolution of the internal chemical structure of the exhalations. Comparatively little is "known" about that. At a first glance, one may assert that the initial chemical and isotopic stratigraphies are swiftly ironed out by the violence of the explosion. However, the deep interior of the exhalations is a very dense, albeit also hot, totally ionized plasma with considerable rigidity. It is perhaps not unreasonable to assume that the convective pattern of an exhalation is similar to that of its progenitor star, at least for a while. An exhalation has a substantial heat source (^{56}Ni + ^{56}Co) in Zone I, but heat sources analogous to those of shell-burning in the progenitor star are missing. Perhaps one may expect intrazonal mixing, but comparatively little mixing between Zones I, II, and III, or even of their subzones, for a while. Quantitative studies of this issue are greatly needed.

This is the basic assumption of the studies of Lattimer *et al.* (1978) and Clayton (1977, 1978, 1979, 1981). Lattimer *et al.* (1978) have treated the development of T and ρ, which decrease in lockstep, at two interior locations. They have concluded that favorable conditions for the condensation of dust eventually do occur, provided that the deduced condensation sequences are not truncated or bypassed owing to heating of matter by $^{56}Ni + ^{56}Co$ in Zone I. However, the failure of condensation to occur does not imply absence of chemical processing in the gas phases where molecules, molecular ions, and radicals can form. The presence of ^{56}Co and ^{26}Al (but also of other species such as ^{36}Cl) will keep Zones I and II partially ionized for long times, hence ionic species could be very significant in chemical processing. The presence of magnetic and electrical fields could further complicate chemical processing in an environment that is a far cry from any laboratory-based analogs.

Lattimer *et al.* (1978) and Clayton (1981) make frequent references to minerals such as corundum, melilite, spinel, enstatite, periclase, etc., but it is misleading to think of solid condensates as euhedral crystals of the uncontaminated minerals. The condensates are probably riddled with lattice defects, and the minerals are probably quite "dirty" owing to the occlusion of "non-formula" elements in their lattices.

Despite the multitude of uncertainties, one can assume with considerable confidence that the initial chemical processing takes place in the chemical zones of Fig. 1 regardless of whether condensation does or does not occur. The next phase is likely to be chemical "reprocessing" when the zones begin to mix. From then on, "reprocessing" is a recurrent theme as the matter wanders through various interstellar media and ultimately into meteorites (see Clayton, 1981 and references in that paper).

MEMORIES

Cosmochemists have long known that chemical processing in distinctly different physical environments tends to lead to remarkably similar results. Thus, it is usually impossible to select one environment exclusively, because there are always several rational choices. For example, oldhamite (CaS) occurs in enstatite chondrites, but it would be rash to identify this occurrence exclusively with Zone I of a supernova exhalation, even though CaS is deduced to occur transiently in its gas phase (see Lattimer *et al.*, 1978). Likewise, it would be rash to associate the well-known trends of total Fe-content and metal content among H-, K-, and LL-chondrites with chemical processing at the top of Zone I and the base of Zone II after some mixing, even though the appropriate gradients of Fe and O might occur there. Much stronger, albeit seldomly exclusionary cases can be made by combining chemical and isotopic observations, but one must then distinguish between isotopic variations due to radioactive transformations and isotopic anomalies for which there is no radioactive parentage. No geochemist will claim, for example, that $^{87}Sr/^{86}Sr$ variations in terrestrial samples are pre-solar memories, but he may argue about the solar system homogeneity of initial values of that ratio. Isotopic variations, due to the decay of ^{26}Al, ^{107}Pd, and ^{129}I are less clear-cut, because one may argue, as D. D. Clayton has done (see Clayton, 1981), that what appears to be *in situ* decay of these species in the meteorites is actually quantitative decay in interstellar media, hence a "memory" from them.

The ambiguities associated with radioactive decay do not occur for those isotopic anomalies that cannot have a radioactive parentage. The isotopic anomaly is the pre-solar memory that has apparently detectably survived the initial chemical processing and also every subsequent chemical reprocessing as mentioned above.

Supernova exhalations are probably significant contributors to oxygen anomalies as Clayton *et al.* (1973) have first pointed out. Zone II of Fig. 1 is the initial repository of most of the oxygen of the exhalation. All condensates from this zone are endowed with an ^{16}O overabundance. Certain chemical species from Zone I, regardless of whether they are in the gas phase or in condensed phases may become similarly endowed with an ^{16}O overabundance when Zones I and II begin to intermingle. A veritable catalog of $^{16}O/^{17}O/^{18}O$ ratios may occur in chemical species where Zones II and III intermingle. Therefore, what appears to be a small number of "components" in the systematics of oxygen anomalies of the solar system (see Clayton *et al.*, 1976) is likely to be the result of repeated chemical reprocessing of an isotopically greatly diverse initial inventory. It is not unlikely, therefore, that continued studies of oxygen will come up with hitherto undiscovered "components." (However, in terms of total yields, Type II supernovae are probably not greatly enhanced in ^{17}O. Their bulk $^{18}O/^{16}O$ ratios vary greatly, decreasing, as they do, with increasing mass.)

The suspicion that *r*-nuclei are grossly underabundant in supernova exhalations implies that the exhalations must be reprocessed with matter that is overabundant in these nuclei. It is therefore unlikely that the excesses of the *r*-isotopes of Ba, Nd, and Sm in the EK-1-4-1 inclusion of Allende (see McCulloch and Wasserburg, 1978a,b; Lugmair *et al.*, 1978) are due to a "spike" of *r*-matter from one or several supernovae, but rather are due to variable chemical reprocessing of *s*-matter from their exhalations. However, the *r*-anomalies of Kr and Xe could be surviving memories from mini *r*-processes of exploding stars (Heymann and Dziczkaniec, 1979, 1980). These must then be memories from very early chemical processing, perhaps ion implantation in the first crop of condensates.

The *p*-nuclei have received comparatively little attention. It is fairly reasonable to assert that isotopic *p*-anomalies are likely to occur in conjunction with ^{16}O anomalies and that they are not necessarily coherently correlated with *r*-anomalies. When one considers the isotopic stratigraphies of Zone II of Figs. 1 and 2, one concludes that the abundances of *p*-isotopes other than those of Kr and Xe in meteoritic phases with ^{16}O excess could be below normal, normal, or above normal. One cannot deny the possibility that all three cases will eventually be found.

Let me close with a case for which chemical processing in a supernova exhalation might result in a violation of commonsense geochemical rules. It has long been known that the isotopic variations of the *p*-isotopes of Kr and Xe are coherently anti-correlated in carbonaceous chondrites (see Frick, 1977). Lewis *et al.* (1975) have argued that the anti-correlation is due to endogenic (i.e., in the solar system) mass fractionation. Heymann and Dziczkaniec (1980) have argued for memory from a supernova exhalation. Figure 2 shows that the locations of maximum overproduction of the *p*-isotopes of these geochemically very similar elements are substantially separated physically. However, the memory seen in the meteorites must then have been established during the earliest chemical processing. The most attractive hypothesis is that *p*-Kr has either remained in the gas phase, or if trapped in dust, has been quantitatively returned to the gas phase by later chemical processing, whereas *p*-Xe has not. When one considers that the location of maximum overabundance of *p*-Kr is either in Zone I, or else closer to Zone I than that of *p*-Xe, which is in Zone II, and that Zone I contains the substantial heat source of ^{56}Co, one can understand how the commonsense geochemical rule that "Kr goes with Xe" might become violated. If my interpretation is correct, then excess *p*-Xe in chondritic meteorites must either be contained in or else be associated with tiny *inorganic* grains.

ADDENDUM

Hans Suess (reviewer) has brought to my attention a manuscript submitted to *Science* by Thiemens and Heidenreich, entitled: "A novel isotope effect: the mass-independent fractionation of oxygen and its possible cosmochemical implications," which was not available to me at the time of this conference. In the laboratory, the authors have discovered a mass-independent oxygen isotopic fractionation whose trend is essentially identical to the mixing line of the anhydrous phases from the C2 and C3 meteorites. Thiemens and Heidenreich cautiously offer this type of fractionation as an alternative for current nucleogenetic models of oxygen isotopic variations in meteorites.

This suggestion does not change the substance of my contribution to these Proceedings. I have reviewed conditions of *early* chemical processing of matter for the specific case of supernova exhalations. I have pointed out that memories from early processing are vulnerable to erasure during later processing. Clayton (1977, 1978, 1979, 1981) has amply written on that topic. The process of Thiemens and Heidenreich must be added to the catalog of processes that may erase records from early processing, but it might also have some significance at its latest stages when an exhalation, or else regions thereof, have cooled sufficiently. Nevertheless, if condensation occurs in Zone II, and if Zones I and II mix significantly, then a substantial mass of probably primary isotopically anomalous oxygen is initially stored in solid phases. These anomalies may become erased later, but it is difficult to see how *that* oxygen could significantly participate in a process *à la* Thiemens and Heidenreich.

Acknowledgments. This is the last paper supported by NASA grant NGL44-006-127. I thank A. Elsweiler for her help in preparing this manuscript.

REFERENCES

Anders E. and Heymann D. (1969) Elements 112 to 119: Were they present in meteorites? *Science* **164**, 821–823.

Arnould M. (1982) Explosive nucleosynthesis. Cahier N°8 of *Physique Nucleaire Theorique U.L.B.* Universite Libre de Bruxelles, Belgium. In press.

Black D. (1972) On the origins of trapped helium, neon and argon isotopic variations in meteorites–II. Carbonaceous meteorites. *Geochim. Cosmochim. Acta* **36**, 377–394.

Chevalier R. A. and Kirshner R. P. (1978) Spectra of Cassiopeia A. II. Interpretation. *Astrophys. J.* **219**, 931–941.

Chevalier R. A. and Kirshner R. P. (1979) Abundance inhomogeneities in the Cassiopeia A. Supernova remnant. *Astrophys. J.* **233**, 154–162.

Clayton D. D. (1977) Solar system isotopic anomalies: Supernova neighbor or presolar carriers. *Icarus* **32**, 255–269.

Clayton D. D. (1978) Precondensed matter: key to the early solar system. *Moon and Planets* **19**, 109–137.

Clayton D. D. (1979) Supernovae and the origin of the solar system. *Space Sci. Rev.* **24**, 147–226.

Clayton D. D. (1981) Key issues in isotopic anomalies. *Proc. Lunar Planet. Sci. 12B*, pp. 1781–1801.

Clayton R. N., Grossman L., and Mayeda T. K. (1973) A component of primitive nuclear composition in carbonaceous meteorites. *Science* **182**, 485–488.

Clayton R. N., Onuma N., and Mayeda T. K. (1976) A classification of meteorites based on oxygen isotopes. *Earth Planet. Sci. Lett.* **30**, 10–18.

Frick U. (1977) Anomalous krypton in the Allende meteorite. *Proc. Lunar Sci. Conf. 8th*, pp. 263–272.

Heymann D. and Dziczkaniec M. (1979) Isotopic composition of xenon from intermediate zones of supernova. *Proc. Lunar Planet. Sci. Conf. 10th*, pp. 1943–1960.

Heymann D. and Dziczkaniec M. (1980) A first roadmap for kryptology. *Proc. Lunar Planet. Sci. Conf. 11th*, pp. 1179–1213.

Howard W. M., Arnett W. D., Clayton D. D., and Woosley S. E. (1972) Nucleosynthesis of rare nuclei from seed nuclei in explosive carbon burning. *Astrophys. J.* **175**, 201–216.

Lamb S. A., Howard W. M., Truran J. W., and Iben I., Jr. (1977) Neutron capture nucleosynthesis in helium-burning cores of massive stars. *Astrophys. J.* **217**, 213–221.

Lattimer J. M., Schramm D. N., and Grossman L. (1978) Condensation in supernova ejecta and isotopic anomalies in meteorites. *Astrophys. J.* **219**, 230–249.

Lee T., Schramm D. N., Wefel J. P., and Blake J. B. (1979) Isotopic anomalies from neutron reactions during explosive carbon burning. *Astrophys. J.* **232**, 854–862.

Lewis R. S., Srinivasan B., and Anders E. (1975) Host phase as a strange Xe component in Allende. *Science* **190**, 1251–1262.

Lugmair G. W., Marti K., and Scheinin N. B. (1978) Incomplete mixing of products from *r*-, *p*-, and *s*-process nucleosynthesis: Sm-Nd systematics in Allende inclusion Ek-1-04-1 (abstract). In *Lunar and Planetary Science IX*, pp. 672–674. Lunar and Planetary Institute, Houston.

McCulloch M. T. and Wasserburg G. J. (1978a) Barium and neodymium isotopic anomalies in the Allende meteorite. *Astrophys. J. Lett.* **220**, 115–119.

McCulloch M. T. and Wasserburg G. J. (1978b) More anomalies from the Allende meteorite: Samarium. *Geophys. Res. Lett.* **5**, 599–602.

Morgan J. W. (1979) Nucleosynthesis in supernovae. Ph.D. thesis, Rice University. 158 pp.

Reeves H. (1978) The "bing-bang" theory of the origin of the solar system. In *Protostars and Planets* (T. Gehrels, ed.), pp. 399–426. Univ. of Arizona Press, Tucson.

Reynolds J. (1960) Determination of the age of the elements. *Phys. Rev. Lett.* **4**, 8–10.

Truran J. W. and Arnett W. D. (1970) Nucleosynthesis in explosive oxygen burning. *Astrophys. J.* **160**, 181–192.

Truran J. W., Cowan J. J., and Cameron A. G. W. (1978) The helium-driven *r*-process in supernovae. *Astrophys. J. Lett.* **222**, 163–167.

Weaver T. A. and Woosley S. E. (1980) Evolution and explosion of massive stars. *Ann. N. Y. Acad. Sci.* **336**, 335–357.

Weaver T. A., Zimmerman G. B., and Woosley S. E. (1978) Presupernova evolution of massive stars. *Astrophys. J.* **275**, 1021–1029.

Woosley S. E. and Howard W. M. (1978) The *p*-process in supernovae. *Astrophys. J. Suppl. Ser.* **36**, 785–804.

Woosley S. E. and Weaver T. A. (1980) Explosive neon burning and ^{26}Al gamma-ray astronomy. *Astrophys. J.* **238**, 1017–1025.

Woosley S. E., Arnett W. D., and Clayton D. D. (1973) The explosive burning of oxygen and silicaon. *Astrophys. J. Suppl. Ser.* **26**, 231–312.

On the Alteration of Allende Chondrules and the Formation of Matrix

R. M. Housley and E. H. Cirlin
Rockwell International Science Center, 1049 Camino dos Rios, Thousand Oaks, California 91360

We report new scanning electron microscope observations of Allende chondrules and matrix. The most striking new feature we describe is a peculiar intergrowth of Fe-rich olivine ($\sim Fa_{50}$) and clinoenstatite ($\sim Fs_2$). All clinoenstatite we have observed is involved in these associations along the outer margins, grain boundaries, and internal crack surfaces. Most of the chondrules also contain abundant opaque inclusions, which we show must have been strongly altered and partially replaced subsequent to chondrule formation. We believe that the Fe-rich olivine mentioned above has formed from the clinoenstatite by a reaction of the type $Fe + 1/2O_2 + MgSiO_3 \rightarrow (Fe,Mg)_2SiO_4$. The observed intergrowth textures require that Fe was able to move readily along internal cracks. Since the strains associated with the reaction cause the clinoenstatite to cleave, all Allende matrix olivine could have been produced from initially much more abundant enstatite-rich chondrules and Fe metal by this reaction. We argue that a variation of the small internally heated planetesimal as discussed by Fish *et al.* (1960) can provide the most plausible environment in which these reactions could have occurred. In such a setting conditions can be more oxidizing and have a higher gas pressure at a given temperature than they could in the solar nebula.

INTRODUCTION

The Allende CV3 fall of February 8, 1969 greatly increased the total amount of carbonaceous chondrite material in the world's meteorite collections. It has now been extensively studied and continues to be the object of much work. A thorough basic petrographic and chemical description has been provided by Clarke *et al.* (1970).

Unusual, sometimes centimeter-sized, Ca,Al-rich inclusions constituting about 9% of the volume (Clarke *et al.*, 1970) provide a primary focus of attention. Initially, they stimulated excitement because of their similarity in composition to material predicted to condense from the solar nebula at high temperature. Although it was soon recognized that some have melt textures, it can still be argued that others may be direct high temperature condensates. All have been altered subsequent to formation; some were altered extensively. Recognized alterations include the formation or deposition of grossular, wollastonite, nepheline, sodalite, and anorthite in interior cavities and of nepheline, sodalite, diopside, hedenbergite, and andradite in rims. Most of the known characteristics of these Ca,Al-rich inclusions and their various suggested interpretations have been carefully reviewed by Grossman (1980).

Matrix constitutes 57% of the section point counted by Clarke *et al.* (1970). Allende matrix is unusual, perhaps even unique, in character. It consists predominantly of 2–10-μm-sized olivine grains with a composition near Fa_{50}. Some of these grains are subhedral or even euhedral.

Somewhat smaller grains of pentlandite, Ni-rich metal, high Ca pyroxene, nepheline, and sodalite are dispersed through this matrix in subordinate amounts. They generally appear to have grown in contact with the olivine.

Most of the known constituents of Allende have been nicely described and illustrated by Wark (1979), who in particular shows a scanning electron microscope (SEM) picture of a typical matrix area. In this paper Wark offers a well-developed, logically consistent interpretation of these components and much of the known information concerning Allende, in terms of nebular condensation processes.

Another important general petrographic study of Allende was reported as part of a comprehensive survey of CO and CV meteorites by McSween (1977a,b). He regards all components, either rounded or irregular in shape, that show evidence of solidification from a melt as chondrules, and we retain this terminology. Chondrules are futher subdivided into four distinct categories. Type I, which is by far the most abundant in CO and CV meteorites, consists predominantly of forsteritic olivine, usually granular, in the composition range Fa_0–Fa_{10}, with or without clinoenstatite Fs_0–Fs_7 frequently poikilitically enclosing olivine, and with or without abundant opaques. Type II consists predominantly of more iron-rich olivine Fs_{15}–Fs_{50} with chrome-spinel and little, if any, magnetite, metal, or sulfide. Type III consists of excentroradial pyroxene chondrules and although it is a minor component in CO chondrites, it has not been observed in CV meteorites. Type IV consists of Ca,Al-rich particles showing a distinct melt texture. We will adopt these categories in this paper. Type I chondrules are the dominant type in Allende. In a random sampling of 1572 points, McSween (1977b) found only trace abundances of type II and IV chondrules. Matrix constitutes 38% of the area surveyed in this case.

Following the observational work mentioned above, McSween (1979) reviewed the available evidence concerning alteration in carbonaceous chondrites. He suggested that starting with CO3 or CV3 meteorites, one basic alteration path produced equilibration of Fe/Mg ratios in silicates, as is believed to occur in the ordinary chondrites, while another produced various degrees of hydrothermal alteration and led to the formation of CM and CI meteorites.

He also noted another type of alteration that occurs fairly generally and is especially pronounced in the CV meteorites (McSween, 1977b, 1979). It is the oxidation of various amounts of metal in the chondrules to magnetite, accompanied by the enrichment of remaining metal and sulfides in Ni. Based on this, he subdivided CV meteorites into oxidized and reduced groups. Allende is the most oxidized one known.

McSween (1977b) describes opaques in CV3 meteorites as generally occurring in subrounded grains inside chondrules. He indicates that in the oxidized subgroup many of these consist of magnetite or magnetite with metal cores. The opaques in Allende have been more fully described by Haggerty and McMahon (1979), who show reflected light micrographs of several of these magnetite-metal complexes. Magnetite in chondrites has long been interpreted, as McSween (1977b) interpreted it, as evidence of secondary alteration (Ramdohr, 1963).

This is a necessary interpretation even if chondrules formed by the melting of heterogeneous solid precursors. The phase equilibria studies of Muan and Osborn (1956) show that magnetite would readily dissolve in an ultramafic melt of chondrule composition and that upon cooling a corresponding amount of magnesioferrite would crystallize. Therefore, rounded magnetite and magnetite-metal complexes must have formed by subsolidus oxidation of primary metal or sulfide, as is generally believed.

Since the magnetite contains little Ni this alteration leads to enrichment of the remaining metal in Ni. When the Ni content in this metal gets high enough, a reaction between metal and troilite to form pentlandite becomes favorable below about 900°K (Kullerud, 1963). Therefore, it is entirely possible, even probable, that the original opaque assemblages in Allende consisted primarily of metal and troilite as in the less oxidized CV3 meteorites and the ordinary chondrites.

Thus both Ca,Al-rich inclusions and opaque assemblages in chondrules are known to have been altered subsequently to formation in Allende. We present here evidence, largely obtained by scanning electron microscopy, that clinoenstatite in chondrules has been partially altered to Fe-rich olivine. The sharp contacts between phases with highly disequilibrium Fe/Mg ratios requires that the Fe participating in the alteration must have been transported by a process other than solid state diffusion. We also present evidence of significant post-chondrule formation transport of Na, Si, P, S, Cl, Ca, and Cr.

We can find no diagnostic difference between the Fe-rich olivine in Allende matrix and that observed in chondrules in a texture that requires formation by alteration of clinoenstatite. Therefore, we argue that Allende matrix probably formed by the reaction of the appropriate amounts of metal and pyroxene-rich chondrules that previously existed, but are now consumed.

We will argue that the environment in which these reactions could have most easily occurred would have been provided by small internally heated planetesimals as discussed by Fish *et al.* (1960) and further developed by DuFresne and Anders (1962) and Armstrong *et al.* (1982).

EXPERIMENTAL METHODS

All reported SEM work was done with an ETEC Autoscan ideally capable of 50 Å resolution, equipped with a Princeton Gamma-Tech energy dispersive X-ray analyzer having a resolution of 160 eV, and interfaced with a Kevex Quantex-Ray data analysis system. It was periodically calibrated to demonstrate an actual resolution of 150 Å or better. In order to obtain good X-ray count rates and cathodoluminescence images, most work was done under compromise conditions, but the operating resolution was always better than 1000 Å. Phases were identified from energy dispersive X-ray spectra obtained under standardized geometrical conditions. Semi-quantitative analyses were made by comparing peak intensities to those obtained from known standards.

Samples, including one polished thin section, were all prepared from a single small commercially obtained Allende stone. About half a gram of material free of fusion crust and obvious white inclusions was gently crushed. Five largely matrix-free chondrules and several small matrix fragments, one with white material on one edge, were removed. One chondrule with a chip broken off and the matrix fragments were mounted without further treatment for SEM study.

The remaining gently crushed material was further disaggregated by ultrasonically agitating in ethanol for about ten hours. Suspended material was periodically poured off and replaced with fresh ethanol. When little more fine-grained material came off, the remainder was dried and sieved. One hundred and twenty-one

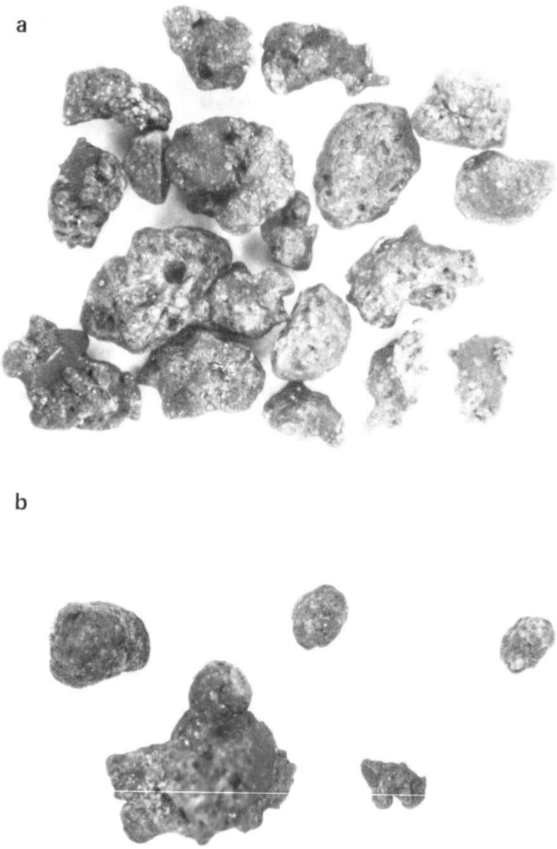

Fig. 1 Ultrasonically disaggregated ≥420 μm Allende chondrules. **(a)** Typical collection illustrating irregular shapes and dark cavities. **(b)** Five of the most nearly spherical ones, two still firmly attached to irregular fragments.

particles larger than 420μm were obtained. All still have adhering matrix, most have complex shapes, and many have large, nearly spherical cavities. A typical collection of them is shown in Fig. 1a. Five of the most spherical particles, three free and two attached to irregular chunks, are shown in Fig. 1b. Nine whole particles were directly mounted for SEM study and another nine randomly selected particles were prepared as polished mount. Based on our subsequent study of these mounted particles, it seems that at least most of the separated particles are chondrules according to our melting criterion.

All SEM samples were carbon coated. The polished samples were coated with an empirically determined thickness that allowed good contrast between minerals to be obtained in the secondary electron imaging mode. A gold coating on the polished section from previous work (Housley, 1981) was removed with very dilute, cold KCN solution. Cathodoluminescence imaging in the SEM was used to detect and map sodalite occurrences.

For incidental reasons we have largely used secondary electron imaging in this study, although backscattered electron images of the polished section would have been more esthetically pleasing. The main difference is that in these images cracks and holes appear bright white rather than dark.

OBSERVATIONS

Evidence for clinoenstatite alteration

Of the nine particles in the polished mount, we have made detailed SEM studies of six. They are all opaque-rich Type I chondrules as defined by McSween (19771a). Five are composed largely of forsteritic olivine with minor clinoenstatite, while one is largely clinoenstatite with minor ovoids of forsterite. We have also carefully studied four similar chondrules in the polished section.

In all nine of the largely forsterite chondrules studied, the subordinant clinoenstatite ($\sim Fs_2$) is preferentially located around the periphery, where it frequently poikilitically encloses angular anhedral to ovoid forsterite grains. A similar observation was mentioned by Clarke *et al.* (1970).

The clinoenstatite itself does not show any crystal faces. On the exposed sides, it instead makes sharp, continuous, but irregular contacts with Fe-rich olivine of matrix composition ($\sim Fa_{50}$). This, in turn, frequently grades into aggregates of Fe-rich olivine blades indistinguishable from the matrix in a distance on the order of 10 μm. In *all* ten cases we have studied, Fe-rich olivine is also distributed throughout the clinoenstatite crystals along grain boundaries and along abundant internal fractures in a texture strongly suggesting formation by an alteration reaction.

The above features are illustrated in Figs. 2 and 3. Figure 2a is typical of the forsteritic chondrules. It consists largely of angular forsterite blocks cemented together along a network of sealed cracks containing, among other things, chromite and sulfides (Housley, 1981). Near the right edge there is some clinoenstatite, much of which cannot be distinguished from the forsterite in this picture. Then, right next to the edge, there is a border of lighter gray Fe-rich olivine ($\sim Fa_{50}$).

Figure 2b is an enlargement of the region in the box. Notice the sharp contacts between clinoenstatite and Fe-rich olivine and the fractured and porous nature of the Fe-rich olivine. Also note stringers of sulfides, as are better illustrated in Fig. 2c from a nearby region. Figure 2d from another chondrule shows an enstatite grain ($\sim Fs_2$) surrounded by forsterite ($\sim Fa_5$). This enstatite, in turn, contains micrometer-sized regions of Fe-rich olivine along internal cracks.

Figures 3a, b, and c are from an interior region of the pyroxene chondrule. Figure 3a shows typical intergrowths of Fe-rich olivine ($\sim Fa_{50}$) and clinoenstatite ($\sim Fs_2$). Several ovoids of forsterite ($\sim Fa_5$) are present. The three in the box are enlarged in Figs. 3b and c. Rims of Fe-rich olivine about 1 μm thick surround the forsterite ovoids. Figure 3d is a typical enstatite, Fe-rich olivine intergrowth from yet another chondrule.

Even though the Fe/Mg ratio of near one in the Fe-rich olivine is dramatically out of equilibrium with those of the forsterite and enstatite, we frequently have not been able to detect composition gradients even on a micrometer scale.

It naturally seems surprising that such a ubiquitous and unusual feature as these intergrowths of Fe-rich olivine in chondrule clinoenstatite should not have been reported and discussed previously. However, even knowing exactly where to look, we have tried to recognize them using reflected light microscopy without any success; this suggests at least a plausible reason.

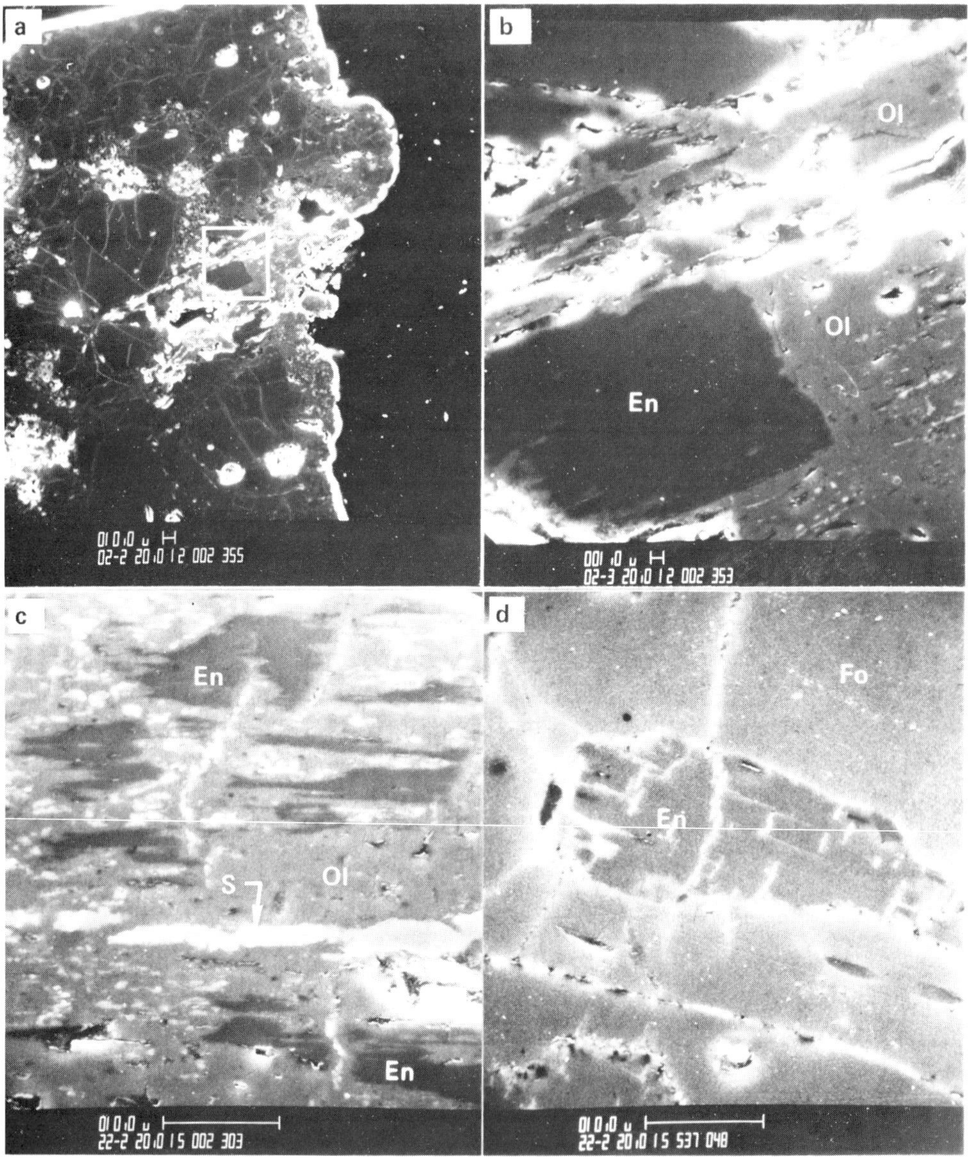

Fig. 2 Secondary electron images of polished particles. They differ from backscattered electron images in that cracks and pores show very bright edges and dark interiors. Enstatite is labeled En, forsterite (~Fa_5) Fo, olivine (~Fa_{50}) Ol, and sulfides S. **(a)** Low magnification image showing porosity and network of sealed cracks. Fe, Cr, and S are strongly enriched along cracks. **(b)** Enlargement of center of region in box showing enstatite in sharp contact with Fe-rich olivine of matrix composition. **(c)** Nearby region showing enstatite, Fe-rich olivine, and abundant fine sulfide stringers. **(d)** Enstatite crystal in separate chondrule enclosed in forsteritic olivine, but showing Fe-rich olivine along border and internal cracks.

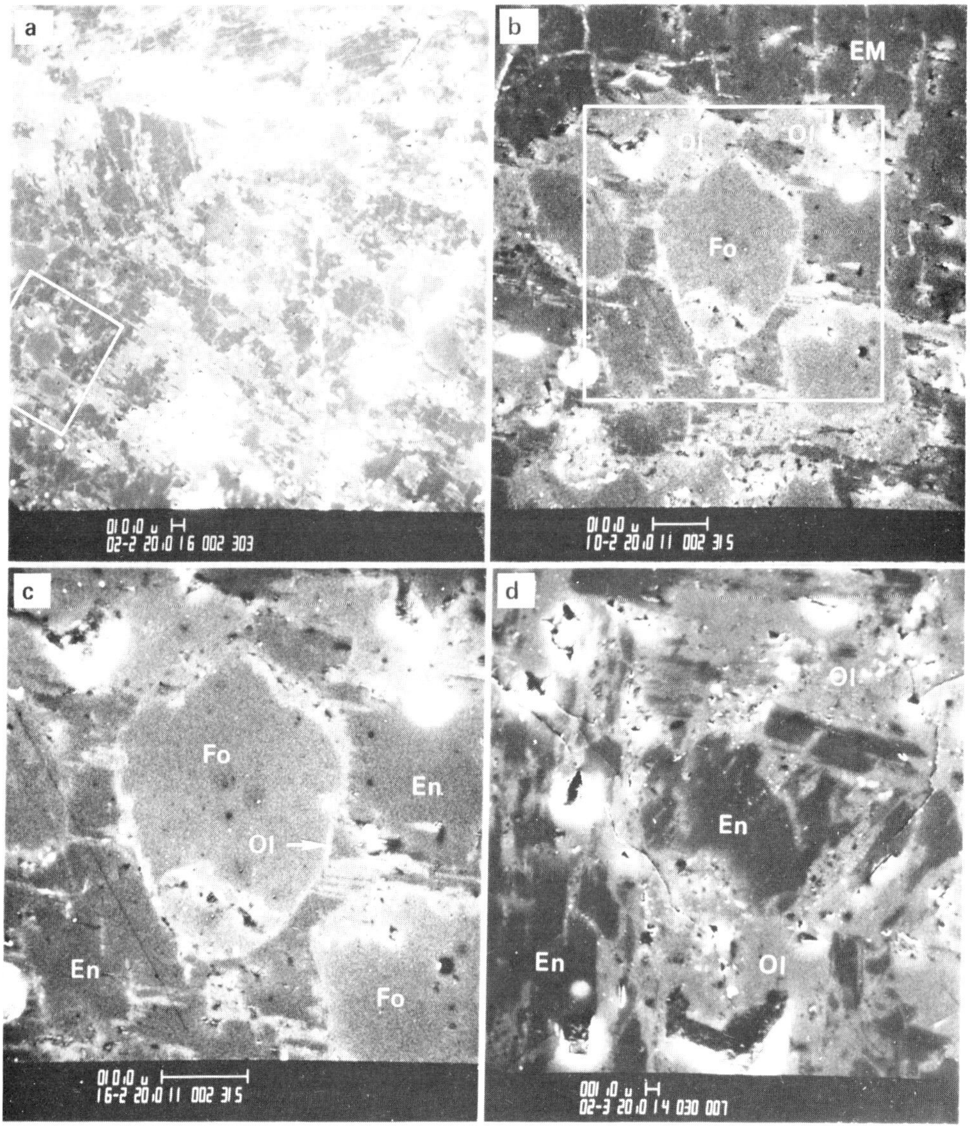

Fig. 3 Secondary electron images of polished particles. **(a)** Low magnification SEM image of pyroxene chondrule. Dark areas are enstatite. Intermediate areas are olivine, mostly of matrix composition. **(b)** Enlarged and rotated view of indicated region. Three rounded objects near the center are forsteritic olivine grains that were included in the enstatite. Fe-rich olivine now forms a border around these Fe-poor olivines. **(c)** Same region at higher magnification showing that rim is less than 1 μm thick. **(d)** A region of intergrown Fe-rich olivine and enstatite in another chondrule.

It is worth noting that nepheline and sodalite have a strong tendency to be associated with the Fe-rich olivine-clinoenstatite intergrowths, and are frequently completely surrounded by Fe-rich olivine. This suggests that the Fe-rich olivine formation and much of the alteration in Ca,Al-rich inclusions (as described, for example, by Wark, 1981) took place at the same time in the same environment.

Chondrule-matrix interfaces

As already described, the separated chondrules in the polished mount frequently grade into material similar to the matrix around their peripheries. In the polished section with either SEM or reflected light microscopy, it can be seen that boundaries between chondrules and matrix are frequently very gradational. This is illustrated by the chondrule in the upper left of Fig. 4a, which is a reflected light micrograph. Magnified SEM images of the region in the box, Fig. 4b and c, show intergrowths of clinoenstatite and Fe-rich olivine similar to those previously described. Figure 4d illustrates typical matrix between chondrules. It consists largely of subhedral Fe-rich olivine with, in this case, some intergrown sodalite, especially in the group in the middle left. Pentlandite and Ni-rich metal are also common in the matrix.

In SEM and reflected light surveys of our whole polished section, we did not find any matrix qualitatively different from that described and illustrated in Fig. 4. Frequently, boundaries between chondrules and matrix are gradational in terms of the size and packing density of Fe-rich olivine blades, but not in their composition.

Extensive SEM study of chondrules in our unpolished mount has also shown that most of the surface area exposed consists of subhedral Fe-rich olivine blades or fine fragments. Frequently the blades occur in subparallel arrangements. Typical examples are shown in Figs. 5a and 5b.

Matrix olivine

Inspection of thousands of matrix olivine grains during SEM study of the matrix fragments showed the overwhelming majority of them to be either subhedral plates without any terminations or completely irregular fragments. Euhedral crystals are very rare in the matrix. This is, of course, in agreement with what we see in the polished section when using either reflected light or the SEM.

Another rare constituent of the matrix is enstatite, although we have found a few relic grains in the polished section. An example of two grains completely surrounded by Fe-rich olivine is shown in Fig. 5c.

In the one matrix fragment that had adhering white inclusion material, we do see beautiful euhedral Fe-rich olivine crystals in a narrow border zone. Such zones have been described by MacPherson and Grossman (1981) and are also reported by Wark (1979). They are distinctly different from typical matrix.

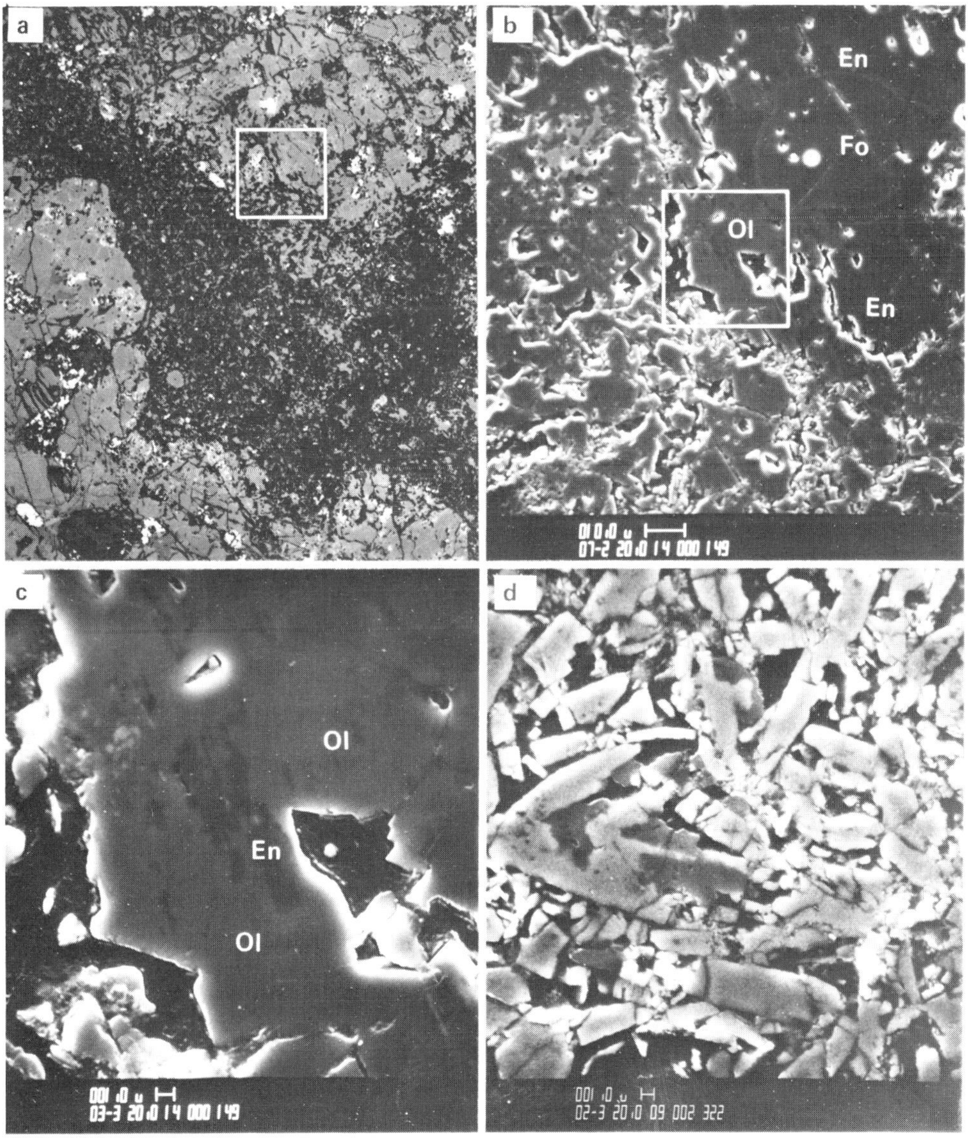

Fig. 4 Secondary electron images of polished particles. **(a)** Reflected light micrograph of matrix between two chondrules. Note gradual transition between matrix and chondrule at upper right. **(b)** SEM image of indicated area at the beginning of transition region. The upper right is largely enstatite with an included forsterite oval. The rest is largely Fe-rich matrix olivine, with some pentlandite in the upper left. **(c)** Higher magnification of the center of **(b)** showing residual enstatite in matrix olivine. **(d)** Typical Allende matrix. Fe-rich olivine with sodalite, dark gray.

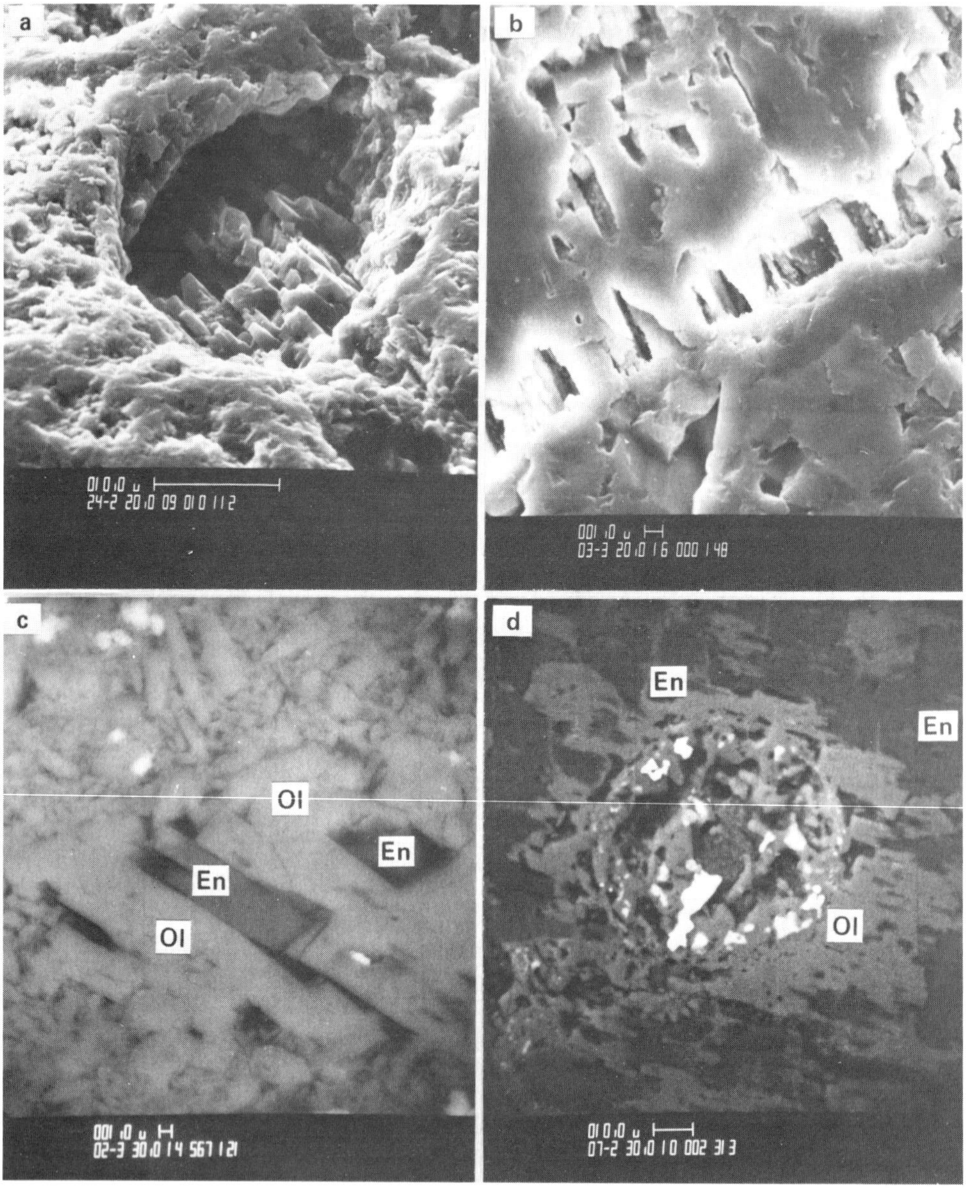

Fig. 5 **(a)** and **(b)** Parallel intergrowths of Fe-rich olivine exposed on the surfaces of chondrules separated by ultrasonic agitation. In **(a)** delicate structure was protected in cavity. **(c)** Backscattered electron (BSE) image showing two rare relict enstatite grains in matrix olivine. White areas are pentlandite. **(d)** BSE image. White regions are Ni-rich metal and pentlandite outlining a highly altered circular inclusion. Enstatite around the inclusion has been partially altered to Fe-rich olivine, which also now fills much of the circle.

Opaque inclusions and cavities in chondrules

All the type I chondrules we have examined in Allende contain abundant subspherical to irregular opaque inclusions and cavities. There seems to be a continuous gradation from solid inclusions to porous inclusions, to partly full cavities, to nearly empty cavities. These inclusions and cavities are apparent in Figs. 2a and 4a. The most abundant minerals in them are troilite, pentlandite, magnetite, Ni-rich metal, and Fe-rich olivine, followed by Ca phosphate, which is common, and by chromite and heazlewoodite, which are rare. The associations and relative abundances of these minerals vary greatly from one inclusion or cavity to another. For example, some inclusions consist largely of magnetite with small amounts of Ni-rich metal, while others consist largely of troilite or largely of pentlandite. Many typical inclusion assemblages have been described by Haggerty and McMahon (1979). A common type that they did not mention in which Fe-rich olivine in continuity with that in the surrounding altered enstatite partially or completely refills a former cavity outlined by Ni-rich metal and pentlandite is shown in Fig. 5d.

Some examples of the contents of small cavities in forsterite are shown in Fig. 6. The porous pentlandite shown in Fig. 6a is very common. Notice the euhedral crystal shapes in the other cases. Fortunately, the broken chondrule that was mounted without any exposure to liquids contains a number of these cavities, including two large ones on the fresh fracture surface. Part of the contents of one of these large ones are illustrated in Fig. 7. It also contains some very fine-grained Cr-rich phase, probably chromite, and some other unidentified fine-grained material.

DISCUSSION

Fe-rich olivine-clinoenstatite intergrowths

The textures of the Fe-rich olivine-clinoenstatite intergrowths in chondrules illustrated in Figs. 2 and 3 do not resemble any familiar igneous textures in rocks of comparable composition. The igneous deposition of Fe-rich olivine along cracks and grain boundaries would require the unlikely existence of considerable, very fluid fayalite-rich residual melt down to the temperatures at which the cracks formed due to differential thermal contraction. Therefore we rule out an igneous origin for these intergrowths. The highly disequilibrium Fe/Mg ratios and the general lack of composition gradients also rule out any simple thermal metamorphic origin.

The textures shown in Figs. 2 and 3 seem to resemble most closely some seen in partially weathered terrestrial rocks. Here the mechanism, though perhaps more complicated in reality, can be simply visualized. A chemical reaction takes place at an exposed grain surface. The second phase produced has a different volume than the first. Thus, the reaction produces strains. These, in turn, lead to the initiation and propagation of additional cracks. The reaction then proceeds along the new crack surfaces.

In the Allende chondrules, a reasonably expected reaction consistent with the observed compositions is the same one assumed to control the Fe content of olivine in nebular condensation

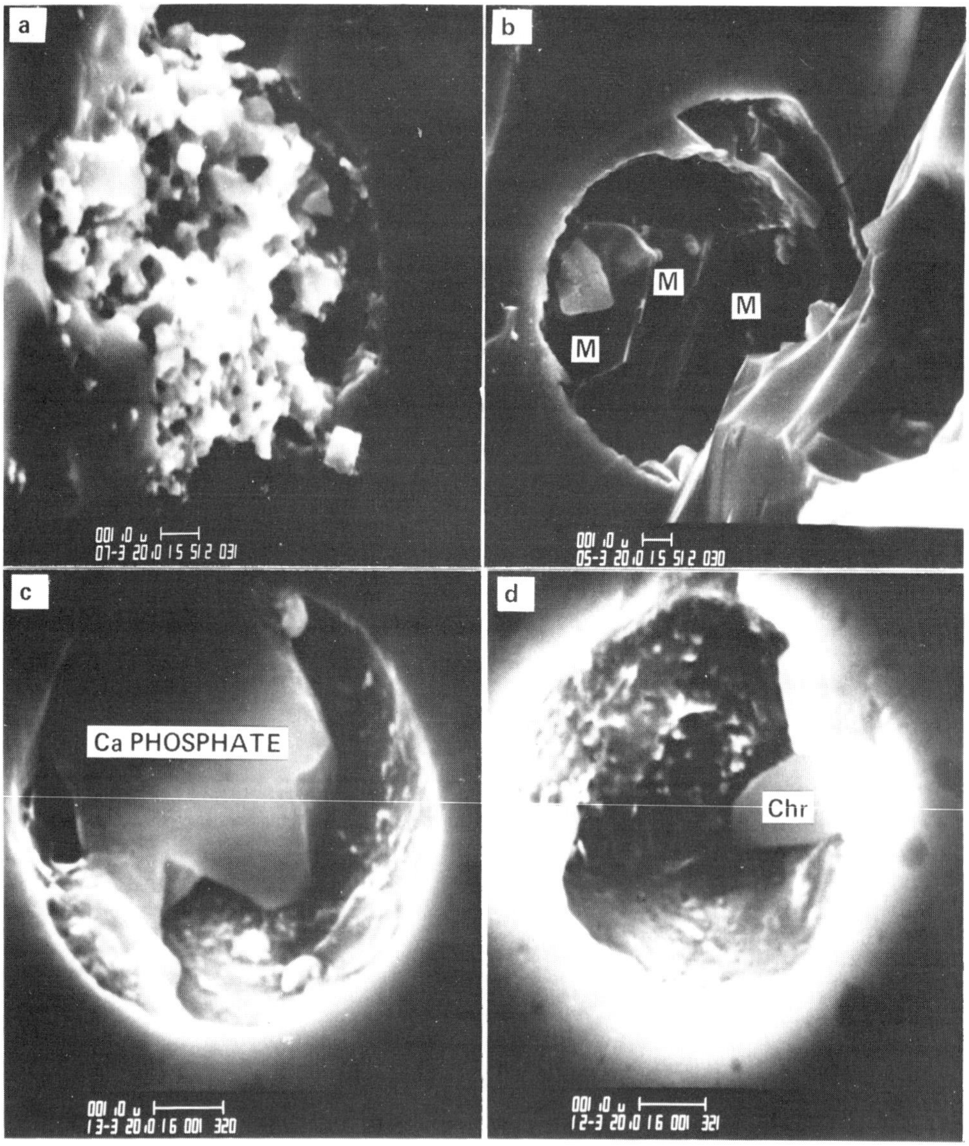

Fig. 6 SEM photographs of typical small cavities and their contents. **(a)** Porous fine-grained pentlandite aggregate in broken chondrule. This is the most common cavity-filling material. **(b)** Ni-rich metal crystals in bottom of cavity in broken chondrule. **(c)** Ca phosphate crystal in polished section cavity. **(d)** Chromite crystal in polished section cavity.

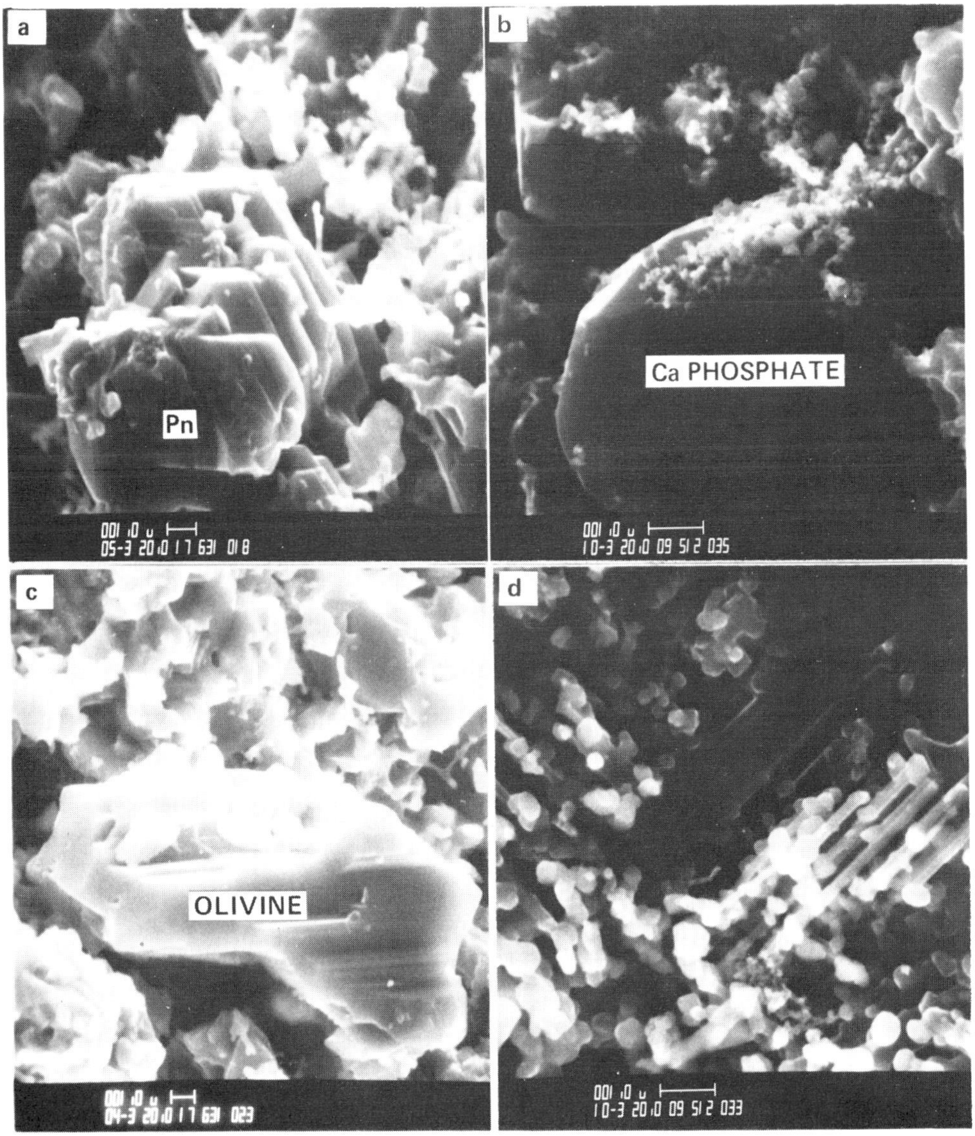

Fig. 7 Partial contents of large cavity in broken chondrule. **(a)** Pentlandite Pn crystal with finer-grained pentlandite. **(b)** Ca phosphate crystal partly coated with very fine material. **(c)** Olivine crystal of matrix composition. **(d)** Very fine needles of unidentified Fe-rich silicate mineral growing into cavity.

models (Grossman, 1972):

$$Fe + 1/2O_2 + MgSiO_3 \rightarrow (Fe,Mg)_2 SiO_4. \quad (1)$$

During this reaction there is a volume change from 31 cm^3 per mole for the enstatite to about 45 cm^3 per mole for the olivine. To account for the absence of observed composition gradients, it is necessary that Fe diffuse through the newly formed Fe-rich olivine, perhaps about an order of magnitude faster than cation diffusion in the clinoenstatite or forsterite. Since cation diffusion depends strongly on the defect density in the crystal, this is not unreasonable.

To account for the observed distribution of Fe-rich olivine along grain boundaries and internal cracks, it is also necessary to assume some process that transports Fe rapidly through cracks and pores. In the case of terrestrial weathering, the transport medium is generally water. Small amounts of presumably hydrated layer lattice silicate related to serpentine have been reported from cavities in Allende by Tomeoka and Buseck (1982a), who have also reported mica and montmorillonite from the Ca,Al-rich inclusions (1982b). Evidence of limited water transport has been reported for the CO meteorite, Warrenton, by Kerridge (1972). Despite all this, it seems clear from the phase equilibria studies of Bowen and Tuttle (1949) that liquid water did not play any significant role in the alteration of enstatite that we describe. In the presence of water enstatite is always unstable toward the formation of forsterite plus talc.

In principle, the reactions we observe could possibly have taken place in a nebular gas, although we consider it very unlikely. Fe-rich olivine of the observed composition would only become stable below about 500°K (Grossman, 1972). At that temperature reaction rates would be very low and gaseous species available to transport Fe very limited.

Planetesimal model

We believe a model of small internally heated planetesimals can most easily provide a plausible environment in which the observed alterations could have taken place. Many of the general characteristics and possibilities expected of such planetesimals were described by Fish *et al.* (1960). DuFresne and Anders (1962) considered a special class of such planetesimals in which liquid water was trapped inside an ice shell as a plausible setting for the formation of CI and CM meteorites. Recently Armstrong *et al.* (1982) have considered such planetesimals as the most likely settings in which alterations of the type they have observed in Murchison Ca,Al-rich inclusions could have occurred. They have further discussed how many of the features commonly observed in meteorites could easily be explained by the processes expected to have occurred in such planetesimals.

To see how such a model can easily explain our observations we will specifically consider a planetesimal with the rock to H_2O ratio, surface temperature, and heat generation adjusted so that an icy shell surrounds a rocky interior without an intervening liquid water layer. A significant H_2O vapor pressure could have still been present in the hot interior, depending on the temperature at the ice boundary. At a given temperature conditions would have been much more oxidizing than in the nebula because of the initial absence of hydrogen and its episodic escape whenever the shell was disrupted.

Substantial pressures of many gases (including, for example, CO, HC1, H_2S, SO_2, and NH_3) could have been present in such a planetesimal without condensing out at the ice interface. Such gases could have acted as catalysts to speed local equilibrium, for example, by transporting cations or carbonyls or chlorides. Because of the temperature gradient in the body they could also have produced net transport of material by cyclical disequilibrium chemical vapor transport reactions (Cirlin and Housley, 1980).

Opaque inclusions and cavities

It is clear that the magnetite, chromite, Ca phosphate, and Fe-rich olivine now in the opaque inclusions must have resulted from alteration or deposition reactions. They cannot be primary phases since they are soluble in a melt of chondrule composition and would not have remained in an immiscible opaque liquid phase, even if the chondrules formed by melting heterogeneous solid precursors.

In other CV meteorites, including the type example, Vigarano, metal and sulfide inclusions are common in the type I chondrules. We have argued that the textures in the Fe-rich olivine-clinoenstatite intergrowths in Allende can be most satisfactorily interpreted if relatively free migration of Fe took place along cracks. This leads naturally to the idea that cavities in the Allende chondrules are holes left by the partial or complete removal of Fe metal and are not vesicles at all. Thus, we suggest that much of the original Fe metal migrated out to participate in reaction (1). Some remained but was largely oxidized to magnetite, leaving a residue of Ni-rich metal. Various minerals, including Ca phosphate and Fe-rich olivine, were then deposited in the cavities.

Matrix

Alteration of clinoenstatite in Allende chondrules produces Fe-rich olivine identical in composition to matrix olivine. The boundary between chondrules and matrix is frequently gradational. Parallel groups of subhedral Fe-rich olivine blades frequently occur adjacent to chondrules. Enstatite blades completely enclosed in Fe-rich olivine rarely occur in the matrix.

Based on the above observations, we see no reason why all matrix olivine in Allende could not have been produced from initially more abundant enstatite-rich chondrules and metal by the alteration reaction (1). The strains associated with the reaction generally seem to cause the clinoenstatite to cleave into blades a few micrometers thick. These continue to react until they generally are completely Fe-rich olivine. It is only necessary to disaggregate them further to produce typical matrix.

The widespread distribution of fine-grained pentlandite throughout the matrix appears to require that it was produced at temperatures below 900°K.

SUMMARY

We have shown that partial alteration to Fe-rich olivine is a common characteristic of clinoenstatite in those Allende chondrules containing it. We have shown that these chondrules have gradational boundaries with the matrix.

The evidence presented suggests the view that all matrix olivine in Allende was produced from initially more abundant pyroxene-rich chondrules and metal by reaction (1). We argued that the most plausible environment in which this reaction could have taken place would have been provided by a small internally heated planetesimal.

Acknowledgments. This work was supported by NASA contract NAS9-11539. We gratefully acknowledge a very helpful review by David Wark that contributed considerably to the improvement of this manuscript.

REFERENCES

Armstrong J. T., Meeker G. P., Huneke J. C., and Wasserburg G. J. (1982) The Blue Angel: I. The mineralogy and petrogenesis of a hibonite inclusion from the Murchison meteorite. *Geochim. Cosmochim. Acta* **46**, 575–595.

Bowen N. L. and Tuttle O. F. (1949) The system $MgO\text{-}SiO_2\text{-}H_2O$. *Geol. Soc. Amer. Bull.* **60**, 439–460.

Cirlin E. H. and Housley R. M. (1980) Redistribution of volatiles during lunar metamorphism. *Proc. Lunar Planet. Sci. Conf. 11th,* pp. 349–364.

Clarke R. S., Jarosewich E., Mason B., Nelen J., Gomez M., and Hyde J. R. (1970) The Allende, Mexico, meteorite shower. *Smithson. Contrib. Earth Sci.* **5**, 1–53.

DuFresne E. R. and Anders E. (1962) On the chemical evolution of the carbonaceous chondrites. *Geochim. Cosmochim. Acta* **26**, 1085–1114.

Fish R. A., Goles G. G., and Anders E. (1960) The record in the meteorites. III. On the development of meteorites in asteroidal bodies. *Astrophys. J.* **132**, 243–258.

Grossman L. (1972) Condensation in the primitive solar nebula. *Geochim. Cosmochim. Acta* **36**, 597–619.

Grossman L. (1980) Refractory inclusions in the Allende meteorite. *Ann. Rev. Earth Planet. Sci.* **8**, 559–608.

Haggerty S. E. and McMahon B. M. (1979) Magnetite-sulfide-metal complexes in the Allende meteorite. *Proc. Lunar Planet. Sci. Conf. 10th*, pp. 851–870.

Housley R. M. (1981) SEM, optical and Mössbauer studies of submicrometer chromite in Allende. *Proc. Lunar Planet. Sci. 12B,* pp. 1069–1077.

Kerridge J. F. (1972) Iron transport in chondrites: evidence from the Warrenton meteorite. *Geochim. Cosmochim. Acta* **36**, 913–916.

Kullerud G. (1963) The Fe-Ni-S system. *Carnegie Inst. Wash., Yearb.* **62**, 175–189.

MacPherson G. J. and Grossman L. (1981) Clastic rims on inclusions: Clues to the accretion of the Allende parent body (abstract). In *Lunar and Planetary Science XII*, pp. 646–647. Lunar and Planetary Institute, Houston.

McSween H. Y. (1977a) Carbonaceous chondrites of the Ornans type: A metamorphic sequence. *Geochim. Cosmochim. Acta* **41**, 477–491.

McSween H. Y. (1977b) Petrographic variations among carbonaceous chondrites of the Vigarano type. *Geochim. Cosmochim. Acta* **41**, 177–179.

McSween H. Y. (1979) Are carbonaceous chondrites primitive or processed? A review. *Rev. Geophys. Space Phys.* **17**, 1059–1078.

Muan A. and Osborn E. F. (1956) Phase equilibria at liquidus temperatures in the system MgO-FeO-Fe_2O_3-SiO_2. *J. Amer. Ceram. Soc.* **39**, 121–140.

Ramdohr P. (1963) The opaque minerals in stony meteorites. *J. Geophys. Res.* **68**, 2011–2035.

Tomeoka K. and Buseck P. R. (1982a) An unusual layered mineral in chondrules and aggregates of the Allende carbonaceous chondrite. *Nature* **299**, 327–329.

Tomeoka K. and Buseck P. R. (1982b) Intergrown mica and montmorillonite in the Allende carbonaceous chondrite. *Nature* **299**, 326–327.

Wark D. A. (1979) Birth of the presolar nebula: The sequence of condensation revealed in the Allende meteorite. *Astrophys. Space Sci.* **65**, 275–295.

Wark D. A. (1981) Alteration and metasomatism of Allende Ca-Al-rich materials (abstract). In *Lunar and Planetary Science XII*, pp. 1145–1147. Lunar and Planetary Institute, Houston.

Conditions and Time of Chondrule Accretion

R. Hutchison and A. W. R. Bevan

Department of Mineralogy, British Museum (Natural History), London SW7 5BD, England

We studied a suite of ten H-group chondrites as coexisting mineral assemblages that, at least in part, shared the same history. Textural observations indicate that Tieschitz (H3) accreted when some chondrule liquids were still plastic; dark rim material was already present and deformed with the chondrules. We argue that dark rims (dark matrix) probably represent quenched liquids, and that the ordinary chondrites accreted hot, at 800° ± 100°C. Most chondrules, clasts, and glassy objects in ordinary chondrites have modal or normative feldspar or feldspathoid that is soda-rich, but in petrologic types 3 and 4 some of the silicate components have modal or normative calcic plagioclase. Thus it appears that a minority of chondritic silicates had a calcium-rich, sodium-poor "noritic" source distinct from that of most chondrules. Individual chondrules and glassy objects in Manych (L3) (Dodd, 1978a,b) have Ca/Al atomic ratios ranging from almost zero to over 1.6, indicating that vapor fractionation was not a major factor in producing chemical variation among chondrules. Instead, we argue that crystal-liquid fractionation was a more probable mechanism, and that some planetary differentiation preceded accretion of the ordinary chondrite parent bodies. Metal in unreheated H-group chondrites comprises α-kamacite, "zoneless plessite," γ-taenite, and γ-tetrataenite, in order of increasing Ni content. In Kernouve (H6) observed zoneless plessites have structures that coarsen as their bulk Ni content decreases from 15 to 10 wt.%. Zoneless plessites did not exchange Fe and Ni with neighboring metal, and each formed by internal decomposition of martensite. Formation of martensite in the observed composition range normally requires rapid cooling to below 500°C. Solid state diffusion of Ni in ordinary chondrites was more localized than the Wood (1967) cooling model requires. A mechanism is necessary to produce compositional variation among the metal particles inherited by these meteorites before the onset of slow cooling. Vapor fractionation (condensation) seems less likely than crystal-liquid fractionation, which is compatible with the formation of silicate chondrules. The ordinary chondrites apparently accreted after some planetary objects had formed and differentiated. If this is so, the meteorites may be able to provide only indirect evidence on the former presence of a solar nebula.

INTRODUCTION

It is often argued that the ordinary chondrites formed by cold accretion of chondrules, metal, sulfide, and low temperature matrix, followed by thermal metamorphism under either closed (Larimer and Anders, 1967; Larimer, 1973) or open (Wasson, 1974) system conditions. In either case the presence of a "solar nebula" during chondrule formation and/or accretion is usually postulated (e.g., Larimer and Anders, 1967; Dodd, 1978b; Clayton, 1981; Grossman and Wasson, 1981; Scott *et al.*, 1982; Taylor *et al.*, 1982). For the past six years, in collaboration with S. O. Agrell, J. R. Ashworth, H. J. Axon, and M. Christophe Michel-Lévy, we have tried to take a fresh look at the petrography and mineralogy of a suite of H-group chondrites. We attempted to observe them as rocks, coexisting assemblages of silicate, metal, and sulfide minerals that, at least in part, shared the same history. Our conclusions favor the formation of ordinary chondrites by the hot accretion of chondrules and imply that no direct evidence of a solar nebula may exist. As our collaborators may not share our more radical views, we accept sole responsibility for the conclusions stated here.

TEXTURAL OBSERVATIONS

We define chondrules as objects that were undoubtedly liquid before or during the accretion period(s) that led to the formation of chondrites, and recognize as chondrules subspherical masses of mainly silicate, ranging through silicate plus metal-sulfide, to metal-sulfide (Bevan and Axon, 1980). In addition, we also recognize irregularly-shaped objects as chondrules if they satisfy the "liquid before or during accretion" criterion. Except for Dodd's specification that chondrules are "chiefly silicate," we generally accept his definition as stated at the Conference on Chondrules and their Origins.

In Tieschitz, one of the least metamorphosed chondrites studied by us, round chondrules and angular clasts have dark rims (dark matrix) on most surfaces (Christophe Michel-Lévy, 1976). Between chondrules and clasts, channels up to some tens of micrometers wide are sometimes present; these are filled with "white matrix" (Christophe Michel-Lévy, 1976), which is dominantly composed of nepheline plus albite in clastic porous aggregates (Hutchison *et al.,* 1979). In rare instances there is evidence that chondrules aggregated as groups or clusters when one or more of the objects was molten or plastic. For example, the 1-mm-long object in Fig. 1 has its irregular outline (shown by arrows) surrounded by a dark rim, although in places it is very thin. Five more regularly-shaped silicate chondrules (four are visible in Fig. 1) and three metal grains abut the object, which is composed of a lilac-colored, partially devitrified glass with olivine dendrites and phenocrysts. The olivines are concentrated around the margins, and some dendrites extend inward from them, indicating that cooling and crystallization occurred from the presently observed surface. Furthermore, the shape of the object, which we call a chondrule, must have been determined before the onset of crystallization of an optically continuous olivine dendrite 0.2 mm long. It seems inescapable that the irregular outline of the chondrule was determined, while it was molten and after it had picked up its dark rim, by the impingement of solid silicate and metal objects upon it. A second silicate object, regarded by us as a chondrule, had a similar agglomeration history (Fig. 4 of Hutchison *et al.,* 1979). This is a microcrystalline pyroxene chondrule whose dark rim is folded around a tongue-like extension apparently trapped between two impinging, regular silicate chondrules (Fig. 2). Bevan and Axon (1980) recognized clusters of metal-sulfide chondrules that have dark matrix (rim) material between the components (their Figs. 1a and 1b). In these cases aggregation occurred after dark rims had been acquired, but before solidification was complete everywhere. The aggregating objects could not have been thoroughly molten on accretion, otherwise they would have coalesced (Bevan and Axon, 1980).

These textural observations indicate that dark rim material was added to at least some silicate and metal-sulfide chondrules when they were hot. The ambient temperature during the aggregation of some groups of chondrules must have been high, otherwise millimeter-sized droplets of liquids would have quenched before plastic deformation occurred.

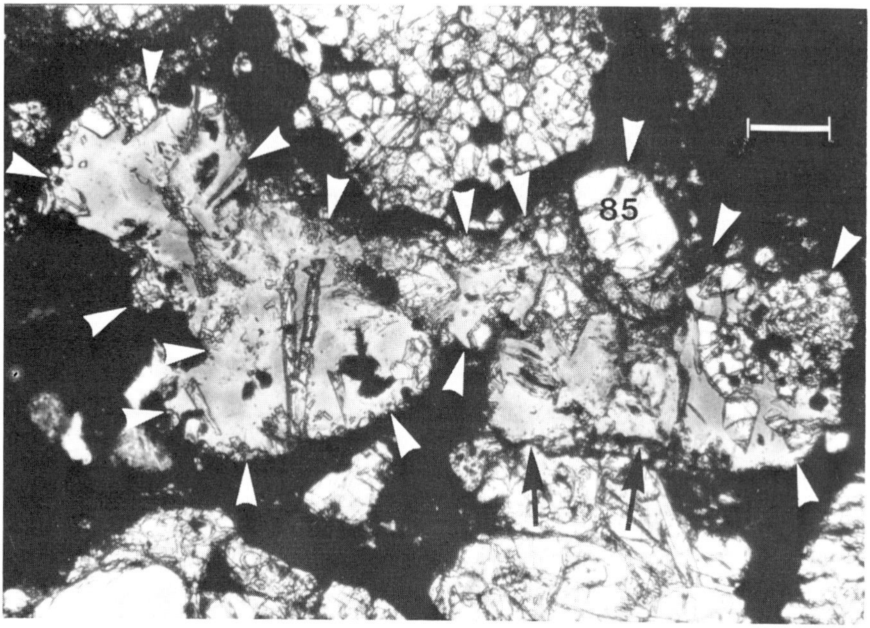

Fig. 1 Tieschitz deformed chondrule, C6. Scale bar = 100 μm. The margins are indicated by arrows. Olivine phenocrysts/dendrites are set in a sodium-rich, nepheline normative glass and are Fo_{98-91} in composition, except for a few more iron-rich crystals, one of which is indicated (Fo_{85}). The 200-μm-long bifurcating dendrite, like an elongate letter "Y", is in optical continuity, testifying to its crystallization after the chondrule had been deformed. It ranges from Fo_{97} near the margin to Fo_{89}.

Ashworth (1977, 1981) used high-voltage transmission electron microscopy (TEM) to examine the fine structure in the dark rims of chondrules. He found these "non-clastic" rims to be compact, with few voids in comparison with the coarser-grained "clastic matrix" between chondrules (cf. Huss *et al.,* 1981, p. 45). Ashworth found the dark rims on Chainpur chondrules to be uniformly very fine-grained (Ashworth, 1977), whereas some of those in Tieschitz contain microphenocrysts of olivine (Ashworth, 1981). Very fine-grained (submicrometer) rim material always makes sharp contact with chondrule margins or microphenocrysts. More recently, Ashworth (personal communication, 1982) found that dark matrix in Bishunpur (LL3) contains subspherical vugs that occur singly or in groups. Their size (in the micrometer range) contrasts with the grain size of the matrix, which ranges from about 0.1 μm to amorphous. In each of the three type 3 ordinary chondrites, metal-sulfide-rich areas of dark matrix were observed.

Because of their very fine grain size, Ashworth originally (1977, p. 31) likened dark rims to rapidly crystallized melts or glasses ("impact splashing?"); this conclusion seems to be reinforced by his observations on Bishunpur. The amorphous areas in the matrix may, in fact, be glasses, and

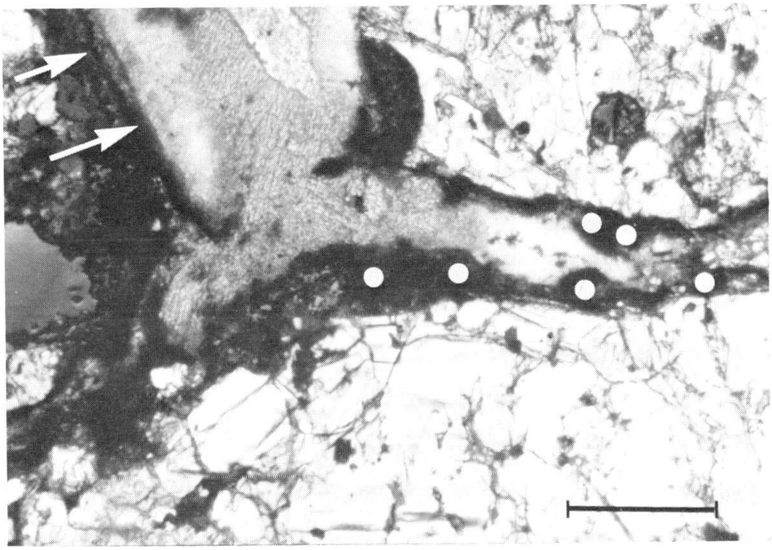

Fig. 2 Tieschitz deformed("squeezed") chondrule. Scale bar = 50 μm; interference contrast. The dark rim has deformed around a tongue-like extension of the chondrule on the right. The spots indicate the approximate positions of points taken for analysis no. (4), Table 1; the arrows, of no. (3).

it seems impossible to produce subspherical vugs of diameter much greater than the average grain size by a process other than solidification of gas "bubbles" in a melt. Accretion of dust grains onto chondrule surfaces, a process favored by Allen *et al.* (1980), seems unable to account for the observed textures unless special conditions prevailed, such as solution to produce the vugs, deposition or recrystallization to seal intergranular pores, and hydrothermal alteration to produce amorphous areas without attacking metal or chondrule margins. Under reflected light, the dark matrix in Bishunpur is seen to fill deep embayments in silicate chondrules, metal, and sulfide (Fig. 3), indicating that it may have been intrusive. Taking all the textural evidence together, we suggest that the dark matrix was liquid when added to chondrules (silicate and metal-sulfide) and clasts. It was, therefore, probably hotter than 900°C, the Fe,Ni-S eutectic temperature (Smith and Goldstein, 1977), and some of the still-molten chondrule precursors must have been even hotter.

CHEMICAL OBSERVATIONS

Here we emphasize chemical differences within the various components of some ordinary chondrites. In contrast, previous workers tended to use average compositions of suites of chondrules (e.g., Gooding *et al.*, 1980; Lux *et al.*, 1981), or trends within suites of chondrules (e.g., Dodd, 1978b), in their efforts to determine chondrule origins, and Huss *et al.* (1981) quoted mean

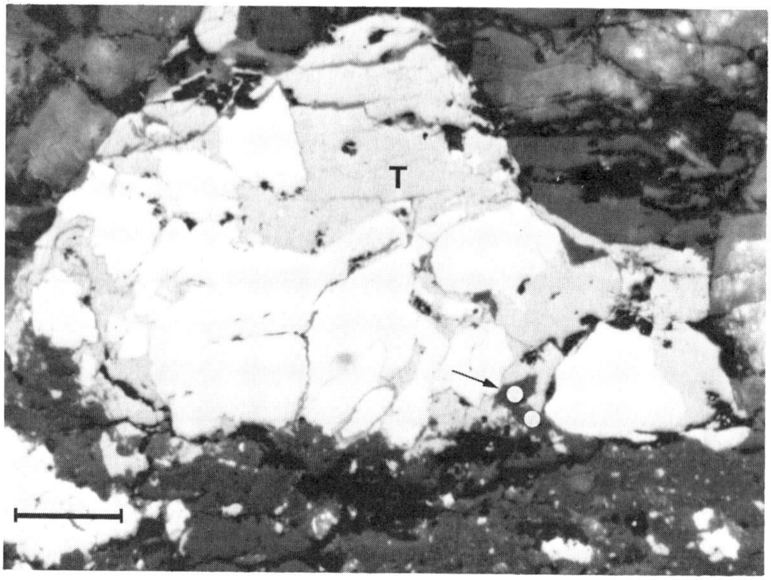

Fig. 3 Bishunpur metal-sulphide intergrowth "intruded" by black matrix. Scale bar = 50 μm; interference contrast. Two white spots indicate the positions of analyses of black matrix; the spot designated by the arrow shows no. (4), Table 2. Note that analysis no. (3) was performed on a point some 1.5 mm distant.

bulk compositions for "recrystallized" matrix and "opaque" matrix in an attempt to understand the genesis of chondrites.

Silicate liquids

Dodd (1978b, p. 77) remarked on the "wide scatter" in the Ca/Al ratios of Manych droplet chondrules and glasses, and hence concluded from that and Na/Al relationships that vapor fractionation was only a minor factor in producing the "diversification of chondrules." More specifically, Ca and Al should not fractionate from each other in a gas of solar composition below about 1550°K, a temperature at which Na would be wholly gaseous (Grossman, 1975; Ganapathy and Anders, 1974). By using Dodd's (1978a,b) data in plotting Na/Al (atomic) against CaO for Manych microporphyritic chondrules, droplet chondrules, and irregular glassy objects (Fig. 4a), it is shown that some fractionation process had operated. There is a hiatus between a group of droplet chondrules and microporphyritic chondrules with Na/Al below 0.31, and a larger group that includes glasses with Na/Al greater than 0.62. In this second group there is no correlation between Na/Al ratio and CaO content, but in the group with low Na/Al, there is a weak negative correlation (r = 0.75) with CaO. Of the ten members of the low Na/Al group,

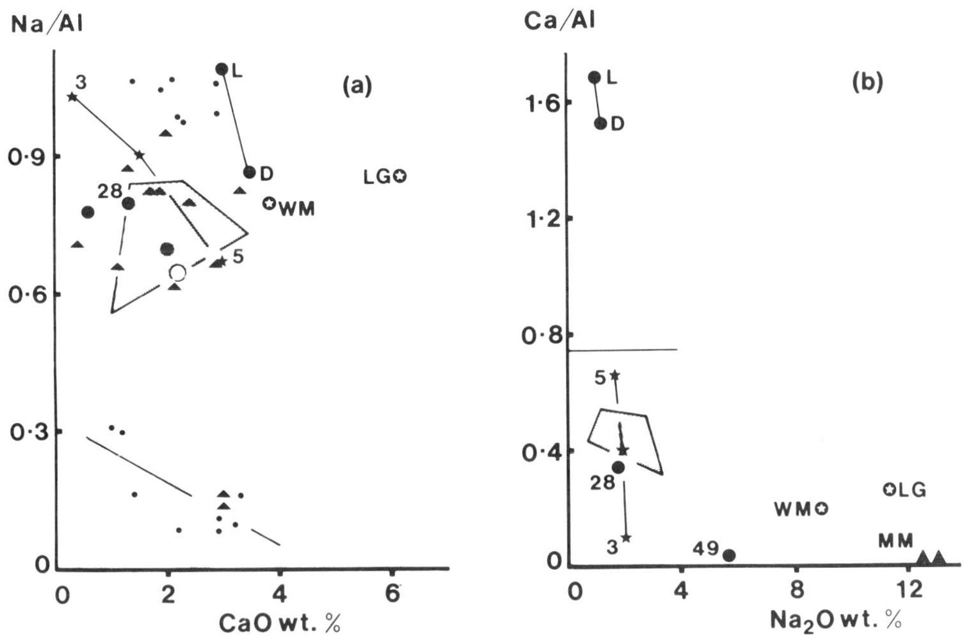

Fig. 4 **(a)** Na/Al atomic ratio vs. CaO in wt.%. Dots: Manych droplet chondrules; solid circles: Manych glasses (Dodd, 1978b); triangles: Manych microporphyritic chondrules (Dodd, 1978a). Shaded area: Bishunpur dark matrix (this work) includes average opaque matrices of Tieschitz and Bishunpur (Huss *et al.*, 1981); the open circle on the lower margin is the L-group mean (Dodd, 1978b). Solid stars, (3)–(5): Tieschitz black matrix, Table 1. WM and LG = white matrix and lilac glass (Tieschitz, Table 1). 28 = glass selvage No. 28 from Dodd (1978b). L and D = light and dark glasses (see No. 36 in Dodd, 1978b). The line, lower left, is the best fit to the low Na/Al points. **(b)** Ca/Al atomic ratio vs. Na_2O in wt.% shows extremes in Ca/Al ratios in Manych chondrules and glasses (Dodd, 1978a,b) and extreme Na_2O enrichment coupled with very low Ca/Al in the mesostasis of a Mezö-Madaras chondrule (MM), No. XXI of Kurat (1967). 49 = Manych nepheline normative droplet chondrule (Dodd, 1978b); remaining symbols as in **(a)**. Horizontal line represents the mean ordinary chondritic Ca/Al, 0.74.

seven have normative hypersthene and normative calcic plagioclase, and an additional two have normative anorthite dominant over minor normative nepheline; the tenth member is too undersaturated to have normative anorthite. The calculation of the norm is a convention for expressing a bulk analysis as a theoretical mineral assemblage, and the small degree of silica undersaturation implied by two of the norms may not be significant. These two objects would then be troctolitic in composition (No. 60 and 1A of Dodd, 1978b) and so could be genetically related to the seven objects whose normative mineralogy is noritic. The tenth object is a barred olivine chondrule that may be unrelated to the others, or else it could be a rare example of a chondrule that lost silica by vapor fractionation. Whatever its origin, Manych certainly contains chondrules of noritic affinity that are distinct from those with more normal, higher Na/Al ratios.

A chondrule with noritic modal mineralogy was found in Goodland (L4) (Hutchison and Graham, 1975), and several occur in Kramer Creek (L4) (Gibson *et al.*, 1977). A further example of noritic material in an ordinary chondrite is a crystalline norite clast in Tieschitz (Hutchison, 1982) that is partly rimmed and embayed by dark matrix (Fig. 5). Finally, the significance of an anorthositic inclusion in Bovedy (L3) has already been discussed (Graham *et al.*, 1976; Hutchison, 1982). Such materials, though uncommon, seem to have been distributed among type 3 and 4 ordinary chondrites.

Figure 4b illustrates the extremes of Ca/Al variation and Na_2O enrichment among the components of unequilibrated ordinary chondrites. In Manych, the highest Ca/Al ratios (about 1.6 atomic) in the objects studied by Dodd (1978b) occur in devitrified glass and are attributable to calcic pyroxene. In contrast, a barred olivine chondrule has the lowest Ca/Al ratio coupled with the highest Na_2O content (No. 49 of Dodd, 1978b). This chondrule is silica-undersaturated, there being 26% nepheline in the norm. Other examples of Na_2O and Al_2O_3 enrichment coupled with depletion in CaO are apparent in chondrule mesostases, such as the lilac-colored glass in the deformed Tieschitz chondrule (Table 1). An even more extreme case occurs in a Mezö-Madaras chondrule (Kurat, 1967) that is also highly enriched in FeO relative to MgO. [The 100 FeO/(MgO + FeO) atomic ratio is about 90.] Both mesostases are highly nepheline normative, and other similar examples exist (Kurat, 1967, 1971). Dodd (1978b) pointed out that Ca/Al increases with SiO_2 content, ascribing this to an increasing Ca-pyroxene component in Manych liquids. The opposite trend of Ca-pyroxene removal may have operated to enhance SiO_2 and CaO depletion in nepheline normative chondrule liquids observed in Tieschitz and Mezö-Madaras by Kurat (1967, 1971).

From the foregoing discussion we are led to conclude that chondrules and their associated glasses comprise three broad petrologic groups: (1) "noritic," with low Na/Al; (2) hypersthene normative with high Na/Al; and (3) nepheline normative, also with high Na/Al. Dodd (1978b) did not recognize the groups. His use of the relative proportions of normative olivine, pyroxene, and feldspar (both calcic and sodic), omitting nepheline, in all his analyzed chondrules and glasses together seems to have obscured trends within the groups. We believe that his dismissal of crystal-liquid fractionation as an important mechanism is not justified. Theories for the origin of chondrules must account for the chemical differences between the groups. In the nepheline normative group the more volatile Na_2O seems to correlate with refractory Al_2O_3, so we reject vapor fractionation or recondensation during chondrule formation. In this respect we agree with Dodd (1978b) rather than Kurat (1967, 1971, 1982). However, Dodd's (1978b) theory of a single olivine porphyry parent, if it has a chondritic Na/Al ratio, could not yield the "noritic" suite of chondrules by selective sampling of different proportions of its component minerals. Partial melting and removal of a Na_2O-rich liquid followed by melting of the CaO-enriched residue would be required. Furthermore, without vapor fractionation or high pressure, nepheline norma-

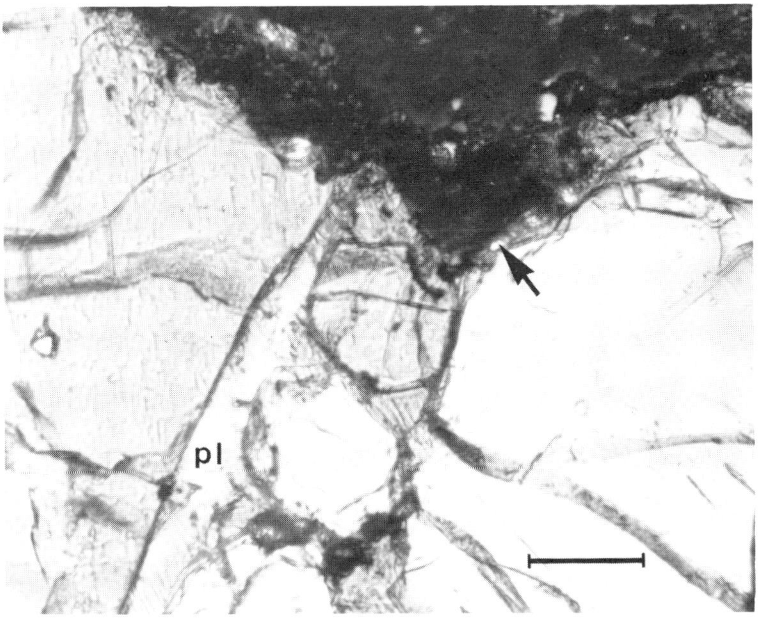

Fig. 5. Detail of margin of norite clast in Tieschitz (see Hutchison, 1982, Fig. 4). Scale bar = 50 μm; interference contrast. Note the penetrating wedge of dark matrix (shown by arrow). An elongate, anhedral calcic plagioclase crystal is labeled "pl."

tive liquids cannot be produced from a hypersthene normative parent. Under these circumstances each of the three groups of chondrules requires its own chemically distinct source.

The case is strongest for the "noritic" suite of chondrules and clasts. These have lunar highland or eucritic affinities and testify to a parentage of volatile-poor, differentiated planetary material. The noritic clast in Tieschitz predates its dark rim, and the meteorite has an internal Rb-Sr isochron age of 4.53 Ga (Minster and Allègre, 1979), so it is not a "young" regolith breccia. The anorthositic inclusion, with a high positive Eu anomaly, in Bovedy (L3) is also taken as evidence that planetary differentiated material was available at the time of accretion of the ordinary chondrites (Graham *et al.*, 1976; Hutchison, 1982).

Inter-chondrule matrices

We have already argued that the white matrix in Tieschitz represents expelled, fractionated, nepheline normative chondrule liquids (Hutchison *et al.*, 1979). A representative analysis is given in Table 1 and plotted in Fig. 4, with an analysis of the lilac glass in the deformed chondrule (Fig.

Table 1. Compositions of Tieschitz chondrule glass and matrix.

	Lilac glass	White matrix	Dark matrix		
	(1)	(2)	(3)	(4)	(5)
SiO_2	51.3	56.2	29.4	45.2	40.0
TiO_2	0.97	0.50	—	—	—
Al_2O_3	21.7	18.3	3.26	3.56	4.2
Cr_2O_3	0.37	—	0.50	0.42	—
FeO	2.10	4.75	22.1	22.8	23.5
MnO	0.09	—	0.32	0.56	—
MgO	3.80	3.32	17.0	21.4	17.4
CaO	6.17	3.85	0.37	1.57	3.0
Na_2O	11.3	8.89	2.05	1.96	1.7
K_2O	0.60	0.40	0.18	0.15	0.25
P_2O_5	—	—	0.16	0.17	—
FeS	—	—	21.5	—	—
Sum	98.40	96.21	96.84	97.79	90.05
Ca/Al	0.26	0.19	0.10	0.40	0.65
F/FM	24	44	42	37	43

Ca/Al = the atomic ratio. F/FM = 100 (FeO + MnO):(FeO + MnO + MgO) molecular ratio.

(1) Lilac-colored glass, chondrule C6 (Fig. 1). Analyzed by S. O. Agrell and R. Hutchison.

(2) White matrix. Analyzed by S. O. Agrell and R. Hutchison.

(3)–(5) Dark matrix. (3) and (4), means of 8 points each, "squeezed chondrule" (Fig. 2). Analyzed by R. Hutchison. All of the above are energy dispersive microprobe analyses.

(5) Wavelength dispersive analysis, based on Christophe Michel-Lévy (1976, Table 1).

1) for comparison. Tieschitz black matrix (Table 1 and Fig. 4; Huss *et al.*, 1981) has a variable but low Ca/Al ratio, a variable but high Na/Al ratio, and a uniform Na_2O content (1.5–2.05 wt.%). The FeO/(FeO + MgO) molecular ratio ranges upwards from 0.37 (this work and Christophe Michel-Lévy, 1976), the mean figure of Huss *et al.* (1981), 0.64 for "opaque" matrix, being at the high end of our range. From only eight analyses, four of which appear in Table 2 and are plotted in Fig. 4, the black matrix in Bishunpur has a more variable Na_2O content than that in Tieschitz, but a more uniform Ca/Al ratio. The FeO/(FeO + MgO) ratio extends from 0.35 to 0.70. The mean composition of Bishunpur "opaque" matrix of Huss *et al.* (1981) lies within the range of our few analyses. From their chemistry we cannot infer with certainty the origin of black matrices, but from Ashworth's observations (already discussed) we suggest that liquids were involved. In Bishunpur, one variety of black matrix is fairly rich in Al_2O_3, CaO, and Na_2O, and the silicate portion could represent a chondrule liquid depleted in olivine. This liquid could be related to a devitrified glass selvage (Table 2, Fig. 4) around a porphyry clast in Manych; clearly Dodd (1978b) interprets the selvage (No. 28) as a solidified liquid. Even if we are ultimately proven wrong, it seems premature to dismiss the possibility that black matrix was added to chondrules and clasts as a liquid; certainly, we have already demonstrated two instances in which chondrules and dark matrix rims were deformed when hot (Figs. 1 and 2).

Table 2. Composition and norms, Bishunpur black matrix.

	Black matrix				Glass No. 28		Molecular norms		
	(1)	(2)	(3)	(4)	(5)		(3)	(4)	(5)
SiO_2	37.8	35.9	47.0	47.6	40.5	Qtz	3.7	—	3.7
TiO_2	0.05	0.03	0.07	0.04	0.15	Or	3.0	3.0	1.5
Al_2O_3	2.29	2.07	6.50	6.12	3.5	Ab	35.0	30.0	7.5
Cr_2O_3	0.39	0.23	0.36	0.46	0.5	Ab	35.0	30.0	7.5
FeO	36.2	43.7	12.1	10.0	24.1	Ne	—	—	4.5
MgO	14.1	9.90	12.0	9.66	27.5	Di	9.6	14.0	4.0
CaO	1.34	0.99	2.28	3.46	1.3	Hyp	35.2	44.6	—
Na_2O	1.17	0.70	3.35	2.72	1.7	Ol	15.0	—	80.7
K_2O	—	—	0.45	0.44	0.25	Cm	0.5	0.6	0.5
P_2O_5	0.18	0.12	0.03	0.03	—	Il	0.2	0.1	0.2
FeS	0.62	0.44	3.26	4.66	—				
Fe,Ni*	1.16	1.14	2.68	3.32	—				
Sum	95.30	94.23	90.08	88.51	99.9†		100.0	100.0	99.9
Ca/Al	0.53	0.43	0.32	0.51	0.34				
F/Fm	59	71	36	37	33				

*Fe,Ni calculated from Ni content assuming 50% Ni in metal.

†Total includes 0.4% MnO. Ca/Al = the atomic ratio. F/FM = 100 FeO/(FeO + MgO) molecular ratio.

(1)–(4) Wavelength dispersive microprobe analyses of Bishunpur black matrix. For location of analysis no. 4, see Fig. 3. Analyzed by R. Hutchison.

(5) Microprobe analysis of glass selvage no. 28, in Manych (from Dodd, 1978b). Molecular norms are used because they approximate closely to volume proportions; they are calculated from the atomic proportions of the constituent oxides when recalculated to 100%.

Based on Fig. 4, we conclude that dark matrices cannot have been parental to chondrules by melting and reduction followed by removal of metal, as proposed by Scott *et al.* (1982). All dark matrix compositions have Ca/Al ratios lower than the H- and L-group averages, 0.74 atomic (from 1.10 wt.; Ahrens, 1970), and the spread of ratios is more limited than that among chondrules and related glasses. We reiterate that chondrules apparently require two stages of chemical fractionation for their production.

Metal

Bevan and Axon (1980) showed that metal and sulfide in Tieschitz (H3) retain textural evidence of a primary molten stage in their history. Furthermore, through their interpretation of polycrystalline γ-taenite, they suggested that heterogeneities were produced in the primary metal population, during or before accretion, by non-equilibrium solidification. Hutchison *et al.* (1981)

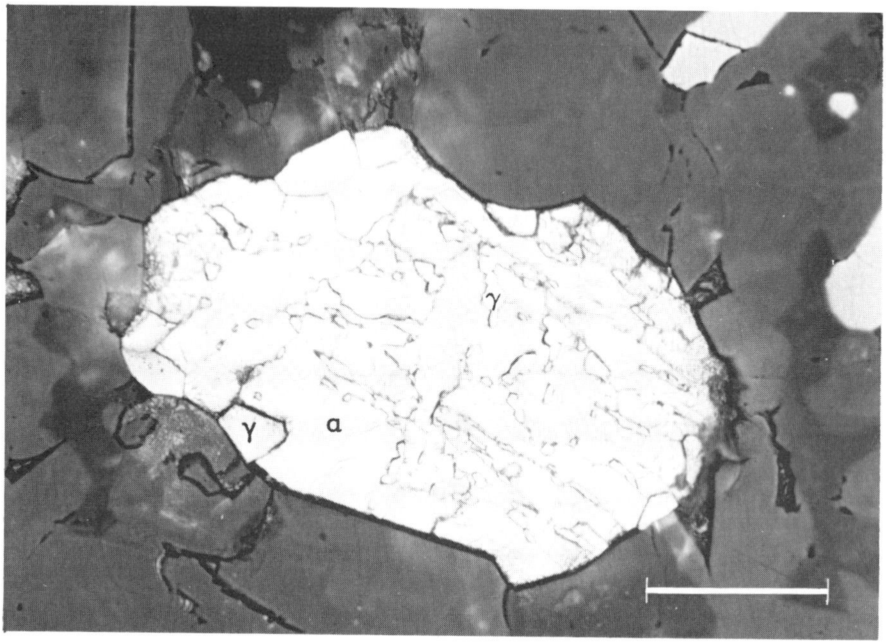

Fig. 6 Zoneless plessite ($\alpha + \gamma$) Fe,Ni grain in Kernouve. Scale bar = 25 μm; 2% nital etch. Note the blocky texture. One α-and two γ-crystallites are labeled, one of the latter belonging to a group of γ-taenite crystallites developed along metal/silicate interfaces.

extended this interpretation to the equilibrated H-group chondrites, although here crystallization and solid state diffusion during cooling from high temperatures have all but obliterated textural evidence of a solidification history in their metallic minerals. However, it is worth remembering that, in parallel with the metal, most of the evidence of silicate solidification has also been destroyed in type 6 chondrites, leaving rare relic chondrules as the only testimony to an earlier molten stage.

In addition to α-kamacite, zoned γ-taenite, and γ-tetrataenite, unreheated equilibrated (types 4–6) ordinary chondrites also contain small (25–100 μm) two-phase ($\alpha + \gamma$) particles lacking an overall Ni zonation and γ-taenite rims. Instead, they consist of continuous micrometer-sized intergrowths of $\alpha + \gamma$ showing either blocky morphologies (Fig. 6) or α-kamacite and γ-taenite blebs and lamellae in regular orientation (Fig. 7). Occasionally, small areas of γ-taenite occur along metal-silicate grain boundaries. These particles were noted by Sears and Axon (1975) and labeled "zoneless plessite." In section, zoneless plessite grains always appear to be enclosed in silicate chondrules or in large crystals of silicate in matrix, which may be recrystallized chondrule fragments.

Zoneless plessite occurs in all ten H-group chondrites studied by Hutchison *et al.* (1981). Here we present metallographic and microprobe data on zoneless plessite grains in Kernouve

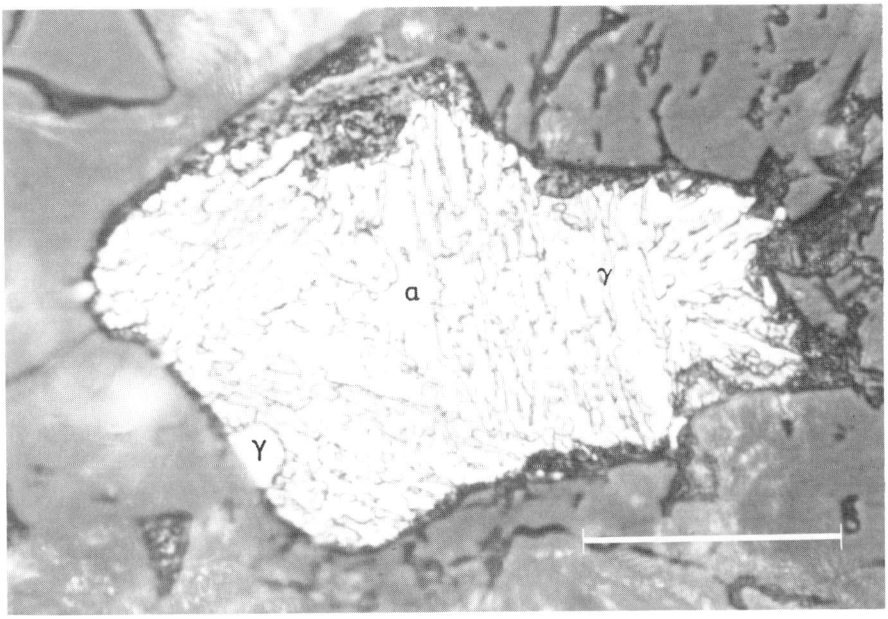

Fig. 7 Zoneless plessite ($\alpha + \gamma$) grain in Kernouve. Scale bar = 25 μm; 2% nital etch. Note the pseudo-Widmanstätten pattern.

(H6). This meteorite is unbrecciated, unshocked, and was not reheated subsequent to its cooling from peak ("metamorphic") temperatures, which allowed the crystallization of its metal and silicate. Many of the other features of its metallic minerals are typical of unreheated H-group and other ordinary chondrites.

In fine-structured zoneless plessite grains it was impossible to resolve their α and γ phases by microprobe. Instead, spot analyses yielded compositions approximating the bulk composition of the particle. More coarsely-structured zoneless plessite particles in Kernouve have α-kamacite with 5–6 wt.% Ni (0.6–0.8% Co) and γ-taenite with up to 51% Ni (Fig. 8), indicative of equilibrium at 350°C. Analyzed zoneless plessite grains in Kernouve have variable (10–15 wt.%) bulk Ni contents and are enriched in Ni and Co relative to the whole meteorite bulk metal composition obtained by wet chemistry (9 wt.% Ni; Hutchison *et al.*, 1981; see also Fig. 8). However, all the particles analyzed have bulk Co/Ni ratios close to the cosmic ratio of 1:20 (Cameron, 1973).

Bulk compositions, which lie close to the whole meteorite bulk metal, and the absence of both continuous γ-taenite rims and an overall Ni zonation clearly distinguish them from plessites formed from γ-taenites, which are zoned and have higher (25–30 wt.%) Ni contents (Wood, 1967). Zoneless plessite grains did not undergo extensive diffusive exchange of Fe and Ni with neighboring metal particles during slow, post-accretion cooling. In this respect they deviate markedly from the cooling model of Wood (1967).

The regular, pseudo-Widmanstätten arrangement of α-kamacite and γ-taenite within zoneless plessite grains is consistent with formation by decomposition of martensite (α_2). Metal with 10–15 wt.% Ni, rapidly cooled from temperatures of γ stability, undergoes diffusionless transformation to martensite in the temperature range 500°–350°C (Kaufman and Cohen, 1956). If subsequent cooling is slow enough, martensite can isothermally decompose to $\alpha + \gamma$ by the reaction $\gamma \rightarrow \alpha_2 \rightarrow \alpha + \gamma$. From laboratory experiments, martensite transformation morphologies vary from blocky irregular structures to acicular platy structures as the Ni content of the alloy increases (Speich and Swann, 1965), and this would be reflected in the structure of any subsequent decomposition product. It is significant that analyzed zoneless plessites with blocky $\alpha + \gamma$ structures have lower bulk Ni contents than those displaying pseudo-Widmanstätten patterns, consistent with the experimental results.

There appear to be a number of explanations as to how such particles may have originated. One possibility is that they had a different origin and thermal history from the remainder of the metal in the chondrites in which they occur and were formed in assemblages that pre-dated the aggregation of the chondrites. Gooding *et al.* (1979; Gooding, 1982) noted "martensite" particles with varying bulk compositions (10–20 wt.% Ni) in chondrites of types 3 and 4; these may be the same as zoneless plessite of Sears and Axon (1975). The former authors suggest that some angular metallic grains represent unmelted, pre-existing materials: it seems impossible that they could have survived alteration during incorporation into molten silicate chondrules. Alternatively, all the metal in the chondrite shared the same thermal history. In the highly equilibrated Kernouve, zoneless plessites coexist with α-kamacite, zoned γ-taenite, and γ-tetrataenite with compositions and structures indicative of an undisturbed cooling history. In addition, the bulk Co/Ni ratios tend to support a common origin for all the metal (Fig. 8). In any case, zoneless plessite grains must have been effectively isolated during post-accretion cooling until a late stage when, below 350°C, localized grain boundary or surface diffusion might have occurred along opening cracks in the surrounding silicates to produce small areas of γ-taenite along interfaces of zoneless plessite with silicate. However, this appears to have had a negligible effect on the bulk composition of the particles.

Zoneless plessite particles are largest in unreheated type 6 ordinary chondrites. In Kernouve they have the same size range as zoned γ-phases and appear to be as abundant, although the actual proportion of the whole meteorite metal represented by zoneless plessite is unknown. The cooling rate model for the chondrites used by Wood (1967, 1979) assumes that α-kamacite was formed from γ-taenite of uniform composition by the process of solid state diffusion. Zoneless plessite testifies to the existence of metal of varying composition before the onset of slow cooling, and suggests that some of the metal did not undergo diffusive exchange of Fe and Ni. What effect this has on the estimation of cooling rates of the ordinary chondrites has yet to be determined, but rapid cooling to below 500°C to produce martensite is consistent with the model of hot accretion during

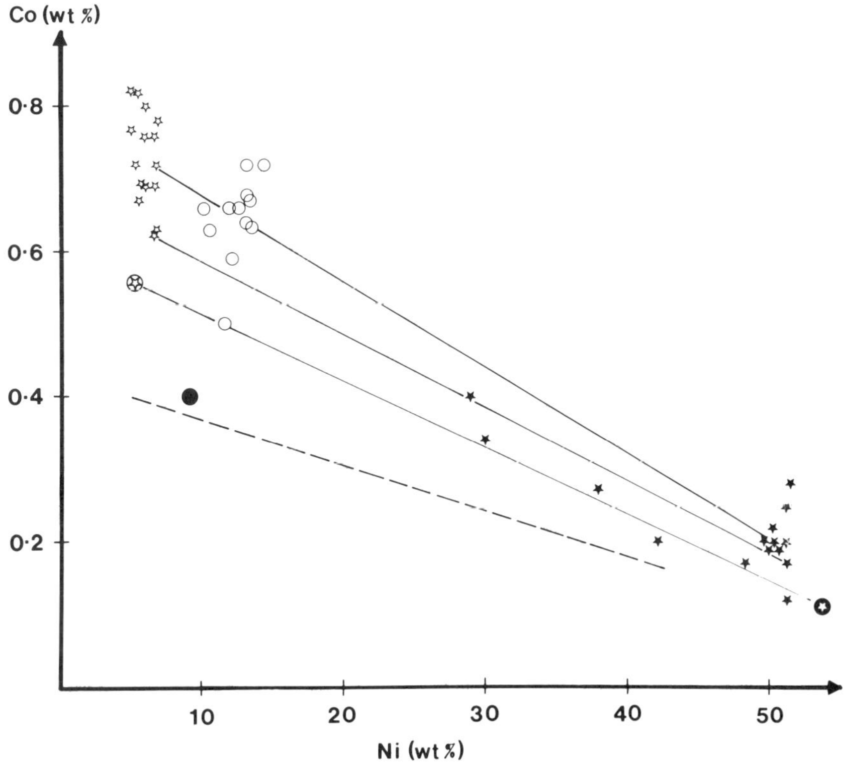

Fig. 8 Ni-Co relationships of metal in Kernouve. Open stars represent resolved α-kamacite, solid stars resolved γ-taenite, both in zoneless plessite grains. Contiguous phases are indicated by tie-lines. Open circles are bulk analyses of zoneless plessite grains. Circled stars are contiguous, massive α-kamacite (open) and γ-taenite (dark border) in the matrix of the chondrite. The dotted line represents the lowest limit of Co within the composition range of contiguous α/zoned γ-taenite in the matrix. The closed circle represents the whole meteorite metal, from "wet" chemical analysis (Hutchison *et al.*, 1981).

rapid cooling proposed by Hutchison *et al.* (1980, 1981). It appears that before slow cooling, the ordinary chondrites acquired a chemically heterogeneous population of metal grains whose origin is still uncertain. A condensing solar nebula could have produced metal with up to 22 wt.% Ni and about 1% Co (Kelly and Larimer, 1977). Subsequent oxidation of Fe could have achieved even higher Ni and Co abundances. Alternatively, some form of crystal-liquid fractionation ("non-equilibrium solidification"), as proposed by Bevan and Axon (1980) for Tieschitz metal, cannot be ruled out; instead, it seems consistent with a simultaneous origin of chondritic silicate and metal-sulfide liquids.

CONCLUSIONS

Christophe Michel-Lévy (1981) argued that Tieschitz accreted below 350°C, but that H4, H5, and H6 chondrites accreted at successively higher temperatures up to 1000°C. Her estimated accretion temperature for Tieschitz is based on the form of free carbon in the meteorite. However, if we are correct in our interpretation of the dark matrix as solidified liquids, the carbon must have been introduced at a cooler, post-accretion stage. From our own observations and those of Rambaldi and Rajan (1982), phosphate in type 3 ordinary chondrites is usually associated with metal, suggesting that elemental P expelled from cooling metal became oxidized to phosphate. The source of oxygen could have been gaseous CO or CO_2, and graphite could have been produced by the reaction.

Many workers [e.g., Dodd (1978b), Gooding *et al.* (1980), and Christophe Michel-Lévy (1981)] have argued that chondrules formed by the melting of pre-existing differentiated solids. In general, differentiation is invoked only to account for metal-silicate fractionation. [In Dodd's (1978b) model, the olivine microporphyry parental to chondrules is formed from a melt after removal of immiscible Fe,Ni-S liquid.] Our conclusions go further, favoring fractionated planetary precursors for chondrules and their associated clasts and liquids, and hence support the models of Zook (1980) for chondrule and chondrite formation and of Smith (1979) for planetary formation. A two-component model is certainly demanded by the oxygen isotopic data (Clayton, 1981). Clayton argues that ordinary chondrites formed from two reservoirs, one of solids enriched in ^{16}O, the other gaseous and ^{16}O depleted. Other gaseous reservoirs are required to account for isotopic mixing among other groups of meteorites. These reservoirs, he suggests, could have been produced in giant gaseous protoplanets.

The "noritic" component in ordinary chondrites seems to require the existence of a differentiated, volatile-poor lunar or eucritic type of planetary body before the accretion of the ordinary chondrites. Such a body could also have been a source of heterogeneous Fe,Ni-S liquids. In addition, a volatile-bearing source is required, so that some model based on collision between a partly molten, differentiated volatile-poor body and a gaseous protoplanet seems appropriate for the genesis of chondrules and the ordinary chondrites. The model could include aerodynamic sorting among the metal and silicate components of chondrites (Dodd, 1976) and the addition of a mist of liquid droplets to solid and plastic chondrules and clasts to produce dark matrix. Material that accumulated on a substrate (planetary fragment?) soon after the collision would have been buried and more slowly cooled (type 6) than the material that arrived later (types 3 and 4). The later accreted material would have been near the planetary surface and so could have received components (C-chondrite, "noritic" material?) from exotic sources.

The existence of differentiated precursors to ordinary chondrites was argued by Hutchison (1982), partly on the basis of age and isotopic data on differentiated meteorites. More recently it was shown that isotopic closure of Ag and Pd occurred in Cape York (Chen and Wasserburg, 1982) and in Grant (Kaiser and Wasserburg, 1983), while the short-lived ^{107}Pd was extant. Cape York certainly solidified from a melt (Esbensen and Buchwald, 1981), and the new data strengthen the view that differentiation in planetary bodies occurred very early. The implication is that Urey (1959) may have been correct in maintaining that the solar system comprises primary and secondary planetary objects, although many of the interpretations on which his idea was based are no longer tenable. If ordinary chondrites really are reworked planetary materials, they cannot provide direct evidence for the former existence of a solar nebula, so other hypotheses of planetary origins ought to be reconsidered. For example, Woolfson's (1978) capture theory of the origin of planets demands a collision between gaseous protoplanets, but assumes that there was never a solar nebula.

Acknowledgments. We thank Andrew Graham for reading and discussing the manuscript and for his helpful suggestions. R. T. Dodd's constructive but critical review is much appreciated.

REFERENCES

Ahrens L. H. (1970) The composition of stony meteorites; IX, Abundance trends of the refractory elements in chondrites, basaltic achondrites and Apollo 11 fines. *Earth Planet. Sci. Lett.* **10**, 1–6.

Allen J. S., Nozette S., and Wilkening L. L. (1980) A study of chondrule rims and chondrule irradiation records in unequilibrated ordinary chondrites. *Geochim. Cosmochim. Acta* **44**, 1161–1175.

Ashworth J. R. (1977) Matrix textures in unequilibrated ordinary chondrites. *Earth Planet. Sci. Lett.* **35**, 25–34.

Ashworth J. R. (1981) Fine structure in H-group chondrites. *Proc. Roy. Soc. London* **A374**, 179–194.

Bevan A. W. R. and Axon H. J. (1980) Metallography and thermal history of the Tieschitz unequilibrated meteorite—metallic chondrules and the origin of polycrystalline taenite. *Earth Planet. Sci. Lett.* **47**, 353–360.

Cameron A. G. W. (1973) Abundances of elements in the solar system. *Space Sci. Rev.* **15**, 121–146.

Chen J. H. and Wasserburg G. J. (1982) Isotopic composition of Ag, Cd and Pb at ultralow levels in iron meteorites (abstract). In *Papers Presented to the 45th Annual Meteoritical Society Meeting*, p. VI-2. Lunar and Planetary Institute, Houston.

Christophe Michel-Lévy M. (1976) La matrice noire et blanche de la chondrite de Tieschitz (H3). *Earth Planet. Sci. Lett.* **30**, 143–150.

Christophe Michel-Lévy M. (1981) Some clues to the history of the H-group chondrites. *Earth Planet. Sci. Lett.* **54**, 67–80.

Clayton R. N. (1981) Isotopic variations in primitive meteorites. *Phil. Trans. Roy. Soc. London* **A303**, 339–349.

Dodd R. T. (1976) Accretion of the ordinary chondrites. *Earth Planet. Sci. Lett.* **30**, 281–291.

Dodd R. T. (1978a) The composition and origin of large microporphyritic chondrules in the Manych (L3) chondrite. *Earth Planet. Sci. Lett.* **39**, 52–66.

Dodd R. T. (1978b) Compositions of droplet chondrules in the Manych (L3) chondrite and the origin of chondrules. *Earth Planet. Sci. Lett.* **40**, 71–82.

Esbensen K. H. and Buchwald V. F. (1981) Late crystallisation of the natural Fe-Ni-S-P system—Evidence from Cape York troilite inclusions (abstract). *Meteoritics* **16**, 313.

Ganapathy R. and Anders E. (1974) Bulk compositions of the moon and earth, estimated from meteorites. *Proc. Lunar Sci. Conf. 5th*, pp. 1181–1206.

Gibson E. K., Jr., Lange D. E., Keil K., Schmidt T. E., and Rhodes J. M. (1977) The Kramer Creek, Colorado meteorite: A new L4 chondrite. *Meteoritics* **12**, 95–107.

Gooding J. L. (1982) Origin of chondrules in ordinary chondrites by incomplete melting of geochemically fractionated solids (abstract). In *Papers Presented to the Conference on Chondrules and their Origins*, p. 20. Lunar and Planetary Institute, Houston.

Gooding J. L., Keil K., Fukuoka T., and Schmitt R. A. (1979) The metal components of chondrules (abstract). *Meteoritics* **14**, 404–407.

Gooding J. L., Keil K., Fukuoka T., and Schmitt R. A. (1980) Elemental abundances in chondrules from unequilibrated chondrites: Evidence for chondrule origin by melting of pre-existing materials. *Earth Planet. Sci. Lett.* **50**, 171–180.

Graham A. L., Easton A. J., Hutchison R., and Jérome D. Y. (1976) The Bovedy meteorite; mineralogy and chemistry of its Ca-rich glass inclusions. *Geochim. Cosmochim. Acta* **40**, 529–535.

Grossman J. N. and Wasson J. T. (1981) Evidence for primitive nebular components in the chondrules from the Chainpur chondrite. *Geochim. Cosmochim. Acta* **46**, 1081–1099.

Grossman L. (1975) Petrography and mineral chemistry of Ca-rich inclusions in the Allende meteorite. *Geochim. Cosmochim. Acta* **39**, 433–454.

Huss G. R., Keil K., and Taylor G. J. (1981) The matrices of unequilibrated ordinary chondrites: implications for the origin and history of chondrites. *Geochim. Cosmochim. Acta* **45**, 33–51.

Hutchison R. (1982) Meteorites—evidence for the interrelationships of materials in the solar system of 4.55 Ga ago. *Phys. Earth Planet. Inter.* **29**, 199–208.

Hutchison R. and Graham A. L. (1975) Significance of calcium-rich differentiates in chondritic meteorites. *Nature (London)* **255**, 471.

Hutchison R., Bevan A. W. R., Agrell S. O., and Ashworth J. R. (1979) Accretion temperature of the Tieschitz, H3, chondritic meteorite. *Nature (London)* **280**, 116–119.

Hutchison R., Bevan A. W. R., Agrell S. O., and Ashworth J. R. (1980) Thermal history of the H-group of chondritic meteorites. *Nature (London)* **287**, 787–790.

Hutchison R., Bevan A. W. R., Easton A. J., and Agrell S. O. (1981) Mineral chemistry and genetic relations among H-group chondrites. *Proc. Roy. Soc. London* **A374**, 159–178.

Kaiser T. and Wasserburg G. J. (1983) The isotopic composition and concentration of Ag in iron meteorites and the origin of exotic silver. *Geochim. Cosmochim. Acta* **47**, 43–58.

Kaufman L. and Cohen M. (1956) The martensitic transformation in the iron-nickel system. *Trans. AIME* **206**, 1393–1401.

Kelly W. R. and Larimer J. W. (1977) Chemical fractionations in meteorites—VIII. Iron meteorites and the cosmochemical history of the metal phase. *Geochim. Cosmochim. Acta* **41**, 93–111.

Kurat G. (1967) Einige chondren aus dem Meteoriten von Mezö-Madaras. *Geochim. Cosmochim. Acta* **31**, 1843–1857.

Kurat G. (1971) Die chemische Zusammensetzung von Gläsern und Chondrenmatrizes im Chondriten von Tieschitz. *Chem. Erde* **30**, 235–249.

Kurat G. (1982) Impact origin of chondrules (abstract). In *Papers Presented to the Conference on Chondrules and their Origins*, p. 36. Lunar and Planetary Institute, Houston.

Larimer J. W. (1973) Chemical fractionations in meteorites—VII. Cosmothermometry and cosmobarometry. *Geochim. Cosmochim. Acta* **37**, 1603–1623.

Larimer J. W. and Anders E. (1967) Chemical fractionation in meteorites—II. Abundance patterns and their interpretation. *Geochim. Cosmochim. Acta* **31**, 1239–1270.

Lux G., Keil K., and Taylor G. J. (1981) Chondrules in H3 chondrites: textures, compositions and origins. *Geochim. Cosmochim. Acta* **45**, 675–685.

Minster J. F. and Allègre C. J. (1979) ^{87}Rb-^{87}Sr chronology of H chondrites: Constraint and speculations on the early evolution of their parent body. *Earth Planet. Sci. Lett.* **42**, 333–347.

Rambaldi E. R. and Rajan R. S. (1982) Evidence for primitive phosphates in highly unequilibrated ordinary chondrites (abstract). In *Papers Presented to the 45th Annual Meteoritical Society Meeting*, p. XIV-3. Lunar and Planetary Institute, Houston.

Scott E. R. D., Taylor G. J., Rubin A. E., Keil K., and Kracher A. (1982) Origin of chondrules by preaccretionary melting of silicate matrix material (abstract). In *Papers Presented to the Conference on Chondrules and their Origins*, p. 54. Lunar and Planetary Institute, Houston.

Sears D. W. and Axon H. J. (1975) The metal content of common chondrites (abstract). *Meteoritics* **10**, 486–487.

Smith B. A. and Goldstein J. I. (1977) The metallic microstructures and thermal histories of severely reheated chondrites. *Geochim. Cosmochim. Acta* **41**, 1061–1072.

Smith J. V. (1979) Mineralogy of the planets: a voyage in space and time. *Mineral. Mag.* **43**, 1–89.

Speich G. R. and Swann P. R. (1965) Yield strength and transformation substructure of quenched iron-nickel alloys. *J. Iron Steel Inst. (London)* **203**, 480–485.

Taylor G. J., Scott E. R. D., and Keil K. (1982) Cosmic setting for chondrule formation (abstract). In *Papers Presented to the Conference on Chondrules and their Origins*, p. 58. Lunar and Planetary Institute, Houston.

Urey H. C. (1959) Primary and secondary objects. *J. Geophys. Res.* **64**, 1721–1737.

Wasson J. T. (1974) *Meteorites*. Springer, Berlin. 316 pp.

Wood J. A. (1967) Chondrites: Their metallic minerals, thermal histories, and parent planets. *Icarus* **6**, 1–49.

Wood J. A. (1979) Review of the metallographic cooling rates of meteorites and a new model for the planetesimals in which they formed. In *Asteroids* (T. Gehrels, ed.), pp. 849–891. Univ. of Arizona Press, Tucson.

Woolfson M. M. (1978) The capture theory and the origin of the solar system. In *The Origin of the Solar System* (S. F. Dermott, ed.), pp. 179–198. Wiley, Chichester.

Zook H. A. (1980) A new impact model for the generation of ordinary chondrites (abstract). *Meteoritics* **15**, 390–391.

Reduction, Partial Evaporation, and Spattering: Possible Chemical and Physical Processes in Fluid Drop Chondrule Formation

Elbert A. King

Department of Geosciences, University of Houston, Houston, Texas 77004

The major chemical differences between fluid drop chondrules and their probable parent materials may have resulted from the loss of volatiles such as S, H_2O, Fe, and volatile siderophile elements by partial evaporation during the chondrule-forming process. Vertical access solar furnace experiments in vacuum and hydrogen have demonstrated such chemical fractionation trends using standard rock samples. The formation of immiscible iron droplets and spherules by *in situ* reduction of iron from silicate melt and the subsequent evaporation of the iron have been observed directly. During the time that the main sample bead is molten, many small spatter spherules are thrown off the main bead, thereby producing many additional chondrule-like melt spherules that cool rapidly and generate a population of spherules with size frequency distribution characteristics that closely approximate some populations of fluid drop chondrules in chondrites. It is possible that spatter-produced fluid drop chondrules dominate the meteoritic fluid drop chondrule populations. Such meteoritic chondrule populations should be chemically related by various relative amounts of iron and other volatile loss by vapor fractionation.

INTRODUCTION

It has been demonstrated that some large populations of meteoritic chondrules are depleted in Fe, Co, Ni, and Ir compared with bulk meteorite chemistry (Osborn and Schmitt, 1971; Osborn *et al.*, 1973). However, the petrographic nature of these chondrules was only generally described. A more comprehensive study of a much smaller but well-characterized population of fluid drop chondrules (for definition of "fluid drop chondrule" see King and King, 1978) by Gooding *et al.* (1980) confirmed the earlier result and extended the data to show depletion of iron and siderophile elements relative to bulk CI chondrites. Gooding *et al.* concluded that metal/silicate fractionation must have occurred prior to or during chondrule formation, but no specific mechanism for this fractionation was proposed. A discussion of the depletion of iron in chondrules relative to bulk chondrites and presumed solar system bulk composition is included in Dodd (1981). Scott *et al.* (1982) have provided additional documentation of iron depletion in chondrules relative to silicate matrix in chondrites and suggest that the melting of the matrix-like material occurred in the solar nebula. It should be noted that fluid drop chondrules comprise only a minor fraction of the chondrule populations in many meteorites (King and King, 1978).

A series of vertical access solar furnace experiments and subsequent analyses of refractory furnace residues was undertaken, in part, for the purpose of investigating fractional vaporization as a possible mechanism to account for siderophile loss during chondrule formation (King,

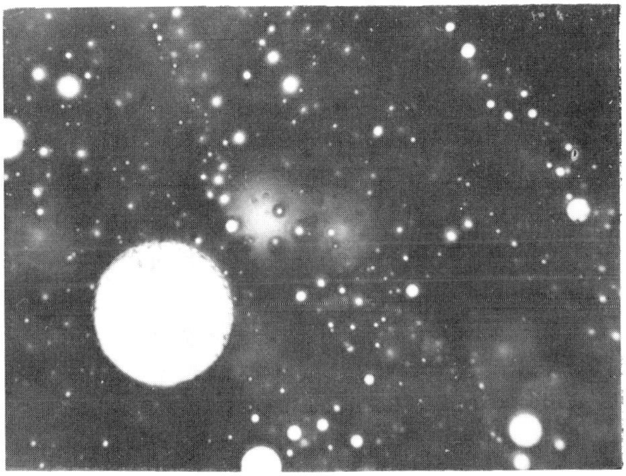

Fig. 1a Immiscible metal spherules composed almost entirely of iron in a matrix of basaltic glass. Fusion of the powdered basalt (BCR-1) occurs within a few seconds and *in situ* reduction of the iron from the melt occurs within a few minutes of heating in vacuum (<1 mm Hg) at maximum temperatures approximately 3000°C. The largest metal spherule visible is approximately 0.15 mm in diameter, reflected light.

1982a,b,c,d; MacPherson *et al.*, 1982). An unanticipated result of these experiments was the observation of a spatter mechanism, which may have been the dominant physical process for the formation of meteoritic fluid drop chondrules.

EXPERIMENTS AND OBSERVATIONS

During the summers of 1981 and 1982, a suite of fusion and partial evaporation experiments using basalt standard BCR-1, bulk Allende, and bulk Murchison were accomplished at the Centre National de Researche Scientifique, Laboratory for Ultrarefractories in Odeillo (Font-Romeu), France. This laboratory presently contains the only available operating vertical access solar

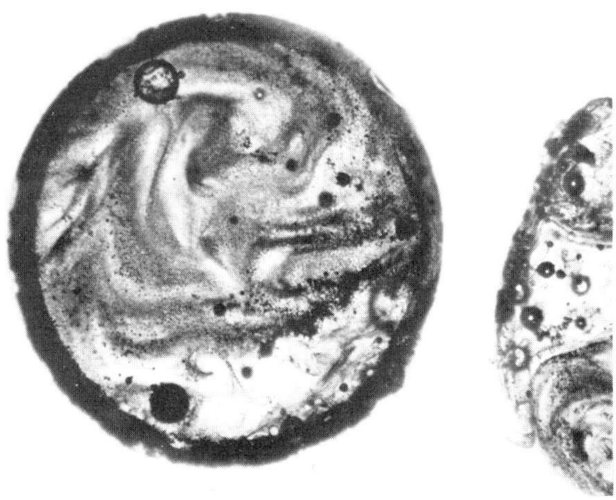

Fig. 1b A basalt spatter spherule approximately 4 mm in diameter containing abundant tiny and a few larger immiscible spherules of metallic iron. The tiny, dusty metal spherule inclusions are responsible for the easy visibility of the *schlieren*, transmitted light.

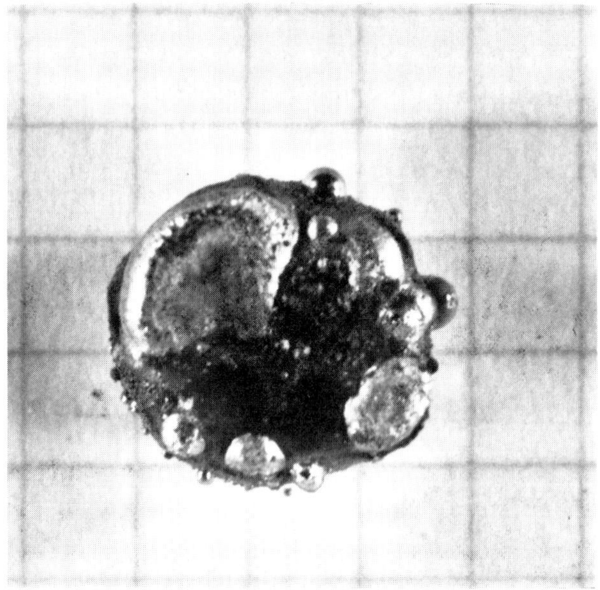

Fig. 2 Bottom of the residual melt bead formed from the fusion and partial evaporation in the solar furnace in vacuum ($<$1 mm Hg) of powdered bulk Allende showing numerous large iron spherules on the periphery of the bead. Spacing of the background grid is approximately 2.5 mm between adjacent parallel lines.

furnace facilities. The advantages in using this facility and the general geometry of the experiments and associated apparatus have been described by King (1982a,d).

Reduction and partial evaporation

Melting and continued heating above the liquidus of powdered BCR-1 samples *in vacuum* at maximum temperatures ranging from approximately 2400°C to 3000°C for minutes to tens of minutes in all cases resulted in the *in situ* reduction of iron from the silicate melt (Fig. 1). Some of these immiscible beads either formed on the surface of the main silicate melt bead and/or formed in the interior of the bead and reached the surface by convective circulation and can now be observed on and in the surfaces of the cooled melt beads. However, the production of iron metal beads is much more evident in bulk samples of Allende and Murchison that are heated *in vacuum* to maximum temperatures ranging from just above the liquidus to as much as 3000°C (Fig. 2). These metal spherules are so large (as much as 4 mm in diameter) that it was possible to observe the formation of the individual iron spherules on the active surface of the silicate melt beads while the beads were in the focus of the furnace. More importantly, it was possible to observe the growth of individual iron spherules during the first few minutes of heating above the liquidus and, under continued observation, see the same individual spherules shrink in diameter during the following few minutes to tens of minutes until no metal spherules were visible. Later examination of the cooled melt beads pulled from the furnace at various stages of iron spherule formation indicates

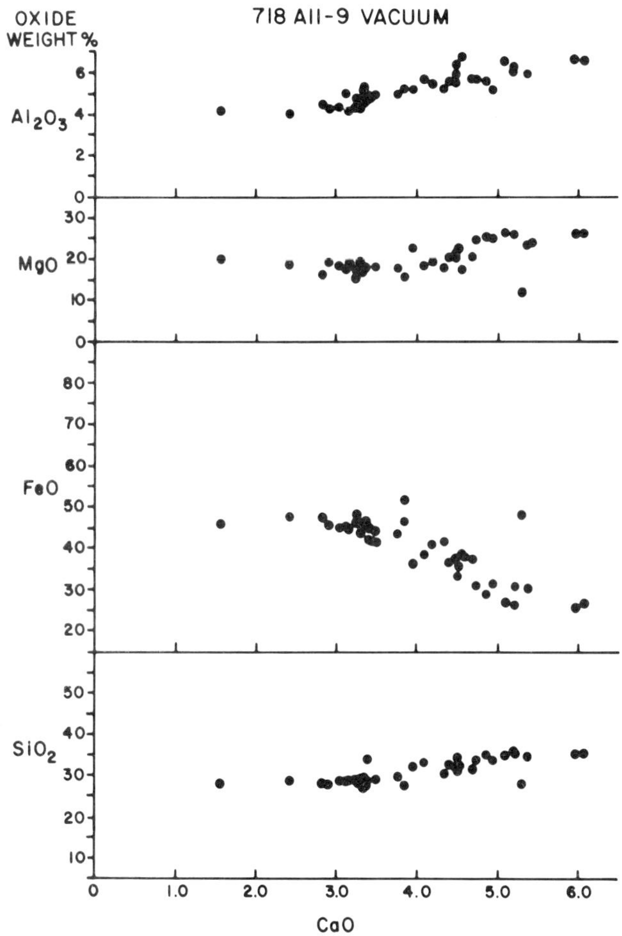

Fig. 3 Some major elemental correlations of randomly selected spatter spherules from solar furnace heating of bulk Allende in vacuum. Analyses are by broad beam energy dispersive X-ray technique. The maximum sustained temperature obtained is estimated to be 2300° ± 100°C, total heating time 27 minutes. Spherules plotting to left presumably were spattered from main melt bead earlier than those that plot to right. Note the trends that show calcium, magnesium, aluminum, and silicon enrichment, but a depletion of iron.

that the spherules form and grow both on the surfaces of the melt beads and in the interiors. However, once a metal spherule reaches the surface of a silicate melt bead, it tends to stay there and eventually visibly reduces in size due to evaporation to the point that it is no longer visible to the unaided eye within a few minutes to tens of minutes, provided that a temperature several hundred degrees above the liquidus is maintained. The chief difference observed as a function of maximum temperature attained is the rate of reduction and evaporation. In fact, at temperatures just above the liquidus, only reduction was observed directly. However, evaporation surely also occurred as amorphous iron-bearing silicate was observed as condensate on the interior of the vacuum globe in the initial stages of all sample runs in which condensate was collected.

At maximum iron spherule formation, prior to substantial evaporation of the metallic iron, there is a concentration of metal spherules and blebs about the periphery of the silicate melt beads.

This texture is very similar to that of many chondrules in chondrites, e.g., Renazzo, in which silicate chondrules with metal spherule concentrations about their rims are especially prominent. If the high temperature heating is continued, the reduction/evaporation process will remove virtually all of the iron from a silicate melt. This has been verified by energy dispersive X-ray analyses of some residual beads in which there is no detectable iron present. Figure 3 shows the general trends of enrichment and depletion of some major elements in spatter spherules that serve as consecutive samples of a residual Allende melt bead heated to a maximum temperature of approximately 2300°C and held at that temperature for 27 minutes.

Energy dispersive X-ray analyses of metallic iron spherules from the surfaces of solar furnace residual beads from heated samples of both bulk Allende and bulk Murchison demonstrate that they range in composition from almost pure iron to iron containing minor and variable amounts of nickel, silicon, and chromium. Minor element contents are especially variable adjacent to contacts with the silicate melt. No systematic correlation of the amount of iron reduction and evaporation with the nickel or chromium content of the metal spherules has yet been observed.

If the silicate melt bead with the metal spherules on its surface is pulled from the furnace, the difference in thermal contraction between the silicate and the metal permits some of the iron spherules to be easily dislodged from the surface. Grossman (1982) independently has suggested this mechanism as an important one for iron loss from fluid drop chondrules, based chiefly on his observations of "craters" in chondrule rims and surfaces that he interprets as sites where metal-rich spherules have been lost. He further suggests that this mechanism may not account for all of the iron variation in chondrules. However, many fluid drop chondrules that are depleted in iron relative to presumed parent materials do not have cavities in their surfaces that might have been occupied by metal spherules, and unattached iron spherules are not very common in chondrites. If Grossman's mechanism is correct, the iron spherules must somehow be removed and separated from the chondrules so that they do not accumulate with them, or they must be evaporated after separation from their parent chondrule. The writer prefers the interpretation that most of the iron was evaporated from the surfaces of the fluid drop chondrules themselves either from iron spherules, iron-rich rims, or directly from the silicate melt, but variations in the total iron content of fluid drop chondrule parent material certainly cannot be ruled out. In any event, the iron reduction/evaporation mechanism is more than efficient enough to provide the magnitude of iron and siderophile element loss observed in fluid drop chondrules, *provided that* a sustained intense heat source is available. The necessity for a sustained heat source (King, 1982a,d) is a serious constraint on the origins proposed for many, if not most, fluid drop chondrules. Sudden heating events followed by rapid cooling do not provide the time needed for the kinetics of substantial iron reduction and evaporation. Heating in vacuum at temperatures in excess of approximately 1800°C for period of minutes to tens of minutes is required, depending on the total pressure and partial pressures of various gases present.

Spattering

During the fusion and continued heating of powdered BCR-1, bulk Murchison, and bulk Allende samples, dozens to hundreds of fluid drop chondrule-like spatter spherules were formed in all sample fusion runs regardless of heating rates or maximum temperatures attained (King, 1982a,c,d). The spatter mechanism is most efficient in the production of chondrules at high heating rates and high temperatures in volatile-rich samples, but also occurs in samples that are heated very slowly to melting and in samples of refractory compositions. Many of the synthetic spatter chondrules have textures and mineralogies that are similar to some meteoritic chondrules, in particular non-porphyritic fluid drop chondrules. Such similarities in textures and mineralogies have been observed by many previous workers, who produced chondrule-like spherules from molten drops by a variety of experimental processes (e.g., Rinne, 1897; Fredriksson and Ringwood, 1963; Nelson *et al.*, 1972). However, in addition to similarities in textures and mineralogies, at least some populations of synthetic spatter chondrules have been produced in the solar furnace that closely resemble some populations of meteoritic fluid drop chondrules in grain size frequency distribution characteristics, including sorting (King, 1982a,c,d). The problem of generating populations of meteoritic fluid drop chondrules with the observed grain size frequency distributions and sorting has been a serious constraint on proposed mechanisms of chondrule formation and history (e.g., Dodd, 1976; King and King, 1978, 1979). However, if the spatter process is the *dominant* process of fluid-drop chondrule formation (King, 1982c), there is no necessity for a complex two-step process to yield the observed sorting and other grain size frequency characteristics. It should be noted that only an average of *more than one* spatter chondrule need be produced for each fluid drop chondrule formed by some other process (e.g., direct melting) for the spatter mechanism to numerically dominate the fluid drop chondrule populations. Such spatter chondrule dominated populations, if derived from the same or chemically similar primary melts, should be chemically related chiefly by a range in the amount of refractory element enrichment and volatile loss due to partial evaporation. More specifically, there should be differences in the amounts of enrichment in refractory lithophile elements and depletion of iron and siderophile elements including iridium (Grossman *et al.*, 1982) as a result of partial evaporation and vapor fractionation. This appears to be a possible interpretation of at least two of the best available data sets, Osborn *et al.* (1973) and Gooding *et al.* (1980).

CONCLUSIONS

It appears that most fluid drop chondrules have originated by melting of primitive solar system dust by a sustained heat source. Spattering is an experimentally demonstrated mechanism that may have produced the vast majority of fluid drop chondrules. The high temperature and its minimum duration of minutes to tens of minutes (and possibly much longer) caused partial evaporation of iron and volatile siderophile elements in chondrule parent melts and the individual spatter-produced fluid drop chondrules. This can account for the broad patterns, as well as many (but not all) of the details of fluid drop chondrule chemistry relative to presumed parent materials.

Acknowledgments. This work is supported by a grant from the National Aeronautics and Space Administration (NAGW-178). I am especially indebted to the staff of the CNRS Laboratory, Odeillo, for their help and assistance, in particular Dr. J. P. Coutures. Jac Cashore performed the energy dispersive X-ray analyses. This manuscript was substantially improved in response to comments by J. L. Gooding, E. R. D. Scott, and an anonymous reviewer.

REFERENCES

Dodd R. T. (1976) Accretion of the ordinary chondrites. *Earth Planet Sci. Lett.* **30**, 281–291.

Dodd R. T. (1981) *Meteorites.* Cambridge University Press, N. Y. 368 pp.

Fredriksson K. and Ringwood A. E. (1963) Origin of meteoritic chondrules. *Geochim. Cosmochim. Acta* **27**, 639–641.

Gooding J. L., Keil K., Fukuoka T., and Schmitt R. A. (1980) Elemental abundances in chondrules from unequilibrated chondrites; evidence for chondrule origin by melting of pre-existing materials. *Earth Planet. Sci. Lett.* **50**, 171–180.

Grossman J. N. (1982) Mechanisms for the depletion of metal in chondrules (abstract). In *Papers Presented to the Conference on Chondrules and their Origins*, p. 23. Lunar and Planetary Institute, Houston.

Grossman L., Kawabe I., Ekambaram V., and King E. A. (1982) Chemical studies of evaporation residues produced in a solar furnace (abstract). *Meteoritics* **17**, 226.

King E. A. (1982a) Refractory residues, condensates and chondrules from solar furnace experiments (abstract). In *Lunar and Planetary Science XIII*, pp. 389–390. Lunar and Planetary Institute, Houston.

King E. A. (1982b) Iron loss from fluid drop chondrules by partial evaporation (abstract). In *Papers Presented to the Conference on Chondrules and their Origins*, p. 33. Lunar and Planetary Institute, Houston.

King E. A. (1982c) Spattering: The dominant process of fluid drop chondrule formation (abstract). In *Papers Presented to the Conference on Chondrules and their Origins*, p. 34. Lunar and Planetary Institute, Houston.

King E. A. (1982d) Refractory residues, condensates and chondrules from solar furnace experiments. *Proc. Lunar Planet. Sci. Conf. 13th*, in *J. Geophys. Res.* **87**, A429–A434.

King T. V. V. and King E. A. (1978) Grain size and petrography of C2 and C3 carbonaceous chondrites. *Meteoritics* **13**, 47–72.

King T. V. V. and King E. A. (1979) Size frequency of fluid drop chondrules in ordinary chondrites. *Meteoritics* **14**, 91–96.

MacPherson G. J., Kawabe I., Grossman L., and King E. A. (1982) A solar-furnace study of the formation of refractory residues by evaporation (abstract). In *Lunar and Planetary Science XIII*, pp. 459–460. Lunar and Planetary Institute, Houston.

Nelson L. S., Blander M., Skaggs S. R., and Keil K. (1972) Use of a CO_2 laser to prepare chondrule-like spherules from supercooled molten oxide and silicate droplets. *Earth Planet. Sci. Lett.* **14**, 338–344.

Osborn T. W. and Schmitt R. A. (1971) Elemental abundances in meteoritic chondrules (abstract). *Meteoritics* **6**, 297–298.

Osborn T. W., Smith R. H., and Schmitt R. A. (1973) Elemental composition of individual chondrules from ordinary chondrites. *Geochim. Cosmochim. Acta* **37**, 1909–1942.

Rinne F. (1897) Kugelrunde Eiskrystalle und Chondren von Meteoriten. *Neues Jahr. Min. Geol. u. Paläon.*, ser. 11, **1**, 259–261.

Scott E. R. D., Taylor G. J., Rubin A. E., Keil K., and Kracher A. (1982) Origin of chondrules by preaccretionary melting of silicate matrix material (abstract). In *Papers Presented to the Conference on Chondrules and their Origins*, p. 54. Lunar and Planetary Institute, Houston.

Chondrule Formation by Impact? The Cooling Rate

F. Kluger, H. H. Weinke, and W. Kiesl
Institute for Analytical Chemistry, University of Vienna, A-1094 Wien, Austria

Different sets of initial conditions for impacts of stray objects on solid surfaces of planetary bodies or asteroids have been computed for cooling times of chondrules. The results of the calculations indicate cooling times shorter than 60 seconds for condensed matter from 10^4-10^{15}g generated by impact and initial temperatures of the gas phase formed up to presumed degeneracy limit of 150,000 K. To retain solid impact products originated that way on meteorite parent bodies by gravitational forces, a target diameter of at least 800 km is required. It is shown that chondrule formation by impact seems to be an unlikely event.

INTRODUCTION

A number of different origins for chondrules have been proposed during the past decades. Among these, the so-called impact theory was stressed first by Fredriksson (1963). Thereafter the theory has been modified more or less until the present, e.g., by Kurat (1982). The main process, however, is an impact occurring on the surface of a parent body. The proponents of the impact theory point to abundant collisions, producing melt and vapor, during the early accretion stage of planets and planetesimals. A serious problem is the definition of what particle is called a chondrule (Dodd, 1982; King and King, 1978). From the literature it is evident that most investigators, when speaking of chondrules, mean "barred olivine chondrules," "radiating pyroxene chondrules," and "glassy chondrules" after Van Schmus (1969), or the so-called "true chondrules" of Grossman and Olsen (1974). Whatever the definition might be, in the paper presented here we show that a collision between objects of a certain size must fail in producing "true chondrules."

THEORY

Since impacts on the surface of parent bodies during the early accretion stage of planets and planetesimals should be responsible for the formation of chondrules, we describe the mechanism with respect to the cooling rate of a gas phase containing condensed matter after a cosmic impact. The early thermal history of the chondrules (formed out of the condensed fraction mentioned) is clearly controlled by the thermal behavior of the surrounding gas phase.

The computation of the cooling time of gas as well as of condensed matter is based on the law of Stefan-Boltzmann:

$$\frac{dt}{dT} = -\frac{m \cdot C_p \cdot 1.76 \cdot 10^4}{F \cdot (T_m^4 - T_s^4)} \tag{1}$$

with T = temperature (K), T_m = matter (T_g = gas, T_c = chondrules), and T_s = temperature of the surrounding environment. For the chondrules the environment is the cooling gas, thus variable in temperature, but for the gas the environment is the interplanetary space with a constant temperature of 200 K, which corresponds to the black body temperature in the asteroidal region, considering the present value of the solar constant $S = 1.36 \cdot 10^6 \text{erg cm}^{-2} \text{sec}^{-1}$ (e.g., Unsöld, 1974); t = time (sec), m = mass (g), F = surface (cm^2), C_p = specific heat; for two-atomic gases 9R/2, for mono-atomic gases 5R/2. Taking the universal gas constant $R = 8.31 \cdot 10^7 \text{erg Mol}^{-1} K^{-1}$, a mean molecular weight of 40 and a mean atomic weight of 20 yields $C_{p_g} = 9.87 \cdot 10^6 \text{ erg g}^{-1} K^{-1}$. This value corresponds to the high temperature limit of fluid $MgSiO_3$ or Mg_2SiO_4 (JANAF, 1968).

The relief of pressure after the collision defines the zero point of time scale and the initial temperature T_i of the gas phase. Our calculations had been carried out for T_i between 2000 K and a presumed degeneracy limit of 150,000 K, because for high speed impacts (i.e., greater than 3.5 km/sec) the shock temperature may well be above the temperature of fusion and vaporization but remains below the limit of degeneracy of matter, as has been calculated for example by Öpik (1958a). Thus we exclude plasma, an excellent thermal conductor. At most the gas expands without interference to interplanetary space with the speed of sound: $dr/dt = (2\, C_{p_g} \cdot T_i)^{1/2}$; the corresponding surface becomes $F_g = 2\, C_{p_g} \cdot T_i\, t^2\, G$ with G as an undimensional factor depending on the geometry of expansion and the simplifying assumption that at t = 0 the surface is zero. The cooling of the gas from T_i down to 1500 K is calculated pre-supposing thermal equilibrium within the gas and radiative energy loss at the surface only, unaffected by the condensed matter. Allowing the black body radiation to come from a gray and semi-transparent object by introducing an efficiency factor for hot spheres of gas (Unsöld, 1974), $E_g = 1/2$ (thus the effective temperature of $T_{eff}^4 = T_g^4/2$), eq. (1) becomes:

$$\frac{t^2 dt}{dT_g} = -\frac{8818 m_g}{G \cdot T_i (T_g^4/2 - T_s^4)} \,. \tag{2}$$

During the cooling time of the gas, chondrules are formed by melting of the precursor material, evaporation, or condensation until the entire mass has reached a temperature of 1500 K. This limit has been chosen with respect to rare relict inclusions and liquidus temperature (Nagahara, 1981; Rambaldi, 1981; Rubin *et al.*, 1982) of chondrules. Further cooling of chondrules is described by eq. (1), accounting for non-ideal radiation with efficiency $E_c = 2^{-1/4}$ (Unsöld, 1974):

$$\frac{dt}{dT_c} = -\frac{2.7 \cdot 10^{11} m_c}{E_c (T_c^4 - T_g^4)} \,. \tag{3}$$

Since the cooling of the chondrules increases the thermal energy of the gas, eq. (2) has to be extended; furthermore, allowing the gas to approximate the temperature of T_s of the surrounding space, E_g has to be replaced by:

$$E_g' = \frac{(1500-T_s)}{(1500-2T_s + T_g)} . \tag{4}$$

Thus the cooling of the gas for temperatures less than 1500 K finally is described:

$$\frac{t^2 dt}{dT_g} = - \frac{m_g \cdot C_{p_g} + \dfrac{(dt/dT)_c}{(dt/dT)_g} \cdot m_c \cdot C_{p_c}}{1.13 \cdot 10^{-4} \cdot G \cdot T_i \cdot C_{p_g} (E_g' T_g^4 - T_s^4)} . \tag{5}$$

Rather extensive estimates indicate that the influence of physical or chemical parameters, like thermal aspects of dissociation, recombination, crystallization, or the temperature dependence of the C_p data may be neglected. Convective and/or turbulent motion of the gas tends to reduce all cooling times when high total masses are involved, especially in combination with high initial temperatures.

RESULTS

Equations (3) and (5) are linked by time and temperature. The calculations have been realized iteratively inserting the following constants:

- The geometry of expansion is a spherical sector with $\sin\beta = 4/5$ ($2\beta = 106°$), thus $G = 5.03$;
- The density of chondrules is 3 g cm^{-3} neglecting temperature dependence and variations in solid or fluid matter;
- $(m/F)_c$ is found to be 0.021 g cm^{-2} with respect to spherical shape and a mean diameter of 0.042 cm (Dodd, 1976); and
- The impact-produced ratio of masses gas/(liquid + solid) is 4/3. This value is given by Öpik (1958b) for impact velocities greater than 15 km/sec as an upper limit. Of course, impacts may be less effective in the production of vapor, but any reduction of the fraction of vapor will result in reduced cooling times of chondrules too, as is evident from eq. (2) and eq. (5).

The results for some selected values are presented in Fig. 1 and listed in Table 1.

Table 1. Input parameters and cooling periods for three selected cases of impact.*

		Case 1	Case 2	Case 3
Initial temperature T_i of gas	[K]	2,000	20,000	150,000
Speed of expansion	[km sec^{-1}]	1.73	5.47	14.98
Mass of condensed matter	[g]	10,000	10^{11}	10^{15}
Mass of gas	[g]	13,300	$1.33 \cdot 10^{11}$	$1.33 \cdot 10^{15}$
Cooling of gas from T_i to 1500 K, eq. (2)	[sec]	0.016	1.91	21
Cooling of gas from 1500 K to 700 K, eq. (4)	[sec]	0.038	2.65	30.2
Cooling of gas from 700 K to 400 K, eq. (4)	[sec]	0.012	3.14	35.6
Cooling of chondrules from 1500 K to 700 K, eq. (3)	[sec]	5.0	6.77	36.6
Cooling of chondrules from 700 K to 500 K, eq. (3)	[sec]	9.8	10	25.6

* Black body temperature of the surrounding space is 200 K.

CONCLUSION

If chondrule formation by impact on the surface of a parental body would have been an abundant event, any chemical, petrological, or experimental study has to be in agreement with the cooling time of these particles. In any case, it is evident from Fig. 1 that the cooling period is extremely short, although depending on several parameters like impact velocity and masses of colliding bodies.

With respect to Table 1 the temperature interval from 1500 K, the lower liquidus temperature of chondrules, down to 500 K, the temperature where diffusion within the chondrules is more or less hampered, is passed very rapidly in less than 60 seconds. From the studies of Nagahara (1981), Rambaldi (1981), and Rubin *et al.* (1982), it seems evident that some chondrules were not necessarily crystallized from a completely molten liquid. This means for an impact model that the temperature of the impact products did not exceed the liquidus of the bulk chemistry of these chondrules, or the high temperature period of the gas has been too short. This is consistent with our calculations for the quick cooling of the gas from T_i down to 1500 K (Table 1; 0.016 to 21 seconds). Blander *et al.* (1976) and Tsuchiyama *et al.* (1980) compared the textures of natural chondrules with synthetic spherules. The results point to cooling rates of chondrules 50–120 K min^{-1} when crystallization occurred. Hewins *et al.* (1981) reported on experiments with synthetic spherules and concluded that pyroxene excentroradial chondrules could have formed by cooling from the liquidus at rates of about 10 K/hr to greater than 3000 K/hr.

Planner and Keil (1982) proposed a three-stage model for the cooling history of olivine-porphyritic fluid droplet chondrules found in ordinary chondrites. Experimental investigations of thermal parameters in the pre-accretionary chondrules environment point to an initial continuous cooling from at least liquidus temperature at rates of 300–4000 K/hr to about 1600 K, a subsequent duration isothermal event at that temperature, and finally quenching. Results of these experiments are not in agreement by at least one order of magnitude with the cooling history of impact products.

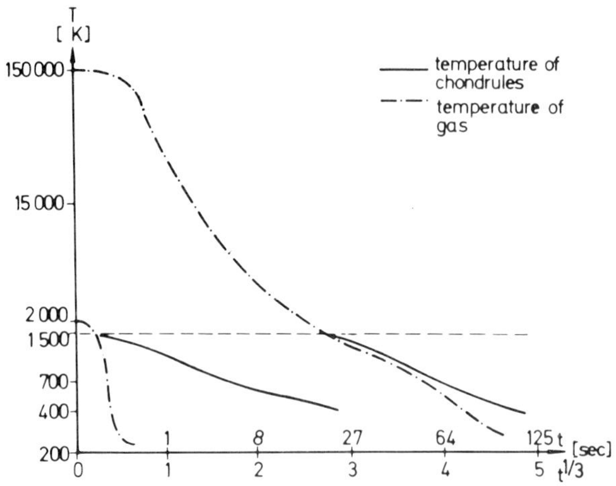

Fig. 1 Time-temperature dependence for two selected cases of Table 1 at T_s of gas is 200 K. Upper curves: $T_i = 2{,}000$ K, $m_g = 13{,}300$ g, $m_c = 10{,}000$ g; lower left curves: $T_i = 150{,}000$ K, $m_g = 1.33 \cdot 10^{15}$ g, $m_c = 10^{15}$ g.

The oxygen isotope ratios of chondrules have been used to support the impact history (Kurat, 1982). The oxygen isotope ratios of the precursor materials may vary as much as 10‰. Since the gas cools faster than the condensed matter, there is no possibility of establishing strictly limited values of oxygen isotope ratios for chondrules of various classes (Clayton *et al.*, 1976). Recently it has been shown by Gooding *et al.* (1980) that isotopic variations might be explained by mixing of mafic and felsic components if chondrules form by melting of pre-existing, heterogeneous materials. Caffee *et al.* (1982) also point in this direction, without specifying in detail the different precursor materials. It should be noted here that these authors state that most tests to find trends between isotopic structures and petrological parameters are of questionable significance.

Concerning a pressure range of 10^{-3} to 10^{-7} atm for the impact gas cloud (T_g less than 1500 K), O'Keefe (1976) has demonstrated that evaporation of FeO is faster than diffusion within a chondrule. As a consequence glass chondrules, produced by impact and having FeO homogeneously distributed, should exhibit a depletion of FeO content at the rims. Since hydrosilicates and/or crystal water cannot be ruled out either as constituents of the projectiles or as parts of the surface layers of the target bodies, the marginal water content of chondrules has to be inspected in the light of the impact model. Water will be dissolved in the melt during the impact peak pressure of up to and greater than 1 Mbar (Cintala *et al.*, 1979). Following O'Keefe (1964) a temperature of 2500 K should be effective for at least 60 seconds to drive out water by diffusion.

To avoid complications due to the postulation of an additional event of accretion, the formation of chondrules by impact is further restricted by the need to retain the impact products on the meteorite parent body. Experiments with impact velocities of more than 6 km/sec show no significant melting of the colliding materials (Gault *et al.*, 1963). A steel projectile impacting the Bjurböle chondrite with 4.5 km/sec (Fredriksson and De Carli, 1964) produced veins by partial melting of compressible materials, e.g., feldspar and troilite. Allowing scaling difficulties in extrapolation and missing knowledge of target strength a mean collision speed for asteroids of 5 km/sec (Davis *et al.*, 1979) may be high enough to produce vapor and melt. Nevertheless, there is no question that, at these impact speeds, targets with diameters less than 800 km are eroded (Marcus, 1969; Davis *et al.*, 1979). As a consequence a comparably thin layer of chondrules may cover a chemically related meteorite parent body essentially free of chondrules; however, this is not likely.

Properly setting the $(m/F)_c$ value the model calculation is useful to evaluate the cooling history of lunar type "chondrules," glassy spheres, or agglutinates. Finally, we agree with Taylor *et al.* (1982) that some molten droplets, glass spheres, and quench-crystallized impact melts in lunar soils, regolith breccias, and even meteorites can certainly form by impact; however, most chondrules did not.

Acknowledgments. We thank D. D. Clayton, R. H. Hewins, and A. Kracher for useful suggestions concerning the manuscript, as well as valuable scientific comments. Parts of this article were presented at the 43rd Annual Meteoritical Society Meeting, San Diego, 1980.

REFERENCES

Blander M., Planner H. N., Keil K., Nelson L. S., and Richardson N. L. (1976) The origin of chondrules: experimental investigation of metastable liquids in the system Mg_2SiO_4-SiO_2. *Geochim. Cosmochim. Acta* **40**, 889–896.

Caffee M. W., Hohenberg C. M., and Swindle T. D. (1982) Isotopic measurements in individual chondrules (abstract). In *Papers Presented to the Conference on Chondrules and their Origins*, pp. 6–8. Lunar and Planetary Institute, Houston.

Cintala M. J., Head J. W., and Wilson L. (1979) The nature and effects of impact cratering on small bodies. In *Asteroids* (T. Gehrels, ed.), pp. 579–600. Univ. of Arizona Press, Tucson.

Clayton R. N., Onuma N., and Mayeda T. K. (1976) A classification of meteorites based on oxygen isotopes. *Earth Planet. Sci. Lett.* **30**, 10–18.

Davis D. R., Chapman C. R., Greenberg R., Weidenschilling S. J., and Harris A. W. (1979) Collisional evolution of asteroids: populations, rotations, and velocities. In *Asteroids* (T. Gehrels, ed.), pp. 528–557. Univ. of Arizona Press, Tucson.

Dodd R. T. (1976) Accretion of the ordinary chondrites. *Earth Planet. Sci. Lett.* **30**, 281–291.

Dodd R. T. (1982) Objects we call chondrules (abstract). In *Papers Presented to the Conference on Chondrules and their Origins*, p. 15. Lunar and Planetary Institute, Houston.

Fredriksson K. (1963) Chondrules and the meteorite parent bodies. *Trans. N.Y. Acad. Sci.* **25**, 756–769.

Fredriksson K. and De Carli P. (1964) Shock emplaced argon in a stony meteorite 1. Shock experiment and petrology of sample. *J. Geophys. Res.* **69**, 1403–1406.

Gault D. E., Shoemaker E. M., and Moore H. J. (1963) Spray ejected from the lunar surface by meteoroid impact. *NASA Tech. Note D-1767*. 28 pp.

Gooding J. L., Keil K., Mayeda T. K., Clayton R. N., Fukuoka T., and Schmitt R. A. (1980) Oxygen isotopic compositions of petrologically characterized chondrules from unequilibrated chondrites. *Meteoritics* **15**, 295.

Grossman L. and Olsen E. (1974) Origin of the high-temperature fraction of C2 chondrites. *Geochim. Cosmochim. Acta* **38**, 173–187.

Hewins R. H., Klein L. C., and Fasano B. V. (1981) Conditions of formation of pyroxene excentroradial chondrules (abstract). In *Lunar and Planetary Science XII*, pp. 448-450. Lunar and Planetary Institute, Houston.

JANAF (1968) *Thermodynamical tables, Third addendum*. Dow Chemical Company, Midland, Michigan.

King T. V. V. and King E. A. (1978) Grain size and petrography of C2 and C3 carbonaceous chondrites. *Meteoritics* **13**, 47–72.

Kurat G. (1982) Impact origin of chondrules (abstract). In *Papers Presented to the Conference on Chondrules and their Origins*, p. 36. Lunar and Planetary Institute, Houston.

Marcus A. H. (1969) Speculations on mass loss by meteoroid impact and formation of the planets. *Icarus* **11**, 76–87.

Nagahara H. (1981) Evidence for secondary origin of chondrules. *Nature* **292**, 135–136.

O'Keefe J. A. (1964) Water in tektite glass. *J. Geophys. Res.* **69**, 3701–3707.

O'Keefe J. A. (1976) *Tektites and their Origin*. Elsevier, N.Y. 254 pp.

Öpik E. J. (1958a) Meteor impact on solid surface. *The Irish Astron. J.* **5**, 14–33.

Öpik E. J. (1958b) On the catastrophic effect of collision with celestial bodies. *The Irish Astron. J.* **5**, 34–36.

Planner H. N. and Keil K. (1982) Evidence for a three-stage cooling history of olivine-porphyritic fluid droplet chondrules. *Geochim. Cosmochim. Acta* **46**, 317–330.

Rambaldi E. R. (1981) Relict grains in chondrules. *Nature* **293**, 558–561.

Rubin A. E., Scott E. R. D., Taylor G. J., and Keil K. (1982) Silicate matrix material in type 3 ordinary chondrites: implications for the origin of chondrules. *Meteoritics* **17**, 275–276.

Taylor G. J., Keil K., and Scott E. R. D. (1982) Chondrules did not form by impact on parent bodies. *Meteoritics* **17**, 288–289.

Tsuchiyama A., Nagahara H., and Kushiro I. (1980) Experimental reproduction of textures of chondrules. *Earth Planet. Sci. Lett.* **48**, 155–165.

Unsöld A. (1974) *Der neue Kosmos*. Springer, Berlin. 438 pp.

Van Schmus W. R. (1969) Mineralogy, petrology and classification of type 3 and 4 carbonaceous chondrites. *Symp. Meteorite Res.* (P. M. Millman, ed.), pp. 480–491. Reidel, Dordrecht.

Ferromagnesian Chondrules in Carbonaceous Chondrites

Harry Y. McSween, Jr., A. Kem Fronabarger, and Steven G. Driese
Department of Geological Sciences, University of Tennessee, Knoxville, Tennessee 37996

Carbonaceous chondrites (groups CO and CV) contain three types of ferromagnesian chondrules that can be distinguished on the basis of mineralogy and texture. Statistical analysis of the bulk chemical compositions of these chondrule types suggests that two of the three types are genetically related. The chondrules in carbonaceous chondrites can thus be separated into two populations. These populations are also apparently distinct in terms of redox state and oxygen isotopic composition. The more abundant population (I) is clearly different from chondrules in unequilibrated ordinary chondrites, but the minor population (II) is similar to them in many respects. These chondrule populations may have formed by melting of solid, locally homogeneous precursor materials in different regions of the nebula, and were mixed after solidification was complete.

INTRODUCTION

Within the past few years, there has been a resurgence of interest in meteoritic chondrules. This research has focused almost exclusively on chondrules in ordinary chondrites, but ferromagnesian chondrules are the most abundant compositional types in both ordinary and carbonaceous chondrites. Although the chondrules in carbonaceous chondrites have many similarities to those in ordinary chondrites, they differ in some important respects. The only recent systematic studies of ferromagnesian chondrules in carbonaceous chondrites were published by McSween (1977a) and Simon and Haggerty (1980). This paper will reevaluate the same chemical data set utilized in McSween's previous study. This data set was previously treated only in graphical form; the application of statistical methods in this paper has resulted in some new insights into chondrule populations in carbonaceous chondrites, as well as new conclusions about chondrule origin at variance with the previous work.

ANALYTICAL METHODS

Some carbonaceous chondrites have experienced aqueous alteration, as in the case of all CM chondrites, or thermal metamorphism, as in the case of a few CO and CV chondrites (McSween, 1979). Such meteorites have been excluded from the present study. Only chondrule analyses from the following relatively unaltered type 3 carbonaceous chondrites (with respective numbers of chondrules in parentheses) are included in this compilation: CV [Allende (14), Arch (8), Bali (17), Efremovka (17), Grosnaja (16), Kaba (14), Leoville (13), Mokoia (6), Vigarano (12)], and CO [Felix (30), Kainsaz (48), Lancé (43), Ornans (29)]. A tabulation of individual analyses may be requested from the senior author.

Chondrule analyses were obtained using an electron microprobe defocused beam procedure described by McSween (1977a). The compositions of glass chondrules were determined from simple averages of random spot analyses covering the chondrules. A modified data reduction procedure (Bower *et al.*, 1977) was employed for coarse-grained chondrules. After each iteration, a provisional norm for the target material was calculated, providing estimates of the compositions of the mineral phases in which each element was cited; in succeeding iterations, corrections were calculated for each element on the basis of this heterogeneous mixture of hypothetical minerals. This procedure is not rigorously correct, but for heterogeneous targets it yields accurate analyses if the normative and modal mineralogies are similar (Albee *et al.*, 1977). The only significant difference between normative and modal phases in ferromagnesian chondrules is that normative feldspar components are actually cited in interstitial glass. However, because these glass compositions are generally similar to feldspars (McSween, 1977b), this is not a significant source of error. Lux *et al.* (1980) tested the accuracy of this procedure against chondrules analyzed by INAA and found good agreement.

Chondrule bulk compositions (361 chondrules in 13 meteorites) obtained by the procedure above were previously presented only in graphical form (McSween, 1977a). In this paper various statistical methods have been employed. Programs by Barr *et al.* (1976) and Jennrich and Sampson (1981) were utilized for factor analysis and stepwise discriminant analysis, respectively.

PETROGRAPHIC FEATURES OF CHONDRULES IN THIS STUDY

Chondrule types

The chondrules considered here are distinct from inclusions and mineral fragments in that all are polymineralic and have internal textures that testify to solidification of molten material, but some are drop-formed and others have irregular shapes. McSween (1977a) devised a working classification for chondrules in carbonaceous chondrites, based on texture and mineralogy. A slight modification of this classification, summarized in Table 1, is adopted here. More detailed

Table 1. Petrologic classification of ferromagnesian chondrules in carbonaceous chondrites.

	Texture	Mineralogy	Modal Abundance Ranges*	
			CO	CV
Type I	Granular	Olivine (Fo_{90-100}) ± pyroxene (En_{90-100}) ± spinel ± metal or magnetite with colorless interstitial glass	25–39	35–51
Type II	Microporphyritic or barred	Olivine (strongly zoned from Fo_{85-50}) + Cr-rich hercynite with dark brown interstitial glass	<3	<1
Type III	Excentroradial or glassy	Pyroxene (En_{90-100}) + light brown to yellow glass	<2	Trace

*Modal data in vol.% of the meteorite, determined from point counts on polished thin sections.

descriptions of the chondrule types were published by McSween (1977c, d). The textural and mineralogical properties of each chondrules type provide sufficient differences to allow petrographic classification. The type I chondrules were previously subdivided into those that contained opaque minerals (metal or magnetite) and those that did not. Chemical data cannot differentiate between these groups, and because this study deals with chondrule chemistry, the distinction between opaque-bearing and opaque-absent type I chondrules has not been continued. We also note in Table 1 that type II chondrules have two very different textures: microporphyritic or barred. It is possible that these may represent two distinct types, but their mineralogies and bulk compositions are indistinguishable. Microporphyritic chondrules occur more commonly in CO and barred occur more commonly in CV chondrites. McSween (1977a) also recognized type IV chondrules, which have refractory compositions similar to Ca, Al-rich inclusions and probably represent their melted equivalents; this paper will focus only on ferromagnesian chondrules.

Chondrule abundances

The modal abundance ranges for various chondrule types are presented in Table 1. Abundance data for nine CV chondrites are from McSween (1977c). The ranges for nine CO meteorites include new data for the ALHA 77003, 77029, and 77307 chondrites (McSween, unpublished data), newly discovered Antarctic representatives of this group (Scott *et al.*, 1981), as well as previously published CO data (McSween, 1977d). There are significant differences between the relative abundances of the three chondrule types in carbonaceous chondrites. However, the relative abundances may have been changed prior to or during accretion, as chondrules in these meteorites show clear evidence of sorting by size (King and King, 1978).

Additional petrographic observations

Some double and even triple chondrules have been observed in carbonaceous chondrites (McSween, 1977a). These were welded together while still hot or in some cases partially molten, as the join may serve as a point of crystal nucleation for both droplets. When multiple chondrules occur, they are almost invariably of the same type. Similar relationships have been observed among chondrules in ordinary chondrites (Lux *et al.*, 1981), although a few exceptions have been noted (Gooding and Keil, 1981).

Relict grains of olivine, or less commonly orthopyroxene, that survived the chondrule formation process without melting have been described in chondrules of ordinary chondrites (Nagahara, 1981; Rambaldi, 1981). These grains typically are dusted with metal inclusions and exhibit reverse Mg-Fe zonation. Kracher *et al.* (1983) described several similar grains in chondrules of the Vigarano CV chondrite. However, relict grains in the chondrules of carbonaceous chondrites must be much less common than in ordinary chondrites or most must have a different appearance.

CHONDRULE COMPOSITIONS

Chemical and normative compositions

Average chemical compositions and standard deviations for each chondrule type are presented in Table 2. The various types of chondrules in carbonaceous chondrites appear to have distinctive compositional ranges, as illustrated by variations in MgO, FeO, and SiO_2 (Fig. 1). However, some overlap occurs among the three groups, particularly between types I and III. These data include some multiple chondrules, which are always of the same type and composition.

Figure 2 shows analyses of chondrules for Ni, Mn, and Fe (calculated as oxides). These three elements can exist in several oxidation states and thus respond differently to changes in redox conditions. Nickel contents of type I chondrules are highest, as expected for chondrules that contain metal, and type II chondrules lie off the NiO/FeO trend of the other two types. The MnO/FeO pattern for type II chondrules is also distinct, but type I and III chondrules overlap. These differences are interpreted as fractionations that occurred under different redox conditions. [These elements also have different condensation temperatures, and therefore could be fractionated during condensation. However, separation of Fe and Ni (Fig. 2) in different chondrule types

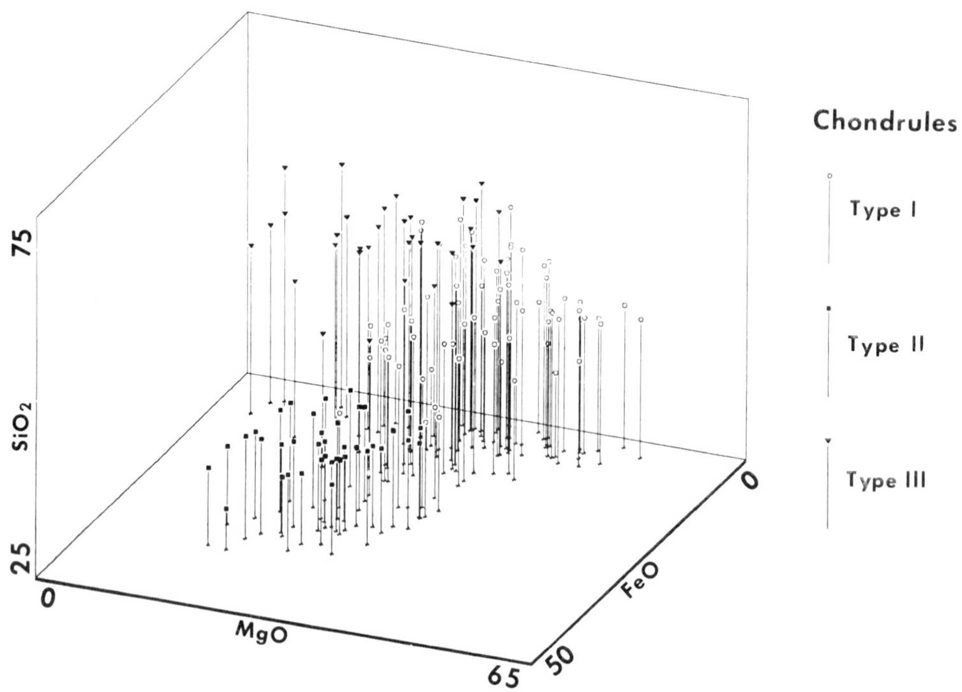

Fig. 1 FeO vs. MgO vs. SiO_2 (wt.%) for chondrules in carbonaceous chondrites.

may require condensation under different redox conditions.] Variations in oxidation state are also reflected in the mineralogy of type I and II chondrules (McSween, 1977b). Metal is absent, and olivines and spinels contain distinctly more Fe^{2+} in type II chondrules, reflecting their higher

Table 2. Average chondrule compositions in carbonaceous chondrites.

Wt. % oxides	Type I	Type II	Type III
SiO_2	44.3 (47)†	36.8 (21)	52.4 (41)
TiO_2	0.18 (10)	0.12 (7)	0.28 (17)
Al_2O_3	4.24 (261)	3.29 (183)	7.65 (493)
Cr_2O_3	0.42 (20)	0.39 (23)	0.54 (24)
FeO*	9.25 (594)	27.8 (61)	7.45 (377)
MnO	0.14 (10)	0.24 (7)	0.29 (17)
MgO	38.2 (74)	27.0 (69)	23.0 (91)
CaO	2.89 (172)	1.91 (143)	7.26 (380)
Na_2O	0.60 (54)	1.35 (98)	1.81 (150)
K_2O	0.05 (5)	0.06 (5)	0.08 (11)
Ni	0.32 (35)	0.12 (12)	0.09 (12)
S	0.48 (95)	0.19 (22)	0.14 (19)
Mg/(Mg + Fe)‡	0.88	0.63	0.85
No. analyses	282	42	37
Norms§			
or	0.29	—	0.47
ab	5.05	—	15.17
an	8.68	2.78	12.39
ne	—	6.28	—
lc	—	0.28	—
di	4.41	1.96	18.53
hy	10.49	—	29.31
ol	69.22	87.52	22.55
ilm	0.34	0.23	0.53
chr	0.62	0.59	0.79
py	0.89	0.36	0.26
Normative Mineral Compositions			
plag	Or_2Ab_{36}	An_{100}	Or_2Ab_{54}
di	$Wo_{53}En_{41}$	$Wo_{51}En_{28}$	$Wo_{52}En_{37}$
hy	En_{86}	—	En_{82}
ol	Fo_{85}	Fo_{54}	Fo_{80}

* Total Fe as FeO.
† Numbers in parentheses represent one standard deviation in terms of least unit given, i.e., 44.3 (47) is ± 4.7.
‡ Molar ratios.
§ CIPW norms in wt.%.

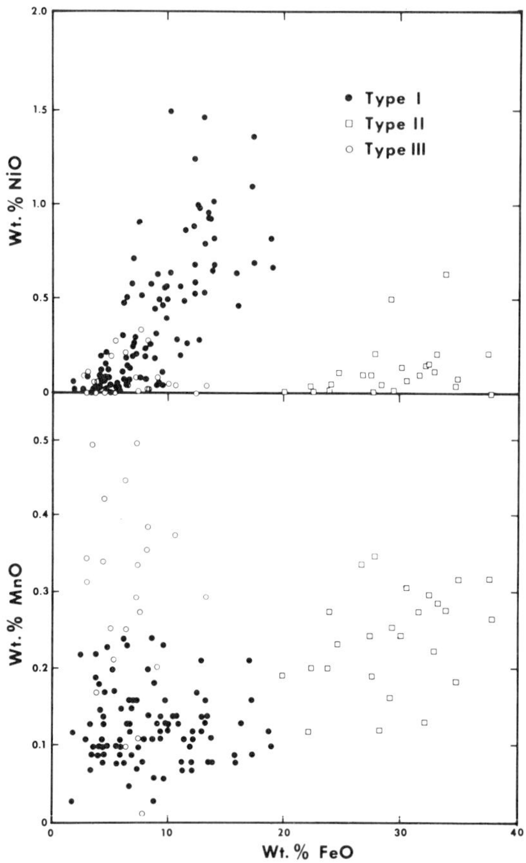

Fig. 2 NiO and MnO vs. FeO (wt.%) for chondrules in carbonaceous chondrites.

oxidation state. Differences in redox conditions established during droplet crystallization might produce different mineral compositions, but it is difficult to see how this could alter bulk compositions. Thus fractionations under different redox conditions must have occurred in the chondrule precursor materials prior to the chondrule-forming event. Type II chondrules must have formed from more highly oxidized material than types I or III.

Another way of emphasizing the differences between the compositions of chondrule types is by comparison of calculated CIPW norms (Table 2). It should be noted that these are "oxidized" norms, as a portion of Fe has not been allotted to metal in type I chondrules. This was not done because the proportion of Fe in metal is unknown, and because we wish to compare the compositions of these chondrules on equal bases. The entire chondrule population in carbonaceous chondrites varies in normative olivine/pyroxene ratio at roughly constant feldspar plus feldspathoid content (Fig. 3). However, there are significant differences between types I, II, and III. Type I chondrules show a much wider range of normative mineralogy than other types. This is consistent with their highly variable modal proportions of olivine and pyroxene. Relative to type I,

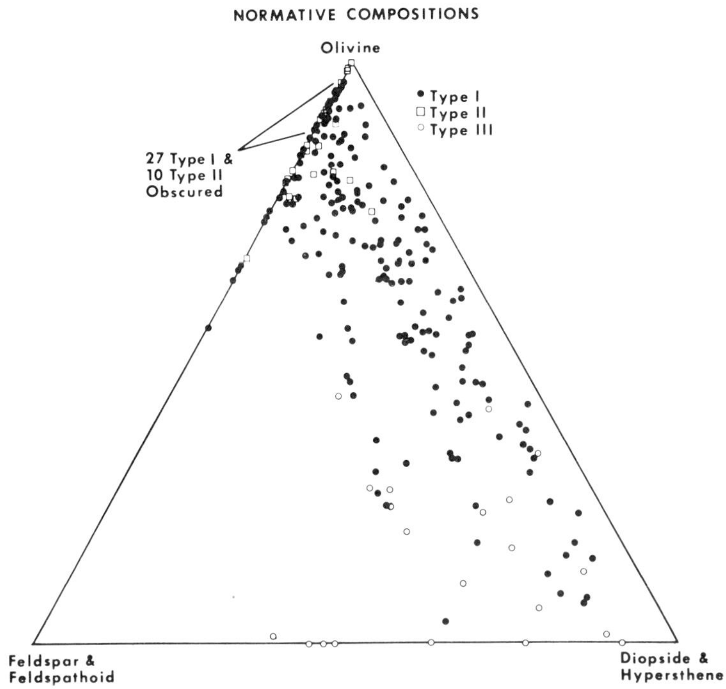

Fig. 3 CIPW normative components for chondrules in carbonaceous chondrites.

most type II chondrules contain almost no pyroxene component and appreciable amounts of feldspathoid, reflecting their low silica contents. These chondrules contain no modal pyroxene, and normative feldspathoid and plagioclase are represented modally by interstitial glass. Type III chondrules contain the highest proportions of pyroxene component, but have some similarities to type I. These chondrules contain abundant modal pyroxene, but no modal olivine.

Statistical analysis

In order to assess the significance of differences in all the analyzed elements in types I, II, and III chondrules, both factor analysis and stepwise discriminant analysis were employed. Factor analysis attempts to reduce the number of variables needed to define a population. This is accomplished by combining the most highly correlated variables into simple linear equations, or factors. No initial judgement is made about whether a specific chondrule belongs to one type or another, and chondrules of the same type should have similar factor scores. Factor analysis is deemed successful if a large number of variables can be reduced to a few factors, and each individual variable is associated with a minimum number of factors. Discriminant analysis, on the other hand, maximizes the difference between pre-defined populations by developing linear equations for each population. Thus, each chondrule type is defined by a discriminant function, and an assessment is made of how well these functions classify each chondrule. Discriminant analysis is judged successful if only a few samples are incorrectly classified by their respective discriminant functions.

Prior to applying these multivariant analysis techniques, we ascertained that no systematic increase in coefficient of variance occurred with increasing arithmetic mean for each oxide. For this reason, arithmetic (unnormalized) rather than geometric (normalized) means were calculated for average chondrule compositions (Table 2). Because our results of factor and discriminant tests using unnormalized data were successful, additional analyses employing normalized data were judged unnecessary.

The results of R-mode factor analysis are shown in Table 3. The twelve analyzed elements are reduced to four rotated factors with eigenvectors greater than 1.0, and there are noticeable variations among element loadings for the different factors. F1 is dominated by MgO, TiO_2, and feldspar components (Na_2O, CaO, and Al_2O_3); F2 is controlled by FeO and SiO_2; F3 is most affected by MnO and Cr_2O_3; and F4 is controlled by S and Ni. These four factors account for 73.7% of the total variance (Table 3). A plot of factor scores for F1 vs. F2 is presented in Fig. 4. F1 partially discriminates between the fields of chondrule types I and III, and F2 differentiates the field of type II chondrules from I and III. Plots involving F3 and F4 provide less resolution than F1 or F2.

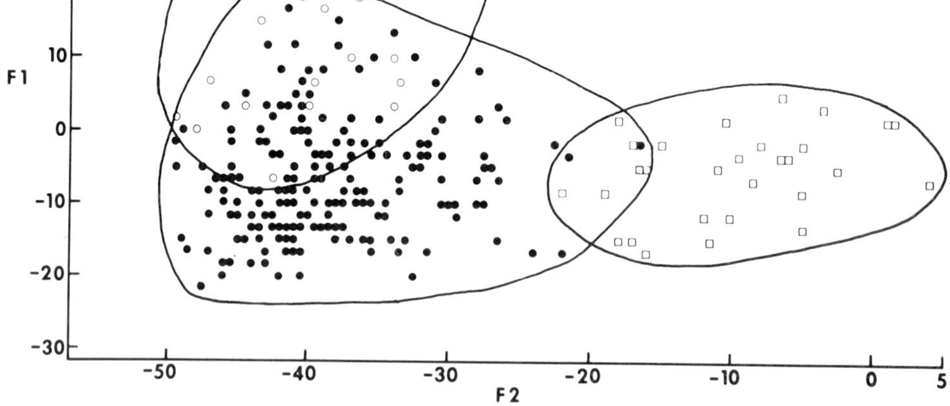

Fig. 4 An example of factor analysis for chondrules. The coefficients for factors F1 and F2 are defined in Table 3.

Table 3. Factor and discriminant analyses for chondrules.

	Na_2O	MgO	Al_2O_3	SiO_2	S	K_2O	CaO	TiO_2	Cr_2O_3	MnO	FeO	Ni	%Total Variance Explained
						*Factor Analysis for Chondrules**							
F1	0.6452	−0.6211	0.9156	0.2860	0.0608	0.5099	0.8695	0.8390	−0.0455	0.0499	−0.1977	−0.0308	31.9
F2	0.4681	−0.3433	−0.0315	−0.6928	0.0996	0.5161	−0.1707	−0.1891	−0.1071	0.1196	0.8359	0.0401	17.1
F3	0.2637	−0.5897	−0.0023	0.3839	0.0140	−0.0602	0.2326	0.0008	0.7273	0.8149	0.3055	−0.0077	13.3
F4	−0.1812	−0.0108	−0.0149	−0.2105	0.8132	−0.0053	−0.0070	0.0520	0.2640	−0.2490	0.1079	0.8218	11.4
						Discriminant Analysis for Chondrules†							
	Constant												
D1	−148.39	5.02					11.13	92.35	27.65	69.70	3.65	−2.06	
D2	−165.85	4.97					10.70	100.71	17.88	77.40	5.00	−13.24	
D3	−140.79	4.70					11.92	77.98	25.18	89.93	3.41	−4.50	
						Canonical Equation Coefficients							
C1	1.123	0.005					0.086	−1.674	1.717	−1.128	−0.244	1.968	
C2	−3.609	0.088					−0.231	4.165	0.592	−5.650	−0.081	0.567	

* Equimax rotated principal axis factor method, analysis program by Barr *et al.* (1976).
† Stepwise discriminant analysis program by Jenrich and Sampson (1981).

There are some inherent difficulties in the interpretation of factor analyses. Because chemical analyses theoretically add to 100%, any one of the oxides in an analysis is a dependent variable of the others. Koch and Link (1971) showed that this "constant sum" problem does not appreciably affect factor analysis because the method, by its very nature, is searching for independent linear combinations of variables. Le Maitre (1982) discussed problems in making petrogenetic interpretations of R-mode factors. As there is no unique solution to the initial factor matrix, an indeterminate set of factor scores and loadings can be generated. Factor analysis is a valuable tool, however, in that it examines the degree of correlation among the variables, identifying the minimum number of independent variables needed to characterize the data. For these reasons, we have not attempted to assign mineralogic or petrologic significance to the individual factors, but have used two-dimensional plots of factor scores only to separate the data into chondrule types.

Stepwise discriminant analysis results are also shown in Table 3. The order in which oxides are admitted into the discriminant functions is predicated on their ability to differentiate between the three chondrule types. This order of entrance was limited by a cutoff value for each element which we arbitrarily set so as to provide a confidence level of > 98% for the last element admitted. The components meeting this screen were FeO, MgO, Ni, MnO, Cr_2O_3, TiO_2, and CaO; the remaining oxides were not used in the analysis. The discriminant functions were effective in classifying each individual chondrule as to type I, II, or III. Assignment of chondrules to type I was correct in 95% of all cases, to type II in 93%, and to type III in 88%. Most of the incorrect classifications were between chondrules of types I and III. An F-matrix was also generated, as another means of testing the confidence level at which the discriminant functions classified the chondrule types. These data indicated that all three types were discriminated above the 99% confidence level. The F-matrix scores were greatest for types I vs. II or II vs. III, indicating that type II chondrules are most clearly resolved from types I and III.

Two canonical functions (Table 3) that maximized the differences between chondrule types in two-dimensional space were calculated from the discriminant function data. In a plot of these functions (Fig. 5), type I and III chondrules define a continuous, slightly overlapping population, whereas type II chondrules are clearly resolved. Even though the canonical variables do not completely separate types I and III, the discriminant functions accurately differentiate between them in most cases.

Chondrule populations

Factor analysis and discriminant analysis both lead to the same conclusion: that carbonaceous chondrites contain three recognizable types. However, these three types can probably be grouped into two independent *populations* of chondrules, as the two types have many similarities. One population, consisting only of type II chondrules and thus called population (II), is clearly resolved by both statistical methods (Figs. 4 and 5). The other consists of poorly resolved types I and III chondrules (Figs. 4 and 5) and henceforth will be called population (I). Types I and III, i.e., population (I), chondrules also appear to be related in terms of the redox state of their precursor

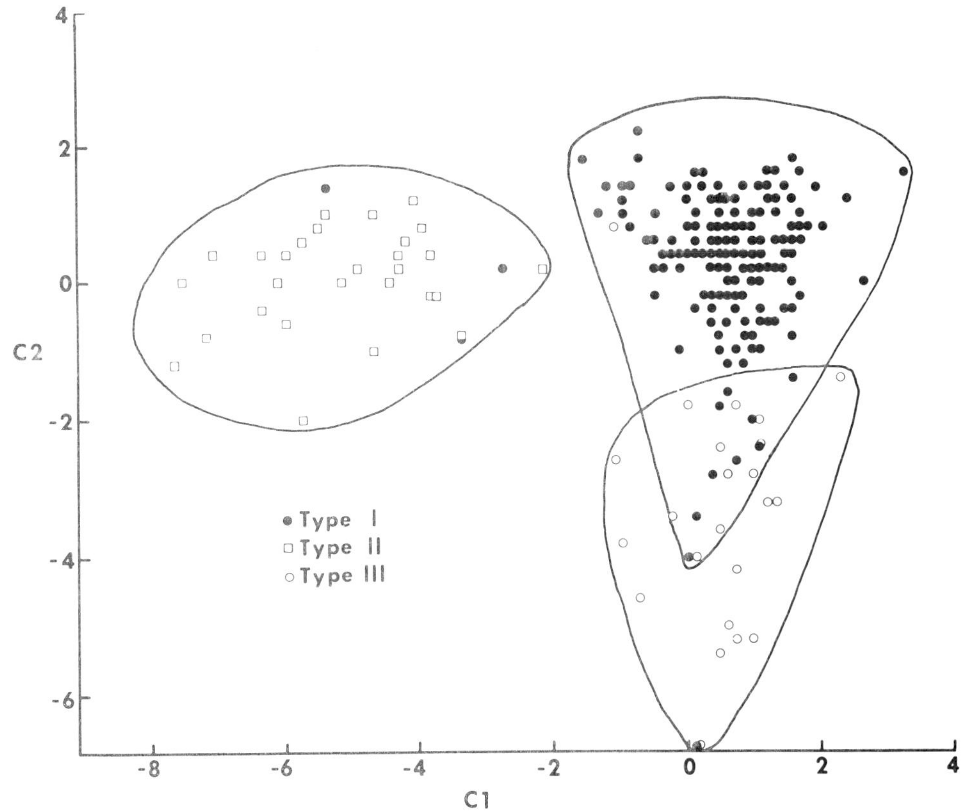

Fig. 5 Results of discriminant analysis for chondrules. The canonical variables C1 and C2 are calculated as the sum of the constants and oxide-coefficient products given in Table 3.

materials, and these are clearly distinct from the more oxidized population (II). The petrographic distinction between types I and III is primarily textural, but a textural change from excentroradial to granular may occur within a continuous compositional sequence of chondrules. Comparison of the normative mineralogy of types I and III chondrules (Table 2) suggests that type III chondrules could have formed from more highly differentiated but related protolithic material relative to type I.

COMPARISON OF CHONDRULES IN CARBONACEOUS AND ORDINARY CHONDRITES

The weighted average compositions for the two chondrule populations in thirteen carbonaceous chondrites are compared to the average chondrule compositions for unequilibrated ordinary chondrites in Table 4. The latter represent data from three H3 chondrites by Lux *et al.*

Table 4. Comparison of chondrule populations in carbonaceous and ordinary chondrites.

Wt. % Oxides	Carbonaceous Population (I)*	Carbonaceous Population (II)	H3 Chondrites†	L3 chondrites ‡
SiO_2	45.2	36.8	50.3	46.0
TiO_2	0.19	0.12	0.14	0.13
Al_2O_3	4.64	3.29	3.23	3.35
Cr_2O_3	0.43	0.39	0.54	0.60
FeO§	9.04	27.8	8.82	16.4
MnO	0.16	0.24	0.33	0.41
MgO	36.4	27.0	29.7	28.0
CaO	3.40	1.91	2.18	2.05
Na_2O	0.74	1.35	1.40	1.07
K_2O	0.05	0.06	0.18	0.22
No. analyses	319	42	87	23

* Weighted average of type I and III chondrules in Table 2.
† Data from the Sharps, Tieschitz, and Bremervörde chondrites by Lux *et al.* (1981).
‡ Data from the Manych chondrite by Dodd (1978).
§ Total Fe as FeO.

(1981) and one L3 chondrite by Dodd (1978). These two data sets were chosen for comparison because they included most of the same elements analyzed in chondrules of carbonaceous chondrites. Both data sets were also obtained by microprobe defocused beam analysis, though Dodd (1978) did not apply corrections for heterogeneous targets to his data. Both H3 and L3 chondrule data sets are very similar, in agreement with the conclusion of Gooding *et al.* (1980) that unequilibrated ordinary chondrites of different chemical groups (H, L, LL) contain indistinguishable chondrule populations. The only noticeable differences in the data for H and L chondrites is in FeO content (Table 4). Gooding *et al.* (1980) reported a mean FeO content of 13.9 wt.% for chondrules from four H, L, LL3 chondrites, intermediate between that of Lux *et al.* (1981) and Dodd (1978), and their reported standard deviations were higher for iron than for any other major element. Neither of these data sets was obtained using random sets of chondrules, and this may account for the difference in FeO values. Within each ordinary chondrite, there may also exist two distinct chondrule populations, one porphyritic and one non-porphyritic (Gooding *et al.*, 1980); however, the chemical differences between these are mainly in refractory and siderophile elements, many of which are in trace quantities. Since corresponding data are not available for chondrules in carbonaceous chondrites, we have not considered these possible groups in ordinary chondrites separately.

Scott *et al.* (1982) have emphasized the similarities between the components (including chondrules) of carbonaceous and unequilibrated ordinary chondrites. However, some important differences are apparent in Table 4. A comparison of the compositions of population (I) chondrules in carbonaceous vs. ordinary chondrites indicates the following: both have equivalent amounts of feldspar components, but CaO contents in carbonaceous chondrite population (I) are higher and alkalis are lower; and chondrules in population (I) have much higher MgO/FeO ratios than those in ordinary chondrites.

A comparison of population (II) chondrules with those in ordinary chondrites indicates some strong similarities (Table 4). Most oxide concentrations are very similar, except for lower FeO and higher SiO_2 in ordinary chondrites. The difference in FeO contents could be explained if chondrules in ordinary chondrites and population (II) formed from similar materials, but chondrules in ordinary chondrites experienced reduction during melting and subsequent fractionation of metal. Grossman (1982) suggested that reduced Fe metal beads have been lost from chondrules in ordinary chondrites, and a general depletion of siderophile elements in chondrules relative to cosmic values (Gooding *et al.*, 1980) is consistent with this idea. Loss of Fe would also possibly explain the higher contents of oxides other than FeO (especially SiO_2 and MgO) in the chondrules of ordinary chondrites. Reduction and loss of FeO from population (II) to ordinary chondrite levels (assumed 13.9 wt.%, Gooding *et al.,* 1980) would cause increases of 5.9% in SiO_2, 4.4% in MgO, and only modest increases in other oxides to approximate the values in the chondrules of ordinary chondrites.

Gooding *et al.* (1982) reported oxygen isotopic compositions of individual chondrules in unequilibrated ordinary chondrites. They found that such chondrules scatter about a line with slope ~1 on a three-isotope diagram, and that the isotopic compositions bear no relationship to the chemical group of the parent meteorite. Clayton *et al.* (1983) reported new isotopic data for chondrules in the Allende carbonaceous chondrite. Chondrules whose textural and mineralogical descriptions suggest they are of population (I) (one type III and numerous type I chondrules) also define a line of slope ~1, but it is significantly displaced from the regression line for chondrules in ordinary chondrites. Barred olivine, i.e., type II or population (II), chondrules in Allende define a distinct grouping that plots between the two previously described populations. These trends might be explained by partial equilibration of solid chondrule precursor materials rich in ^{16}O with two different gaseous reservoirs depleted in ^{16}O. One of these reservoirs would plot between the chondrules in ordinary chondrites and population (II), and the other on an extension of the population (I) trend. Thus population (II) and chondrules in ordinary chondrites, which also share some chemical similarities, may have equilibrated with the same isotopic reservoir and thus formed in similar environments. In this regard, population (II) chondrules may be considered as interlopers that were transported into the carbonaceous chondrite accretion zone, within which the more abundant population (I) chondrules were typical.

CONSTRAINTS ON CHONDRULE ORIGIN

In carbonaceous chondrites, there appear to be two distinct populations of chondrules, differing in chemical and mineralogical composition. Variations in Fe, Ni, and Mn contents may reflect fractionations at different oxidation states that were established in the precursor materials before chondrule formation. These differences are difficult to reconcile with a condensation origin (e.g., McSween, 1977a), though condensation of metastable liquids (Blander and Katz, 1967) might be possible. Chondrules were more likely produced by melting solid precursor materials that in some cases had been processed from their presumed nebular starting compositions.

The present populations of chondrules in carbonaceous chondrites experienced mechanical sorting prior to assembly (King and King, 1978). Such sorting is unlikely to have occurred during accretion, as fine-grained matrix was accreted along with coarser chondrules. No genetic significance can be attached to the limited size range or relative abundances of chondrule types, because these were probably affected by the sorting process. The distinction between droplet and lithic chondrules in these meteorites may also be meaningless because lithic chondrules may have been larger droplet chondrules that were broken to meet size-sorting requirements.

Multiple chondrules that obviously formed in close proximity are almost always of the same type and composition (McSween, 1977a). Thus, the chondrule-forming process probably did not sample a coarse-grained, heterogeneous protolith on a small (non-representative) scale (e.g., Dodd, 1978). Such a process would produce chondrules of varying compositions, but these would form together and thus occasionally be attached as multiple chondrules. Another argument against this process is the observation that differences in chondrule compositions within a population cannot be explained by differences in the proportions of melted phases of fixed composition. For example, type I and III chondrules within population (I) differ not only in the proportions of normative feldspar, pyroxene, and olivine components, but also in the normative compositions of these minerals (Table 2). These arguments mitigate against formation of chondrules by impacts into regoliths, unless these were chemically and isotopically homogeneous on a small scale but heterogeneous on a large scale. More likely, chondrules probably formed by melting of locally homogeneous (i.e., fine-grained) material in the nebula, before being mixed with other populations during accretion. This interpretation may conflict with the observation of coarse relict grains in chondrules, but such large grains may have been unusual in this material (possibly explaining their survival during melting). Rubin *et al.* (1982) have argued from the observation of matrix lumps in chondrules that fine-grained matrix was the chondrule precursor. Although it seems possible that many of these lumps are of secondary origin, having filled chondrule surface irregularities and interior voids connected to the surface, something like matrix would have been a suitable precursor material.

Population (I) chondrules are apparently chemically and isotopically distinct from population (II) and the chondrules in ordinary chondrites. The latter two chondrule groups may have formed in proximity to each other in a heterogeneous nebula, as they can be interpreted as having partially equilibrated with the same gaseous reservoir. An alternative is that both formed at similar times in a homogeneous nebular gas with an evolving isotopic composition. Clayton *et al.* (1983) suggested that the porphyritic (population I) chondrules may have been only partially molten and the barred (population II) chondrules were completely molten, allowing the latter to experience more complete isotopic exchange with the same gas. The high FeO/MgO contents of population (II) chondrules would lower their solidus temperatures and facilitate melting. However, it is not known whether such partial equilibration took place with the solid precursor materials or while the chondrules were molten. It is difficult to envision how chondrule melts suspended in reasonable nebula gas compositions could become so oxidized as population (II), or even remain so oxidized during melting of an oxidized solid precursor. It is clear, however, that the nebula in which chondrules formed was a complex place, containing solids of different chemical composition and redox state and possibly gases of different isotopic composition. The many similarities between all of these chondrule populations suggest that all formed by a similar process, but different solid precursor materials were clearly involved.

Acknowledgments. We are grateful to J. L. Gooding, M. R. Smith, and J. N. Grossman for thoughtful reviews. This work was partly supported by NASA grant NSG 7413.

REFERENCES

Albee A. L., Quick J. E., and Chodos A. A. (1977) Source and magnitude of errors in "broad-beam analysis" (DBA) with the electron probe (abstract). In *Lunar Science VIII*, pp. 7–9. The Lunar Science Institute, Houston.

Barr A. J., Goodnight J. H., Sall J. P., and Helwig J. T. (1976) *A User's Guide to SAS 76*, pp. 112–119. SAS Institute, Raleigh.

Blander M. and Katz J. L. (1967) Condensation of primordial dust. *Geochim. Cosmochim. Acta* **31**, 1025–1034.

Bower J. F., Wood J. A., Richardson S. M., McSween H. Y., and Ryder G. (1977) Rock compositions by defocussed beam analysis (abstract). In *Abstracts of VIII International Conf. on X-ray Optics and Microanalysis*, Boston.

Clayton R. N., Onuma N., Ikeda Y., Mayeda T. K., Hutcheon I. D., Olsen E. J., and Molini-Velsko C. (1983) Oxygen isotopic compositions of chondrules in Allende and ordinary chondrites. This volume.

Dodd R. T. (1978) Compositions of droplet chondrules in the Manych (L-3) chondrite and the origin of chondrules. *Earth Planet. Sci. Lett.* **40**, 71–82.

Gooding J. L. and Keil K. (1981) Relative abundances of chondrule primary textural types in ordinary chondrites and their bearing on conditions of chondrule formation. *Meteoritics* **16**, 17–43.

Gooding J. L., Keil K., Fukuoka T., and Schmitt R. A. (1980) Elemental abundances in chondrules from unequilibrated chondrites: evidence for chondrule origin by melting of pre-existing materials. *Earth Planet. Sci. Lett.* **50**, 171–180.

Gooding J. L., Mayeda T. K., Clayton R. N., Keil K., Fukuoka T., and Schmitt R. A. (1982) Oxygen isotopic compositions of chondrules in unequilibrated chondrites: Further petrological implications (abstract). In *Lunar and Planetary Science XIII*, pp. 271–272. Lunar and Planetary Institute, Houston.

Grossman J. N. (1982) Mechanisms for the depletion of metal in chondrules (abstract). In *Papers Presented to the Conference on Chondrules and their Origins*, p. 23. Lunar and Planetary Institute, Houston.

Jennrich R. and Sampson P. (1981) P7M, Stepwise discriminant analysis. In *BMDP Statistical Software* (W. J. Dixon, ed.), pp. 519–537. Univ. of California Press, Berkeley.

King T. V. V. and King E. A. (1978) Grain size and petrography of C2 and C3 carbonaceous chondrites. *Meteoritics* **13**, 47–72.

Koch G. S., Jr. and Link R. F. (1971) *Statistical Analysis of Geological Data, Volume II*. Dover, N. Y. 438 pp.

Kracher A., Scott E. R. D., and Keil K. (1983) Dusty olivines in the Vigarano (CV3) chondrite: Evidence for an ubiquitous reduction process (abstract). In *Lunar and Planetary Science XIV*, pp. 407–408. Lunar and Planetary Institute, Houston.

Le Maitre R. W. (1982) *Numerical Petrology, Statistical Interpretation of Geochemical Data*. Elsevier, N. Y. 281 pp.

Lux G., Keil K., and Taylor G. J. (1980) Metamorphism of the H-group chondrites: implications from compositional and textural trends in chondrules. *Geochim. Cosmochim. Acta* **44**, 841–855.

Lux G., Keil K., and Taylor G. J. (1981) Chondrules in H3 chondrites: textures, compositions and origins. *Geochim. Cosmochim. Acta* **45**, 675–685.

McSween H. Y. (1977a) Chemical and petrographic constraints on the origin of chondrules and inclusions in carbonaceous chondrites. *Geochim. Cosmochim. Acta* **41**, 1843–1860.

McSween H. Y. (1977b) On the nature and origin of isolated olivine grains in carbonaceous chondrites. *Geochim. Cosmochim. Acta* **41**, 411–418.

McSween H. Y. (1977c) Petrographic variations among carbonaceous chondrites of the Vigarano type. *Geochim. Cosmochim. Acta* **41**, 1777–1790.

McSween H. Y. (1977d) Carbonaceous chondrites of the Ornans type: a metamorphic sequence. *Geochim. Cosmochim. Acta* **41**, 477–491.

McSween H. Y. (1979) Are carbonaceous chondrites primitive or processed? A review. *Rev. Geophys. Space Phys.* **17**, 1059–1078.

Nagahara H. (1981) Evidence for secondary origin of chondrules. *Nature* **292**, 135–136.

Rambaldi E. R. (1981) Relict grains in chondrules. *Nature* **293**, 558–561.

Rubin A. E., Scott E. R. D., Taylor G. J., and Keil K. (1982) Silicate matrix material in type 3 ordinary chondrites: implications for the origin of chondrules. *Meteoritics* **17**, 275–276.

Scott E. R. D., Taylor G. J., and Keil K. (1982) Origins of ordinary and carbonaceous chondrites and their components (abstract). In *Lunar and Planetary Science XIII*, pp. 704–705. Lunar and Planetary Institute, Houston.

Scott E. R. D., Taylor G. J., Maggiore P., Keil K., McKinley S. G., and McSween H. Y. (1981) Three CO3 chondrites from Antarctica—comparison of carbonaceous and ordinary type 3 chondrites. *Meteoritics* **16**, 385.

Simon S. B. and Haggerty S. E. (1980) Bulk compositions of chondrules in the Allende meteorite. *Proc. Lunar Planet. Sci. Conf. 11th*, pp 901–927.

Chondrules Formed Through Incomplete Melting of the Pre-existing Mineral Clusters and the Origin of Chondrules

Hiroko Nagahara

Geological Institute, University of Tokyo, Hongo, Tokyo 113, Japan

Relic olivine found in some type 3 ordinary chondrites was the first direct evidence for the secondary origin of chondrules. Olivine grains enclosed in pyroxene phenocrysts of poikilitic chondrules are also considered to be relic minerals. The proportion of those relic-bearing chondrules is about half of the porphyritic chondrules, indicating that nearly half of the porphyritic chondrules were formed through incomplete melting of the precursor mineral clusters. Chemical and petrographic similarities between relic-bearing and relic-free chondrules suggest that all the chondrules were formed from pre-existing minerals through complete or incomplete melting. The poikilitic texture was successfully reproduced by experiments. The charges heated at 15°C below the liquidus temperature for 2 minutes and cooled at a rate of 5°C/hr contain porphyritic pyroxene with poikilitically enclosed olivine which resembles the poikilitic texture in natural chondrules. This result supports the subliquidus heating of the poikilitic chondrules. Compositional and oxygen-isotopic diversities among individual chondrules require the clusters of minerals (mainly olivine, pyroxene, and plagioclase) as the precursor of chondrules, and their formation would have been at the early stage of the solar system. In order to give the wide range of cooling rate estimated from recent experimental studies, the nebula, with abundant clusters and fine-grained materials (dust), must have been present in the optically opaque nebula, which played a role of insulation for the cooling chondrules. The mechanism of heating the nebula with clusters is still unknown.

INTRODUCTION

Many recent petrological and chemical studies suggest that chondrules were formed from pre-existing minerals (i.e., Gooding *et al.*, 1980a; Gooding and Keil, 1981; Grossman and Wasson, 1982; Lux *et al.*, 1981). The presence of relic olivine in some chondrules of type 3 ordinary chondrites was the first direct evidence for the secondary origin of chondrules (Nagahara, 1981; Rambaldi, 1981), and the oxygen-isotopic diversity of individual chondrules (Gooding *et al.*, 1980b, 1982; Onuma *et al.*, 1982) is also strong evidence of their secondary origin.

The texture of chondrules is one of the important manifestations of their origin and condition of formation. As Dodd (1971, 1978a,b) and Gooding and Keil (1981) have pointed out, porphyritic or microporphyritic chondrules are much more abundant than those of other textures such as barred olivine, radial pyroxene, and glassy (or cryptocrystalline) chondrules. Dodd (1978a,b) suggested that the microporphyritic chondrules are not the crystallization products of molten droplets, but are fragments of magmatic rocks larger than 10 cm because of irregular outline and low CaO contents of olivine. On the other hand, Gooding and Keil (1981) considered that the surface roughness of chondrules would have been dominated by later processes and that all chondrules have the same origin. Thus, the origin of porphyritic chondrules and the difference between porphyritic and non-porphyritic chondrules are important problems of chondrule study.

Experimental studies on reproducing the textures of chondrules and estimating the conditions of chondrule formation have continued since the early study of Nelson *et al.* (1972), who duplicated barred or radial olivine texture. Tsuchiyama *et al.* (1980) reproduced barred and porphyritic olivine textures from completely or incompletely melted starting materials of three different compositions. Tsuchiyama and Nagahara (1981) and Hewins *et al.* (1981) reproduced radial pyroxene texture with cooling rates of 1°–100°C/hr and 50°–300°C/hr, respectively, a slower cooling rate than that for barred olivine texture (several thousand degrees per hour) by Tsuchiyama *et al.* (1980). Planner and Keil (1982) produced porphyritic olivine and pyroxene texture by three-stage cooling which includes an isothermal plateau; the cooling rate from superliquidus temperature to 1300°C was estimated to be 300°–4000°C/hr. These experiments show that barred or radial texture can be easily formed from completely melted starting materials, but porphyritic texture, which is most common in the natural chondrules, was difficult to obtain, except for complex three-stage cooling.

In this paper, the importance of incomplete melting of the pre-existing mineral clusters and the origin of chondrules are discussed based on petrographical observations and experimental results. Details of petrography and experiments will be given in the future as separate papers.

SAMPLES AND METHODS OF STUDY

The samples investigated in the present study were ALH-77015 (L 3.4), 77278 (LL 3.6), 77299 (H 3.7), and 77304 (L 3.9). Brief descriptions of these samples are given in Score *et al.* (1981). ALH-77015 is considered to be paired with the other 33 Antarctic L3 chondrites (McKinley *et al.*, 1981). The compositions of constituent minerals of chondrules were analyzed by the focused beam of the electron microprobe and the bulk compositions of individual chondrules by the defocused beam.

The Pt-wire loop method was used for the dynamic crystallization experiments. The starting material was a mixture of bronzite (70 wt.%), olivine (20%), and plagioclase (10%), which is nearly the same as sample 3 of Tsuchiyama *et al.* (1980). The sample was ground into several microns in size, pressed into pellets, attached to the Pt-wire, and put into the furnace, which had already been heated at the fixed temperature. A vertical siliconit furnace was used, where oxygen fugacity was controlled to be 10^{-9}atm at 1600°C and 10^{-12}atm at 1200°C by constant H_2/Co_2 gas mixing. The sample was cooled from the fixed temperature to 1200°C at a linear rate automatically. After the run, the charges were mounted with epoxy and sliced into thin sections for petrographical observations.

RELIC OLIVINE-BEARING CHONDRULES

Nagahara (1981) and, independently, Rambaldi (1981) and Michel-Lévy (1981), have found relic olivine in chondrules of different chondrites. Relic olivine, which has numerous dusty inclusions and shows a dirty appearance, is always present in the centers of chondrules as large or small anhedral grains (Fig. 1a,b). The newly crystallized olivines are small euhedral clear grains,

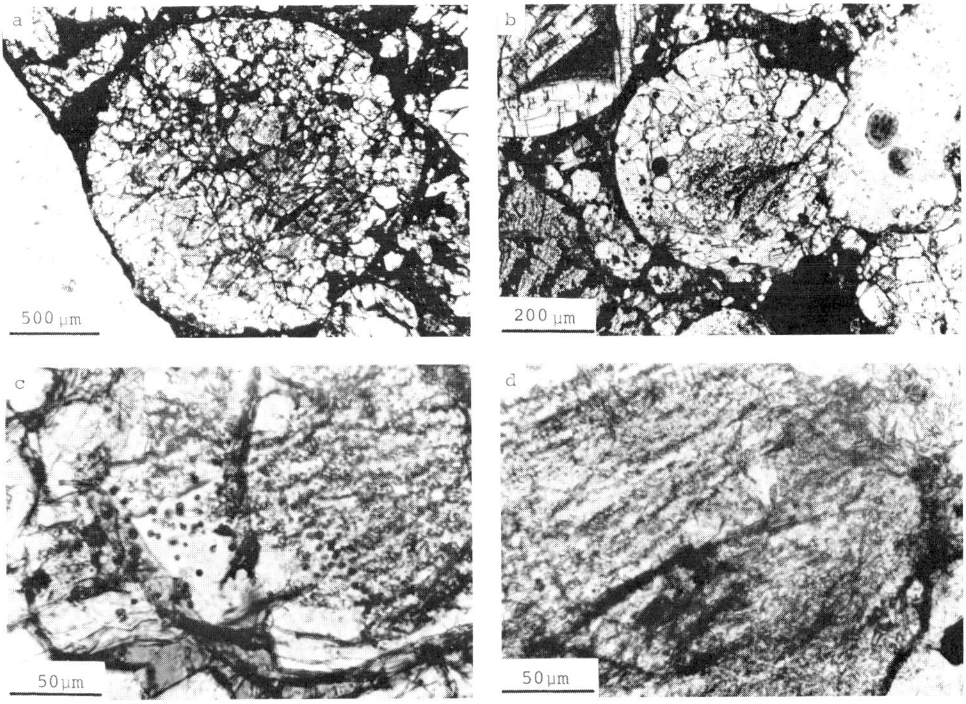

Fig. 1 **(a)** Relic olivine-bearing porphyritic chondrule in ALH-77015 chondrite. **(b)** Relic olivine-bearing porphyritic chondrule in ALH-77299 chondrite. **(c)** Overgrowth of clear olivine around the dusty relic olivine in a porphyritic chondrule of ALH-77015 chondrite. **(d)** Enlargement of dusty inclusions in relic olivine in **(a)**.

which exist in the outer portion of the chondrules. Clear olivine also occurs as overgrowth on relic olivine (Fig. 1c). The inclusions of the relic olivine often show alignment in one direction (Fig. 1d).

Rambaldi and Wasson (1982) investigated the dusty inclusions in the relic olivines, and showed that most are small grains of almost pure metallic iron; they said that reduction of iron-rich olivine to metallic iron and more magnesian olivine had occurred at the time of chondrule formation.

The relic and newly crystallized olivines have different chemical compositions. The relic olivine is rich in FeO and shows "reverse" zoning, with the core more enriched in FeO than the rim. In contrast, the newly crystallized olivine is more magnesian and shows "normal" zoning. The rim of normally zoned, crystallized olivine grains and reversely zoned, relic olivine grains have the same composition.

The relic and newly crystallized olivines are also distinguished by their CaO content. Figure 2 shows the CaO content versus Mg/Mg + Fe ratio of the two olivines in four different chondrules of ALH-77015 and two of ALH-77299 chondrite. The newly crystallized olivine contains more

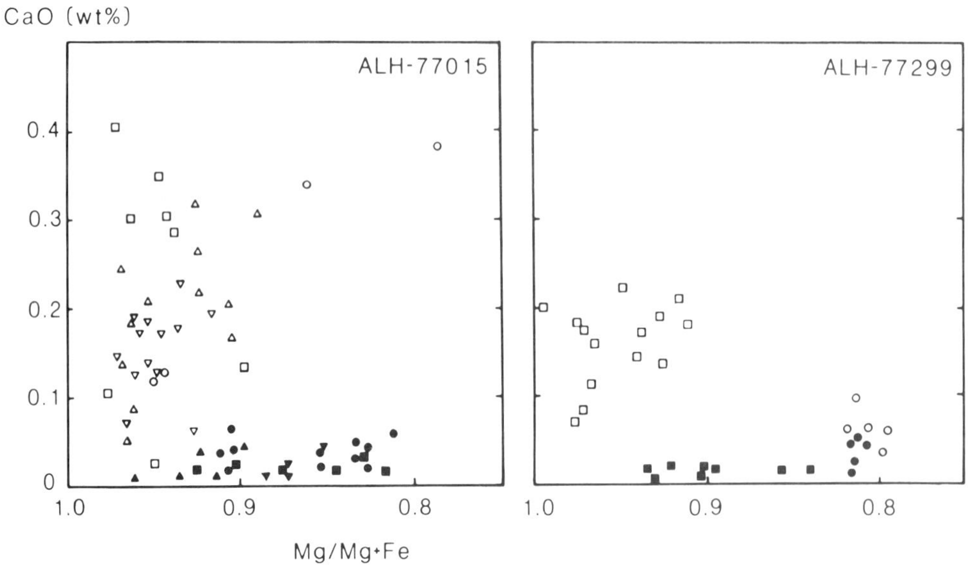

Fig. 2 Relationship between Mg/Mg + Fe ratios and CaO contents of relic and crystallized olivines in four chondrules in ALH-77015 chondrite (left) and two chondrules in ALH-77278 chondrite (right). Solid symbol: relic. Open symbol: crystallized. Solid and open marks of same symbols are in a single chondrule.

CaO than the relic olivine. In general, olivine, crystallized at higher temperatures or at rapid growth rates, contains more CaO (i.e., Donaldson *et al.*, 1975). This fact supports the idea that the clear olivine crystallized from liquid when chondrules were heated, and that the dusty olivine is relic olivine that had been equilibrated at lower temperatures or under a more oxidized atmosphere before chondrule formation.

As already mentioned, porphyritic olivine and pyroxene chondrules are the most abundant among all textural types of chondrules. Three different modes of occurrence of olivine and pyroxene are recognized: (1) olivine is euhedral and interstitial pyroxene is subhedral or anhedral; (2) pyroxene is euhedral and olivine is poikilitically enclosed in pyroxene; and (3) both olivine and pyroxene are euhedral and exist as discrete grains. Cases 1 and 3 are less common than case 2, where olivine is enclosed in pyroxene.

In case 2, discrete euhedral olivine grains are often present, in addition to olivine poikilitically enclosed in pyroxene (Fig. 3a). These two different olivines show the same features as the above described relic and newly crystallized olivines; the poikilitically enclosed olivine (Fig. 3b) often contains many dusty inclusions in its center (Fig.3c), while the discrete olivine is clear (Fig. 3d). The dusty inclusions resemble those in the relic olivine (Fig. 1d), suggesting that the poikilitically enclosed olivine is also relic.

These two kinds of olivine in poikilitic chondrules also chemically resemble previously described relic and newly crystallized olivines. Figure 4 shows the CaO content of the poikilitically

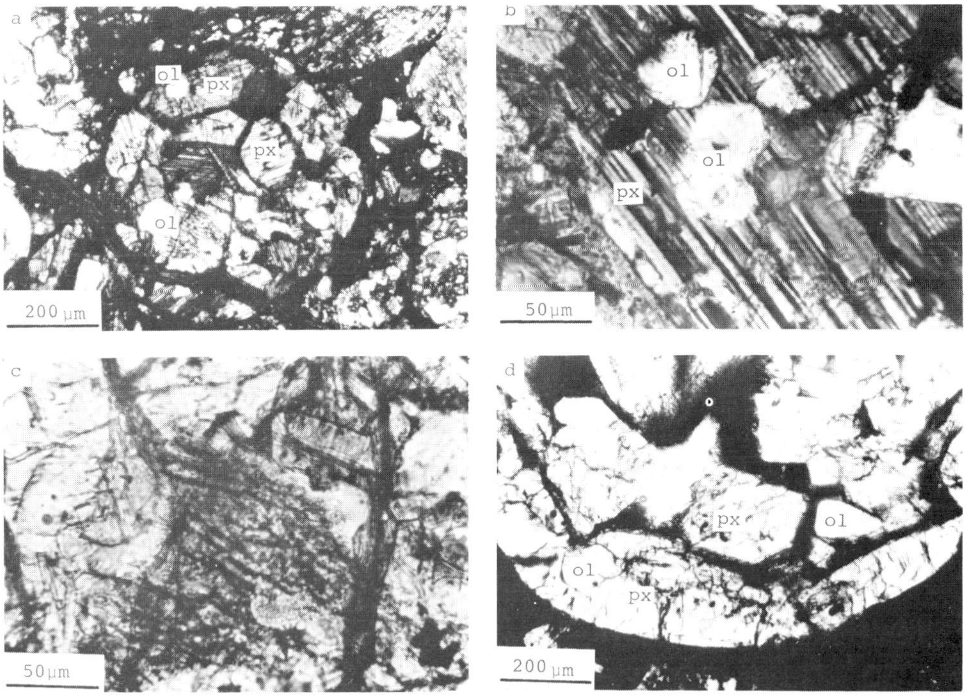

Fig. 3 **(a)** Poikilitic chondrule in ALH-77015 chondrite. **(b)** Poikilitically enclosed olivine in clinoenstatite. **(c)** Dusty inclusions in poikilitically enclosed olivine in **(a)**. **(d)** Discrete clear euhedral olivine and pyroxene in poikilitic chondrule; ol=olivine, px=pyroxene.

enclosed olivine and the euhedral discrete olivine in the porphyritic chondrule of ALH-77278 chondrite. Poikilitically enclosed olivine has a low Mg/Mg + Fe ratio and CaO content, while the clear olivine has a higher ratio and content. These textural and chemical similarities show that the enclosed olivine is relic and that the euhedral olivine crystallized from a liquid at the time of chondrule formation.

Figure 5 shows the CaO content of poikilitically enclosed large (more than 100 μm) and small (less than 50 μm) olivines in three different chondrules in ALH-77015 chondrite. The CaO content of enclosed olivine varies depending on the grain size; larger olivine contains less CaO than the smaller ones. This would be a result of diffusion of Ca at or after crystallization of surrounding pyroxene. Small olivine that was originally Ca-poor would have been surrounded by pyroxene with higher CaO and would become rich in CaO by diffusion at higher temperatures, even during short duration. On the other hand, large olivine was much less affected.

Poikilitically enclosed olivine in pyroxene does not always contain dusty inclusions, though it is plausible that most of the poikilitic chondrules were formed by the same processes; that is, they would also be the products of incomplete melting of the pre-existing minerals. If this is the case,

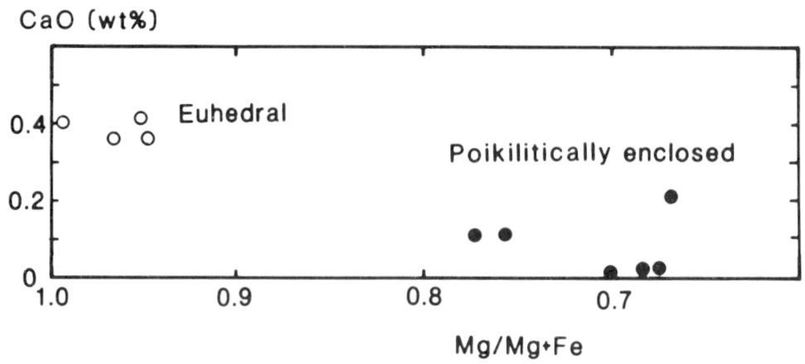

Fig. 4 Relationship between Mg/Mg + Fe ratios and CaO contents of poikilitically enclosed and discrete clear olivines in a porphyritic chondrule of ALH-77278 chondrite. Solid circle: poikilitically enclosed olivine. Open circle: discrete olivine.

poikilitically enclosed olivine with or without dusty inclusions is relic, and pyroxene and clear discrete olivine crystallized from partially molten liquid.

Proportions of those relic-bearing and poikilitic chondrules were measured in two chondrites, and the results are shown in Fig. 6. In ALH-77015 chondrite, about 10% of all chondrules have

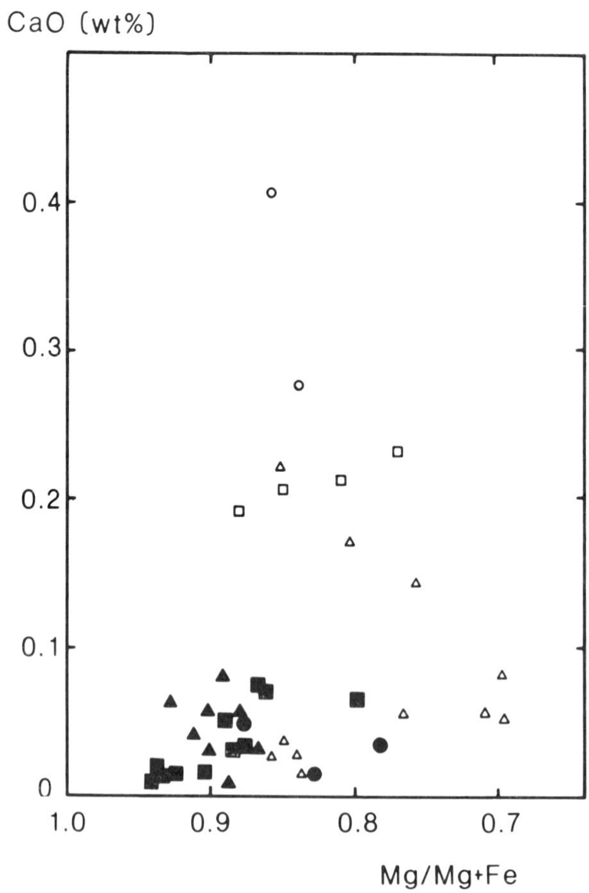

Fig. 5 Relationship between Mg/Mg + Fe ratios and CaO content of poikilitically enclosed olivine in three chondrules in ALH-77015 chondrite. Closed symbols: olivine larger than 100 μm. Open symbols: olivine smaller than 20 μm. Solid and open marks of same symbols are in a single chondrule.

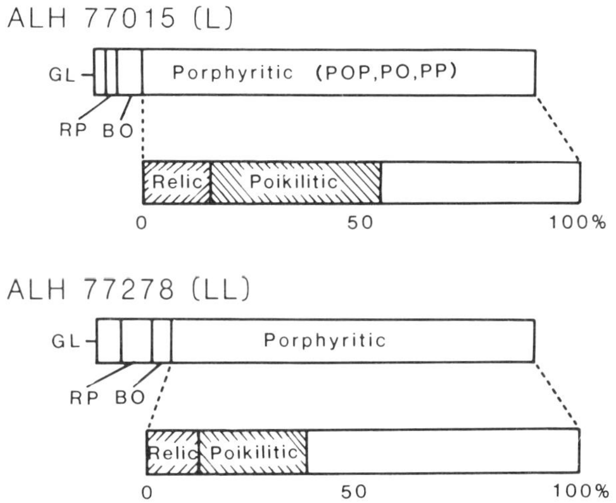

Fig. 6 Proportion of relic-bearing and poikilitic chondrules in ALH-77015 and ALH-77278 chondrites. POP=porphyritic olivine and pyroxene chondrule. PO=porphyritic olivine chondrule. PP=porphyritic pyroxene chondrule. BO=barred olivine chondrule. RP=radial pyroxene chondrule. GL=glassy or cryptocrystalline chondrule.

olivine with dusty inclusions and 32% show poikilitic texture. In the poikilitic chondrules, 10% have dusty inclusions, and the rest (22%) show poikilitic texture without dusty inclusions. As a result, about half of the porphyritic chondrules and 40% of all chondrules were formed through incomplete melting of the precursor mineral clusters.

EXPERIMENTAL REPRODUCTION OF POIKILITIC TEXTURE

Factors which determine the texture of chondrules are thought to be original bulk chemical composition, cooling rate, maximum temperature of heating, and duration of heating according to petrographical observations of natural chondrules (Tsuchiyama *et al.*, 1980) and the results of dynamic crystallization experiments (i.e., Lofgren, 1980). In the course of systematic experiments to reproduce the chondrule textures, poikilitic texture was successfully reproduced. Charges cooled from a temperature 15°C below the liquidus at rates smaller than 100°C/hr contained large euhedral clinoenstatite phenocrysts that poikilitically enclosed small olivine grains. The result of cooling with a rate of 5°C/hr is shown in Fig. 7, which is similar to the natural poikilitic chondrules shown in Fig. 3. The size and the proportion of enclosed olivine grains would depend on the original grain size of the starting material, maximum heating temperature, heating duration, and original bulk chemical composition (especially modal ratio of olivine and pyroxene).

The present experimental result supports the inference from petrographic observations that poikilitic texture was formed through incomplete melting of the pre-existing mineral clusters. It further shows that the cooling rate of poikilitic chondrules was smaller than those of barred olivine (Tsuchiyama *et al.*, 1980) and radial pyroxene (Tsuchiyama and Nagahara, 1981; Hewins *et al.*, 1981). As a result, the range of cooling rates for various textures of chondrules is very wide, ranging more than four orders of magnitude (10^4–1°C/hr).

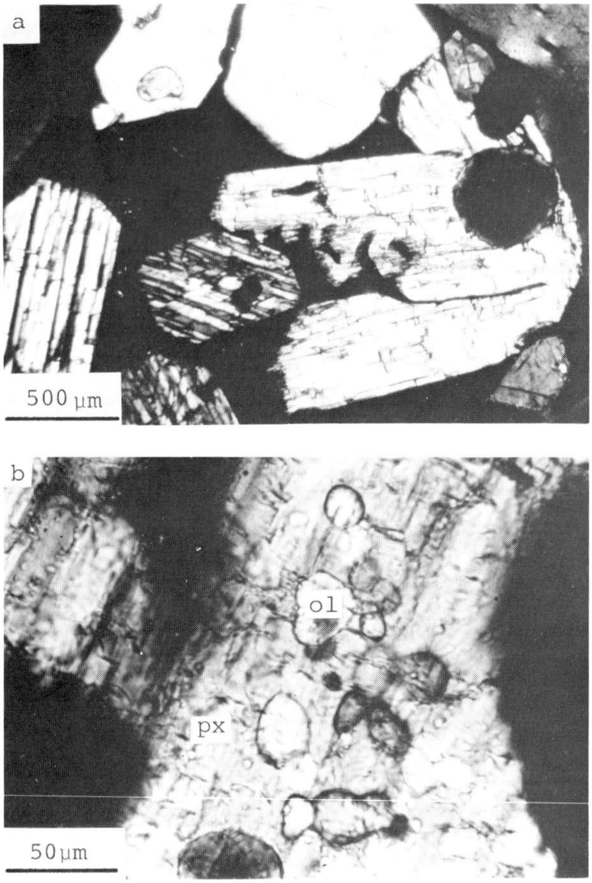

Fig. 7 **(a)** Experimentally reproduced porphyritic pyroxene texture with poikilitically enclosed olivine. The charge was heated at temperature 15°C below the liquidus temperature for 2 minutes and cooled at a rate of 5°C/hr. **(b)** Poikilitically enclosed olivine in pyroxene shown in Fig. 6. Ol=olivine. Px=pyroxene.

DISCUSSION

Petrographic observations and experimental results reveal that the poikilitically enclosed olivine grains in olivine and pyroxene porphyritic chondrules are also relic, like previously reported relic olivine in olivine porphyritic chondrules (Nagahara, 1981; Rambaldi, 1981). Figure 6 shows that at least about half of the porphyritic chondrules were formed through incomplete melting of the precursor mineral clusters.

There is no direct evidence for the formation of other chondrules (barred olivine, radial pyroxene, relic-free porphyritic chondrules without poikilitic texture, and glassy or cryptocrystalline chondrules) from pre-existing mineral clusters. However, the following indirect evidences suggest that all chondrules were formed from precursor mineral clusters: (1) bulk chemical compositions of relic-bearing and relic-free chondrules are indistinguishable; (2) the size ranges of both types of chondrules are the same; and (3) both chondrules are well-mixed and show no difference in occurrence in all chondrites. Consequently, it is suggested that all chondrules were formed from mixtures of pre-existing minerals. This exclusively eliminates the possibility of direct liquid condensation from primitive solar gas for the origin of chondrules (Wood and McSween, 1977; Blander, 1982); condensed liquid cannot contain relic minerals.

The presence of relic olivine in various chondrules indicates that the heating temperature of these chondrules was between liquidus and solidus temperatures. Most chondrules have a liquidus temperature of 1400°–1700°C, and the solidus approximately 1000°–1200°C based on the bulk composition and the melting experiments of chondrule-like materials and terrestrial ultramafic rocks. Thus, the heating temperatures of relic-bearing chondrules should have been between these temperatures.

Mineral clusters that were heated at temperatures lower than those for relic-bearing chondrules could not form round or ellipsoidal shapes because of the presence of small amounts of liquid between liquidus and solidus temperatures. The amount of liquid decreases with decreasing temperature, so those heated to about 1200°C or below contained small amounts of liquid and would have failed to form round shapes; that is, they are not "droplet" chondrules. Those heated to subsolidus temperatures, of course, contained no liquid. They formed irregular-shaped materials such as "lithic fragments" or "inclusions" with or without small amounts of liquid (glass or devitrified glass), and/or fine-grained minerals. Thus, the shape of chondrules would depend on the maximum temperature of heating or the amount of liquid. The shape continuously varies from a completely-rounded one to an irregularly-shaped one, and therefore cannot be one of the defining factors for chondrules.

Dodd (1978a,b) has discussed that the microporphyritic chondules were formed by fragmentation of rocks that crystallized from liquids, and that they were different in origin from the remelted chondrules, such as barred olivine or radial pyroxene chondrules. Though his model seems to explain the pattern of chemical variation among chondrules with different textures, microporphyritic chondrules with completely rounded shapes are difficult to form by fragmentation of bodies crystallized from wholly-melted liquid. Furthermore, rocks, once wholly melted, would have easily exchanged the oxygen isotopes and become equilibrated with the ambient gas, from which isotopically heterogeneous chondrules could not be formed. Because of their same size distribution, different origins for microporphyritic and droplet chondrules seem difficult to explain. Crystallization from incompletely melted pre-existing mineral clusters, that is, crystallization in the presence of relic minerals as nuclei, is an effective explanation for the formation of microporphyritic chondrules and the difference between (micro-)porphyritic and droplet chondrules.

The present experimental result indicates that chondrules containing porphyritic pyroxene with enclosed olivine cooled at slow rates. The slow cooling rate seems to be inconsistent with the presence of considerable amounts of Na_2O in most chondrules in type 3 ordinary chondrites. Tsuchiyama *et al.* (1981) investigated the volatilization of Na_2O from the liquid droplets of chondrule compositions and revealed that Na_2O is easily vaporized from the liquid under low oxygen partial pressure. They further showed that if chondrules were heated to about 2200°C or above, all Na_2O would have been lost during heating and cooling even at very rapid rates.

Grossman *et al.* (1979) also discussed that sudden melting and rapid cooling could prevent the volatile loss.

The above discussion, however, applies only to liquid droplets. The present petrographic observations and the experimental results show that at least half of the chondrules were not entirely melted. The bulk chemical composition of chondrules consists mainly of olivine, pyroxene, and plagioclase components, and the partially melted liquid that crystallized olivine and pyroxene must be enriched in plagioclase component. Such a liquid should be more polymerized than the completely melted chondrules, which are more enriched in olivine and pyroxene components, and the volatiles, especially Na_2O, would hardly be vaporized from the liquid. Thus, slow cooling of some kinds of chondrules does not necessarily conflict with the presence of considerable amounts of volatiles.

Oxygen isotopic composition for individual chondrules in the type 3 chondrites is variable (Gooding *et al.*, 1980a, 1982; Onuma *et al.*, 1982), and it requires the presence of isotopically different solid precursors of the solar system (Onuma *et al.*, 1982; Wood, 1981). Further, they would have been incorporated into the precursor materials of chondrules. Formation of chondrules in the early solar system from chemically and isotopically heterogeneous mineral clusters seems to be the most plausible explanation for the following problems: (1) mechanically aggregated precursor materials can explain the chemical and isotopic heterogeneity; (2) abundant dust and dense gas, which were present in the early solar system, could have been an insulator for the cooling chondrules; and (3) the difference of optical depth has caused the wide range of cooling rates of chondrules.

Though the mechanism of heating the nebula where clusters and abundant fine-grained dust were present is still unknown, formation of chondrules at the early stage of the solar system eliminates the usual several models for chondrule origin. Formation of chondrules by collision between small bodies (Lange and Larimer, 1973; Kieffer, 1975; Kerridge and Kieffer, 1977) cannot be accepted because high speed collision (more than several km/sec) to form silicate liquid ($\gtrsim$1200°C) was impossible in dense nebula. Impact origin on the large body (or bodies) (Fredriksson and Ringwood, 1963; Fredriksson, 1969; Kurat, 1967; Wlotzka, 1969; King *et al.*, 1972; Dodd, 1978a,b, 1981) is also difficult because the large body could not be present in the early solar system. The model of heating by internal energy of accumulation of interstellar particles into large particles (Clayton, 1980) does not need any external energy supply, but the assumption that the amorphous grains became crystals is uncertain. Thus, these models cannot satisfactorily explain the origin of chondrules. Therefore, another new model of heating the nebula, which contained mineral clusters as the precursor of chondrules, abundant fine-grained dust, and dense gas by a certain event, should be considered.

Acknowledgments. I would like to thank Prof. E. King of the University of Houston and the Program Committee for providing me with the chance to attend the conference from which this volume resulted. Profs. I. Kushiro and H. Takeda of the University of Tokyo kindly made arrangements for it. I am also grateful to Drs. R. T. Dodd of the State University of New York, G. Lofgren of NASA/JSC, and I. Kushiro for a critical reading of the manuscript, and Dr. A. Tsuchiyama of NASA/JSC for discussions and help with the experiments. The samples were supplied by the National Institute of Polar Research, Japan.

REFERENCES

Blander M. (1982) Condensation of chondrules (abstract). In *Papers Presented to the Conference on Chondrules and their Origins*, pp. 1–2. Lunar and Planetary Institute, Houston.

Clayton D. D. (1980) Chemical energy in cold-cloud aggregates: the origin of meteoritic chondrules. *Astrophys. J.* **239**, 37–41.

Dodd R. T. (1971) The petrology of chondrules in the Sharps meteorite. *Contrib. Mineral. Petrol.* **31**, 201–227.

Dodd R. T. (1974) The petrology of chondrules in the Hallingeberg meteorite. *Contrib. Mineral. Petrol.* **47**, 97–112.

Dodd R. T. (1978a) The composition and origin of large microporphyritic chondrules in the Manych (L-3) chondrite. *Earth Planet. Sci. Lett.* **39**, 52–66.

Dodd R. T. (1978b) Compositions of droplet chondrules in the Manych (L-3) chondrites and the origin of chondrules. *Earth Planet. Sci. Lett.* **40**, 71–82.

Donaldson C. H., Usselman T. M., Williams R. J., and Lofgren G. F. (1975) Experimental modeling of the cooling history of Apollo 12 olivine basalts. *Proc. Lunar Sci. Conf. 6th,* pp. 843–869.

Fredriksson K. (1969) The Sharps chondrite: a new evidence on the origin of chondrules and chondrites. In *Meteorite Research* (P. M. Millman, ed.), pp. 155–165. Reidel, Dordrecht.

Fredriksson K. and Ringwood A. E. (1963) Origin of meteoritic chondrules. *Geochim. Cosmochim. Acta* **27**, 639–641.

Gooding J. L. and Keil K. (1981) Relative abundance of chondrule primary textural types in ordinary chondrites and their bearing on conditions of chondrule formation. *Meteoritics* **16**, 17–43.

Gooding J. L., Keil K., Fukuoka T., and Schmitt R. A. (1980a) Elemental abundances in chondrules from unequilibrated chondrites: evidence for chondrule origin by melting of pre-existing materials. *Earth Planet. Sci. Lett.* **50**, 171–180.

Gooding J. L., Keil K., Mayeda T. K., Clayton R. N., Fukuoka T., and Schmitt R. A. (1980b) Oxygen isotopic compositions of petrologically characterized chondrules from unequilibrated chondrites (abstract). *Meteoritics* **15**, 295.

Gooding J. L., Mayeda T. K., Clayton R. N., Keil K., Fukuoka T., and Schmitt R. A. (1982) Oxygen isotopic compositions of chondrules in unequilibrated chondrites: Further petrological interpretations (abstract). In *Lunar and Planetary Science XIII*, pp. 271–272. Lunar and Planetary Institute, Houston.

Grossman J. N. and Wasson J. T. (1982) Evidence for primitive nebular components in chondrules from the Chainpur chondrite. *Geochim. Cosmochim. Acta* **46**, 1081–1089.

Grossman J. N., Kracher A., and Wasson J. T. (1979) Volatiles in Chainpur chondrules. *Geophys. Res. Lett.* **6**, 597–600.

Grossman L. (1972) Condensation in the primitive solar nebula. *Geochim. Cosmochim. Acta* **36**, 597–619.

Hewins R. H., Klein L. C., and Fasano B. V. (1981) Conditions of formation of pyroxene excentroradial chondrules. *Proc. Lunar Planet. Sci. 12B*, pp. 1123–1133.

Kerridge J. F. and Kieffer S. W. (1977) A constraint on impact theories of chondrule formation. *Earth Planet. Sci. Lett.* **35**, 35–42.

Kieffer S. W. (1975) Droplet chondrules. *Science* **189**, 333–340.

King E. A., Jr., Carman M. F., and Butler J. C. (1972) Chondrules in Apollo 14 samples: Implications for the origin of chondritic meteorites. *Science* **175**, 59–60.

Kurat G. (1967) Zur Entstehung der Chondren. *Geochim. Cosmochim. Acta* **31**, 491–502.

Lange D. E. and Larimer J. W. (1973) Chondrules: An origin by impacts between dust grains. *Science* **182**, 920–922.

Larimer J. W. (1967) Chemical fractionations in meteorites—I. Condensation of the elements. *Geochim. Cosmochim. Acta* **31**, 1215–1238.

Lofgren G. E. (1980) Experimental studies on dynamic crystallization of silicate melts. In *Physics of Magmatic Processes* (R. B. Hargraves, ed.), pp. 487–551. Princeton University Press, New Jersey.

Lux G., Keil K., and Taylor G. J. (1981) Chondrules in H3 chondrites: textures, compositions and origins. *Geochim. Cosmochim. Acta* **45**, 675–685.

McKinley S. G., Scott E. R. D., Taylor G. J., and Keil K. (1981) A unique type 3 ordinary chondrite containing graphite-magnetite aggregates—Allan Hills A77011. *Proc. Lunar Planet. Sci. 12B*, pp. 1039–1048.

Michel-Lévy M. C. (1981) Some clues to the history of the H-group chondrites. *Earth Planet. Sci. Lett.* **54**, 67–80.

Nagahara H. (1981) Evidence for secondary origin of chondrules. *Nature* **292**, 135–136.

Nelson L. S., Blander M., Skaggs S. R., and Keil K. (1972) Use of CO_2 laser to prepare chondrule-like spherules from supercooled molten oxide and silicate droplets. *Earth Planet. Sci. Lett.* **14**, 338–344.

Onuma N., Clayton R. N., Ikeda Y., Hutcheon I. D., Mayeda T. K., and Molini-Velsko C. (1982) Oxygen isotopes in chondrules of ordinary and carbonaceous chondrites (abstract). Paper given at the Conference on Chondrules and their Origins, November 15–17, 1982. Lunar and Planetary Institute, Houston.

Planner H. N. and Keil K. (1982) Evidence for three-stage cooling history of olivine-porphyritic fluid droplet chondrules. *Geochim. Cosmochim. Acta* **46**, 317–330.

Rambaldi E. R. (1981) Relict grains in chondrules. *Nature* **293**, 558–561.

Rambaldi E. R. and Wasson J. T. (1982) Fine, nickel-poor Fe-Ni grains in the olivine of unequilibrated ordinary chondrites. *Geochim. Cosmochim. Acta* **46**, 929–939.

Score R., Schwarz C. M., King T. V. V., Mason B., Bogard D. D., and Gabel E. M. (1981) Antarctic meteorite descriptions 1976–1977–1978–1979. *Antarctic Meteorite Newsletter* **4**, no. 1. Lunar Curatorial Branch, NASA Johnson Space Center. 144 pp.

Tsuchiyama A. and Nagahara H. (1981) Effect of pre-cooling thermal history and cooling rate on the texture of chondrules: a preliminary report. *Mem. Natl. Inst. Polar Res., Spec. Issue No. 20* (T. Nagata, ed.), pp. 175–192.

Tsuchiyama A., Nagahara H., and Kushiro I. (1980) Experimental reproduction of textures of chondrules. *Earth Planet. Sci. Lett.* **48**, 155–165.

Tsuchiyama A., Nagahara H., and Kushiro I. (1981) Volatilization of sodium from silicate melt spheres and its application to the formation of chondrules. *Geochim. Cosmochim. Acta* **45**, 1357–1367.

Wlotzka F. (1969) On the formation of chondrules and metal particles by "shock melting." In *Meteorite Research* (P. M. Millman, ed.), pp. 174–184. Reidel, Dordrecht.

Wood J. A. (1981) The interstellar dust as a precursor of Ca,Al-rich inclusions in carbonaceous chondrites. *Earth Planet. Sci. Lett.* **56**, 32–44.

Wood J. A. and McSween H. Y. (1977) Chondrules as condensation products. In *Comets, Asteroids, and Meteorites* (A. H. Delsemme, ed.), pp. 365–373. Univ. of Toledo, Ohio.

SiO_2-Bearing Chondrules in the Murchison (C2) Meteorite

Edward J. Olsen
Department of Geology, Field Museum of Natural History, Roosevelt Road at Lake Shore Drive
Chicago, Illinois 60605

Two chondrules in the Murchison C2 meteorite have been found that contain segregations of glass that are essentially pure silica, associated with low FeO crystalline phases and other (possibly cryptocrystalline) glasses. In each case the compositions of crystalline phases and glasses indicate an equilibrium assemblage that can be interpreted in terms of a magmatic cooling trend. This indicates that the cooling rate was slow enough to permit nucleation and crystallization of equilibrium phases, and that part of the cooling history involved an isothermal state for some period of time before a quench took place. This is consistent with the three-stage cooling process determined experimentally by Planner and Keil (1982). Slow cooling suggests these chondrules were surrounded by an insulating medium, possibly dense gas or dust.

INTRODUCTION

Much discussion has appeared during the past decade over the nature of discrete inclusions in carbonaceous chondrites. Some inclusions have clearly had histories that involved their total or partial melting into obvious melt-droplet chondrules. The processes involved in the formation of other kinds of inclusions, especially those with baroque shapes and absence of glass, are not fully agreed upon and are the subjects of ongoing discussion.

Fuchs *et al.* (1973) made distinctions between several morphological types of inclusions and this was amplified by Olsen and Grossman (1978). These workers have concluded that among the numerous inclusions, melt-droplet chondrules are relatively rare; most of the inclusions result from histories that do not involve melting. Other workers, however, have taken the view that virtually all of the inclusions they have seen in thin sections of carbonaceous chondrites (especially the much-studied Allende and Murchison chondrites) are melt-droplet chondrules or the processed remnants thereof (McSween, 1977a,b; Richardson and McSween, 1978; McSween, 1979; Desnoyers, 1980; Roedder, 1981). During this decade of active discussion about these features, very little detailed work has actually been done on those inclusions in carbonaceous chondrites that no one would argue are not melt-droplet chondrules—objects that are morphologically round and contain both crystalline phases and glass.

During the past year, I have begun a project to examine melt-droplet chondrules from the C2 meteorite, Murchison. The first object of this work is to determine how similar clear-cut melt-droplet chondrules in Murchison are in their mineralogies and petrographic relationships to the much-studied melt-droplet chondrules in the ordinary chondrites. The second object is to determine whether there exists within this single C2 meteorite any genetic relationship between obvious melt-droplet chondrules and the other types of inclusions for which origins are not necessarily simple and obvious.

MELT-DROPLET CHONDRULES IN MURCHISON

Olsen and Grossman (1978) illustrated a histogram of diameters of melt-droplet chondrules in Murchison (their Fig. 10D). Their sample was obtained by hand-picking from the surfaces of fragments of the meteorite and, to a small extent, from chondrules encountered in thin sections. They reported data on only 27 chondrules that ranged in diameter from 175 micrometers to 1081 micrometers. Their data show a mode around 300 micrometers and a computed mean of 420 micrometers.

Since the time of that work, freeze-thaw methods have been employed on Murchison. About 20 grams of randomly chosen fragments have been disaggregated, yielding an impressive population of objects, some of which had never been seen by any workers in the course of extensive thin section examinations over many years (e.g., corundum-melilite-anorthite-diopside aggregates; Bar-Matthews *et al.*, 1982). They also yielded a population of obvious melt-droplet chondrules that were set aside for study. The real diameters of these have been measured and an analysis of their size distribution, compared to populations of other kinds of inclusions found in Murchison, has been made (Olsen, unpublished data).

A total of 434 chondrules were released by freeze-thaw methods on about 20 grams of Murchison. If these 434 chondrules represent the total chondrule population in the approximately 20 grams of Murchison that were disaggregated, then they comprise less than 0.001% by volume of the meteorite. [Olsen and Grossman (1978) estimated less than 2% by volume, clearly a large overestimate.] Even if the estimate of less than 0.001% is off by many orders of magnitude, which is highly unlikely, the absolute abundance of melt-droplet chondrules in Murchison is very low. It is probable that only those chondrules with unaltered glass in them survive freeze-thaw extraction. There is, however, no reason why alteration should be size-selective. Thus, the population obtained here can be considered to be a representative sample.

Approximately 100 chondrules were randomly selected and prepared for study: half in polished section, and half in thin sections. Thus far, 20 have been studied in great detail by optical and scanning electron microscopy. Two of them were found to bear free silica; this paper deals with the results on those two.

METHODS

These chondrules have been studied by scanning electron microscope (SEM) in both secondary electron and back-scattered electron modes. Analyses were performed using a solid-state detection system with the output corrected by an on-line, dedicated computer. Some analyses were repeated by electron microprobe (EMP) (solid-state detection, using a different correction algorithm) to test consistency. EMP, however, lacks the visual resolution necessary to observe certain small features; the SEM method was superior.

CHONDRULE 3A

This chondrule (Fig. 1) is 280 micrometers in diameter as it appears in polished section. It is a porphyritic chondrule and consists of an interior of blocky to rounded anhedral crystals of enstatite (Fs 1.0) that range from 30 to 100 micrometers in their longest dimensions. These are separated from each other by linear to cuspate patches of glass, within which are microlites of augite (Fs 0.0) that are as large as 2 micrometers; the majority, however, are much smaller, down to less than a

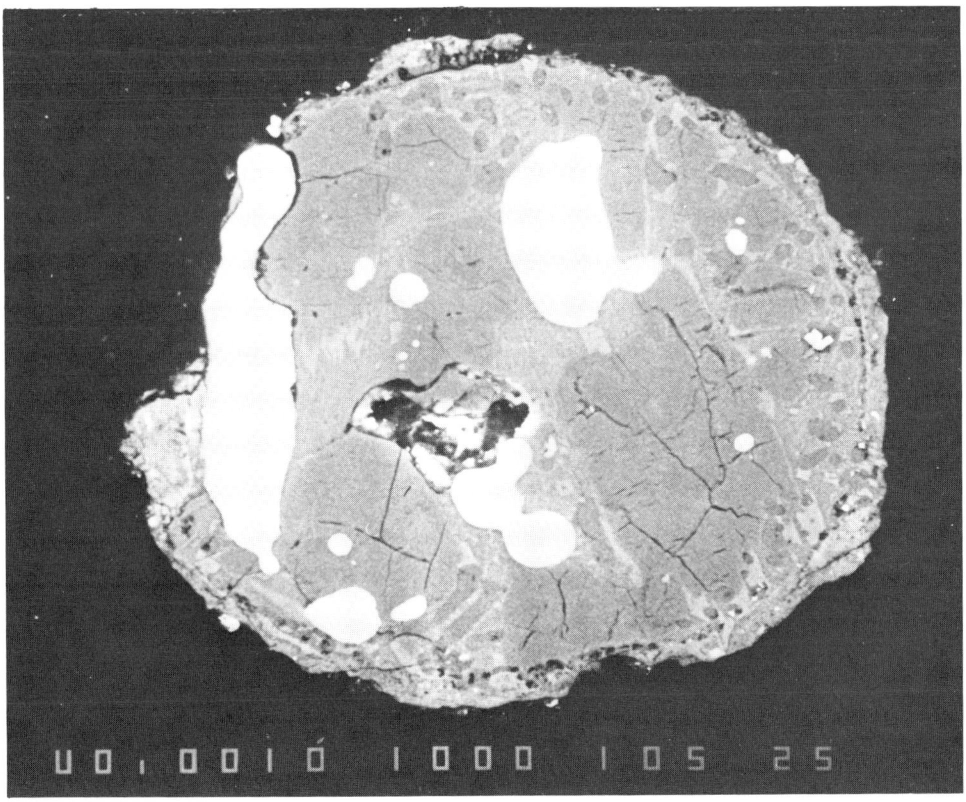

Fig. 1 Chondrule 3A, 280 micrometers in diameter. Silica pods are the darkest shade of gray. Pale gray patches are residual glasses (col. 5 in Table 2). Medium gray shades are enstatite. Small, whitish-gray grains are calcic-pyroxene. Bright white is metal. (During section preparation a grain was plucked out of the approximate center of the chondrule, leaving a hole.)

few hundredths of a micrometer. Because of these augite microlites, the glass appears to vary somewhat in composition from point to point, depending on the varying abundance of them.

Glass is most abundant around the outer portion of the chondrule, forming a "crust" from 10 to 30 micrometers thick (as it appears in polished section) that encloses the more porphyritic interior. Within this "crustal" region are numerous round to elliptical pods, 5 to 15 micrometers in longest dimension, of essentially pure silica.

Within the interior, over twenty irregular-lobate to circular patches of metal occur, ranging from 8 to 32 micrometers in their longest dimensions. The largest of these occurs as an elongated mass partially along one edge of the chondrule, overlaid in part by a thin layer of glass. The metal is low in Ni (5%) and contains up to 1% Cr (Table 1). One small anhedral grain of forsterite (Fa 0.8) was found enclosed in enstatite on three sides and by glass on the fourth side.

Analyses of crystalline phases are given in Table 1. The bulk composition of the chondrule was measured by three methods: (1) the beam was enlarged and moved along a zig-zag pattern until the entire area was scanned; (2) the same method was used with a small beam size (2 micrometers); and (3) a small beam was stepped at 5-micrometer intervals along a grid pattern over the entire area, and counting was done for a fixed time at each point. All three methods gave essentially the same result (Table 2). Since the metal contains virtually all of the iron in the chondrule, the bulk composition of the non-metal portion was easily computed (Table 2).

Table 1a. Mineral analyses of chondrule 3A.

	PYROXENES						OLIVINE
SiO_2	57.26	56.22	58.77	57.99	54.88	55.94	43.30
Al_2O_3	0.89	1.72	1.20	3.74	5.35	5.79	0.00
Cr_2O_3	0.33	0.49	0.32	0.00	0.00	0.00	0.00
FeO	0.69	1.35	0.47	0.00	0.00	0.00	0.83
MgO	37.98	36.60	39.32	23.71	18.99	19.32	54.88
CaO	0.26	0.37	0.26	15.71	20.86	18.92	0.00
Sum	97.41	97.39	100.35	101.16	100.08	100.00	99.01
Wo	0.5	0.7	0.5	32.4	44.1	41.3	—
Fs	1.0	2.0	0.7	0.0	0.0	0.0	—
En	98.5	97.3	98.9	67.6	55.9	58.7	—
Fa	—	—	—	—	—	—	0.8

	METAL		
Fe	93.45	93.65	92.69
Ni	5.01	5.26	5.00
Cr	0.69	1.09	0.78
Sum	99.15	100.00	98.47

Table 1b. Mineral analyses of chondrule 4B.

PYROXENES						
SiO_2	51.60	53.49	54.87	54.83	58.60	58.51
TiO_2	3.49	2.11	1.25	2.11	0.00	0.00
Al_2O_3	4.41	4.20	3.78	3.58	1.52	1.87
FeO	0.94	0.67	0.92	0.00	0.00	0.00
MgO	18.44	19.10	20.55	20.46	37.57	37.07
CaO	19.25	20.35	18.62	18.93	0.36	0.32
Sum	98.13	99.92	99.98	99.92	98.99	97.77
Wo	42.2	42.8	38.9	39.9	0.7	0.6
Fs	1.6	1.1	1.5	0.0	1.3	0.0
En	56.2	56.1	59.6	60.1	98.0	99.4

METAL			
Fc	94.00	93.82	94.29
Ni	5.30	5.46	5.35
Cr	0.27	0.27	0.00
Sum	99.57	99.56	99.64

PLAGIOCLASE				
SiO_2	49.50	49.95	48.35	48.96
Al_2O_3	31.08	31.31	30.98	31.31
CaO	18.35	17.84	17.54	18.51
Na_2O	1.05	0.83	0.76	1.23
Sum	100.98	99.94	97.62	100.01
An	91	92	93	89

COOLING HISTORY OF CHONDRULE 3A

The bulk composition of the silicate portion of the chondrule is very closely represented by the quaternary $CaO-MgO-Al_2O_3-SiO_2$, which has been studied experimentally by Osborn *et al.* (1954). The bulk composition plots just inside the primary phase field of forsterite; the first crystalline phase that should appear on cooling of a complete melt is forsterite at approximately 1500°C. Under conditions of equilibrium crystallization, the melt would move to the pyroxene boundary and enstatite would form by reaction of the melt with forsterite, exhausting all of it. Enstatite would be joined by calcic-pyroxene just before the melt hits the pyroxene-tridymite cotectic.

Table 2. Analyses of bulk chondrules and glassses within them.

	Chondrule 3A					Chondrule 4B				
	1	2	3	4	5	6	7	8	9	10
SiO_2	58.24	46.40	98.87	98.19	63.81	56.89	49.38	45.99	50.36	99.93
TiO_2	0.00	0.00	0.00	0.00	0.00	0.00	0.00	0.00	5.39	0.05
Al_2O_3	7.27	5.79	2.27	1.39	16.84	17.04	33.89	27.29	8.83	1.70
Cr_2O_3	0.55	0.59	0.00	0.00	0.18	0.00	0.00	0.00	0.00	0.00
FeO	0.10	20.27	0.00	0.00	0.39	0.80	0.00	0.00	1.29	0.00
MnO	0.00	0.00	0.00	0.00	0.37	0.00	0.00	0.00	0.00	0.00
MgO	28.85	22.98	1.27	0.00	4.19	15.60	0.00	3.43	15.38	0.00
CaO	3.37	2.68	0.00	0.35	10.02	8.71	18.20	12.89	17.66	0.00
K_2O	0.00	0.00	0.00	0.00	0.00	0.11	0.00	0.00	0.00	0.00
Na_2O	1.62	1.29	0.00	0.00	2.45	0.85	0.01	0.00	0.00	0.00
Sum	100.00	100.00	102.42	99.93	98.23	100.00	101.48	98.97	98.91	101.68

1. Chondrule 3A bulk, metal-free (normalized to 100%).
2. Chondrule 3A—All metal oxidized and added to silicates (normalized to 100%).
3. Chondrule 3A—Silica pod.
4. Chondrule 3A—Silica pod.
5. Chondrule 3A—Glass surrounding silica pods (ave. of 4 analyses).
6. Chondrule 4B bulk, metal/sulfide-free (normalized to 100%).

7-10. Chondrule 4B—Glasses within chondrule (see text).

Analyses were made of the glass surrounding the pods of silica (Table 2). The analyses necessarily include microlites of augite, which (as stated earlier) cause small variations in the composition of the glass, depending on their abundance. Since augite was precipitating at the same time as pods of silica separated, the composition of the glass plus any microlites adjacent to the pods must represent the local bulk composition of the glass associated with the silica. A series of points on glass were analyzed around the pods, with large crystallites (micrometer-size) of pyroxene being avoided. When these analyses were averaged and plotted in the quaternary, the point lies on the pyroxene-tridymite cotectic at 1300°C.

The silica pods run as high as 99% SiO_2 and contain mainly only minor amounts of Al_2O_3 and MgO or CaO. As it is now mounted, no thin section is possible without danger of losing the chondrule. The phases referred to as glass all luminesce brilliantly under the electron beam, which is often an indication of a glassy state. It is possible, however, that cryptocrystalline devitrification has occurred. The silica pods, viewed at 8000× in back-scattered electron mode, reveal no noticeable differences in composition within them.

Texturally, the pods could appear to represent liquid immiscibility; however, I can make no judgment about that. It may well be that the pods are masses of tridymite microlites that separated from the liquid.

This chondrule appears to have undergone a cooling history that at least approached equilibrium. (The small grain of forsterite is apparently an unreacted remnant of the earliest crystalline phase.) The last liquid concentrated around the outside as a crustal layer. (Admittedly, the chondrule is seen in only two dimensions and this feature may only be apparent.) Based on its petrography and composition, this chondrule is different from melt-droplet chondrules in the ordinary chondrites, which show nucleation around their exteriors with residual glass primarily in their interiors.

CHONDRULE 4B

This chondrule is 275 micrometers in diameter (Fig. 2). Petrographically, it is different from chondrule 3A. It consists of two rounded crystals of enstatite, each 100 micrometers in the longest dimension and separated from each other at two positions along the edge, and several small crystals of enstatite, 20 to 30 micrometers, along the opposite edge. The interior consists of a filigree of skeletal crystals of augite and plagioclase (An 89–93) with interstitial lenses of glass of three distinct compositions: one glass is composed of essentially Si, Al, and Ca oxides with occasional small amounts of Na (normatively it is plagioclase with excess silica and/or alumina); another glass (found around the outer edge of the chondrule) has, in addition, variable amounts of Ti, Mg, and Fe; and the third, relatively abundant glass, is almost pure silica with minor Al and a trace of Ti. This latter glass occurs in the interior of the chondrule and is always surrounded by skeletal crystals of plagioclase.

Small blebs of metal (5.3% Ni, 0.27% Cr) occur together at one edge and are associated with the enstatite grains. An unusual Ti-Cr oxide was found associated with an augite crystal.

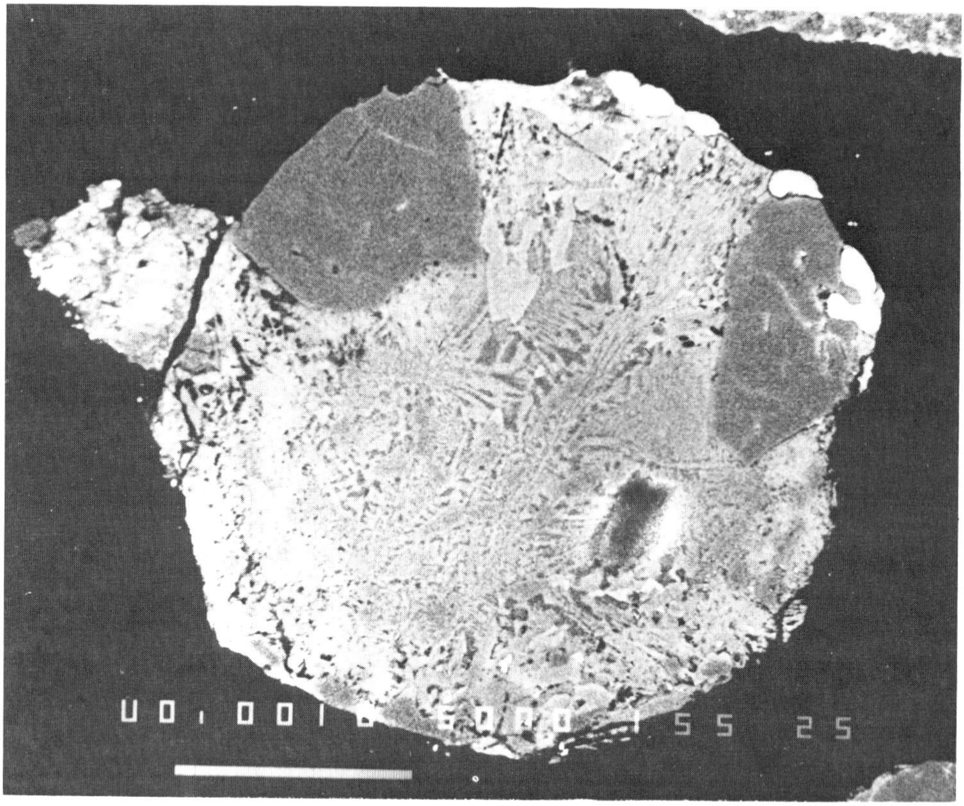

Fig. 2 Chondrule 4B, 275 micrometers in diameter. Enstatites are dark gray and along the edge of the chondrule. White dendrites are plagioclase. Dark gray interior patches are glasses. Metal at edge is bright white. Chondrule has a hole in the southeast quadrant. (Fragment at left edge, outside the chondrule, is Murchison matrix.)

COOLING HISTORY OF CHONDRULE 4B

The bulk composition of this chondrule was measured in the same way as described for chondrule 3A (Table 2). As with 3A, bulk silicate 4B falls into the quaternary CaO-MgO-Al_2O_3-SiO_2 (Osborn *et al.*, 1954). The bulk composition lies in the primary phase field of pyroxene at approximately 1325°C. Equilibrium crystallization begins with pyroxene, is joined by plagioclase at approximately 1275°C, and proceeds to the pyroxene-plagioclase-tridymite eutectic at approximately 1250°C.

The presence of silica surrounded by calcic plagioclase indicates that the cooling path approximated equilibrium. The texture indicates that the cooling was slow enough for nucleation of equilibrium phases, but too fast to permit significant crystal growth. The texture also indicates that initial nucleation and crystallization took place at the outer edge of the chondrule, where the largest crystals are, and proceeded inward. This is the more normal texture observed in chondrules of the ordinary chondrites and experimentally-made chondrules. In experimental systems, plagioclase nucleates with difficulty and may often be bypassed metastably (McSween, 1977a). The fact that the plagioclase did nucleate in this chondrule suggests that the cooling rate was initially relatively slow.

DISCUSSION

Magmatic crystallization to the stage of separation of a silica phase occurs in eucrites. Presumably, the process there involves fairly large melt pools with sufficient insulation to allow cooling slow enough for equilibrium to be approximated. In the instances of these two chondrules, 3A and 4B, the cooling rate must have been slow enough to permit a similar approach to equilibrium. Recently, Brigham *et al.* (1982) have reported three silica-bearing chondrules, one each in three H3 ordinary chondrites: Bremervörde, Dhajala, and Sharps. In those three cases, the bulk compositions are very different from the Murchison chondrules. Their chondrules are all high in FeO. In one case, Sharps, the silica occurs intimately mixed with fayalite. In Dhajala and Bremervörde, silica laths are embedded in pyroxenes of varying Fs content. In addition, the bulk composition of the Bremervörde chondrule is high in silica: $SiO_2 = 70\%$, $FeO = 13\%$, $MgO = 17\%$, and minor CaO. This bulk composition plots essentially in the system MgO-FeO-SiO_2 (Bowen and Schairer, 1935) in the primary phase field of cristobalite at 1700°C. In this chondrule, the silica laths radiate inward from the rim where nucleation must have taken place. All three of these chondrules are low in Ca and Al.

Based on their descriptions of these silica-rich chondrules, it appears that silica is not a phase at the end of a line of magmatic descent but rather an initial phase, due to the siliceous bulk composition of the parent material. Brigham *et al.* (1982) offer the suggestion that this may imply an unusual, unsampled parent body.

Chondrule 3A in Murchison is from parent material essentially free of FeO and unlike any known meteorite in bulk composition. Were we to suppose that the metal now present in the chondrule were due to reduction of FeO during the melting process, the hypothetical parent material would have the computed composition shown in Table 2. This "parent" material is also unlike any known meteorite type.

Arguments of this type are, however, spurious. The volume sampled by any chondrule is very small compared to the crystalline scale of homogeneity of many well-known meteorite types. The chondrules in Murchison have volumes of only 0.011 cubic millimeters, and those described by Brigham *et al.* (1982) range from only 0.034 to 0.113 cubic millimeters.

In order for chondrules 3A and 4B to follow an approximate equilibrium cooling path, it is necessary that they not quench. This suggests that the surrounding "atmosphere" was relatively dense. It could have been dense with other chondrules, each radiating heat and mutually retarding each other's cooling rates. Within each chondrule, nucleation and crystallization of each solid phase creates a retardation of cooling rate as the latent heats of crystallization are released. On the other hand, the "atmosphere" may have consisted of a dense gas or dust, relatively opaque to infrared. In either event, a cooling rate slower than a quench rate is implied by the mineralogy and glass compositions in these two chondrules.

Recent experimental work has come to the same conclusion; in order to realize the textures and mineral compositions in chondrules in unequilibrated ordinary chondrites, Planner and Keil (1982) have demonstrated that cooling must take place in three steps: (1) from the liquidus (or near liquidus) at a rate of 300° to 4000°C per hour to about 1300°C; (2) an isothermal stage at 1300°C for an unknown period of time; and (3) a final stage quench that preserves "residual melts as vitreous or fine-crystalline material" (Planner and Keil, 1982).

Similarly, Nagahara (1982) and Tsuchiyama *et al.* (1980) have concluded that cooling rates must be retarded in order to obtain the textures and compositions observed in chondrules of ordinary chondrites. They postulate a dense gas as an insulating agent.

The isothermal plateau indicated by the work of Planner and Keil (1982) is 1300°C. The temperature of the liquid at which the silica pods separated in Murchison chondrule 3A is also 1300°C. This is probably fortuitous. It is also possible that this is a measure of the real temperature at which an isothermal state of unknown duration set in. Such a temperature may be dictated by a dynamic set of conditions in which gas, dust, and cooling chondrules, each releasing latent heats of crystallization, together cause a halt in cooling until a perturbation causes the system to cool at quench rates. For chondrule 4B the nucleation of plagioclase and the formation of silica indicate an isothermal plateau at about 1250°C prior to a quench.

Experimental workers conclude that some crystallites must be present in order to form the textures seen in chondrules in the ordinary chondrites; that is, total melting cannot take place. This means that chondrules must form from initially crystalline material. As such, they are not primary solids, but secondary. Their precursors may be only dust or dust aggregations, but necessarily crystalline at least at a cryptocrystalline scale. Any kind of rapid input of energy, such as lightning, would be inefficient and tend to blow apart such material (Cameron, 1982). A slow input of energy is indicated. In addition, the isothermal or near isothermal plateau indicated by chondrules 3A and 4B and by experiments indicate a slow initial cooling rate. Both of these conditions might be accommodated by the model of Cameron (1982). Grain clusters migrating into the central region of the primitive solar nebula would undergo partial evaporation of less refractory components and could partially melt. Movement outward along a gradual temperature gradient would permit slow cooling. Acceleration into colder regions with lower grain density would facilitate a final quench.

The compositions of chondrules 3A and 4B (Table 2) are refractory. Based on the Fa and Fs contents measured, their oxidation states are low; $p(O_2)$ is approximately 10^{-15} to 10^{-20} atmospheres at temperatures between 1250°C and 1300°C (Mueller, 1964), the isothermal plateaus indicated earlier in this paper. The liquidus temperatures are, as noted earlier, 1500°C and 1325°C, respectively, although it seems apparent that in order to retain nuclei these temperatures were not reached, or were reached rapidly and cooling ensued before all pre-existing crystalline phases were totally melted. It is also possible that the precursor solid material was richer in less refractory components and these components were evaporated off, leaving a more refractory residuum that almost completely melted, as discussed by Cameron (1982).

CONCLUSIONS

The two Murchison chondrules described in this paper are each approximate equilibrium assemblages of crystalline and glassy phases. Kinetically, these could not have formed by a rapid quench process. They appear to require slow cooling and probably an isothermal stage before quenching. This is consistent with the sequence of events determined experimentally by Planner and Keil (1982).

In studying other Murchison chondrules, they run the gamut from refractory and reduced to less refractory and oxidized (i.e., FeO-rich). Such a range of objects may offer glimpses into all stages of evolution of the enigmatic particles called chondrules.

Acknowledgments. I thank Dr. T. Tanaka and Dr. L. Grossman for making available the population of melt-droplet chondrules separated by freeze-thaw methods from the Field Museum sample of the Murchison meteorite. This work was supported by grant NAGW-23 from the National Aeronautics and Space Administration.

REFERENCES

Bar-Matthews M., Hutcheon I. D., MacPherson G. J., and Grossman L. (1982) A corundum-rich inclusion in the Murchison carbonaceous chondrite. *Geochim. Cosmochim. Acta* **46**, 31–41.

Brigham C., Murrell M. T., and Burnett D. S. (1982) SiO_2-rich chondrules in ordinary chondrites (abstract). In *Papers Presented to the Conference on Chondrules and their Origins*, p. 4. Lunar and Planetary Institute, Houston.

Bowen N. L. and Schairer J. F. (1935) The system MgO-FeO-SiO_2. *Amer. J. Sci.* **229**, 151–217.

Cameron A. G. W. (1982) Chondrule-related processes in the primitive solar nebula (abstract). In *Papers Presented to the Conference on Chondrules and their Origins*, pp. 9–11. Lunar and Planetary Institute, Houston.

Desnoyers C. (1980) The Niger (I) carbonaceous chondrite and implications for the origin of aggregates and isolated olivine grains in C2 chondrites. *Earth Planet. Sci. Lett.* **47**, 223–234.

Fuchs L. H., Olsen E., and Jensen K. J. (1973) Mineralogy, mineral chemistry and composition of the Murchison (C2) meteorite. *Smithson. Contrib. Earth Sci.* **10**. 33 pp.

McSween H. Y., Jr. (1977a) On the nature and origin of isolated olivine grains in carbonaceous chondrites. *Geochim. Cosmochim. Acta* **41**, 411–418.

McSween H. Y., Jr. (1977b) Chemical and petrographic constraints on the origin of chondrules and inclusions in carbonaceous chondrites. *Geochim. Cosmochim. Acta* **41**, 1843–1860.

McSween H. Y., Jr. (1979) Are carbonaceous chondrites primitive or processed? A review. *Rev. Geophys. Space Phys.* **17**, 1059–1078.

Mueller R. F. (1964) Phase equilibria and the crystallization of chondritic meteorites. *Geochim. Cosmochim. Acta* **28**, 189–207.

Nagahara H. (1982) Effect of heating temperature on the texture of chondrules with special reference to the porphyritic chondrules (abstract). In *Papers Presented to the Conference on Chondrules and their Origins*, p. 47. Lunar and Planetary Institute, Houston.

Olsen E. and Grossman L. (1978) On the origin of isolated grains in type 2 carbonaceous chondrites. *Earth Planet. Sci. Lett.* **41**, 111–127.

Osborn E. F., DeVries R. C., Gee K. H., and Kraner H. M. (1954) Optimum composition of blast furnace slag as deduced from liquidus data for the quaternary system CaO-MgO-Al_2O_3-SiO_2. *Trans AIME* **200**, 33–45.

Planner H. N. and Keil K. (1982) Evidence for the three-stage cooling history of olivine-porphyritic fluid droplet chondrules. *Geochim. Cosmochim. Acta* **46**, 317–330.

Richardson S. M. and McSween H. Y., Jr. (1978) Textural evidence bearing on the origin of isolated olivine crystals in C2 carbonaceous chondrites. *Earth Planet. Sci. Lett.* **37**, 485–491.

Roedder E. (1981) Significance of Ca-Al-rich silicate melt inclusions in olivine crystals from the Murchison type II carbonaceous chondrite. *Bull. Minéral.* **104**, 339–353.

Tsuchiyama A., Nagahara H., and Kushiro I. (1980) Investigations on the experimentally produced chondrules: Chemical compositions of olivine and glass and formation of radial pyroxene chondrules. *Mem. Natl. Inst. of Polar Res., Spec. Issue 17* (T. Nagata, ed.), pp. 83–94. National Institute of Polar Research, Tokyo.

Phase Separation in a Chondrule Fragment from the Piancaldoli (LL3) Chondrite

H. N. Planner

Los Alamos National Laboratory, Los Alamos, New Mexico 87545

A chondrule fragment has been found in the Piancaldoli (LL3) chondrite that contains prolate silica spheroids embedded in a matrix with an approximate low-Ca pyroxene (Fs14) composition. Although compositional and textural evidence supports metastable coexistence of two liquid phases, neither phase is preserved as glass. The matrix has a fan-spherulitic texture, and many of the silica spheroids contain dendritic cristobalite(?). Silica interdendrite voids near the two-phase interface do not modify the smooth curvilinear outline of the interface, indicating that the iron magnesium silicate matrix had congealed prior to cristobalite(?) crystallization from the silica-rich phase. Available data suggest that this immiscibility is formed upon cooling from the melt. A temperature near 1275°–1300°C, based on phase composition, is thought to be the temperature at which these coexisting liquids last approached metastable equilibrium.

INTRODUCTION

A nearly rectangular chondrule fragment measuring 0.61×0.43 mm in a thin section of the Piancaldoli (LL3) chondrite has spheroids embedded in a spherulitic host material. Three sides of this fragment are bounded by angular outlines. The fourth side has a smooth curved surface thought to represent the original chondrule surface. Along this curved surface spheroids are closely packed to form a rim ≤ 100 μm thick. Interior to this rim, spheroids are present in trains or are isolated in the matrix. The distribution of spheroids is not unlike enstatite spherulites observed in experimentally devitrified magnesium silicate glass beads heated to temperatures between 800°–925°C for up to 175 hrs. As observed in the chondrule, the experimental charges have a rim along the spherule surface, and spherulites isolated or as trains embedded in glass interior to the rim.

It was hoped that the known thermal conditions used for devitrifying the glass charges could be used to infer that the chondrule had been reheated within a prescribed temperature-time framework. Detailed investigation of the chondrule showed, however, that separation of silica from the melt was responsible for spheroid development, the implication being that a mechanism other than devitrification could produce the textural similarity.

Occurrence of silica in several chondrules from ordinary chondrites has recently been addressed by Brigham *et al.* (1982). With their conclusion that silica-bearing chondrules could be the product of several different physical processes, the present investigation has developed with two objectives. The first is to document phase relations and compositional data that indicate a different mechanism of formation for the chondrule despite overall textural similarity with devitrified samples. The second objective is to identify the mechanism responsible for the chondrule's phase assemblage and estimate thermal conditions consistent with the documented data. These objectives, in conjunction with the work of others, may broaden our understanding of the presence of silica in some chondrules.

DESCRIPTION

The experimentally devitrified charge shown in Fig. 1a was formed from a glass bead heated to 800°C and maintained there for 173.5 hrs. The rim and spheres consist of radially oriented fibrils of orthoenstatite (determined from X-ray diffraction) and the matrix is isotropic glass. The charge is a synthetic magnesium silicate of enstatite composition (Table 1). A silicon X-ray map (Fig. 2b) and electron microprobe data (Table 1) indicate that spherulites, surrounding glass, and bulk sample composition are nearly indistinguishable in composition. The sample texture, consisting of spherulites and spherulitic rim, is reproducible and forms only upon reheating of glass. For comparison, continuously cooled samples with similar composition, which nucleate a

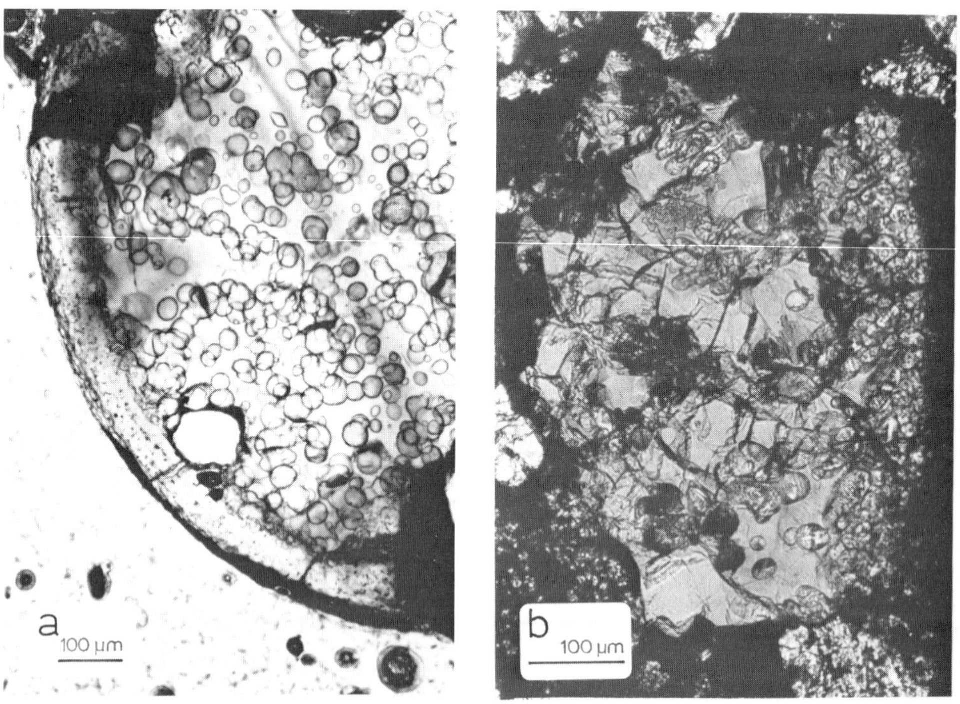

Fig. 1 Rim structure and spheroid distribution in **(a)** experimentally devitrified magnesium silicate glass sample and **(b)** chondrule fragment from the Piancaldoli chondrite. Transmitted light photomicrographs.

Table 1. Electron microprobe analyses.

Wt.%	(1)	(2)	(3)	(4)	(5)
SiO_2	98.87	56.46	60.11	60.01	60.20
TiO_2	0.01	0.08	0.0	0.0	0.0
Al_2O_3	0.23	2.02	0.05	0.05	0.05
Cr_2O_3	n.d.	0.55	0.01	0.06	0.0
FeO	0.64	8.94	0.02	0.01	0.0
MnO	n.d.	0.24	0.0	0.0	0.0
MgO	0.36	28.67	39.04	39.77	39.71
CaO	0.05	1.75	0.11	0.05	0.05
Na_2O	0.16	0.08	0.0	0.0	0.0
K_2O	0.09	0.03	0.0	0.0	0.0
Total	100.41	98.82	99.34	99.95	100.01

1. Silica-rich phase forming spheroids in chondrule fragment. Average of 4 analyses.
2. Iron magnesium silicate with nearly stoichiometric low-Ca pyroxene (Fs_{14}-Wo_4) composition, which forms the matrix phase within the chondrule. Average of 9 analyses.
3. Bulk composition for experimentally devitrified sample. Average of 73 analyses.
4. Average of 11 analyses along traverse across spherulite in experimentally devitrified sample.
5. Glass adjacent to spherulite represented by (4). Average of 11 analyses.

n. d. = not determined.

crystalline phase after being undercooled to $\leq$1000°C, yield a spherulitic texture without rim (Planner, 1974, pp. 10–18).

The Piancaldoli fragment shown in Fig. 1b is assumed to be part of a chondrule because one fragment edge forms a smooth curved surface. A fine-granular rind, similar in composition to a Tieschitz chondrule rind (Kurat, 1970), coats this curved surface and the adjoining angular outline along the fragment's upper surface, as oriented in Fig. 2c. Assuming that the rind coated the chondrule as it remained an isolated object in space (Christophe Michel-Lévy, 1976), presence of the rind along the angular outline is evidence that the chondrule texture, which predates fragmentation, developed while the chondrule resided in space and not after it had agglomerated with surrounding chondritic material. A minimum apparent diameter for the original chondrule is estimated at 1.4 mm. In the plane of the thin section only 14% of the chondrule is represented by the fragment. It is very possible that additional phases present in the original chondrule are not represented in this fragment.

Interior to the densely packed spheroid rim, spheroids have a prolate ellipsoidal form (Fig. 1b). These ellipsoids are arranged as trains centered about double lines separated 25 μm (Fig. 2c) that appear to project inward from the rim. Each double train, in turn, is arranged in parallel and at a regular spacing of about 90 μm. The density of the spheroids drops significantly 300 μm from the chondrule's curved surface.

The spheroids consist of silica (Table 1) with minor amounts of iron (0.53–0.80 wt.% FeO), MgO (0.29–0.43 wt.%), and Al_2O_3 ($\leq$0.40 wt.%). Many display a rib-like structure visible in transmitted light. Electron micrographs, such as Fig. 3, reveal dendrite arms $\leq$5 μm wide with cubic and octahedral orientations typical of cristobalite (Fenner, 1913). These dendrites and their

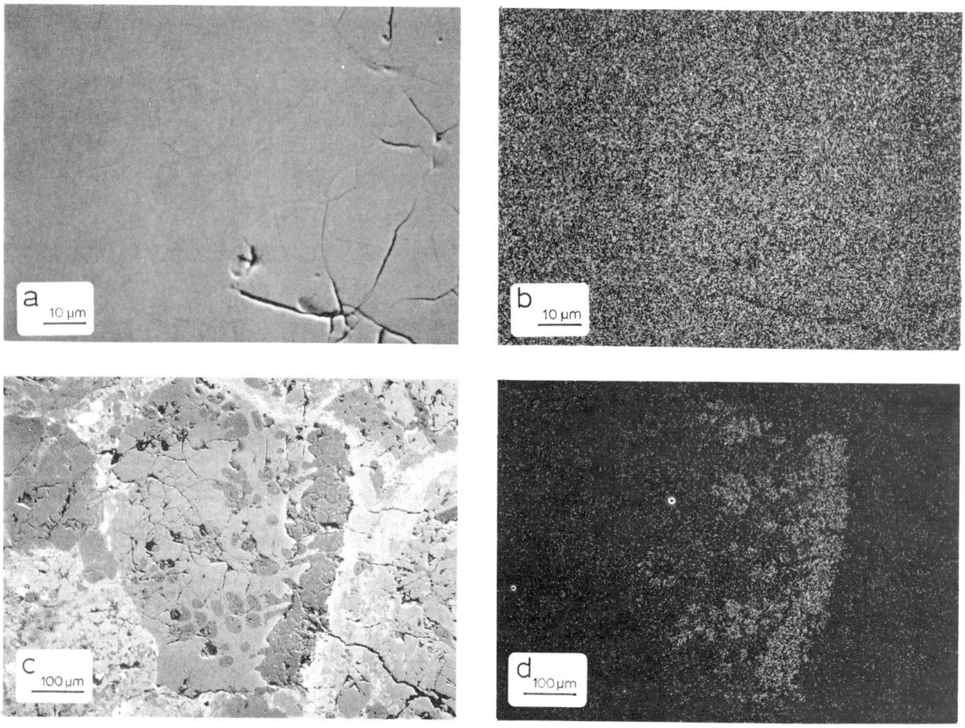

Fig. 2 **(a)** Backscatter electron micrograph for portion of experimentally devitrified glass spherule showing rim, glass, and isolated spherulite. Radiating cracks are associated with spherulites. Botryoidal outline marks devitrification front between rim (on right) and isotropic glass. **(b)** Silicon X-ray map for same area as **(a)**. **(c)** Backscatter electron micrograph of Piancaldoli chondrule fragment showing densely packed silica (dark grey) spheroid rim along curved chondrule outline and spheroid trains. Matrix with approximate low-Ca pyroxene composition identified as medium grey. Fine-granular rind is found along curved surface (on right) and angular outline (top). **(d)** Silicon X-ray map for same area as **(c)**.

interdendrite voids do not perturb the smooth curvilinear outline of ellipsoidal surfaces suggesting cristobalite (?) crystallization from a silica-rich phase after the iron magnesium silicate matrix had congealed.

The matrix to the spheroids has nearly stoichiometric low-Ca pyroxene (Fs_{14}-Wo_4) composition (Table 1). Variability is small with magnesia and silica ranging from 28.0–29.1 wt.% and 55.9–57.4 wt.%, respectively. Compositional gradients near silica spheroids are negligible, except where matrix is tightly confined by spheroid distribution. The matrix textural form is described here as xenomorphic-granular, although the grains do not actually represent individual crystals.

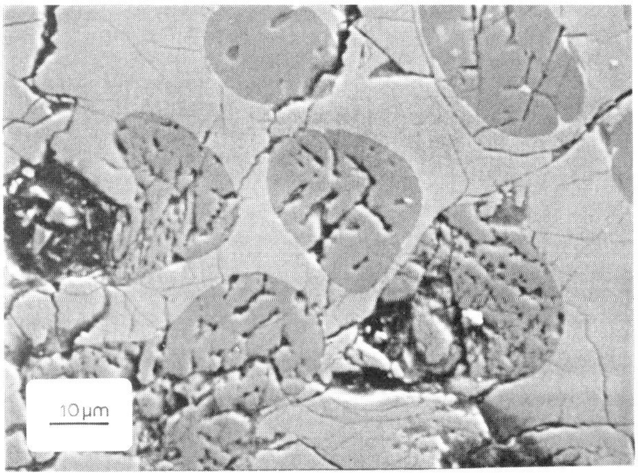

Fig. 3 Backscatter electron micrograph showing cristobalite(?) dendrites and interdendrite voids within silica ellipsoids (dark grey) and the unperturbed interface between spherulitic iron magnesium silicate matrix (medium grey) and silica.

Instead, each grain consists of a fan-spherulite with an off-centered extinction cross visible under crossed nicols. Many of the extinction crosses are centered in proximity to silica spheroids, suggesting that these spheroids acted as nucleation sites for spherulitic growth. All spherulites fan out in one common direction.

Metallic iron-nickel and troilite are present in the chondrule fragment but are restricted in distribution to the outer half of the densely packed spheroid rim (i.e., along the original chondrule periphery). Both phases, generally <2 μm in diameter, are located within silica and not the interstitial magnesium-bearing material.

DISCUSSION

Based on textural appearance, the synthetic magnesium silicate spherule and chondrule fragment shown in Fig. 1 represent a striking side-by-side comparison. Both materials have a rim along a curved peripheral boundary and spherical masses embedded in glassy matrix. It was hoped that the devitrified sample with its known thermal history could be used to explain the chondrule as a glassy chondrule having responded to reheating. Detailed investigation, however, supports an interpretation that the similar appearance between materials is in fact the result of two different phenomena.

A comparison of those features indicating a difference in material processing follows. Spherically-shaped volumes formed by devitrifying the synthetic material consist of radially oriented enstatite fibrils. Spherulite composition differs little from surrounding isotropic glass. Unlike the devitrified sample, the chondrule is compositionally segregated. Spheroids do not consist of fibrils but cristobalite(?) dendrites. Confocal arrangement of dendrites is lacking. The matrix in which the spheroids are embedded consists of fan spherulites instead of isotropic glass. It is interpreted, therefore, that despite similarity in spheroid distribution between synthetic magnesium silicate charge and chondrule, the features in the chondrule do not represent devitrification.

More feasible explanations for the phase relations in the chondrule are cristobalite crystallization from melt or amorphous phase separation. From equilibrium phase diagrams of systems such as MgO-iron oxide-SiO_2, MgO-Al_2O_3-SiO_2, and CaO-MgO-Al_2O_3-SiO_2, it is reasonable to suspect that silica in this chondrule either precipitated at liquidus temperatures or coprecipitated with pyroxene along a cotectic. If either of these conditions occurred, certain interrelationships between silica and the iron magnesium silicate matrix would be expected. First, assuming that silica precipitated with dendritic form and the iron magnesium silicate matrix represents the melt from which silica has crystallized, the interdendrite area within silica spheroids (Fig. 3) should be filled with this matrix material (i.e., the matrix should form reentrants into the silica). Contrary to this, the interdendrite regions are void. Interdendrite voids suggest a phase transformation following formation of the silica-rich spheroids. Prior to this transformation the iron magnesium silicate matrix had congealed because it retains its smooth curvilinear outline (Fig. 3) opposite areas of dendrite growth, which is accompanied by a specific volume decrease. If as expected by dendrite habit (Fenner, 1913) the transformation was in forming cristobalite, then the precursor phase must have been either silica-rich melt or glass.

Secondly, cristobalite precipitated experimentally from magnesium, and iron oxide-silica melts typically exhibit crystal habits described as octahedra, cubes, or stout prisms (Bowen and Andersen, 1914; Van Vlack, 1960). Crystal outlines may be subhedral. The geometric form of silica (Figs. 2c and 3) in contact with matrix in this chondrule does not conform with these observations. Instead, the smooth curvilinear outline suggests that the silica phase was originally non-crystalline.

For the condition that cristobalite and pyroxene coprecipitated from the melt, the matrix surrounding the silica spheroids should be polycrystalline. The matrix in the chondrule does exhibit a grain-like texture; however, the grains are fan-spherulites and not single crystals. Spherulite formation in the matrix is consistent with fibril growth in viscous liquid or a glass.

The chondrule phase relations are more readily explained as derived from two coexisting, originally non-crystalline materials rather than cristobalite crystallizing from melt with or without pyroxene. This phenomenon, known as phase separation, is common to many of the alkaline earth-silicate melts (Greig, 1927; Seward *et al.*, 1968). Under equilibrium conditions it is observed at the silica-rich end of binary systems at temperatures exceeding 1692°C. One of the two melts commonly has >95 wt.% SiO_2, the other varies considerably with temperature. Phase separation can extend to the subliquidus region upon failure to nucleate a crystalline solid (Stevens, 1972), resulting in two metastable liquids.

An evaluation of the thermal conditions necessitating phase separation in this chondrule is obtainable from the matrix composition. Major constituents (>2 wt.%) for the matrix are SiO_2, MgO, and FeO in decreasing proportions. Alumina, at 2 wt.%, is taken into consideration because even this small proportion significantly reduces the immiscibility field (Greig, 1927; Wood and

Hess, 1980). The solvus used is that determined for the $MgO-Al_2O_3-SiO_2$ system by Galakhov *et al.* (1976) as presented by Hess (1980). The appropriate iron-free ternary composition was obtained by recalculating FeO to an equivalent amount of MgO. This has the effect of translating the matrix composition parallel to the silica-poor side of the immiscible dome outline in the system MgO-iron oxide-SiO_2 (Bowen and Schairer, 1935), thus minimizing a temperature error caused by addition of iron oxide.

The solvus temperature corresponding to the adjusted matrix composition in the $MgO-Al_2O_3-SiO_2$ ternary is approximately 1275°C. A less sensitive temperature estimate given by the silica composition is about 1300°C. These temperature-composition relationships represent metastable phase separation according to the data of Galakhov *et al.* (1976) and Ol'shanskii (1951). The low viscosities, extrapolated from magnesium silicate binary data (Bockris *et al.*, 1955; Urbain *et al.*, 1982), indicate that the matrix phase at 1300°C is significantly above its glass transition temperature. Considering that glass transition for pure natural silica is known to be ~1100°C (Brückner, 1970), the phase separation is of the liquid-liquid vs. glass-glass type.

The thermal history for this chondrule cannot be fully evaluated because little of the total chondrule is sampled. The chondrule reported upon by Olsen (1982) from the Murchison (C2) chondrite, which similarly contains silica ellipsoids, contains large volume fractions of enstatite. The phase assemblage in the Murchison chondrule is consistent with a cooling process initiated above the liquidus at 1550°C (Olsen, 1982). In an experimental charge that produced a radial pyroxene chondrule-like texture, Tsuchiyama *et al.* (1980) observed micron-sized spheroids of glass aggregated in evenly dispersed rosettes within a low-Ca pyroxene [crystalline (?)] matrix. This texture resulted from controlled cooling from 1525° to 930°C at 24°C/hr and may indicate liquid-liquid phase separation, although not recognized as such by Tsuchiyama *et al.* (1980).

These observations suggest that the melt now represented by the Piancaldoli chondrule fragment was derived from fractional crystallization. Phases responsible for fractionation could have resided beyond the fragment outline. As the chondrule cooled and fractionation continued, the residual melt became metastable and encroached on the two-liquid solvus, which resulted in phase separation.

CONCLUSIONS

Similarity in spheroid distribution between the synthetic magnesium silicate charge and chondrule fragment shown in Fig. 1 is the result of two different phenomena. Unlike the devitrification features of the synthetic charge, the chondrule has a texture and mineralogy consistent with metastable liquid-liquid phase separation at temperatures near 1275°–1300°C.

Acknowledgments. I thank R. S. Clarke, Jr. of the U.S. National Museum of Natural History for granting access to the Piancaldoli meteorite thin section. I am grateful to P. C. Hess, R. H. Hewins, and M. T. Murrell for constructive reviews of the manuscript, and to K. D. Keefer for stimulating discussion of the topic. I also wish to thank K. Keil for making his analytical facilities available for portions of this research. This work has been performed under the auspices of the U.S. Department of Energy.

REFERENCES

Bockris J. O'M., Mackenzie J. D., and Kitchener J. A. (1955) Viscous flow in silica and binary liquid silicates. *Trans. Faraday Soc.* **51**, 1734–1748.

Bowen N. L. and Andersen O. (1914) The binary system MgO-SiO_2. *Amer. J. Sci.* **187**, 487–500.

Bowen N. L. and Schairer J. F. (1935) The system MgO-FeO-SiO_2. *Amer. J. Sci.* **229**, 151–217.

Brigham C., Murrell M. T., and Burnett D. S. (1982) SiO_2-rich chondrules in ordinary chondrites (abstract). *Meteoritics* **17**, 187–188.

Brückner R. (1970) Properties and structure of vitreous silica. I. *J. Non-Cryst. Solids* **5**, 123–175.

Christophe Michel-Lévy M. (1976) La matrice noire et blanche de la chondrite de Tieschitz (H3). *Earth Planet. Sci. Lett.* **30**, 143–150.

Fenner C. N. (1913) The stability relations of the silica minerals. *Amer. J. Sci.* **186**, 331–384.

Galakhov F. Ya., Aver'yanov V. I., Vailonova V. T., and Aresher M. P. (1976) The metastable liquid phase separation region in the MgO-Al_2O_3-SiO_2 system. *Fiz. Khim. Stekla* (English Trans.) **2**, 405–408.

Greig J. W. (1927) Immiscibility in silicate melts. *Amer. J. Sci.* **213**, 1–44 and 133–154.

Hess P. C. (1980) Polymerization model for silicate melts. In *Physics of Magmatic Processes* (R. B. Hargraves, ed.), pp. 3–48. Princeton University Press, Princeton.

Kurat G. (1970) Zur Genese des kohligen Materials im Meteoriten von Tieschitz. *Earth Planet. Sci. Lett.* **7**, 317–324.

Olsen E. (1982) Melt droplet chondrule in Murchison: Evidence for slow cooling before quench (abstract). In *Papers Presented to the Conference on Chondrules and their Origins*, p. 48. Lunar and Planetary Institute, Houston.

Ol'shanskii Ya. I. (1951) Equilibrium of two immiscible liquid phases in silicate systems containing alkaline earth metals. *Dokl. Akad. Nauk. S.S.S.R.* **76**, 93–96.

Planner H. N. (1974) An experimental investigation of highly undercooled magnesium silicate chondrule-like spherules. M.S. thesis, Univ. of New Mexico, Albuquerque. 89 pp.

Seward T. P., III, Uhlmann D. R., and Turnbull D. (1968) Phase separation in the system BaO-SiO_2. *J. Amer. Ceram. Soc.* **51**, 278–285.

Stevens H. J. (1972) Phase separation of simple glasses. In *Introduction to Glass Science* (L. D. Pye, H. J. Stevens, and W. C. LaCourse, eds.), pp. 197–235. Plenum, N. Y.

Tsuchiyama A., Nagahara H., and Kushiro I. (1980) Investigations on the experimentally produced chondrules: Chemical compositions of olivine and glass and formation of radial pyroxene chondrules. *Mem. Natl. Inst. Polar Res., Spec. Issue 17* (T. Nagata, ed.), pp. 83–94. National Institute of Polar Research, Tokyo.

Urbain G., Bottinga Y., and Richet P. (1982) Viscosity of liquid silica, silicates and alumino-silicates. *Geochim. Cosmochim. Acta* **46**, 1061–1072.

Van Vlack L. H. (1960) Microstructure of silica in the presence of iron oxide. *J. Amer. Ceram. Soc.* **43**, 140–145.

Wood M. I. and Hess P. S. (1980) The structural role of Al_2O_3 and TiO_2 in immiscible silicate liquids in the system SiO_2-MgO-CaO-FeO-TiO_2-Al_2O_3. *Contrib. Mineral. Petrol.* **72**, 319–328.

Can Chondrules Form from a Gas of Solar Composition?

Hans E. Suess
Department of Chemistry, University of California, San Diego, La Jolla, California 92093

William B. Thompson
Department of Physics, University of California, San Diego, La Jolla, California 92093

Chemical composition and petrological texture of individual chondrules as observed by many investigators indicate melting and rapid cooling, as well as relationships with elemental vapor pressures. What would the properties of a dusty gas have to be so that chondrules with the observed properties could form by sudden heating and cooling in a turbulent gas as, for example, in the bow shocks of collisions with larger objects?

GENERAL REMARKS

Chondrules, by definition, are solidified spherical droplets of molten minerals and other substances that occur in meteorites. With the exception of chondrules occurring in enstatite meteorites, chondrules always contain oxidized iron. Therefore, at temperatures above their melting point, they could not have been in equilibrium with a gas containing hydrogen in solar proportions. Oxidized iron may only form below 800° K and in a very tenuous gas, only as a fine dust. Over sufficiently long periods of time, the fine dust particles may agglomerate to produce objects the size of chondrules. In a second high temperature stage, the dust particles must then have been melted to form the chondrules present in meteorites.

It is important to keep in mind that, because of the near constancy of the isotopic composition of the elements, the various genetic components (in particular, the S and R components) must have mixed in a nearly perfect manner before condensates formed out of the primeval gas cloud. It may be mentioned here that oxygen isotopes are a special case: The relatively large non-mass-dependent fractionation of ^{16}O from the heavy oxygen isotopes ^{17}O and ^{18}O (Clayton *et al.*, 1973) can now be understood in the light of the laboratory experiments by Thiemens and Heidenreich (1983). These experiments show that the ratios of the oxygen isotopes in extraterrestrial matter may be the result of photochemical and charged particle reactions. The time at which such fractionation could have occurred is obviously that of the loss of the gases before and during the second high temperature stage.

OBSERVATIONS

In order to explain the presence of both metal and oxidized iron, one must assume that the second high temperature stage must have been either sudden and brief or it must have occurred after nearly all the hydrogen had separated from the condensed particles. In the first case, the hydrogen must not have had enough time to react with the iron oxide to give metal and water. In the second case, a sufficiently large fraction of hydrogen must have left before the second hot period had started, so that only a fraction of the iron was reduced to metal (Herndon and Suess, 1977). No doubt a separation of condensed matter from the gases must have occurred at some time, but no theory is available that would allow us to quantitatively determine the conditions under which this has occurred.

OXIDATION AND EQUILIBRATION

It is not clear yet whether a relatively simple history of typical chondrules can be assumed, *viz.*, condensation at low temperatures from a homogeneous well-mixed gas and subsequent melting of the condensates during a second high temperature stage, or a more complex history involving intermediate parent bodies. However, in either case, it must be assumed that the state of oxidation of the condensate was established by the processes during the second high temperature stage, which one would assume should have led to homogeneous matter and finally to the formation of "equilibrated" chondrites. Inhomogeneous conditions in a highly turbulent gas, however, obviously could have led to the formation of chondrules of varying states of oxidation and to "unequilibrated" chondrites. Some of these appear to have been subjected finally to metamorphic processes and to an "equilibration" of their constituents.

There are, of course, many other ways by which objects with the properties of chondrules could have formed, such as spraying of a melt, collisions, lightning, etc. These possibilities have been discussed extensively in the literature on this subject. In fact, it is quite possible that one of these ways, and not the one we are suggesting, was actually the way in which the chondrules formed. Nevertheless, in order to decide what could and what could not have happened, a more quantitative analysis of the processes assumed by us has been carried out by one of us (W.B.T.). We start out with hot solar matter, completely gaseous so that perfect mixing of genetic components is possible. Then, with decreasing temperature, a fraction of $\sim 10^{-3}$ parts per mass will form dust particles.

SHOCK HEATING

Let us explore the possibility that a pulse of a secondary heating was produced by a strong shock in the gas-dust mixture. There are many possibilities as to how such a shock can be produced. Let us assume that it is produced by the passing of a solid body (meteorite to asteroid size) through the gas at a velocity of 10 km/sec. The post-shock temperature is then of the order of 50 $V^{2\circ}$K, where V is the velocity in km/sec, for a gas primarily of molecular hydrogen. If the initial gas density is $n \simeq 10^{16} cm^{-3}$, grains that are a few microns in size will be heated almost immediately, but will not reach the temperature of the post-shock gas; instead, heat loss by radiation will hold the temperature T_D of the dust at a much lower value:

$$T_D = 122\,[10^{-16} n\, M_G{}^{-1/2}]^{1/2}\, T_G{}^{3/8},$$

where M_G is the molecular weight of the gas, n its density, and T_G its temperature, while the dust grain is assumed to have unit emissivity. For $T_G = 5000°K$, $T_D = 2000°K$, and although the grain remains cooler than the gas, most metals will melt. Since momentum and energy transfer occur at the same high rate, the dust (if of micron size) is swept along with the gas.

CONCLUSIONS AND DISCUSSION

After the shock is past, the gas will cool adiabatically as it reexpands, and its temperature will drop by a factor of 4, carrying the gas temperature with it; however, at this point, cooling becomes much slower and is determined by the radiation loss of the grains. If ~0.1% of the mass is in dust and micron-size grains, the temperature drops below the melting point in a few minutes.

Thus, small dust grains could easily undergo a secondary melting process in a modest shock; however, for the conditions we have considered, collisions between molten grains will be very rare events and coalescence of molten droplets unlikely, unless the dust is already somewhat aggregated.

Acknowledgment. We appreciate financial support through NASA Grant NAGW 103-Supl. 3 for research on meteorites.

REFERENCES

Clayton R. N., Grossman L., and Mayeda T. K. (1973) A component of primitive nuclear composition in carbonaceous meteorites. *Science* **182**, 485–487.

Herndon J. M. and Suess H. E. (1977) Can the ordinary chondrites have condensed from a gas phase? *Geochim. Cosmochim. Acta* **41**, 233–236.

Thiemens M. A. and Heidenreich J. E., III (1983) The mass-independent fractionation of oxygen: a novel isotope effect and its possible cosmochemical implications. *Science* **219**, 1073–1075.

Radiometric Ages of Chondrules

T. D. Swindle, M. W. Caffee, and C. M. Hohenberg
McDonnell Center for the Space Sciences, Washington University, St. Louis, Missouri 63130

Ages of chondrules, as determined by the U-Pb, Rb-Sr (and initial $^{87}Sr/^{86}Sr$), ^{40}Ar-^{39}Ar, and I-Xe chronometers, are reviewed. Due to differences inherent in the dating schemes, they do not always yield the same chronologies, but the specific differences provide supplemental information on the post-accretional histories. Individual chondrules or groups of chondrules seem to be contemporaneous with each other, and with the host chondrites in general, to within the precision of U-Pb and Rb-Sr methods (about 20–30 m.y.). In I-Xe studies, inherently more sensitive to small time differences, age differences between individual chondrules do show up, with a total spread in apparent formation times less than 5 m.y. (10 m.y. if earlier results are included). Although the isotopic structures are complex and interpretation often difficult, ^{40}Ar-^{39}Ar studies of individual chondrules show apparent 200 m.y. age differences between types of chondrules and evidence for excess radiogenic ^{40}Ar. The method is, however, more sensitive to alteration than the I-Xe system and may in fact be reflecting post-accretional thermal metamorphism. The initial strontium isotopic ratios in several Ca-Al-rich Allende chondrules are as primitive as those of any objects yet analyzed, in contrast to the results of I-Xe studies, where the most primitive ages are 5 to 10 m.y. older than those of any of the chondrules. I-Xe studies provide the smallest upper limit to the duration of chondrule formation. Uncertainties in the interpretation of the I-Xe chronometer, due to possible spatial inhomogeneities in the initial iodine isotopic composition, only serve to decrease the spread in real formation ages. Matrix samples give I-Xe ages equal to or older than chondrule ages, but often give Rb-Sr and ^{40}Ar-^{39}Ar ages that are hundreds of millions of years younger, perhaps reflecting the relative sensitivity of these chronometers to subsequent thermal or shock events.

INTRODUCTION

Chondrules are puzzling objects, as evidenced by the multitude of theories about their origin presented in this volume. If actual ages of formation of individual chondrules (as opposed to those of metamorphism) could be determined precisely enough, they would certainly provide a framework with which to evaluate the various models for chondrule formation. Even ages that are established by post-formational metamorphic events can provide some degree of constraint for the evolution of chondrules. In particular, we would like to know the duration of chondrule-forming processes (i.e., what is the maximum difference in age from one chondrule to another), how chondrule ages compare with matrix ages (which formed first and whether one could be formed from the other), whether chondrule formation is dominated by local or global phenomena, and finally how chondrule ages compare with ages of other primitive components like the refractory inclusions of CI and CM meteorites.

In this paper, we will discuss ages of chondrules (either individual chondrules or sets of chondrules) as determined by four isotopic methods widely used in early solar system chronometry: uranium-lead (U-Pb), potassium-argon (as determined by ^{40}Ar-^{39}Ar), rubidium-strontium (Rb-Sr and the associated initial $^{87}Sr/^{86}Sr$), and iodine-xenon (I-Xe). It is not our intent to review these chronometers or the chronology of the early solar system (cf. Wetherill, 1975; Kirsten, 1978), nor will we attempt to resolve the differences among apparently discordant results. What we will do is: (1) summarize chronometric studies of chrondules; (2) discuss the results of each chronometer, identifying where possible current limitations of the method; and (3) attempt to integrate the available data into an overview utilizing inputs from a number of different chronometers.

It should be noted from the outset that, while it is valid to compare different ages determined by the same chronometer, comparisons of ages determined by two different chronometers can be misleading, since very often there are large systematic differences between ages derived by various isotopic systems. While these may be due in part to uncertainties in the decay constants (Podosek, 1979; Minster *et al.*, 1982), the largest effect is due to the different inherent responses of the different chemical/isotopic structures to post-formational metamorphic processes (Nakamura *et al.*, 1976; Unruh *et al.*, 1977).

U-Pb, ^{40}Ar-^{39}Ar, and Rb-Sr are absolute dating schemes, establishing the time of the last major chemical fractionation. Initial strontium and I-Xe, on the other hand, are relative dating schemes, delineating differences in the time interval between two early events: (1) the establishment of an initial reservoir whose isotopic composition is time dependent, and (2) system closure, an event that fixes the parent-daughter patterns, effectively sampling the isotopic composition of the reservoir at that time. The establishment of an isotopic reservoir need not be a singular event clearly defined in time. For example, the I-Xe reservoir was likely established by repetitive r-process additions, hence interpretations of absolute I-Xe ages (or initial iodine isotopic ratios) depend upon the specific formative processes. The utility of the I-Xe system therefore comes from intercomparisons between the times of isotopic closure in different objects. These relative-age schemes are potentially more precise than the absolute-age schemes at determining fine-scale (less than 10 m.y.) chronology in the early solar system. The I-Xe system, for example, is capable of resolving differences in system closure times of less than a hundred thousand years.

URANIUM-LEAD

The two most abundant natural isotopes of uranium (^{238}U and ^{235}U) decay to two different isotopes of lead (^{206}Pb and ^{207}Pb, respectively), providing some unique capabilities for Pb-Pb and U-Pb dating methods. Lead-lead dating, in which variations in the $^{207}Pb/^{204}Pb$ ratio are considered together with variations in the $^{206}Pb/^{204}Pb$ ratio, is a useful absolute chronometer, precise in principle, since it involves measurements of isotopic ratios of one element only. For example, two different groups (Tatsumoto *et al.*, 1976; Chen and Tilton, 1976) have obtained internal isochrons for Allende with stated precisions of 4 m.y. However, the ages they obtain (4553 m.y. and 4565 m.y., respectively) differ by 12 m.y., suggesting that, in spite of the precision, some systematic errors are probably present.

The system is subject to possible contamination with highly radiogenic terrestrial lead (cf. Unruh, 1982) and correction for common lead is not always straightforward. Moreover, lead is sufficiently mobile that the chronometer is relatively easily disturbed (cf. Nakamura *et al.*, 1976; Unruh *et al.*, 1977). Additionally, there are problems of "excess" radiogenic lead in many meteorites (radiogenic lead that cannot be accounted for by the decay of local uranium in 4.5 AE). Disturbances due to these and other effects can often be detected, and themselves dated, in U-Pb "concordia" plots where $^{207}Pb/^{235}U$ is plotted against $^{206}Pb/^{238}U$ after correcting for common lead (cf. Doe, 1970). The beauty of concordia is that all U-Pb systems evolve along the same curve, with the distance along concordia providing the age. Disturbances to the system due to lead loss are reflected by displacements from concordia along a chord toward the origin (zero age). As the chord evolves, the upper intersection with concordia defines the (undisturbed) age and the lower intersection defines the time of disturbance. The presence of "excess" radiogenic lead is demonstrated by displacement to the opposite side of concordia (Wetherill, 1956). Although the source of apparent "excess" lead is not always clear, contributions have been attributed to incorrect choices for the initial lead composition (Tatsumoto *et al.*, 1976), unresolved complexities in the history of the system (Gale, 1979), or terrestrial lead contamination (Unruh, 1982).

The first meteorite isochron was obtained using Pb-Pb dating (Patterson, 1956). This was a composite isochron, suggesting that the ensemble of meteorites represented all formed at approximately the same time. As precision improved, the existence of a single whole rock isochron for all meteorites was brought into question (cf. Gale *et al.*, 1972; Tatsumoto *et al.*, 1973). However, more recent studies again seem to show composite whole-rock isochrons for an ensemble of enstatite chondrites (4577 ± 4 m.y.; Manhes and Allègre, 1978) and equilibrated L (hypersthene) chondrites (4551 ± 7 m.y.; Unruh, 1982), while the model ages of other chondrites usually fall within that range.

Lead isotopes in chondrules or sets of chondrules have been measured in at least four different studies. The 4553 ± 4 m.y. isochron obtained for Allende by Tatsumoto *et al.* (1976) includes data points for three Ca-Al-rich chondrules and six Mg-rich chondrules (although one Mg-rich chondrule was off the isochron). The slightly older (4565 ± 4 m.y.) isochron of Chen and Tilton (1976) includes four chondrule data points, indicating that Allende chondrules of all types appear to have begun retaining lead quantitatively within a few million years of one another. Chondrule samples also fall on internal isochrons for the L5 Barwell (Unruh *et al.*, 1979a) and the H5 Richardton (Unruh *et al.*, 1979b) meteorites. The Barwell isochron gives an age that is, within errors, the same as Allende. Richardton gives a slightly younger Pb-Pb age (4538 ± 7 m.y.), but the data intersect the U-Pb concordia at 4545 ± 13 m.y.

RUBIDIUM-STRONTIUM

Rubidium-strontium dating, based on the decay of ^{87}Rb to ^{87}Sr, has been successfully applied to most classes of chondrites. A series of whole rock studies (cf. Minster *et al.*, 1982, for a summary) indicate that all classes seem to be isochronous to within errors at 4.498 ± 0.015 AE, presumably establishing the time of separation from a common source. The same conclusion was reached a decade earlier in a similar series of experiments with correspondingly lower precision (cf. Wetherill, 1972). Internal isochrons give ages equal to or younger than this age, with the younger ages generally interpreted as metamorphic ages.

Chondrule isochrons (internal isochrons consisting only of data points from chondrules) have been obtained for three meteorites. Those from Parnallee, an LL3 chondrite, define an isochron corresponding to 4.53 ± 0.02 AE (Hamilton *et al.*, 1979), slightly older than the whole rock isochron of Minster *et al.* (1982), although the errors overlap. Meanwhile, chondrules from an LL4, Soko Banja (Minster and Allègre, 1981), and an H5, Richardton (Evensen *et al.*, 1979), gave somewhat younger ages, 4.452 ± 0.020 AE and 4.39 ± 0.03 AE, respectively. In both studies the younger ages were interpreted as reflecting metamorphism rather than the time of formation.

Other work that included chondrules from type 3 chondrites has generally obtained similar results, with most chondrules from Tieschitz (Minster and Allègre, 1979) and Chainpur (Minster and Allègre, 1981) falling on or near the whole rock isochron. However, olivine (Mg-rich) chondrules from Allende (C3V) give model ages ranging from 3.6 to 4.6 AE (Gray *et al.*, 1973; Tatsumoto *et al.*, 1976). The spread is interpreted as evidence for recent (3.6 AE ago or later) disturbance.

Other Rb-Sr studies involving chondrules have also shown signs of disturbances, demonstrating the relative vulnerability of the Rb-Sr system. In a study of Bjurböle (LL4) by Shield *et al.* (1966), none of the components analyzed, including a sample consisting only of chondrules, fell on the 4.5 AE isochron, although the chondrules came the closest. This behavior was attributed to alteration by the seawater in which the meteorite landed, which the authors argued had effected the chondrules less than the rest of the meteorite. Murthy and Compston (1965) analyzed four chondrules from Peace River (L6), and found that they roughly defined an apparent isochron, with an age of 4.5 ± 0.8 AE. Two chondrules from Guidder (LL5) gave model ages of about 4.5 AE, but the system has apparently suffered some disturbance (Minster and Allègre, 1981).

The previous studies all involved whole chondrules (model ages) or whole rock isochrons involving suites of chondrules. To date, no internal Rb-Sr isochrons on individual chondrules have been reported. However, Papanastassiou and Wasserburg (1981) have developed a direct-loading mass spectrometric technique for Rb-Sr analyses that may now make such a study feasible. In a preliminary study, they determined model ages for single chondrules from Olivenza (LL5, 4.48 ± 0.05 AE) and Murchison (CM, about 1.6 AE). The Murchison result is consistent with other studies that suggest that Murchison and other CI and CM meteorites apparently have disturbed isotopic systems (Middlefehldt and Wetherill, 1979). To summarize the Rb-Sr results, chondrules in unequilibrated chondrites appear to be isochronous to within the precision of Rb-Sr dating, typically about 50 million years.

A potentially sensitive variant of Rb-Sr dating involves comparing the initial $^{87}Sr/^{86}Sr$ ratios present in the objects at the time of isotopic closure (Papanastassiou and Wasserburg, 1969). This ratio, which often can be determined to better than one part in 10^4 in rubidium-poor objects, increases with time in any reservoir containing ^{87}Rb. Therefore, if the initial Rb/Sr ratio of the reservoir is known, a time scale can be determined. Fractionation events less than 10 m.y. apart can potentially be resolved with this system (Gray *et al.*, 1973; Birck and Allègre, 1978). The most primitive (lowest) initial $^{87}Sr/^{86}Sr$ yet found is defined by refractory objects in Allende (Gray *et al.*, 1973; Wetherill *et al.*, 1973; Nyquist *et al.*, 1973; Tatsumoto *et al.*, 1976; Nagasawa and Jahn, 1976), including Ca-Al-rich chondrules (Gray *et al.*, 1973; Tatsumoto *et al.*, 1976). Due to their apparent antiquity, it has been proposed that Ca-rich chondrules from unequilibrated ordinary chondrites might provide a better determination of a primitive initial $^{87}Sr/^{86}Sr$ ratio applicable to other meteorites and thus more tightly constrain their internal isochrons (Minster and Allègre, 1981). However, for the time being, it can only be said that no objects have been found in the solar system with initial $^{87}Sr/^{86}Sr$ ratios more primitive than Allende Ca-Al-rich chondrules and inclusions. While this result differs with that obtained from I-Xe studies (described below), the two chronometers differ in their response to disturbance.

^{40}ARGON-^{39}ARGON

^{40}Argon-^{39}argon dating, a variant of K-Ar dating, is based on the decay of ^{40}K to ^{40}Ar. Neutron irradiation converts a fraction of the ^{39}K to ^{39}Ar so that the radiogenic ^{40}Ar still correlated with its potassium parent can be identified. The argon is extracted by heating the sample to progressively higher temperatures (step-wise heating), thereby ordering the released gases on the basis of their thermal properties. The lower temperature extractions, sampling sites with lower argon retentivity, frequently give low apparent model ages, indicating prior loss of radiogenic ^{40}Ar during the intervening aeons. But high-temperature extractions, sampling the more retentive sites, often provide a group of concordant model ages, usually displayed as a "plateau" of uniform age (equivalent to an isochron in other dating methods). This plateau age dates the last thermal event, usually interpreted as either the time of formation or metamorphism strong enough to completely reset the K-Ar clock. Occasionally finer structure is seen in the age plot that can be identified with discrete events that partially reset the K-Ar system (Merrihue and Turner, 1966; Jessberger *et al.*, 1978; Bogard and Hirsch, 1980).

One problem with ^{40}Ar-^{39}Ar techniques in fine-grained material is the effect of recoil in the $^{39}K(n,p)^{39}Ar$ reaction (Huneke and Smith, 1976), which alters the release pattern of ^{39}Ar, thereby disturbing the plateau. However, it is possible that the redistributed ^{39}Ar may not be totally lost from the sample. When the low-temperature fractions show little loss of argon, recoil-redistributed ^{39}Ar might be quantitatively retained (although perhaps no longer associated with the host mineral). In such cases total ^{40}Ar-^{39}Ar ages, equivalent to K-Ar ages, might be preferred; however, the method is now incapable of producing a plateau (isochron), and correspondingly contains no built-in confirmation, so caution must be used in the interpretation of the results.

In an extensive ^{40}Ar-^{39}Ar study of the early solar system, Turner *et al.* (1978) analyzed 16 unshocked chondrites and found plateau ages from 4.42 to 4.52 AE, most of them indistinguishable from 4.48 ± 0.03 AE. Chondrules from Saratov (L4), Barwell (L5), and Olivenza (LL5), also analyzed in this study, all gave ages slightly (less than 1σ) older than the whole rocks. In an earlier study, Podosek (1971) found an ^{40}Ar-^{39}Ar age for Allegan (H5) chondrules of about 4.5 AE, but a sample of Chainpur chondrules gave an unusual release pattern (with no plateau) and a total apparent ^{40}Ar-^{39}Ar age of about 4.7 AE. No explanation was offered for this apparent ancient age, but it is similar to anomalous results obtained for some individual chondrules in a more recent ^{40}Ar-^{39}Ar study that also includes Chainpur, discussed below. I. Flohs (1980) from the Heidelberg group dated feldspar and pyroxene separates from Estacado, Shaw, Kernouve, Guarena, St. Severin, Nadiabondi, and Peetz by the ^{40}Ar-^{39}Ar method, obtaining good age plateaus for every sample except Peetz with ages clustering nicely at 4.48 AE (to a precision of about 10 m.y.).

Most of the ^{40}Ar-^{39}Ar analyses of individual chondrules have been done by the Heidelberg group. In a study of objects from Allende (Jessberger *et al.*, 1980), six groups of chondrules were analyzed, all of which gave ages between 4.51 and 4.63 AE, with 1σ errors that overlapped at about 4.57 to 4.58 AE. Due to the apparent presence of recoil effects, most of the analyses were total ^{40}Ar-^{39}Ar, with only one analysis involving radiogenic argon correlated with potassium, as demonstrated by the usual plateau plot. Several Allende inclusions in the same study gave apparent ages greater than 4.6 AE. Possible interpretations of such anomalously long apparent ages must be considered rather tentative since it is clear that parent and daughter no longer correlate.

In its most recent study, the Heidelberg group has analyzed well-characterized individual chondrules from Chainpur (LL3), Mezö Madara (LL3), Seres (H3,4), Saratov (L4), and Bjurböle (L4), as well as Ca-Al-rich chondrules from carbonaceous chondrites. So far, only preliminary results have been reported (Herrwerth *et al.*, 1982). The chondrules contain little potassium, presumably inhomogeneously distributed among fine-grained phases so that recoil effects during the irradiation are probably responsible for the complex apparent age spectra. No convincing plateaus are observed with model ages for the low-temperature extractions frequently higher than those observed for the more retentive sites, a typical signature of ^{39}Ar recoil effects (Huneke and Smith, 1976). If, in spite of the redistribution, argon losses are small, total ^{40}Ar-^{39}Ar ages (rather than plateau ages) may be significant. Within the framework of this assumption there seems to be a systematic apparent age difference between two morphological types of chondrules. The barred chondrules seem to cluster at about 4.65 AE, roughly 200 m.y. older than porphyritic chondrules. No clear correlations are evident between apparent age and petrologic type or chemical trend. It should be pointed out that I-Xe studies on individual Bjurböle chondrules do detect age differences between chondrules, but such differences are two orders of magnitude smaller (1 or 2 m.y.). Whether this is due to differences in the stabilities of the K-Ar and I-Xe systems or is indicative of difficulties in the method is unclear. In both Heidelberg studies, matrix samples gave ages as young as or younger than that of the youngest chondrule from that meteorite, typically by several hundred million years, in apparent agreement with Rb-Sr studies but, again, not with the I-Xe results.

IODINE-XENON

Iodine-xenon dating is based on the now extinct radionuclide ^{129}I, which, with a half-life of 17 million years, has long since decayed to ^{129}Xe. The presence of radiogenic ^{129}Xe in iodine-bearing mineral sites confirms the presence of live ^{129}I at a time the material stabilized sufficiently to quantitatively retain xenon and maintain its association with iodine. The technique is similar to ^{40}Ar-^{39}Ar in that neutron irradiation converts some ^{127}I to ^{128}Xe, which becomes a tracer for iodine. The gas is released by step-wise heating to progressively higher temperatures. Xenon from the less retentive sites is sampled in the lower temperature fractions, with gas from the more retentive sites released only at the high temperatures, effectively ordering the xenon on the basis of the thermal properties of the host mineral phases. Due to losses from the less tenacious sites, iodine-derived ^{128}Xe released in the lower temperature fractions is typically unaccompanied by the full complement of radiogenic ^{129}Xe. Above a certain temperature, however, the iodine-derived xenon is generally characterized by a constant ratio of reactor-produced ^{128}Xe and radiogenic ^{129}Xe, demonstrating a constant iodine isotopic composition at the time of isotopic closure for xenon in those (higher temperature) mineral sites. Isochronous closure is revealed in the form of an isochron, in this case a mixing line between xenon of normal composition ("trapped" xenon) and iodine-derived xenon of constant composition. The linear isochron actually provides both a measure of the initial iodine isotopic composition and confirmation of its uniformity among all the host mineral sites (cf. Caffee *et al.*, 1982b). Since the initial iodine composition changes with a half-life of 17 m.y., the method is quite sensitive to small differences in xenon closure times. However, the utility of the I-Xe system as a dating scheme, rather than just a measure of the initial iodine isotopic composition, requires assumptions about the spatial homogeneity of the iodine isotopic reservoir.

If the iodine isotopic composition of the early solar system was spatially homogeneous, and changing with time only because of the decay of ^{129}I, then the initial $^{129}I/^{127}I$ ratio is quite a sensitive chronometer, capable of dating events with a precision (1σ) of less than 100,000 years in favorable cases (Caffee *et al.*, 1982b). One must, however, allow for the possibility that some spatial heterogeneity may also be present, in which case the initial iodine isotopic composition (that present at the time of cessation of xenon mobility in retentive trapping sites) may not have exclusively a chronometric correspondence. It does, however, place an upper limit on the actual differences in xenon closure times.

It is easy to demonstrate that if some of the differences observed in initial iodine isotopic compositions are contributed by spatial effects, the differences due to temporal causes alone must be less than the total differences observed. While one may, by chance, have the two effects in opposite directions and roughly equal magnitude in an isolated circumstance, the two effects are in fact independent. Therefore, among any reasonable ensemble of samples, the total observed spread in initial iodine isotopic compositions should equal the square root of the sum of the squares of the contributions due to each of the two independent effects. Consequently, if spatial as well as temporal effects are present, the range of actual xenon closure times must be less than the range inferred by I-Xe dating. This is an important constraint because the precision of I-Xe dating is such that small differences in closure times can be observed, and most observed differences in apparent I-Xe formation times are indeed small.

Iodine-xenon studies have recently been performed on individual chondrules from Allende and Bjurböle. Of ten Bjurböle chondrules investigated (Caffee *et al.*, 1982b), eight yield relatively precise isochrons indicating uniform (to within a percent) but individually distinct initial iodine isotopic compositions for each of the chondrules. The total range in initial compositions, about 6.5%, corresponds to a 1.5 m.y. spread in apparent formation times (Fig. 1). Whereas this time interval is considerably less than differences in formation time observed by other methods, it nevertheless must be considered an upper limit for actual differences in xenon closure times, allowing for the possibility of differences due to non-uniform (spatially heterogeneous) initial iodine.

Isochrons derived in I-Xe dating allow determination not only of the iodine isotopic composition at xenon closure, but also the composition of indigenous "trapped" xenon. A sample of "whole rock" Bjurböle, studied along with the individual chondrules, provides a relatively

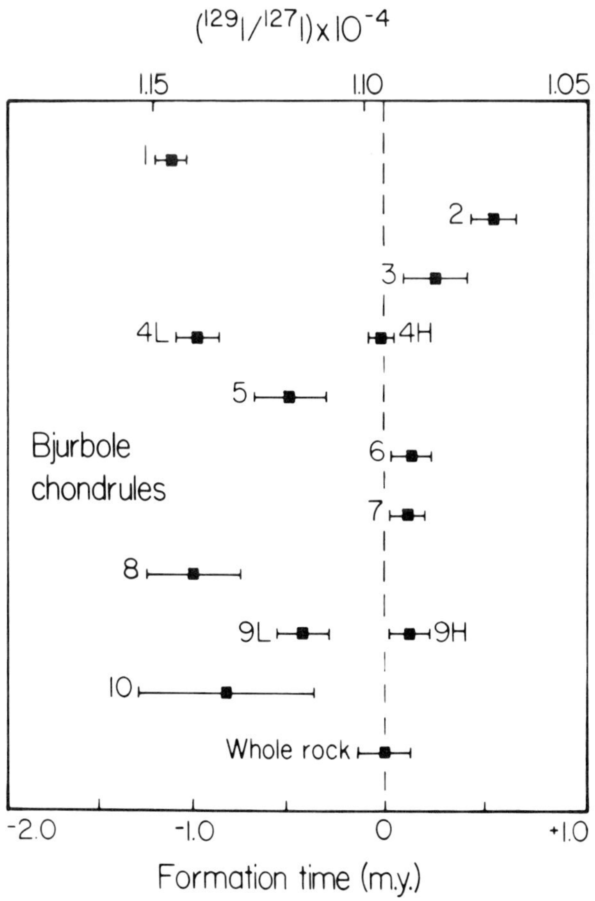

Fig. 1 I-Xe age (bottom) or initial iodine composition (top) of Bjurböle chondrules. Error bars represent 1σ uncertainty. (From Caffee *et al.*, 1982b).

precise composition for "trapped" xenon that is similar to (though resolved from) that representative of Average Carbonaceous Chondrites (AVCC). It is interesting to note that while data for most of the Bjurböle chondrules are consistent with the trapped composition found for whole rock Bjurböle, chondrule 10, which had the smallest xenon concentration, shows evidence for a trapped component substantially (about 30%) enriched in ^{129}Xe. Its initial iodine isotopic composition is, however, within the range of the other chondrules.

The two Bjurböle chondrules that do not yield isochrons have the same unusual isotopic structure. Of the four high temperature extractions (which were isochronous for the other chondrules), the two highest temperature ones had somewhat lower (younger) $^{129}I/^{127}I$ ratios. This is, of course, opposite of what would be expected if the difference were due to partial xenon loss or to the presence of distinctly non-isochronous phases (the higher temperature phase should have formed first, not last). Such reasonably rare (in Bjurböle) effects were attributed to either non-thermal (e.g. shock) disturbances, which can create artifacts not resolvable by step-wise heating (Caffee *et al.*, 1982a), or to an iodine isotopic heterogeneity of about 1.5% (Caffee *et al.*, 1982b).

Of the nine Allende chondrules analyzed (Swindle *et al.*, 1983), only one gives a good isochron (with an age comparable to the most recent of the Bjurböle chondrules). In the other eight chondrules, it was found that model initial iodine compositions for the various extractions correlate quite well with release temperature, with older (richer in ^{129}I) components being released at higher temperatures (Fig. 2). This correlation is consistent with a cooling sequence, but the implied rate of cooling, through the range of temperatures characteristic of xenon closure, is quite slow (100–200K/m.y.). This is orders of magnitude too slow to represent the cooling of an individual chondrule isolated in cold space. It is, however, difficult to see how chondrules would remain

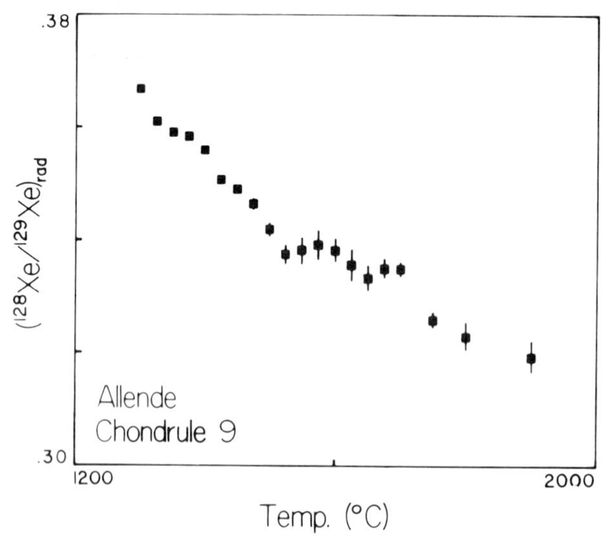

Fig. 2 Apparent model iodine-derived $^{128}Xe/^{129}Xe$ vs. extraction temperature for a 100 mg Allende chondrule (chondrule 9). Temperature given is that of the extraction coil, which is 200°–300° hotter than the chondrule itself. Note that more primitive compositions are richer in ^{129}Xe, and thus plot lower on the figure. (From Swindle *et al.*, 1983).

spherical and mineralogically distinct from their surroundings if they were in thermal contact with a larger object (such as a parent body or hot nebula) at such high temperatures. Furthermore, it is difficult to attribute the pattern to iodine inhomogeneity. Since chondrules apparently went through a molten phase, one would probably expect the iodine isotopes within a single chondrule to have been homogenized. Likewise, mass-dependent fractionation during the experiment cannot account for the structure, since simple Rayleigh distillation is quantitatively unable by several orders of magnitude to produce the observed range of compositions observed.

In all of the Allende chondrules, one or more apparent plateau-like structures can be found in a plot such as Fig. 3, suggesting that the I-Xe system may show the mixing of a few discrete phases or a few types of sites rather than a continuum (Fig. 4). If so, the correlation of initial iodine composition with temperature could indeed be reflecting the thermal history of the chondrules, with the higher temperature plateaus representing one event (such as chondrule formation followed by rapid cooling), and the lower temperature plateaus others (such as reheating during accretion, impact-driven processes strong enough to reset the I-Xe clock at the lower-temperature sites, or slower cooling at the lower temperatures). It should be noted that in I-Xe studies of Allende inclusions, other authors (Zaikowski, 1979, 1980; Podosek and Lewis, 1972) have appealed to multiple phases with different compositions within a single object to explain observations of multiple temperature-ordered isochrons for each inclusion. However, the age differences they obtained were typically about 4 m.y. (larger than those found in the chondrules) and were

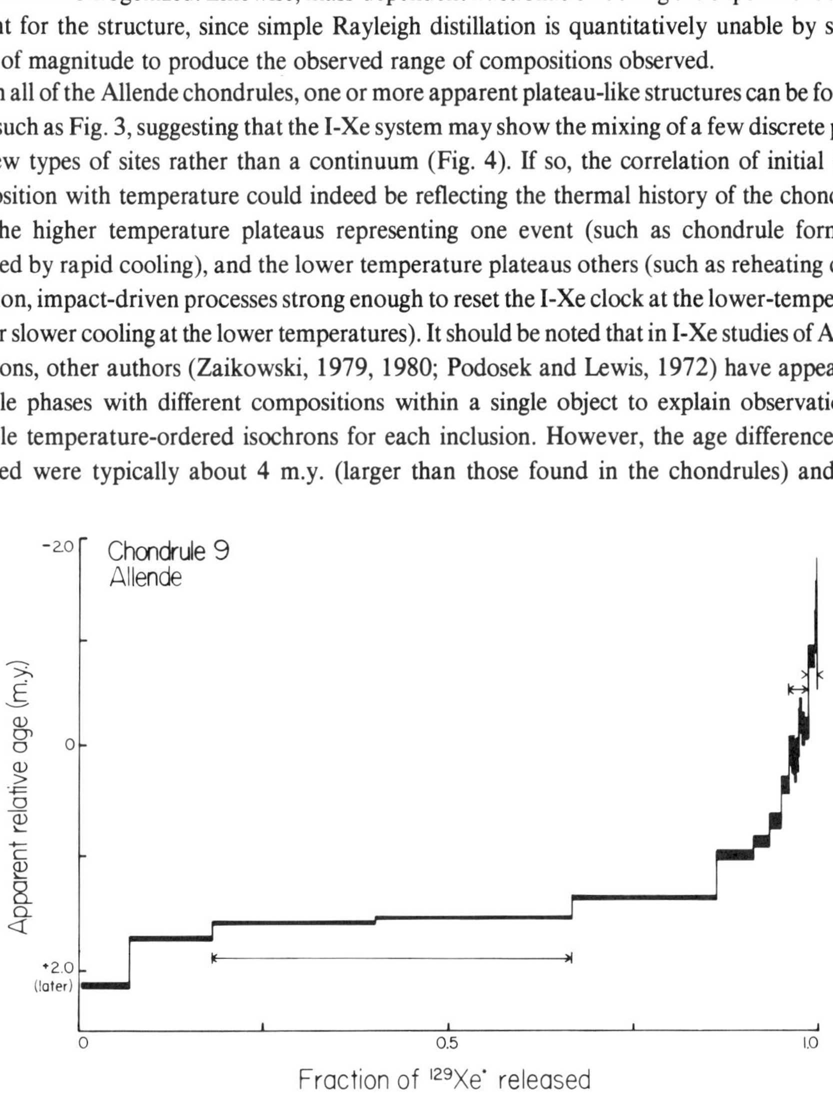

Fig. 3 Apparent age (relative to Bjurböle whole rock) vs. fraction of ^{129}Xe released for Allende chondrule 9 (see Fig. 2). Extents of apparent plateaus are indicated by arrows. (From Swindle *et al.*, 1983).

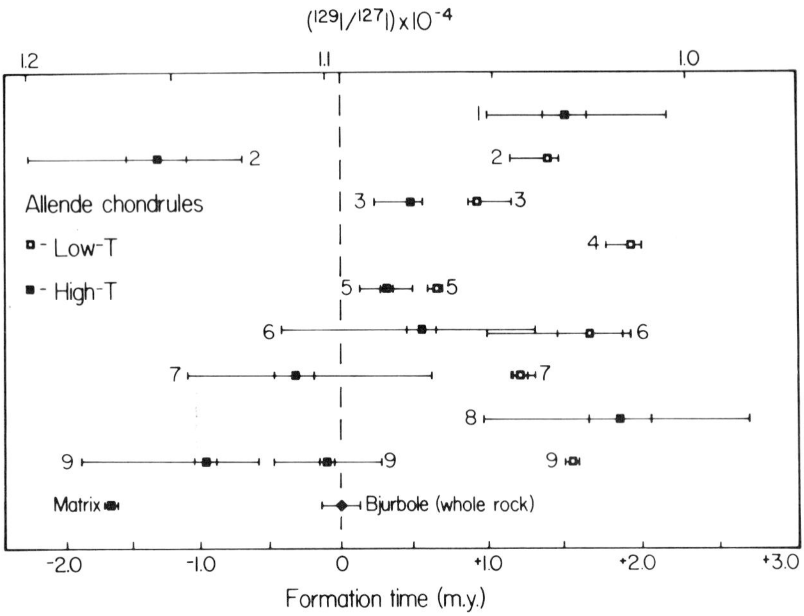

Fig. 4 Apparent plateau ages (and initial iodine compositions) of Allende chondrules. Outer error bars represent extremes of 1σ uncertainties of model ages of temperature extractions making up the plateau, while inner error bars represent 1σ uncertainty of the $^{129}I/^{127}I$ ratio determined from the cumulative gas released in the plateau region. (From Swindle *et al.*, 1983).

isochronous between the two inclusions studied. This parallel, both in age and in isotopic structure, suggests that the chondrules and the inclusions, while not identical, are similar enough to have undergone similar histories.

Podosek's (1970) study of several meteorites by I-Xe techniques included sets of chondrules from Chainpur (LL3) and Allegan (H5). A single chondrule from Bruderheim has been analyzed (Merrihue, 1966; see also Podosek, 1970), giving an age 2 to 10 m.y. later than the whole rock. One sodalite-rich chondrule from Allende was analyzed (Wasserburg and Huneke, 1979), yielding an initial iodine composition comparable to the low-temperature plateaus in the Allende chondrules discussed above (thought also to be due to sodalite phases; Swindle *et al.*, 1983). The ages obtained in all I-Xe studies of chondrules range from about 2 m.y. before Bjurböle whole rock (Allegan and the most primitive compositions in Bjurböle and Allende chondrules) to about 10 m.y. after Bjurböle (Chainpur), although the total range in the more recent Bjurböle and Allende studies is only about 4 m.y. (see Fig. 1 and Fig. 4).

In contrast to the initial strontium studies previously discussed, where chondrules and inclusions from Allende provide the most primitive ratios, initial iodine ratios seemingly more primitive than those observed in chondrules are found in silicate inclusions in iron meteorites

(Niemeyer, 1979), magnetite from CI chondrites (Lewis and Anders, 1975), and some carbonaceous and ordinary chondrite whole rock samples (Crabb *et al.*, 1982; Jordan *et al.*, 1980). However, the Allende chondrules that give the primitive initial strontium ratios are the Ca-Al-rich variety, while the Allende chondrules studies by I-Xe were apparently olivine- or pyroxene-rich (Swindle *et al.*, 1983). These differences, however, may be of lesser importance than the fundamental differences in the various chronometers, discussed in the last section.

Samples of the fine-grained matrix material have also been analyzed in many of the I-Xe studies. A sample of Allende matrix (Swindle *et al.*, 1983) gave an initial iodine ratio comparable to the most primitive ratios seen in the Allende chondrules (Fig. 4); Chainpur matrix (Podosek, 1970) gave an age 7 m.y. before the chondrules, and Allegan matrix (Podosek, 1970) gave an age slightly older (although the same within errors) than the chondrules. This would seem to indicate that matrix material pre-dates chondrules, at least for the I-Xe chronometer. This is seemingly in conflict with results obtained by Rb-Sr and ^{40}Ar-^{39}Ar dating, which both give matrix ages approximately 500 m.y. younger than chondrules ages. The magnitude of the age difference, as well as the sign, tells us that these different chronometers are providing different information.

It is most likely that the discordancy is only an apparent one due to the difference in the chronometers, with I-Xe seemingly the most stable. As an example, Bjurböle, which has a well-studied and well-behaved I-Xe structure with a precise and reproducible isochron (Hohenberg and Kennedy, 1980; Caffee *et al.*, 1982a), shows significant disturbance of its Rb-Sr system (Shields *et al.*, 1966) and its K-Ar system (Hohenberg *et al.*, 1981). The I-Xe clock, which utilizes the correlation of I with ^{129}Xe in high-temperature sites only, has been shown to be insensitive to heating to at least 1000°C on laboratory time scales (Hohenberg and Reynolds, 1969; Rison and Zaikowski, 1980). It has also been shown by studies on lunar troctolite 76535 that xenon trapped in inclusions within high temperature phases can survive extensive metamorphic alteration with its isotopic structure intact (Caffee *et al.*, 1981).

CONCLUSIONS

(1) Probably the most important conclusion is the difference in information provided by the different radiochronometers. Parent-daughter relationships are disturbed by mild metamorphism when one or both of the members are structurally involved in the mineral assemblege. The K-Ar and I-Xe systems both have noble gas daughter products and consequently behave as predicted when driven only by thermal diffusion. They are, however, fundamentally different in most other respects. Potassium participates structurally so that radiogenic ^{40}Ar generally resides within the lattice. It is, however, rather mobile and more susceptible than xenon to diffusion during thermal metamorphism. Iodine, present only in trace amounts in most structures, is more abundant in the interstitials and along grain boundaries. Radiogenic ^{129}Xe produced at the retentive iodine sites is less susceptible to diffusive loss and the I-Xe system has been shown to be rather stable against metamorphic processes (Caffee *et al.*, 1981; Hohenberg and Reynolds, 1969). It is, however, more easily disturbed by non-thermal processes such as shock, presumably due to the compressibility of

some of the less retentive iodine sites (Caffee *et al.*, 1982a; Jessberger and Ostertag, 1982). It may be somewhat surprising but, excluding shock effects, the I-Xe system seems to have the greatest stability against metamorphic change.

(2) The absolute ages of chondrules or groups of chondrules by U-Pb and Rb-Sr techniques are usually indistinguishable from each other and from the host chondrite to within the uncertainties of such absolute-age techniques (about 10 m.y. or more). Younger apparent ages of chondrules or groups of chondrules are seen, but are interpreted as reflecting thermal metamorphism or other disturbance.

(3) Although earlier ^{40}Ar-^{39}Ar studies gave results similar to the above, the most recent experiment, in which many individual chondrules have been analyzed, seems to show apparent 200 m.y. age differences between types of chondrules. Recoil effects were apparently present in this work, which precluded the use of the usual ^{40}Ar-^{39}Ar correlation (plateau) techniques. It is unlikely that both sets of apparent ages in fact correspond to real formation times, since the time span is at least an order of magnitude longer than those obtained by any other chronometer.

(4) Individual chondrules from Allende and Bjurböle do not appear to be isochronous when studied using the I-Xe system, in principle the most sensitive dating scheme presently available. Variations of up to about 1.5 m.y. are seen among Bjurböle chondrules. The I-Xe systems of Allende chondrules seem to record more thermal history, suggesting cooling rates of 100–200 K/m.y. or reheating a few million years after formation. The total range of chondrule ages from all studies is about 10 m.y., while for recent studies, it is less than 5 m.y. In either case, I-Xe gives the smallest upper limit to the duration of chondrule formation. Uncertainties in the chronometer, such as spatial inhomogeneities, only serve to decrease the spread of real formation ages.

(5) Matrix samples give I-Xe ages equal to or older than chondrules ages, but often give Rb-Sr and ^{40}Ar-^{39}Ar ages that are hundreds of millions of years younger. This may simply reflect the relative greater sensitivity of the Rb-Sr and ^{40}Ar-^{39}Ar chronometers to subsequent heating or shock events.

Acknowledgments. We would like to acknowledge Elbert King's role in this paper; not only as organizer of the chondrule conference, but also as one who provided (along with Carleton Moore) the samples we analyzed. We appreciate Laurel Wilkening's patient help as associate editor. We particularly appreciate Elmar Jessberger's willingness to discuss unpublished results and his many helpful discussions. We have also benefited greatly from close association with our colleague Ghislaine Crozaz. This work was supported in part by NASA grant NAG 9-7.

REFERENCES

Birck J. L. and Allègre C. J. (1978) Chronology and chemical history of the parent body of basaltic achondrites studied by the ^{87}Rb-^{87}Sr method. *Earth Planet. Sci. Lett.* **39**, 37–51.

Bogard D. D. and Hirsch W. C. (1980) $^{40}Ar/^{39}Ar$ dating, Ar diffusion properties, and cooling rate determinations of severely shocked chondrites. *Geochim. Cosmochim. Acta* **44**, 1667–1682.

Caffee M. W., Hohenberg C. M., Hörz F., Hudson B., Kennedy B. M., Podosek F. A., and Swindle T. D. (1982a) Shock disturbance of the I-Xe system. *Proc. Lunar Planet. Sci. Conf. 13th,* in *J. Geophys. Res.* **87**, A318–A330.

Caffee M. W., Hohenberg C. M., and Hudson B. (1981) Troctolite 76535: a study in the preservation of early isotopic records. *Proc. Lunar Planet. Sci. 12B,* pp. 99–115.

Caffee M. W., Hohenberg C. M., Hudson B., and Swindle T. D. (1982b) I-Xe ages of individual Bjurböle chondrules. *Proc. Lunar Planet. Sci. Conf. 13th,* in *J. Geophys. Res.* **87**, A303–A317.

Chen J. H. and Tilton G. R. (1976) Isotopic lead investigations on the Allende carbonaceous chondrite. *Geochim. Cosmochim. Acta* **40**, 635–643.

Crabb J., Lewis R. S., and Anders E. (1982) Extinct ^{129}I in C3 chondrites. *Geochim. Cosmochim. Acta* **46**, 2511–2526.

Doe B. R. (1970) *Lead Isotopes*. Springer-Verlag, N. Y. 137 pp.

Evensen N. M., Carter S. R., Hamilton P. J., O'Nions R. K., and Ridley W. I. (1979) A combined chemical-petrological study of separated chondrules from the Richardton meteorite. *Earth Planet. Sci. Lett.* **42**, 223–236.

Flohs I. (1980) ^{40}Ar-^{39}Ar Datierbung von Mineralseparaten gewöhnlicher Chondrite. Diplomarbeit, Max-Planck-Institut für Kernphysik.

Gale N. H. (1979) U-Pb systematics in Allende and a possible explanation of apparent excess radiogenic lead in chondritic meteorites. *Geochem. J.* **13**, 191–199.

Gale N. H., Arden J., and Hutchison R. (1972) Uranium-lead chronology of chondritic meteorites. *Nature Phys. Sci.* **240**, 56–57.

Gray C. M., Papnastassiou D. A., and Wasserburg G. J. (1973) The identification of early condensates from the solar nebula. *Icarus* **20**, 213–239.

Hamilton P. J., Evensen N. M., and O'Nions R. K. (1979) Chronology and chemistry of Parnallee (LL3) chondrules (abstract). In *Lunar and Planetary Science X*, pp. 494–496. Lunar and Planetary Institute, Houston.

Herrwerth I., Jessberger E. K., Muller N., and Kirsten T. (1982) ^{40}Ar-^{39}Ar dating of individual chondrules (abstract). In *Papers Presented to the Conference on Chondrules and their Origins*, p. 24. Lunar and Planetary Institute, Houston.

Hohenberg C. M. and Kennedy B. M. (1980) I-Xe dating: intercomparison of neutron irradiations and reproducibility of the Bjurböle standard. *Geochim. Cosmochim. Acta* **45**, 251–256.

Hohenberg C. M. and Reynolds J. H. (1969) Preservation of the iodine-xenon record in meteorites. *J. Geophys. Res.* **74**, 6679–6683.

Hohenberg C. M., Hudson B., Kennedy B. M., and Podosek F. A. (1981) Noble gas retention chronologies for the St. Severin meteorite. *Geochim. Cosmochim. Acta* **45**, 535–546.

Huneke J. C. and Smith S. P. (1976) The realities of recoil: ^{39}Ar recoil out of small grains and anomalous age patterns in ^{40}Ar-^{39}Ar dating. *Proc. Lunar Sci. Conf. 7th*, pp. 1987–2008.

Jessberger E. K. and Ostertag R. (1982) Shock-effects on the K-Ar system of plagioclase feldspar and the age of anorthosite inclusions from North-Eastern Minnesota. *Geochim. Cosmochim. Acta* **46**, 1465–1471.

Jessberger E. K., Dominik B., Staudacher T., and Herzog G. F. (1980) ^{40}Ar-^{39}Ar ages of Allende. *Icarus* **42**, 380–405.

Jessberger E. K., Staudacher T., Dominik B., and Kirsten T. (1978) Argon-argon ages of matrix samples from Apollo 17 breccia 73255. *Proc. Lunar Planet. Sci. Conf. 9th*, pp. 841–854.

Jordan J., Kirsten T., and Richter H. (1980) ^{129}I/^{127}I: A puzzling early solar system chronometer. *Z. Naturforsch.* **35a**, 145–170.

Kirsten T. (1978) Time and the solar system. In *The Origin of the Solar System* (S. F. Dermott, ed.), pp. 267–346. Wiley and Sons, N. Y.

Lewis R. S. and Anders E. (1975) Condensation time of the solar nebula from extinct ^{129}I in primitive meteorites. *Proc. Nat. Acad. Sci. USA* **72**, 268–273.

Manhes G. and Allègre C. J. (1978) Time differences as determined from the ratio of lead 207 to lead 206 in concordant meteorites (abstract). *Meteoritics* **13**, 543–548.

Merrihue C. (1966) Xenon and krypton in the Bruderheim meteorite. *J. Geophys. Res.* **71**, 263–313.

Merrihue C. and Turner G. (1966) Potassium-argon dating by activation with fast neutrons. *J. Geophys. Res.* **71**, 2852–2857.

Middlefehldt D. W. and Wetherill G. W. (1979) Rb-Sr studies of CI and CM chondrites. *Geochim. Cosmochim. Acta* **43**, 201–206.

Minster J. -F and Allègre C. J. (1979) ^{87}Rb-^{87}Sr chronology of H chondrites: constraint and speculations on the early evolution of their parent body. *Earth Planet. Sci. Lett.* **42**, 333–347.

Minster J. -F. and Allègre C. J. (1981) ^{87}Rb-^{87}Sr dating of LL chondrites. *Earth Planet. Sci. Lett.* **56**, 89–106.

Minster J. -F., Birck J. L., and Allègre C. J. (1982) Absolute age of formation of chondrites studied by the ^{87}Rb-^{87}Sr method. *Nature* **300**, 414–419.

Murthy V. R. and Compston W. (1965) Rb-Sr ages of chondrules and carbonaceous chondrites. *J. Geophys. Res.* **70**, 5297–5307.

Nagasawa H. and Jahn B. M. (1976) REE distribution and Rb-Sr systematics of Allende inclusions (abstract). In *Lunar Science VII*, pp. 591–593. The Lunar Science Institute, Houston.

Nakamura N., Tatsumoto M., Nunes P. D., Unruh D. M., Schwab A. P., and Wildeman T. R. (1976) 4.4-b.y.-old clast in Boulder 7, Apollo 17: a comprehensive chronological study by U-Pb, Rb-Sr and Sm-Nd methods. *Proc. Lunar Sci. Conf. 7th*, pp. 2309–2333.

Niemeyer S. (1979) I-Xe dating of silicate and troilite from IAB iron meteorites. *Geochim. Cosmochim. Acta* **43**, 843–860.

Nyquist L. E., Hubbard N. J., Gast P. W., Bansal B. M., Weismann H. and Jahn B. (1973) Rb-Sr systematics for chemically defined Apollo 15 and 16 materials. *Proc. Lunar Sci. Conf. 4th*, pp. 1823–1846.

Papanastassiou D. A. and Wasserburg G. J. (1969) Initial strontium isotopic abundances and the resolution of small time differences in the formation of planetary objects. *Earth Planet. Sci. Lett.* **5**, 361–376.

Papanastassiou D. A. and Wasserburg G. J. (1981) Microchrons: the ^{87}Rb-^{87}Sr dating of microscopic samples (abstract). In *Lunar and Planetary Science XII*, pp. 802–804. Lunar and Planetary Institute, Houston.

Patterson C. (1956) Age of meteorites and the earth. *Geochim. Cosmochim. Acta* **10**, 230–237.

Podosek F. A. (1970) Dating of meteorites by the high-temperature release of iodine-correlated Xe. *Geochim. Cosmochim. Acta* **34**, 351–365.

Podosek F. A. (1971) Neutron-activation potassium-argon dating of meteorites. *Geochim. Cosmochim. Acta* **35**, 157–173.

Podosek F. A. (1979) Radiometric ages of meteorites. *Meteoritics* **14**, 518–520.

Podosek F. A. and Lewis R. S. (1972) ^{129}I and ^{244}Pu abundances in white inclusions of the Allende meteorite. *Earth Planet. Sci. Lett.* **15**, 101–109.

Rison W. and Zaikowski A. (1980) Proportional retention of I and radiogenic ^{129}Xe in preheated Allende. *Proc. Lunar Planet. Sci. Conf. 11th*, pp. 977–994.

Shields R. M., Pinson W. H., Jr., and Hurley P. M. (1966) Rubidium-strontium analyses of the Bjurböle chondrite. *J. Geophys. Res.* **71**, 2163–2167.

Swindle T. D., Caffee M. W., Hohenberg C. M., and Lindstrom M. M. (1983) Iodine-xenon studies of individual Allende chondrules. *Geochim. Cosmochim. Acta.* In press.

Tatsumoto M., Knight R. J., and Allègre C. J. (1973) Time differences in the formation of meteorites as determined from the ratio of lead-207 to lead-206. *Science* **180**, 1279–1283.

Tatsumoto M., Unruh D. M., and Desborough G. A. (1976) U-Th-Pb and Rb-Sr systematics of Allende and U-Th-Pb systematics of Orgueil. *Geochim. Cosmochim. Acta* **40**, 617–634.

Turner G., Enright M. C., and Cadogan P. H. (1978) The early history of chondrite parent bodies inferred from ^{40}Ar-^{39}Ar ages. *Proc. Lunar Planet. Sci. Conf. 9th*, pp. 989–1025.

Unruh D. M. (1982) The U-Th-Pb age of equilibrated L chondrites and a solution to the excess radiogenic Pb problem in chondrites. *Earth Planet. Sci. Lett.* **58**, 75–94.

Unruh D. M., Hutchison R., and Tatsumoto M. (1979a) U-Th-Pb agc of thc Barwell chondrite: Anatomy of a "discordant" meteorite. *Proc. Lunar Planet. Sci. Conf. 10th*, pp. 1011–1030.

Unruh D. M., Nakamura N., and Tatsumoto M. (1977) History of the Pasamonte achondrite: relative susceptibility of the Sm-Nd, Rb-Sr, and U-Pb systems to metamorphic events. *Earth Planet. Sci. Lett.* **37**, 1–12.

Unruh D. M., Tatsumoto M., and Hutchison R. (1979b) U-Pb study of the Richardton H5 chondrite (abstract). *Meteoritics* **14**, 551–553.

Wasserburg G. J. and Huneke J. C. (1979) I-Xe dating of I-bearing phases in Allende (abstract). In *Lunar and Planetary Science X,* pp. 1307–1309. Lunar and Planetary Institute, Houston.

Wetherill G. W. (1956) Discordant uranium-lead ages. *EOS (Trans. Amer. Geophys. Union)* **37**, 320–326.

Wetherill G. W. (1972) Origin and age of chondritic meteorites. In *Contributions to Recent Geochemistry and Analytical Chemistry* (A. I. Tugarinov, ed.), pp. 21–37 (English translation). Wiley and Sons, N. Y. (1975).

Wetherill G. W. (1975) Radiometric chronology of the early solar system. *Ann. Rev. Nuclear Sci.* **25**, 283–328.

Wetherill G. W., Mark R., and Lee-Hu C. (1973) Chondrites: initial strontium-87/strontium-86 ratios and the early history of the solar system. *Science* **182**, 281–283.

Zaikowski A. (1979) I-Xe chronology of Allende inclusions (abstract). In *Lunar and Planetary Science X*, pp. 1392–1394. Lunar and Planetary Institute, Houston.

Zaikowski A. (1980) I-Xe dating of Allende inclusions: antiquity and fine structure. *Earth Planet. Sci. Lett.* **47**, 211–222.

Cosmic Setting for Chondrule Formation

G. Jeffrey Taylor, Edward R. D. Scott, and Klaus Keil
Institute of Meteoritics and Department of Geology, University of New Mexico, Albuquerque, New Mexico 87131

Impact and igneous processes are sufficiently well-understood to allow us to rule out hypotheses of chondrule origin that involve processes operating on meteorite parent bodies or planetesimals. This leaves the solar nebula as the likely setting for chondrule formation. Properties of lunar, terrestrial, and meteoritic breccias and the lunar regolith are totally inconsistent with the notion that chondrules formed by impact on parent bodies. Impact models cannot explain the high chondrule abundances, the minuscule abundance of agglutinates in chondrites, the absence of hypervelocity impact pits on chondrules, the old ages of chondrules, the presence of chondrule rims, and the diversity of chondrule compositions. Impact during accretion, besides suffering from the same flaws as other impact models, cannot explain quantitatively how impact melts, which require projectile velocities of several km/sec, can be produced when velocities during accretion are less than twice the escape velocity, which is only 0.4 km/sec for a body 300 km in radius. Models depicting chondrules resulting from collisions between molten planetesimals are disproven by the oxygen isotopic heterogeneities of chondrules, the presence of chondrule rims, and the unfractionated bulk-silicate compositions and narrow size distribution of chondrules. Similar arguments rule out volcanism as a means of forming chondrules. A nebula setting is more compatible with the following: (1) The textures of the materials that rim chondrules are not igneous; (2) oxygen isotopic compositions of chondrules can be explained by reaction of solids with nebular gases; and (3) matrix materials in type 3 chondrites vary as widely in composition as do chondrules, demonstrating that compositionally heterogeneous dust was available in the nebula for chondrule formation.

INTRODUCTION

Chondrules are igneous-textured, millimeter-sized, spherical to irregularly-shaped silicate objects that constitute the major component of most chondrites (Dodd, 1982). Researchers agree that chondrules were once molten, but there is no agreement as to how or where the melts formed. We divide models for chondrule origin into two categories, according to the cosmic setting in which chondrules are supposed to have formed. One is a "planetary" setting, which envisages chondrules forming on the surfaces of parent bodies. Melting mechanisms include impact and volcanism. We also include in this category models that imagine chondrules forming by impact during accretion of parent bodies or in collisions between molten or solid planetesimals larger than a few meters in diameter. The other cosmic setting is in the solar nebula, prior to planet formation; we dub this the "nebular" setting. Numerous mechanisms for chondrule melting in the nebula have been proposed, including impact between centimeter-sized objects, condensation as liquids from a vapor, and transient heating events that remelt pre-existing solid materials.

In this paper, we argue that the properties of chondrules and chondrites are drastically different from those of known impact and volcanic deposits on Earth, the moon, and in meteorites. Chondrules, therefore, did not form in a planetary setting, but must have formed in the solar nebula. We also argue that enough is known about differentiated meteorites and igneous processes to rule out the idea that chondrites formed by collisions between molten planetesimals. Some of the arguments have been made before (e.g., Wasson, 1972; Kerridge and Kieffer, 1977; Wood and McSween, 1977; Lux *et al.*, 1981), but the issues have not been discussed as comprehensively as we do here. We address the following specific mechanisms operating in planetary settings: impacts on parent bodies, impacts during accretion, collisions between molten planetesimals, and volcanism. Finally, we point out some virtues of a nebular setting for chondrule formation.

The examples we quote are drawn largely from ordinary chondrites, but there is substantial evidence that chondrules in ordinary, carbonaceous, and enstatite chondrites share similar origins (Scott *et al.*, 1982a; Scott and Taylor, 1983). Specifically, there are similarities in the chemical, mineralogical, and isotopic compositions of chondrules, matrices, and other components in these various chondrite groups.

IMPACT ON PLANETARY SURFACES

The notion that impact on a meteorite parent body could produce chondrules has achieved widespread popularity among meteorite researchers (Urey, 1952; Fredriksson, 1963; Ringwood, 1966, Fredriksson *et al.*, 1970, 1973; Wlotzka, 1969; King *et al.*, 1972a,b; Kurat, 1982). However, the details of impact models tend to be ill-defined. Wlotzka (1969) and Kurat (1982) do not describe the physical setting at all. Urey (1952), Fredriksson (1963), Ringwood (1966) and Fredriksson *et al.* (1973) suggest that chondrules formed when large planetesimals (bigger than asteroids, smaller than the moon) collided. They do not give details about the amount of melt produced, the size distribution of droplets and other blobs of melt, or the compositional diversity of the droplets. King *et al.* (1972a,b) suggest that chondrule-like objects in Apollo 14 breccias formed during the event that produced the breccias, presumed to be the excavation of the gigantic Imbrium basin. By analogy, they suggest that chondrules formed during large impacts on meteorite parent bodies.

All impact models share one common characteristic: chondrules are secondary objects formed by remelting of pre-existing rock. The energy for melting comes from the kinetic energy of impacting projectiles. Such models were given a boost when glass and quench-crystallized spheres of impact origin were discovered in lunar soils and regolith breccias. Impacts certainly cause melting of target materials. However, the detailed characteristics of impact deposits in the lunar highlands and those associated with terrestrial impact craters, the nature of the lunar regolith, and the properties of meteorite regolith breccias demonstrate that few chondrules could have formed by impact on the surfaces of chondrite parent bodies. The following discussion outlines why we believe that the overwhelming majority of chondrules did not form by impact on planetary surfaces.

Chondrule abundances

Glass spheres and other chondrule-like objects constitute merely a few percent of the lunar regolith (e.g., Heiken, 1975), which suggests that the impact process is not efficient at making droplets of melt. This low abundance applies both to loose regolith samples and to regolith breccias (Fig. 1), which are samples of lithified regolith. Achondritic regolith breccias have even lower abundances of chondrule-like objects (Brownlee and Rajan, 1973). Because achondrites formed on small bodies in the asteroid belt, we conclude that the abundance of quenched spheres in regoliths is not greatly dependent on surface gravity or location in the solar system. In contrast, type 3 chondrites consist of 30 to 50% chondrules (Fig. 1); the mechanism that made chondrules must have done so with impressive efficiency. The lack of compound and cratered chondrules on

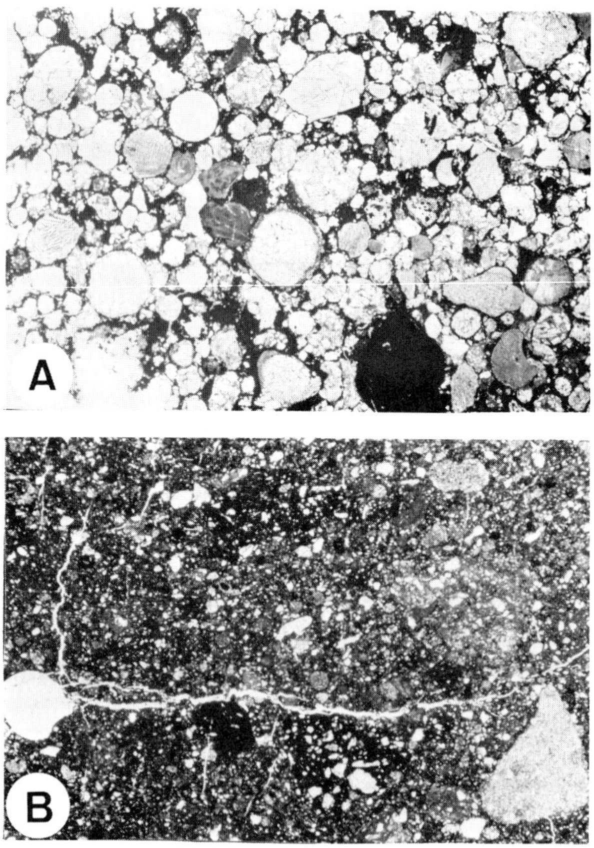

Fig. 1. Comparison of textures of **(a)** type 3 chondrite (Semarkona) and **(b)** lunar regolith breccia (14318); both to the same scale, 15 mm across. Note the great abundance of chondrules in Semarkona and the low abundance in 14318, which consists mostly of broken rock, mineral, and glass fragments.

the moon implies that the number density was small where lunar spheres formed compared to conditions where chondrules formed (Gooding and Keil, 1981).

Agglutinates outnumber glass spheres on the moon

Agglutinates are the single most abundant component in the lunar regolith (Heiken, 1975). These are aggregates of rock, mineral, and glass fragments bound by a brownish, swirly glass. As Kerridge and Kieffer (1977) point out, the abundance of agglutinates (about 50% in most lunar soils) far exceeds the abundance of glass spheres and other chondrule-like objects (a few percent). While discrete agglutinate particles are not generally recognizable in lunar regolith breccias, especially the most coherently lithified ones, the brownish glass characteristic of agglutinates always is (Drozd *et al.*, 1976; Papike *et al.*, 1981). If chondrules formed by impact into a regolith, chondrites ought to contain more agglutinates than chondrules, but in fact, chondrites contain almost no agglutinates. (We have observed one in the Fayetteville chondrite, a regolith breccia.) If chondrules formed in a regolith, chondrites should be chock-full of agglutinates.

Impacts comminute

Comminution is the most efficient process operating in the lunar regolith and during large impacts into bedrock. Although agglutinates compose about 50% of a mature lunar soil (Heiken, 1975), most of the volume of a typical agglutinate is made up of broken rock, mineral, and glass fragments. Studies of terrestrial impact craters (e.g., Pohl *et al.*, 1977) and theoretical studies of cratering dynamics (e.g., O'Keefe and Ahrens, 1975) support these observations: impact melting on a planetary surface is an inefficient process. Impacts break a far greater volume (>97%) of rock than they melt (<3%). Consequently, if chondrules are the products of impact melting, chondrites ought to consist mostly of unmelted, clastic debris, not predominantly of chondrules, the supposed impact melts. Furthermore, chondrite regolith breccias (also called "gas-rich" breccias) attest to the dominance of breaking over melting during impacts: they contain substantial amounts of *comminuted* chondrules and rock fragments (e.g., McSween and Lipschutz, 1980; McSween *et al.*, 1981; Rubin *et al.*, 1983). Far fewer unbroken chondrules are present than in type 3 chondrites, yet if impact could produce chondrules, chondrite regolith breccias ought to have *more* unbroken chondrules than do type 3 chondrites, not fewer. Similarly, achondrite regolith breccias (e.g., the Kapoeta howardite; Fredriksson and Keil, 1963) consist mostly of fragmental material and contain little impact-melted rock.

Hypervelocity impact pits

Microcraters caused by hypervelocity impact of micrometeorites stud the surfaces of rocks, mineral fragments, and glass spheres and fragments in the lunar regolith (e.g., Carter and MacGregor, 1970; McKay *et al.*, 1970; Neukum *et al.*, 1970). Microcraters have small glass-lined pits surrounded by spall zones (Fig. 2a) and range in size from less than a micron to more than a centimeter. It is not uncommon to find several microcraters on a single millimeter-sized soil particle. Microcraters have also been found on the rare chondrule-like objects in Kapoeta, an achondrite regolith breccia (Brownlee and Rajan, 1973). If chondrules formed in planetary regoliths like those on the moon or Kapoeta parent body, one would expect to find hypervelocity impact pits on chondrule surfaces: none have been discovered (Allen *et al.*, 1980; Gooding and Keil, 1981).

Chondrules do have craters, but except for being circular depressions, their morphology is nothing like that of hypervelocity impact craters (Fig. 2b). Instead, they appear to be simple dents caused by low velocity collisions. Das Gupta *et al.* (1978) studied the dents on chondrules from Chainpur and concluded that they formed by "high speed" collisions, citing as evidence the presence of glassy frills and apparent brecciated material within bowl-shaped craters and ejecta surrounding the craters. These features, however, could be formed by low-velocity collisions between partially molten chondrules (Gooding and Keil, 1981). Grossman (1982) suggests that some circular depressions may not be impact pits of any sort, but might instead be places where metallic Fe,Ni droplets were lost after solidification.

Ages

Except for those reheated by shock recently in solar system history (Taylor and Heymann, 1969), chondrites are all 4.4 to 4.5 Gy old. Ages have been determined by Ar-Ar (e.g., Turner *et*

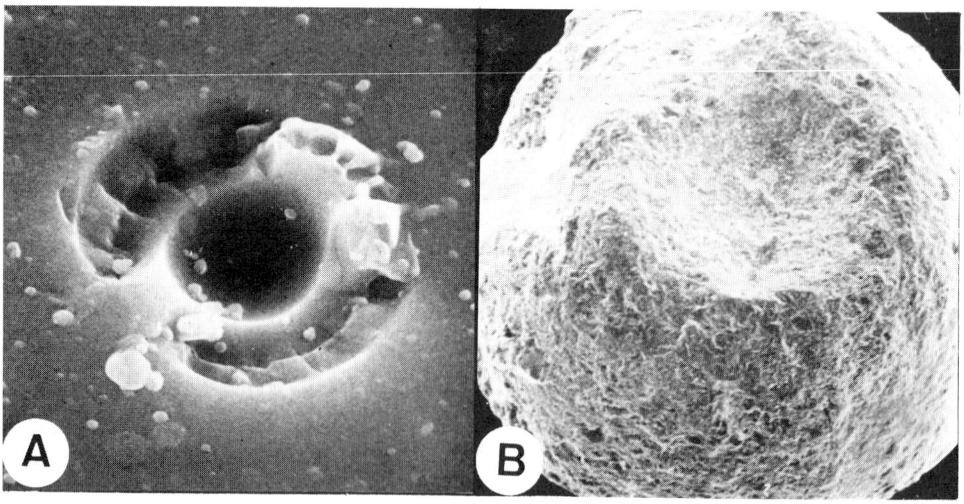

Fig. 2 Comparison of **(a)** hypervelocity impact crater (5 μm in diameter) on a lunar soil particle (photo courtesy of F. Hörz) and **(b)** a crater (640 μm in diameter) on a chondrule from Ochansk (Gooding, 1979). The lunar microcrater has a glass-lined pit surrounded by a spall zone, whereas the chondrule crater appears to be a simple dent caused by a low-speed collision.

al., 1978), Rb-Sr (e.g., Minster and Allègre, 1979), and U-Pb (Unruh and Tatsumoto, 1980). Because chondrites are composed chiefly of chondrules, chondrules must have formed prior to 4.4 Gy ago. The antiquity of chondrules is confirmed by the presence in individual chondrules of ^{129}Xe derived from the decay of extinct ^{129}I (Caffee *et al.*, 1982a,b). If impact was an important chondrule-forming process, surely some chondrules would be younger than 4.4 Gy, as lunar highland bombardment continued until 3.9 Gy.

Absence of pre-chondrule target rock

The formation of chondrules by impact implies the existence of a precursor target rock. If this target was fine-grained nebular dust that had accreted into a parent body, we should have chondrites composed solely of such materials, but we do not. The closest analogous materials are the chondrule-free CI chondrites, but their oxygen isotopic composition is drastically different from those of chondrules (Clayton, 1981). As discussed below, Dodd (1971, 1978) suggests that large sheets of porphyritic rock formed during accretion by impact melting of fine-grained material and were then broken up by subsequent impacts to form porphyritic-olivine chondrules. If this were true, then there ought to be whole chondrites with the composition and texture of porphyritic chondrules, but there are none.

Chondrule rims

Many chondrules in type 3 chondrites are rimmed (Fig. 3) by fine-grained, FeO-rich silicate materials (Allen *et al.*, 1980). These rims have similar compositions and textures as the opaque matrix of ordinary (Huss *et al.*, 1981) and carbonaceous (McSween and Richardson, 1977) type 3 chondrites. They are not compositionally or texturally like the mesostases of chondrules, despite claims to the contrary (Ashworth, 1981; Hutchison, 1982): the rims are much richer in FeO and poorer in feldspathic components than are chondrule mesostases. Any model for chondrule origin must account for the presence of chondrule rims, and the only possibility offered by impact models is rolling around in regolith dust. However, individual mineral grains, rock fragments, or glass spheres in the lunar regolith do not have rims around them. Moreover, the proportion of unequilibrated chondrules with rims in chondrite regolith breccias is smaller than in type 3 chondrites, probably because their rims were abraded away as meteorite impacts reworked the surface of chondrite parent bodies. If chondrules acquired rims by rolling around in a dusty regolith, it is surprising that rim compositions vary drastically from chondrule to chondrule. We have found, for example, that the rims on adjacent chondrules vary by factors of two in their concentrations of FeO and MgO, and by greater amounts in Al_2O_3, Na_2O, K_2O, and CaO.

Chondrule compositions

Chondrules in a given meteorite display a broad range in composition (e.g., Gooding *et al.*, 1980; Lux *et al.*, 1981), yet impact melts tend to be uniform in composition and represent total

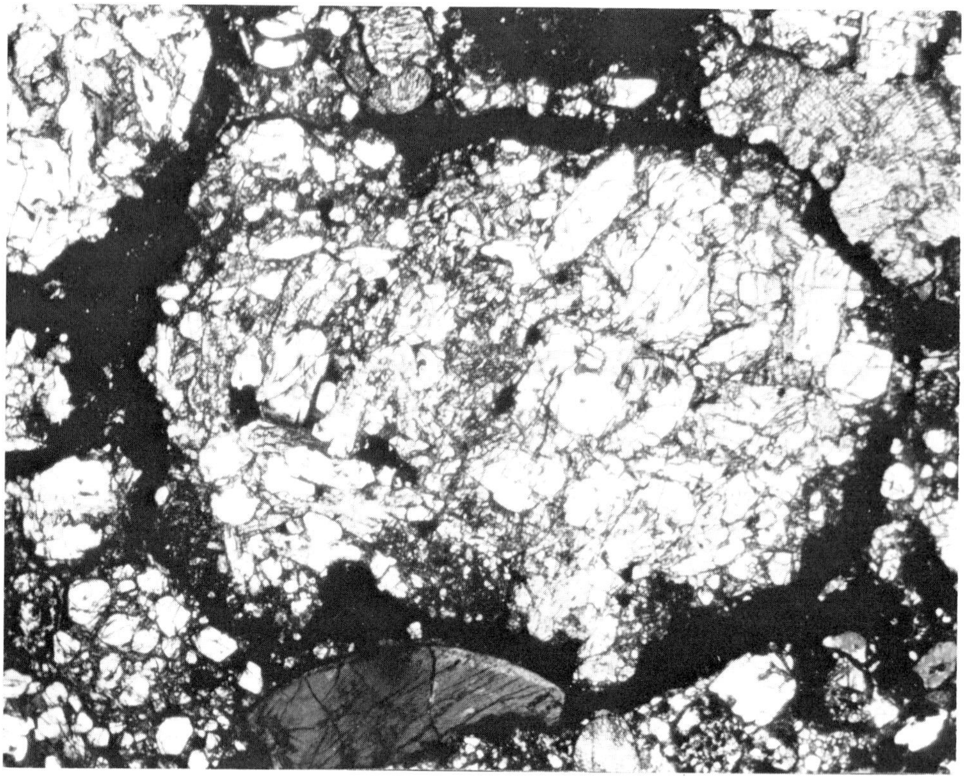

Fig. 3 Chondrule (2 mm across) in Tieschitz surrounded by rim of opaque fine-grained, FeO-rich silicate material. Impact models for chondrule formation cannot account for the presence of such rims.

melts of the target rocks (e.g., Dence, 1971; Floran *et al.*, 1978; Grieve, 1975; Grieve and Floran, 1978). For chondrules to be so variable in composition requires that the chondrite parent bodies were compositionally heterogeneous, so numerous impacts could melt a wide variety of targets. This seems unlikely for chondrite parent bodies. Dodd (1971, 1978, 1981) envisages that chondrule heterogeneity resulted from impact into a coarse-grained target rock. For reasons discussed below, this model no longer seems viable.

Suevitic melts

Terrestrial melt-bearing, fragmental breccias like suevite contain rare chondrule-like, roundish blobs of melt. However, most melt clasts are irregular in shape and appear to have been deformed when the fragmental material was deposited.

COLLISIONS DURING ACCRETION

Dodd (1971, 1978, 1981) proposes an impact origin for chondrules, but unlike the models discussed above, his model depicts chondrule formation during accretion of chondrite parent bodies. Briefly, Dodd proposes that impacts melted dust on growing parent bodies to produce a microporphyritic rock. Later, fragmentation of such melt sheets produced microporphyritic chondrules, and additional impact melting yielded the variety of fluid-drop chondrules observed.

While it seems likely that some chondrite breccias formed during accretion (Wasson and Wetherill, 1979; Scott and Taylor, 1982; Taylor, 1982), it is improbable that more than a tiny proportion of chondrules could have been produced at this time. Most of the arguments listed above also apply to impacts during accretion. Furthermore, it is probably not possible to retain a significant amount of the melt produced during a high-velocity impact (Wasson, 1972). For accretion to take place at all, impacts must be gentle enough to result in a net gain of material to the parent body. Experiments by Gault *et al.* (1963) indicate net mass loss if the impact velocity is more than a few times the escape velocity.

Safronov (1972) and Goldreich and Ward (1973) show that a body accretes from objects whose velocities are not much greater than the escape velocity. The escape velocity of a body 10 km in radius is ~10 m/sec. Even during the terminal stages of accretion of a 300-km-radius body, which has an escape velocity of 0.4 km/sec, the velocities of the accreting projectiles would certainly be <1 km/sec. According to Fig. 2 of Hörz and Schaal (1981), which relates impact velocity to peak shock pressure and the percentage of melt produced, no impact melting would take place at such a low velocity, even in porous, loosely consolidated material.

In summary, the idea that chondrules formed by impact during the accretion of chondrite parent bodies can be ruled out. Velocities during accretion are too low to cause the substantial amounts of melting needed to form chondrules. If the velocities were high enough, accretion would not take place. Even if parent bodies could grow by high velocity collisions, the problems with impact models that we discussed above apply to impacts during accretion.

COLLISIONS BETWEEN MOLTEN PLANETESIMALS

Several authors have proposed that chondrules formed when molten or partially molten planetesimals collided, producing a hot cloud containing droplets of molten silicates (Zook, 1980, 1981, 1982; Wänke *et al.* 1981; Hutchison, 1982; Hutchison and Bevan, 1982; Leitch and Smith, 1981, 1982). The model certainly seems capable of producing vast amounts of molten material, but we believe it is not able to explain most properties of chondrules, as outlined below.

Oxygen isotopic compositions of chondrules

Chondrules in ordinary and carbonaceous chondrites have heterogeneous oxygen isotopic compositions (e.g., Clayton, 1981; Gooding *et al.*, 1982). They tend to plot along a line with a slope of one on the three-isotope plot, probably due to isotopic exchange between silicates and surrounding nebular gas, perhaps during chondrule formation (Clayton, 1981). It would seem that the hypothetical molten planetesimals would have homogeneous oxygen isotopic compositions, and collisions among them would result in a homogeneous distribution of oxygen isotopes in chondrules. Nevertheless, Hutchison and Bevan (1982) argue that the oxygen data can be explained by requiring one of the planetesimals to be richer in ^{16}O than the other so that their collision would result in oxygen isotopic compositions intermediate between the two extremes. However, most chondrules produced in this manner would have oxygen isotopic compositions characteristic of the two molten planetesimals, rather than intermediate compositions. It is difficult to see how kilometer-sized objects could mix on a millimeter scale. Zook (1982) suggests that the mixing line might result from partial melting of the planetesimals, with the most ^{16}O-rich components remaining in the refractory residue. This is not possible, however, as isotopic exchange would occur during melting, leading to compositions lying along a line with a slope of 0.5.

These models are not convincing, but are difficult to refute quantitatively. However, recent results on chondrules separated from the Allende chondrite (Clayton *et al.*, 1983) rule them out. Chondrules from Allende form an array with a slope greater than that of the line defined by most Ca-Al-rich inclusions and matrix. Mixing between two planetesimals of different isotopic compositions cannot explain these trends.

Chondrule rims and matrix lumps

Most chondrules in type 3 chondrites have rims around them (Fig. 3). It is not possible for these rims to be acquired during the collision of two molten planetesimals, though it is possible for this to happen after the resulting cloud of droplets had dispersed. None of the models proposed to date confronts the issue of chondrule rims, though Hutchison and Bevan (1982) suggest that chondrule rims represent splashed liquids, a suggestion that ignores the dramatic differences in chemical composition between chondrule rims and chondrules or chondrule mesostases.

An even more serious problem for the molten planetesimal model is the presence of lumps of dark, fine-grained, FeO-rich matrix material inside chondrules (Rubin *et al.*, 1982a; G. Kurat, personal communication). These lumps of matrix material are clearly contained within some chondrules and are not simply protrusions of rims along cracks in the chondrules. Matrix material must have entered the chondrules before they had crystallized. Because chondrules were partly molten for no more than a few hours (Planner and Keil, 1982; Tsuchiyama and Nagahara, 1981; Hewins *et al.*, 1981), matrix lumps must have been near molten chondrules, probably within 10^4 km. This seems inconsistent with formation of chondrules when two molten bodies splashed into each other.

Chondrule sizes

Chondrules come in a restricted range of sizes (Martin and Mills, 1978; Hughes, 1978; King and King, 1978). Most are 0.5 to 1.5 mm in diameter, though some are as small as one micron (Rubin *et al.*, 1982a); few are larger than 5 mm. How could molten bodies nominally 10 km across collide and produce such a narrow size range of small objects? Presumably the droplets are sorted somehow before accretion, but what happened to the large blobs of melt? Some would probably be hundreds of meters across, and these large blobs would contain most of the mass of the original bodies. Meteorites corresponding to these molten masses should exist, but none have been identified.

Fractionated liquids

A body that heated up gradually would partially melt and produce a range of fractionated liquids. Even if the body ended up totally molten, it would probably be stratifed, with the more refractory components underlying less refractory ones (perhaps with an Fe-Ni-S core). But the silicate portions of chondrules are basically unfractionated in composition relative to CI chondrites (with the obvious exception of Ca-Al-rich chondrules). Their REE patterns are within errors of average chondritic (Gooding *et al.*, 1980); partial melting would almost certainly produce fractionated REE patterns. Similarly, their bulk major element contents, while ranging widely, are on average the same as CI chondrites, minus Fe,Ni metal, trace siderophile elements, and volatiles like H_2O and S (Scott *et al.*, 1982a).

Minor elements in chondrules in E chondrites

Leitch and Smith (1981, 1982) propose that enstatite chondrites formed by the collision and mixing of two planetesimals: one with red-luminescing enstatite, the other with blue-luminescing enstatite. They base the model on the bimodal distribution of enstatite compositions, claiming that red enstatites are consistently higher in Mn, Ti, Cr, and Al than blue enstatites. However, random analyses of enstatite grains in several E chondrites show that compositions and color are not well correlated (McKinley *et al.*, 1982, 1983). For the majority of cases, the composition of enstatite is consistent with crystallization from a liquid and subsequent metamorphism in parent bodies. There are some cases where mixing seems necessary, but these do not require the existence of molten planetesimals.

We do not argue that collisions between partly molten planetesimals never occurred. Such collisions may have been involved in the formation of mesosiderites (Hewins, 1982) and pallasites. But we believe that differentiated (achondrites, irons) and undifferentiated meteorites (chondrites) can be distinguished by their different chemical and petrologic properties. These properties indicate that chondrules did not form from molten or partly molten planetesimals.

VOLCANISM

Volcanism as a mechanism of producing chondrules has been proposed by a number of authors (Tschermak, 1883; Ringwood, 1959; Fredriksson and Ringwood, 1963). A review of the literature reveals that current researchers are not enthusiastic about it being a viable mechanism for chondrule formation. In spite of its current unpopularity, we include a discussion of it here for the sake of completeness. The main problems with the idea are summarized below.

Chondrule compositions

As we noted above when criticizing the molten planetesimal model, bulk silicate compositions of chondrules are, in general, unfractionated compared with solar abundances. Volcanism requires the production of melts. Unless the percentage of partial melting is extremely large (probably greater than 50%), most of the melts would be distinctly non-chondritic in composition. For example, one would expect large variations in FeO/MgO, Al_2O_3, K_2O, Na_2O, and the REE.

Chondrule abundances

Volcanic eruptions on Earth and the moon do not concentrate droplets of melt to the extent that the chondrule forming process did. Although terrestrial volcanoes certainly produce droplets of melt, most of the volume of erupted material forms lava flows, not millimeter-sized spheres. On the moon, pyroclastic deposits like the orange soil at the Apollo 17 landing site (Heiken *et al.*, 1974) are indeed richer in droplets than even chondrites are; such deposits mantle the margins of many maria. Nevertheless, lunar pyroclastic deposits are rare on the moon and are dwarfed in abundance compared to mare basalt lava flows.

No chondrule-composition achondrites

If chondrules formed by volcanic eruptions, then surely lava must have flowed across the surfaces of the chondrite parent bodies, and intrusive rocks must have formed below their surfaces. These igneous rocks would have compositions corresponding to the compositions of chondrules. No such rocks exist among the thousands of meteorites in the world's collection.

Oxygen isotopic compositions

Every model designed to explain some facet of meteoritics has to explain the relevant oxygen isotopic compositions. Volcanism fails the test. Partial melting in a parent body would not yield chondrules that fall along a line with a slope of one on the three-isotope plot. Fractionation of the oxygen isotopes would be along a line with a slope of 0.5. Unless the parent body was astonishingly heterogeneous in its oxygen isotopic composition, all melts produced in one body ought to lie along the same chemical-fractionation line.

VIRTUES OF A NEBULAR SETTING

There is a growing consensus that chondrules formed in the solar nebula by melting of pre-existing solids (Grossman *et al.*, 1979; Gooding *et al.*, 1980; Lux *et al.*, 1981; Wood, 1981; Grossman and Wasson, 1982), rather than by direct condensation from a vapor (Suess, 1949; Wood, 1963; Blander and Katz, 1967; Blander and Abdel-Gawad, 1969; McSween, 1977). The heat sources responsible for the remelting are still debated. Possibilities include: lightning (Whipple, 1966; Cameron, 1966; Rasmussen and Wasson, 1982); heating in shock fronts created during the collapse of the nebula; chemical energy (Clayton, 1980); transient heating events associated with the primitive sun (Wood, 1963; Wood and McSween, 1977); and impact between small (millimeters to centimeters) objects in space (Whipple, 1972; Wasson, 1972; Cameron, 1973; Kieffer, 1975; Grossman *et al.*, 1979). We will not review the pros and cons of these ideas here; adequate discussions appear in Wasson (1972), Dodd (1978), Wood and McSween (1977), and Lux *et al.* (1981). Rather, we wish to point out some features of chondrules that are consistent with conditions thought to have prevailed in the solar nebula, or at least are more consistent with chondrules forming in space rather than on parent bodies.

One argument that persuaded some authors (e.g., Dodd, 1981) that chondrules had formed during planetary processes was the diversity of chondrule compositions. These authors argued that nebular processes could not be responsible. However, recent work on matrix materials in type 3 chondrites (Huss *et al.*, 1981; Ikeda *et al.*, 1981; Rubin *et al.*, 1982b; Scott *et al.*, 1982a,b) demonstrates that matrix materials, including chondrule rims, lumps within chondrules, and matrices of C3 chondrites, vary as widely in composition as do chondrules. Heterogeneous dust *did* exist in the nebula, so there is no need to invoke planetary processing to produce compositionally diverse chondrules. A vexing problem to solve is to elucidate the processes that led to large chemical fractionations of dust in the solar nebula.

Oxygen isotopic variations are more easily understood in a nebular setting than in a planetary one. As Clayton (1981) and Wood (1981) note, the likely carriers of isotopic anomalies in the early solar system were solids, not gases. These solids could exchange oxygen with one or more gaseous reservoirs in the nebula, producing the range of oxygen isotopic compositions observed in chondrules. In contrast, planetary processes tend to homogenize isotopic compositions.

CONCLUSIONS

Properties of lunar, terrestrial, and meteoritic impact breccias and the lunar regolith are totally inconsistent with the notion that most chondrules formed by impact. This conclusion applies whether one considers impact on a parent body, during accretion, or between molten planetesimals. An equally firm case can also be made against a volcanic origin for chondrules. Although there is considerable uncertainty about the kind of nebular processes that could have formed chondrules, there is now some confidence that planetary impact and melting processes are sufficiently understood to be discounted as the major producers of chondrules. Chondrules formed in the solar nebula and, therefore, they contain a wealth of information about conditions in the nebula before planets formed.

Acknowledgments. This work was supported by NASA grant NGL 32-004-064. Comments by Jim Gooding, Larry Grossman, and John Wasson helped clarify the text.

REFERENCES

Allen J. S., Nozette S., and Wilkening L. L. (1980) A study of chondrule rims and chondrule irradiation records in unequilibrated ordinary chondrites. *Geochim. Cosmochim. Acta* **44**, 1161–1175.

Ashworth J. R. (1981) Fine structure in H-group chondrites. *Proc. Roy. Soc. London* **A374**, 179–194.

Blander M. and Abdel-Gawad M. (1969) The origin of meteorites and the constrained equilibrium condensation theory. *Geochim. Cosmochim. Acta* **33**, 701–716.

Blander M. and Katz J. (1967) Condensation of primordial dust. *Geochim. Cosmochim. Acta* **31**, 1025–1034.

Brownlee D. E. and Rajan R. S. (1973) Micrometeorite craters discovered on chondrule-like objects from Kapoeta meteorite. *Science* **182**, 1341–1344.

Caffee M. W., Hohenberg C. M., Lindstrom M. M., and Swindle T. D. (1982a) I-Xe studies of individual Allende chondrules (abstract). *Meteoritics* **17** 190.

Caffee M. W., Hohenberg C. M., Swindle T. D., and Hudson B. (1982b) I-Xe ages of individual Bjurbole chondrules. *Proc. Lunar Planet. Sci. Conf. 13th*, in *J. Geophys. Res.* **87**, A303–A317.

Cameron A. G. W. (1966) The accumulation of chondritic material. *Earth Planet. Sci. Lett.* **1**, 93–96.

Cameron A. G. W. (1973) Accumulation processes in the primitive solar nebula. *Icarus* **18**, 407–450.

Carter J. L. and MacGregor I. D. (1970) Mineralogy, petrology and surface features of some Apollo 11 samples. *Proc. Apollo 11 Lunar Sci. Conf.*, pp. 247–265.

Clayton D. D. (1980) Chemical energy in cold-cloud aggregates: the origin of meteoritic chondrules. *Astrophys. J. Lett.* **239**, L37–L41.

Clayton R. N. (1981) Isotopic variations in primitive meteorites. *Phil. Trans. Roy. Soc. London* **A303**, 339–349.

Clayton R. N., Mayeda T. K., Gooding J. L., Keil K., and Olsen E. J. (1981) Redox processes in chondrules and chondrites (abstract). In *Lunar and Planetary Science XII*, pp. 154–156. Lunar and Planetary Institute, Houston.

Clayton R. N., Onuma N., Ikeda Y., Mayeda T. K., Hutcheon I. D., Olsen E. J., and Molini-Velsko C. (1983) Oxygen isotopic compositions of chondrules in Allende and ordinary chondrites. This volume.

Das Gupta S. P., Sen Gupta P. R., and Dube A. (1978) A SEM study of Chainpur chondrules (abstract). *Meteoritics* **13**, 435–438.

Dence M. R. (1971) Impact melts. *J. Geophys. Res.* **76**, 5552–5565.

Dodd R. T. (1971) The petrology of chondrules in the Sharps meteorite. *Contrib. Mineral. Petrol.* **31**, 201–227.

Dodd R. T. (1978) The composition and origin of large microporphyritic chondrules in the Manych (L-3) chondrite. *Earth Planet. Sci. Lett.* **39**, 52–66.

Dodd R. T. (1981) *Meteorites, A Petrologic-Chemical Synthesis.* Cambridge University Press, Cambridge. 368 pp.

Dodd R. T. (1982) Objects we call chondrules (abstract). In *Papers Presented to the Conference on Chondrules and their Origins*, p. 15. Lunar and Planetary Institute, Houston.

Drozd R. J., Kennedy B. M., Morgan C. J., Podosek F. A., and Taylor G. J. (1976) The excess fission xenon problem in lunar samples. *Proc. Lunar Sci. Conf. 7th*, pp. 599–623.

Floran R. J., Grieve R. A. F., Phinney W. C., Warren J. L., Simonds C. H., Blanchard D. P., and Dence M. R. (1978) Manicouagan impact melt, Quebec, 1. Stratigraphy, petrology and chemistry. *J. Geophys. Res.* **83**, 2737–2759.

Fredriksson K. (1963) Chondrules and the meteorite parent bodies. *Trans. N.Y. Acad. Sci.* **25**, 756–769.

Fredriksson K. and Keil K. (1963) The light-dark structure in the Pantar and Kapoeta stone meteorites. *Geochim. Cosmochim. Acta* **27**, 717–739.

Fredriksson K. and Ringwood A. E. (1963) Origin of meteoritic chondrules. *Geochim. Cosmochim. Acta* **27**, 639–641.

Fredriksson K., Nelen J., and Melson W. G. (1970) Petrography and origin of lunar breccias and glasses. *Proc. Apollo 11 Lunar Sci. Conf.*, pp. 419–432.

Fredriksson K., Noonan A., and Nelen J. (1973) Meteoritic, lunar and Lonar impact chondrules. *The Moon* **7**, 475–482.

Gault D. E., Shoemaker E. M., and Moore H. J. (1963) Spray ejected from the lunar surface by meteoroid impact. *NASA Publ. TN D-1767.* 39 pp.

Goldreich P. and Ward W. R. (1973) The formation of planetesimals. *Astrophys. J.* **183**, 1051–1061.

Gooding J. L. (1979) Petrogenetic properties of chondrules in unequilibrated H-, L-, and LL-group chondritic meteorites. Ph.D. thesis, University of New Mexico, Albuquerque. 392 pp.

Gooding J. L. and Keil K. (1981) Relative abundances of chondrule primary textural types in ordinary chondrites and their bearing on conditions of chondrule formation. *Meteoritics* **16**, 17–44.

Gooding J. L., Keil K., Fukuoka T., and Schmitt R. A. (1980) Elemental abundances in chondrules from unequilibrated chondrites: Evidence for chondrule origin by melting of preexisting materials. *Earth Planet. Sci. Lett.* **50**, 171–180.

Gooding J. L., Mayeda T. K., Clayton R. N., Keil K., Fukuoka T., and Schmitt R. A. (1982) Oxygen isotopic compositions of chondrules in unequilibrated chondrites: Further petrological interpretations (abstract). In *Lunar and Planetary Science XIII*, pp. 271–272. Lunar and Planetary Institute, Houston.

Grossman J. (1982) Mechanisms for the depletion of metal in chondrules (abstract). In *Papers Presented to the Conference on Chondrules and their Origins*, p. 23. Lunar and Planetary Institute, Houston.

Grossman J. N., Kracher A., and Wasson J. T. (1979) Volatiles in Chainpur chondrules. *Geophys. Res. Lett.* **6**, 597–600.

Grossman J. N. and Wasson J. T. (1982) Evidence for primitive nebular components in chondrules from the Chainpur chondrite. *Geochim. Cosmochim. Acta* **46**, 1081–1099.

Grieve R. A. F. (1975) Petrology and chemistry of the impact melt at Mistastin Lake Crater, Labrador. *Geol. Soc. Amer. Bull.* **86**, 1617–1629.

Grieve R. A. F. and Floran R. J. (1978) Manicouagan impact melt, Quebec, 2. Chemical interrelations with basement and formational processes. *J. Geophys. Res.* **83**, 2761–2771.

Heiken G. (1975) Petrology of lunar soils. *Rev. Geophys. Space Phys.* **13**, 567–587.

Heiken G., McKay D. S., and Brown R. W. (1974) Lunar deposits of possible pyroclastic origin. *Geochim. Cosmochim. Acta* **38**, 1703–1718.

Hewins R. H. (1982) The origin of achondrite breccias. In *Workshop on Lunar Breccias and Soils and Their Meteoritic Analogs* (G. J. Taylor and L. L. Wilkening, eds.), pp. 49–53. LPI Tech Rpt. 82-02. Lunar and Planetary Institute, Houston.

Hewins R. H., Klein L. C., and Fasano B. V. (1981) Conditions of formation of pyroxene excentroradial chondrules. *Proc. Lunar Planet. Sci. 12B*, pp. 1123–1133.
Hörz F. and Schaal R. B. (1981) Asteroidal agglutinate formation and implications for asteroidal surfaces. *Icarus* **46**, 337–353.
Hughes D. W. (1978) A disaggregation and thin section analysis of the size and mass distribution of the chondrules in the Bjurböle and Chainpur meteorites. *Earth Planet. Sci. Lett.* **38**, 391–400.
Huss G. R., Keil K., and Taylor G. J. (1981) The matrices of unequilibrated ordinary chondrites: implications for the origin and history of chondrites. *Geochim. Cosmochim. Acta* **45**, 33–52.
Hutchison R. (1982) Meteorites—evidence for the interrelationships of materials in the solar system 4.55 Ga ago. *Phys. Earth Planet. Inter.* **29**, 199–208.
Hutchison R. and Bevan A. W. R. (1982) Conditions and time of chondrule accretion (abstract). In *Papers Presented to the Conference on Chondrules and their Origins*, pp. 32–33. Lunar and Planetary Institute, Houston.
Ikeda Y., Kimura M., Mori H., and Takeda H. (1981) Chemical compositions of matrices of unequilibrated ordinary chondrites. *Mem. Natl. Inst. Polar Res., Spec. Issue No. 20*, pp. 124–144. National Institute of Polar Research, Tokyo.
Kerridge J. A. and Kieffer S. W. (1977) A constraint on impact theories of chondrule formation. *Earth Planet. Sci. Lett.* **35**, 35–42.
Kieffer S. W. (1975) Droplet chondrules. *Science* **189**, 333–340.
King E. A., Jr., Butler J. C., and Carman M. F. (1972a) Chondrules in Apollo 14 samples and size analyses of Apollo 14 and 15 fines. *Proc. Lunar Sci. Conf. 3rd*, pp. 673–686.
King E. A., Jr., Carman M. F., and Butler J. C. (1972b) Chondrules in Apollo 14 samples: Implications for the origin of chondritic meteorites. *Science* **175**, 59–60.
King T. V. V. and King E. A. (1978) Grain size and petrography of C2 and C3 carbonaceous chondrites. *Meteoritics* **13**, 47–72.
Kurat G. (1982) Impact origin of chondrules (abstract). In *Papers Presented to the Conference on Chondrules and their Origins*, p. 36. Lunar and Planetary Institute, Houston.
Leitch C. A. and Smith J. V. (1981) Enstatite chondrites II: Genesis (abstract). In *Lunar and Planetary Science XII*, pp. 610–612. Lunar and Planetary Institute, Houston.
Leitch C. A. and Smith J. V. (1982) Petrography, mineral chemistry and origin of Type I enstatite chondrites. *Geochim. Cosmochim. Acta* **46**, 2083–2098.
Lux G., Keil K., and Taylor G. J. (1981) Chondrules in H3 chondrites: textures, compositions and origins. *Geochim. Cosmochim. Acta* **45**, 675–685.
Martin P. M. and Mills A. A. (1978) Size and shape of near spherical Allegan chondrules. *Earth Planet. Sci. Lett.* **38**, 385–390.
McKay D. S., Greenwood W. R., and Morrison D. A. (1970) Origin of small lunar particles and breccia from the Apollo 11 site. *Proc. Apollo 11 Lunar Sci. Conf.*, pp. 673–694.
McKinley S. G., Keil K., and Scott E. R. D. (1982) Allan Hills A77156, an EH4 enstatite chondrite: some evidence against formation from red and blue luminescing planetesimals (abstract). *Meteoritics* **17**, 251.
McKinley S. G., Scott E. R. D., and Keil K. (1983) Chondrules in enstatite chondrites—Nature and source of enstatite (abstract). In *Lunar and Planetary Science XIV*, pp. 485–486. Lunar and Planetary Institute, Houston.
McSween H. Y., Jr. (1977) Chemical and petrologic constraints on the origin of chondrules and inclusions in carbonaceous chondrites. *Geochim. Cosmochim. Acta* **41**, 1843–1860.
McSween H. Y., Jr. and Lipschutz M. E. (1980) Origin of volatile-rich H chondrites with light/dark structures. *Proc. Lunar Planet. Sci. Conf. 11th*, pp. 853–864.

McSween H. Y., Jr. and Richardson S. M. (1977) The composition of carbonaceous chondrite matrix. *Geochim. Cosmochim. Acta* **41**, 1145–1161.

McSween H. Y., Jr., Biswas S., and Lipschutz M. E. (1981) Petrogenesis of light and dark portions of the Leighton gas-rich chondritic breccia. *Proc. Lunar Planet. Sci. 12B*, p. 1093–1103.

Minster J.-F. and Allègre C. J. (1979) ^{87}Rb-^{87}Sr chronology of H chondrites: Constraint and speculations on the early evolution of their parent body. *Earth Planet. Sci. Lett.* **42**, 333–347.

Neukum G., Mehl A., Fechtig H., and Zahringer J. (1970) Impact phenomena of micrometeorites on lunar surface materials. *Earth Planet. Sci. Lett.* **8**, 31–35.

O'Keefe J. D. and Ahrens T. J. (1975) Shock effects from a large impact on the moon. *Proc. Lunar Sci. Conf. 6th*, pp. 2831–2844.

Papike J. J., Simon S. B., White C., and Laul J. C. (1981) The relationship of the lunar regolith 10 μm fraction and agglutinates. Part I: A model for agglutinate formation and some indirect supporting evidence. *Proc. Lunar Planet. Sci. 12B*, pp. 409–420.

Planner H. N. and Keil K. (1982) Evidence for the three-stage cooling history of olivine-porphyritic fluid droplet chondrules. *Geochim. Cosmochim. Acta* **46**, 317–330.

Pohl J., Stöffler D., Gall H., and Ernstoon K. (1977) The Ries impact crater. In *Impact and Explosion Cratering* (D. J. Roddy, R. O. Pepin, and R. B. Merrill, eds.), pp. 343–404. Pergamon, N. Y.

Rasmussen K. L. and Wasson J. T. (1982) A new lightning model for chondrule formation (abstract). In *Papers Presented to the Conference on Chondrules and their Origins*, p. 53. Lunar and Planetary Institute, Houston.

Ringwood A. E. (1959) On the chemical evolution and densities of the planets. *Geochim. Cosmochim. Acta* **15**, 257–283.

Ringwood A. E. (1966) Genesis of chondritic meteorites. *Rev. Geophys. Space Phys.* **4**, 113–175.

Rubin A. E., Scott E. R. D., and Keil K. (1982a) Microchondrule-bearing clast in the Piancaldoli LL3 meteorite: a new kind of type 3 chondrite and its relevance to the history of chondrules. *Geochim. Cosmochim. Acta* **46**, 1763–1776.

Rubin A. E., Scott E. R. D., Taylor G. J., and Keil K. (1982b) Silicate matrix material in type 3 ordinary chondrites: implications for the origin of chondrules (abstract). *Meteoritics* **17**, 275–276.

Rubin A. E., Scott E. R. D., Taylor G. J., Keil K., Allen J. S. B., Mayeda T. K., Clayton R. N., and Bogard D. D. (1983) Nature of the H chondrite parent body regolith: evidence from the Dimmitt breccias. *Proc. Lunar Planet. Sci. Conf. 13th*, in *J. Geophys. Res.* **87**, pp. A741–A754.

Safronov V. S. (1972) Evolution of the Protoplanetary Cloud and Formation of the Earth and Planets. *NASA TTF-677*. 206 pp.

Scott E. R. D. and Taylor G. J. (1982) Primitive breccias among type 3 ordinary chondrites—origin and relation to regolith breccias. In *Workshop on Lunar Breccias and Soils and Their Meteoritic Analogs* (G. J. Taylor and L. L. Wilkening, eds.), pp. 130–134. LPI Tech. Rpt. 82-02. Lunar and Planetary Institute, Houston.

Scott E. R. D. and Taylor G. J. (1983) Petrologic similarities among chondrules in H, LL, CO, CM, and E chondrites (abstract). In *Lunar and Planetary Science XIV*, pp. 680–681. Lunar and Planetary Institute, Houston.

Scott E. R. D., Taylor G. J., and Keil K. (1982a) Origins of ordinary and carbonaceous type 3 chondrites and their components (abstract). In *Lunar and Planetary Science XIII*, pp. 704–705. Lunar and Planetary Institute, Houston.

Scott E. R. D., Taylor G. J., Rubin A. E., Keil K., and Kracher A. (1982b) Origin of chondrules by preaccretionary melting of silicate matrix materials (abstract). In *Papers Presented to the Conference on Chondrules and their Origins*, p. 54. Lunar and Planetary Institute, Houston.

Suess H. E. (1949) Zur chemie der Planeten und Meteoritenbildung. *Z. Electrochem.* **53**, 237–241.

Taylor G. J. (1982) The three ages of brecciation (abstract). *Meteoritics* **17**, 287–288.

Taylor G. J. and Heymann D. (1969) Shock, reheating, and the gas retention ages of chondrites. *Earth Planet. Sci. Lett.* **7**, 151–161.

Tschermak G. (1883) Beitrug zur classification der Meteoriten. *Sitzber. Akad. Wiss. Wien* **88** (Part 1), 347–371.

Tsuchiyama A. and Nagahara H. (1981) Effects of precooling thermal history and cooling rate on the texture of chondrules: a preliminary report. *Mem. Natl. Inst. Polar Res., Spec. Issue No. 20*, pp. 175–192. National Institute of Polar Research, Tokyo.

Turner G., Enright M. C., and Cadogan P. H. (1978) The early history of chondrite parent bodies inferred from ^{40}Ar-^{39}Ar ages. *Proc. Lunar Planet. Sci. Conf. 9th*, pp. 989–1025.

Unruh D. M. and Tatsumoto M. (1980) A uniform U-Pb age for L chondrites and a method for correcting terrestrial Pb contamination (abstract). *Meteoritics* **15**, 380–381.

Urey H. C. (1952) *The Planets, Their Origin and Development.* Yale University Press, New Haven. 245 pp.

Wänke H., Dreibus G., Jagoutz E., Palme H., and Rammensee W. (1981) Chemistry of the Earth and the significance of primary and secondary objects for the formation of planets and meteorite parent bodies (abstract). In *Lunar and Planetary Science XII*, pp. 1139–1141. Lunar and Planetary Institute, Houston.

Wasson J. T. (1972) Formation of ordinary chondrites. *Rev. Geophys. Space Phys.* **10**, 711–759.

Wasson J. T. and Wetherill G. W. (1979) Dynamical, chemical and isotopic evidence regarding the formation locations of asteroids and meteorites. In *Asteroids* (T. Gehrels, ed.), pp. 926–974. Univ. of Arizona Press, Tucson.

Whipple F. L. (1966) Chondrules: suggestion concerning the origin. *Science* **153**, 54–56.

Whipple F. L. (1972) On certain aerodynamic processes for asteroids and comets. In *From Plasma to Planet* (A. Elvius, ed.), pp. 211–232. Wiley and Sons, N. Y.

Wlotzka F. (1969) On the formation of chondrules and metal particles, by 'shock melting.' In *Meteorite Research* (P. M. Millman, ed.), pp. 174–184. Reidel, Dordrecht.

Wood J. A. (1963) On the origin of chondrules and chondrites. *Icarus* **2**, 152–180.

Wood J. A. (1981) The interstellar dust as a precursor of Ca,Al-rich inclusions in carbonaceous chondrites. *Earth Planet. Sci. Lett.* **56**, 32–44.

Wood J. A. and McSween H. Y., Jr. (1977) Chondrules as condensation products. In *Comets, Asteroids, Meteorites: Interrelations, Evolution, and Origins* (A. H. Delsemme, ed.), pp. 365–373. Univ. of Toledo Press, Ohio.

Zook H. A. (1980) A new impact model for the generation of ordinary chondrites (abstract). *Meteoritics* **15**, 390–391.

Zook H. A. (1981) On a new model for the generation of chondrules (abstract). In *Lunar and Planetary Science XII*, pp. 1242–1244. Lunar and Planetary Institute, Houston.

Experimental Investigation of Electrical Discharge Formation of Chondrules

T. J. Wdowiak

Physics Department, University of Alabama in Birmingham, Birmingham, Alabama 35294

The hypothesis that chondrules are formed in a thermal impulse event with subsequent radiative cooling, as expressed in the Whipple discharge heating of dust balls model and in impact jetting models, has been examined by experiment. The experiment demonstrates a difficulty with those models if the precursor contains volatiles. An apparatus was contructed that heats ground Allende meteorite in a 5 kJ gas discharge and allows the fused material to radiatively cool while free-falling. Spherules produced in the experiment are vesiculated. If chondrules were produced by very short duration impulse heating, as in lightning and hypervelocity impact of precursor material incorporating volatiles as absorbed, structural, or admixture components and cooled rapidly by radiation, they also would be expected to contain vesicles. The absence of vesiculation in chondrules suggests that most are not the product of very short duration impulse heating and radiative cooling.

INTRODUCTION

In 1966 Whipple proposed that chondrules were formed by the fusing of dust in electrical discharges occurring in the primordial solar nebula. He invoked the pinch effect as the mechanism responsible for heating. Cameron (1966) extended the details of the model to include accumulation processes. Experiments by Salisbury involving electrical discharges in an atmosphere containing granite dust were disclosed by Whipple (1975). Those experiments produced spherules with an efficiency of the order of 1%. However, it was also reported that Ursula Marvin and John Wood stated the composition was not like that of chondrules. Because of the interest in the Whipple mechanism, experiments utilizing Allende meteorite material have been carried out.

EXPERIMENT AND ANALYSIS

The experiment consists of apparatus (Fig. 1) capable of discharging 5 kJ of energy, from a 45 microfarad capacitor bank charged to 15 kV, at a pressure of 0.35 torr, through an electrode gap in which a dust ball of ground Allende meteorite is suspended by a thread. The discharge is triggered with an ignitron, which in turn is fired by a thyratron circuit. Equating magnetic and gas pressure provides a calculated estimate of discharge temperature in the range of 10^3–10^4K. The heated material free-falls 1.5 m while radiative cooling takes place, and is collected at the bottom by a funnel and test tube. The dust balls were prepared by wetting the cotton thread with water, dipping

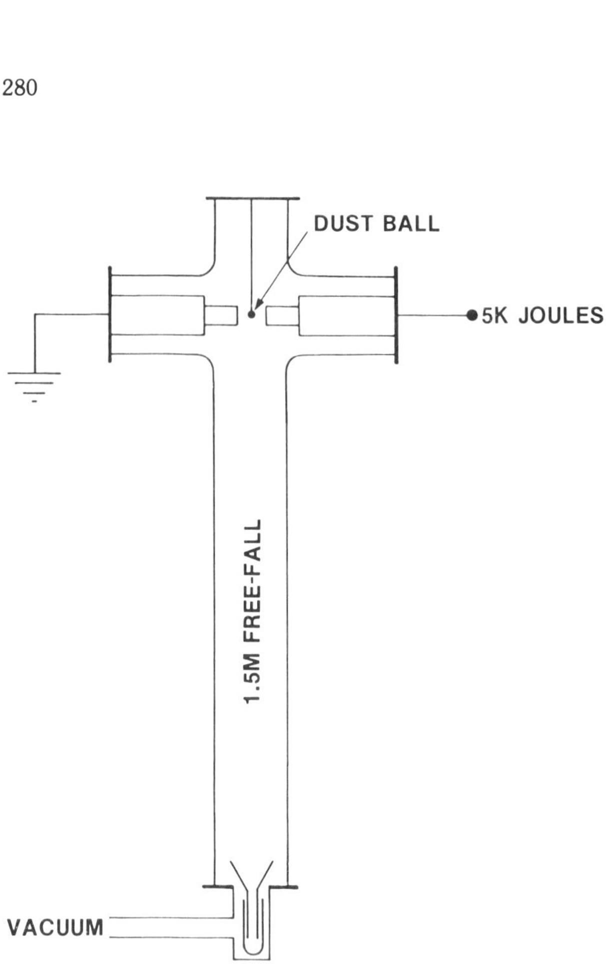

Fig. 1 Experimental apparatus constructed of Kimax with brass electrodes 1 cm in diameter.

it into the dust, and transferring it to the apparatus where it was held at 0.35 torr and room temperature for at least 20 minutes while the capacitor bank was charged.

Microscopic examination revealed that although some of the dust was fused into spherules in the size range of 10 to 200 microns, the bulk experienced melting just sufficient to agglomerate the fines. Spherules were separated out and placed upon Teflon plates where they were potted in thermoplastic for purposes of sectioning. Sectioning was accomplished by using abrasives on glass. Light and SEM microscopy (Fig. 2) revealed interior bubbles. Surface bubbles were also evident for spherules that had not been sectioned. The existence of vesicles is due to the presence of volatiles including sulfur in the meteorite (~100 ppm), residual water from the dust ball preparation process, volatized thread, and atmospheric gases.

It is not unreasonable to expect dust of the primordial solar nebula to contain volatiles such as water, sulfur, and organics as absorbed, structural, or admixture components. The experiment suggests that chondrules produced by the Whipple mechanism would probably contain vesicles as the laboratory-produced spherules do. The absence of vesicles in chondrules indicates it is unlikely that they have been formed by the Whipple mechanism. Impact mechanisms (Kieffer, 1975) have an aspect in common with the discharge mechanism in that both involve impulse heating and rapid

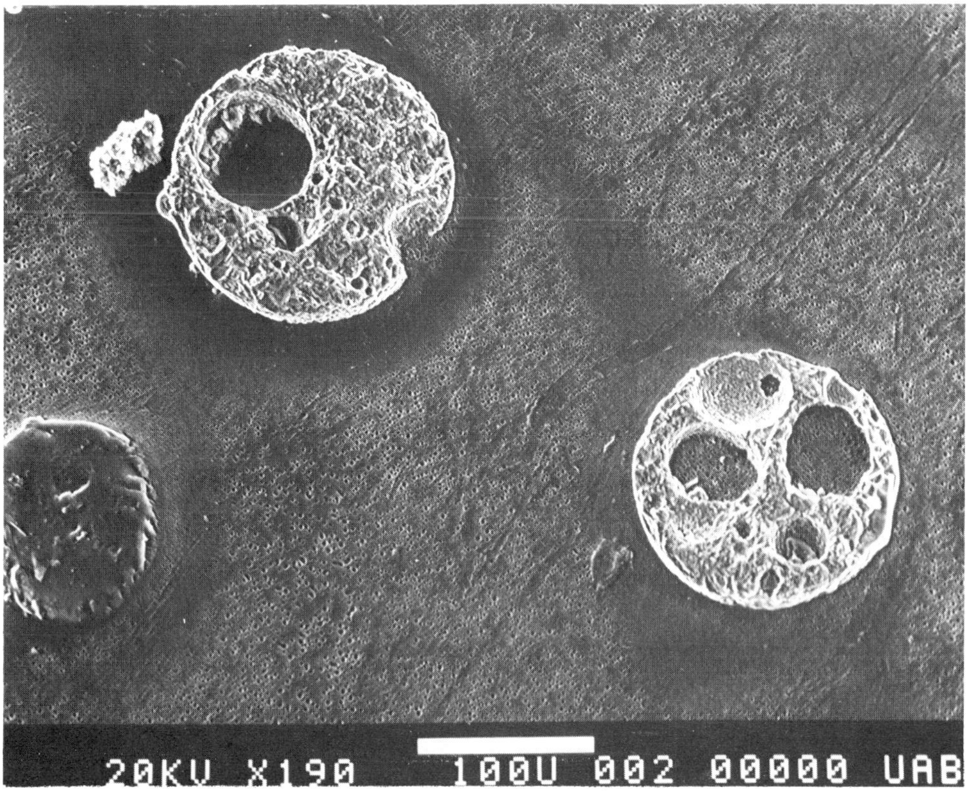

Fig. 2 SEM microphotograph of two spherules sectioned to show interior vesicles. The scale is 100 microns in length.

radiative cooling. Chondrules produced by hypervelocity impact would also be expected to contain vesicles, as do some of the glassy spherules found in the lunar regolith (von Englehardt and Stöffler, 1970). These are thought to be due to meteoroid impacts. Also, the Clayton model (1980) of highly exothermic chemical reactions providing energy for fusion of dust into chondrules could be expected to suffer similar consequences.

The presence of bubbles in spherules that have been produced by impulse heating of precursor solids with subsequent radiative cooling, and why the bubbles would remain, can be understood by considering the effect of decreasing temperature on the volume of a gas bubble and the viscosity of the melt. Examination of viscosity-temperature data (η vs. T) for various basaltic melts (Clark, 1966) indicates a first-order relationship of $\eta \propto T^{-N}$ where $N \geq 10$ and can be as high as 30! This, coupled with a radiative cooling rate of $dT/dt \propto -T^4$, results in bubble entrapment. Indeed the rapid increase in viscosity with decreasing temperature is conducive to retaining the bubble at a greater volume than determined by gas law contraction—bubbles are frozen in!

The result of the experiment is verified by the presence of bubbles in some of the glassy droplets found in lunar regolith samples (von Engelhardt and Stöffler, 1970). Such droplets are quite likely the result of hypervelocity impact of meteoroids on the lunar surface. Droplets not having bubbles can be understood in terms of the lunar precursor material being lacking in volatiles in those cases. If chondrules were produced by very short duration impulse heating and radiative cooling as in the Whipple and impact hypotheses, the precursor material would have to have been devoid of volatiles. The apparent lack of vesiculation in chondrules and its presence in the experimental spherules and in some lunar samples indicate that if chondrules are of secondary origin, the melt must have been held at elevated temperatures longer than radiative cooling in a low temperature environment permits. The recent work of Planner and Keil (1982) and King (1982) suggests that chondrules have been held at elevated temperatures for extended times much longer than radiative cooling times. Other investigations have cast some doubt on primary origin models (Wood, 1981; Rambaldi, 1981; Nagahara, 1981).

CONCLUSIONS

The results of this study and others cited suggests chondrule formation involved melting of precursor solids in a non-impulsive manner with cooling at a slow rate as compared to the radiative process. Perhaps this occurred during the sun's T Tauri stage and may have involved sustained heated gas, ionic bombardment, or electromagnetic heating. The Whipple electrical discharge has also been invoked (Miller *et al.*, 1976) to explain the presence of amino acids in carbonaceous chondrites via a Miller synthesis. Doubt cast upon the reality of electrical discharge heating by this study forces reexamination of the corollary hypothesis of meteoritic amino acid synthesis by discharge.

Acknowledgments. The author acknowledges the support of an institutional faculty research grant. The assistance of Nick Gerontakis, Jerry Sewell, and Frank Egner in constructing and operating the experiment is gratefully acknowledged. Scott Brande, Carl Conger, and Mike Neilson provided valuable assistance in obtaining the necessary photomicrographs. Special thanks are given to John Wood for reading the initial manuscript and making constructive suggestions and to Bert Donn for useful discussions and encouragement while the author was a summer faculty fellow at the Goddard Space Flight Center.

REFERENCES

Cameron A. G. W. (1966) The accumulation of chondritic material. *Earth Planet. Sci. Lett.* **1**, 93–96.

Clark S. P., Jr. (1966) Viscosity. In *Handbook of Physical Constants,* p. 299. Geol. Soc. Amer. Mem. 97.

Clayton D. D. (1980) Chemical energy in cold-cloud aggregates: The origin of meteoritic chondrules. *Astrophys. J.* **239**, L37–L41.

Kieffer S. W. (1975) Droplet chondrules. *Science* **189**, 333–340.

King E. A. (1982) Refractory residues, condensates, and chondrules from solar furnace experiments. *Proc. Lunar Planet. Sci. Conf. 13th*, in *J. Geophys. Res.* **87**, A429–A434.

Miller S. L., Urey H. C., and Ora J. (1976) Origin of organic compounds on the primitive earth and in meteorites. *J. Mol. Evol.* **9**, 59–72.

Nagahara H. (1981) Evidence for secondary origin of chondrules. *Nature* **292**, 135–136.

Planner H. N. and Keil K. (1982) Evidence for the three-body cooling history of olivine-porphyritic fluid droplet chondrules. *Geochim. Cosmochim. Acta* **46**, 317–330.

Rambaldi E. R. (1981) Relict grains in chondrules. *Nature* **293**, 558–561.

von Engelhardt W. and Stöffler D. (1970) *Lunar Rocks Under the Microscope*. No. 09-123-e, Lit. Carl Zeiss, Oberkochen.

Whipple F. L. (1966) Chondrules: Suggestion concerning the origin. *Science* **153**, 54–56.

Whipple F. L. (1975) Comments. In *The Dusty Universe* (G. B. Field and A. G. W. Cameron, eds.), pp. 293–310. Academic, New York.

Wood J. A. (1981) The interstellar dust as a precursor of Ca, Al-rich inclusions in carbonaceous chondrites. *Earth Planet. Sci. Lett.* **56**, 32–44.

Aggregation of Grains in a Turbulent Pre-solar Disk

Bernhard Wieneke and Donald D. Clayton
Department of Space Physics, Rice University, Houston, Texas 77251

The growth and evolution of grains in the protostellar nebula is investigated within the context of turbulent low-mass disk models developed by previous investigators. Because of grain collisions promoted by the turbulent velocities, particles aggregate to millimeter size in times of order 10^3 yrs. During the growth the particles acquire a large inward radial velocity due to gas drag (Weidenschilling, 1977) and spiral into the sun. The calculations indicate that the final size of the particles does not exceed a few centimeters. This result is not very sensitive to the specific nebula parameters. For all conditions investigated it seems impossible to grow meter- or kilometer-sized bodies that could decouple from the gas motion. An additional argument is given that shows that only particles smaller than centimeter size can survive drift into the growing sun by being transported radially outward by turbulent mixing. This agrees well with the maximum size of inclusions and chondrules. Since sedimentation of grains and subsequent dust disk instability is effectively inhibited by turbulent stirring, the formation of planetesimals and planets cannot be explained in the above scenario without further assumptions.

I. INTRODUCTION

In a turbulent pre-solar disk, turbulent viscosity provides a means of transporting angular momentum radially outward. Mass flows inward in the inner part and outward in the outer part of the disk. Turbulence is probably driven by a meridional circulation current (Cameron, 1978) or by a convective instability in the radial direction or in the direction parallel to the axis of rotation (Cameron and Fegley, personal communication; Lin, 1981). As long as turbulent transport is efficient enough, the disk mass never exceeds a few percent of the solar mass. Lynden-Bell and Pringle (1974) and Lin and Bodenheimer (1982) have shown that the inner region of the disk evolves toward a quasi-steady state with an inward mass flux changing on a timescale of $R^2/\upsilon \sim 10^5$–10^6 yrs (υ = turbulent viscosity, R = radius of disk), which is considerably larger than the other dynamical timescales [the turbulent coherence time ($\sim$ 10–100 yrs) or the time for establishing hydrostatic equilibrium in the vertical direction ($\sim$ few yrs)].

The disk is heated by energy dissipation due to viscious friction, but only in about the innermost 1 AU is the temperature above 1000 K. It falls off rapidly with radial distance from the protosun. Therefore in most parts of the disk we expect the condensable matter to reside in the form of solids.

Because of turbulent convection, particles are mixed throughout the disk. In superposition they have a systematic drift inward owing to both the average inward gas motion and gas drag (Adachi *et al.*, 1976; Weidenschilling, 1977). Condensable vapor is mixed outward from the innermost hot region, where grains evaporate partially or completely, and recondenses onto existing grains or forms new grains by nucleation (Morfill, 1983; Cameron and Fegley, personal commmunication). Interesting chemical effects may result that might also explain the rim structure of chondrules and inclusions (Morfill, 1983). Further out in the disk, at lower temperatures, the situation is somewhat simpler. Neglecting the interaction of the grains with the gas phase, the main growth mechanism for grains is collisional aggregation promoted by turbulence-induced velocities. Using representative disk models developed by previous investigators, we investigate the aggregation of grains in a region between 1 and 10 AU. We follow the growth and trajectory of a test particle by making further simplifying assumptions concerning the background size spectrum and its change with time. Our major conclusion, that particles larger than a few centimeters cannot be aggregated in the simple model adopted before radial drift takes them into the sun, has been reached by others (Weidenschilling, 1980; Völk, 1982; Morfill, 1983). Our purpose is to report the present status of our own calculations of the situation in a simplified way, in order that any relevance to the chondrule problem can be found within the present volume. It is our intent to later refine the calculations to a much more physically complete theory.

II. MODEL

A. Disk model

We have used two different steady-state disk models as settings for our calculations. The first one is a minimum-mass disk by Weidenschilling (1980) calculated from the present distribution of mass in the solar system (after adding the solar complement of light elements to the planet masses). The surface density falls off in radial directions as $R^{-1.5}$. At 1 AU the temperature is 600 K and decreases with radius as $R^{-1.0}$. The disk height is taken to be proportional to the radius. We used an inward mass flux in the inner disk of $10^{-6} M_{\odot}/yr$.

The second model is a steady-state convective accretion disk model by Lin (1981). The disk is unstable to vertical convection. Turbulent convection, in return, is the cause for a viscous heat dissipation, which is the energy source for the convection. Taking an inward mass flux of 10^{-6} $M_{\odot}/yr$ the surface density is 320 g/cm^2 out to a radius of 6.9 AU where it drops to 40 g/cm^2. This discontinuity is due to a change in grain opacity at 160 K. It was assumed that the icy components of the grains evaporate at 160 K. This leads to a large drop in the opacity. The temperature falls off roughly as $R^{-0.75}$ from 640 K at 1 AU while the disk height increases as about $R^{0.75}$. Both density distributions are illustrated in Fig. 1.

The basic parameters of the two disks are summarized in Table 1. For both disks the total mass within 10 AU is less than 0.03 $M_{\odot}$. We assume that the turbulent velocity is 0.3 c_s (c_s = sound speed).

B. Motion of grains

As first reported by Whipple (1964) grains drift radially inward due to gas drag. This is due to the fact that the gas disk is partially supported by a radial pressure gradient. Thus the disk rotates

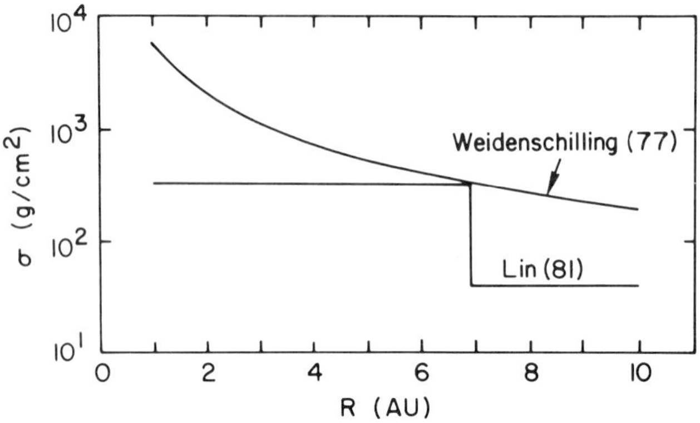

Fig. 1 Surface density vs. radius for the two disk models used in these calculations.

with sub-Keplerian velocities, whereas solid bodies, which try to move on Kepler orbits, experience a gas drag and spiral toward the protosun. This has been calculated quantitatively by Adachi *et al.* (1976) and Weidenschilling (1977).

Also taking into account the average radial gas motion v_g due to viscous transport, $v_g \equiv F/2\pi R\sigma$ (F = radial mass flux, σ = surface density), we show in Appendix A that the radial drift velocity u of a particle is given to first order in $\Delta g/g$ by:

$$u = (\Delta g\ t_e + v_g)\left(1 + \frac{4\pi^2 t_e^2}{T^2}\right)^{-1}, \qquad (1)$$

Table 1. Disk parameters.

Parameter		Model 1*	Model 2†
$T = T_o(R/AU)^{n}T$	n_T:	-1	-0.75
	T_o[K]	600	640
$\sigma = \alpha_o(R/1\ AU)^{n}S$	n_S:	-1.5	0
	$\sigma_o[g/cm^2]$	6,000	320 for $R > 6.9$ AU
			40 for $R < 6.9$ AU
$H = H_o(R/1\ AU)^{n}G$	n_H:	1	0.75
	H_o[AU]	0.125	0.14
V_t/c_s		0.3	0.3
$F[M_{\odot}/yr]$		10^{-6}	10^{-6}
$v_r \equiv \frac{\lvert F\rvert}{2\pi R\sigma}$[cm/sec]	at 1 AU	112	2,100
	at 10 AU	355	1,680

*Weidenschilling (1980).
†Lin (1981).

where Δg is the "residual gravity" due to the pressure gradient, $\Delta g = 1/\rho\, dP/dR$, $g = GM/R^2$, and T and t_e denote respectively the Keplerian orbital period and the particle stopping time, $t_e \equiv F_D/mv_{rel}$ (F_D = gas drag force, m = particle mass, v_{rel} = particle velocity relative to gas). In the Epstein drag regime for small particles ($\lesssim$1 cm) of radius *a* and density ρ_s the stopping time is:

$$t_e = \frac{\rho_s a}{\rho \bar{v}}, \tag{2}$$

where ρ and $\bar{v}$ designate the gas density and the mean thermal velocity of the gas molecules, respectively.

The drift velocity is a maximum ($\sim 10^4$ cm/sec) at $2\pi t_e/T \simeq 1$, which corresponds to a particle size of about 1 m for the above disk models at 1 AU. For the much smaller particles in our calculations, the radial drift is increasing with particle sizes.

Small particles with $2\pi\, t_e/T \ll 1$ are closely coupled to the gas motion and drift inward to first order in t_e/T as:

$$u \simeq v_g - 2\Delta v \frac{2\pi t_e}{T}, \tag{3}$$

where $\Delta v = \frac{|\Delta g|}{2g} v_\phi$ is the difference between the Kepler velocity V_ϕ and the azimuthal gas velocity. Large bodies ($2\pi\, t_e/T \gg 1$) decouple from the gas motion and move on only slightly perturbed Kepler orbits.

Superposed on this systematic inward drift is the random motion due to turbulence. As long as the particle stopping time, t_e, is less than the turbulent coherence or mixing time, τ_t($\tau_t = l_t/v_t$, v_t = turbulent velocity, l_t = size of largest eddy ~ mixing length), the particle partakes in turbulent transport.

The radial motion of the test particle is calculated according to eq. (1), excluding the random walk due to turbulence. This is a simplifying assumption that is not totally correct; nonetheless, our graphical results give a meaningful visual representation of grain growth. A more correct approach, as pointed out by an anonymous referee, would be to solve the coagulation equation in conjunction with the transport equation, as a coupled system, subject to the physical boundary conditions. Although we eventually hope to achieve this, our present simplified results are useful for their direct understandability.

C. Growth of grains

The growth rate due to collisional aggregation for a particle of radius *a* is given by:

$$\frac{da}{dt} = \frac{Q\rho_d}{4\rho_s} \frac{(a + a_o)^2}{a^2} v_{rel}, \tag{4}$$

where Q, ρ_d, and ρ_s denote the sticking probability, the nebular mass density in the form of grains, and the specific density of the solid particles, respectively. We assume that grains at a temperature below 160 K are composed of a fluffy mixture of icy and rocky materials. For simplicity we assume that all icy components evaporate at 160 K. In these two regions we use a specific density of 0.5 or 1.0 g/cm^3 (Weidenschilling, 1980) and a mass ratio of grain to gas equal to 0.012 or 0.0034 (Podolak and Cameron, 1974) for $T < 160$ K and $T > 160$ K, respectively. For simplicity we assume all background particles have the same size a_o, although we will make differing assumptions about that size.

The relative velocity v_{rel} between the test particle and the background particles is calculated by taking into account Brownian motion, differences in the drift velocities, and turbulence-induced relative velocities. Brownian motion leads to a relative velocity between particles of mass m_1 and m_2 equal to:

$$v_{br} = \sqrt{kT}\left(\frac{1}{m_1} + \frac{1}{m_2}\right)^{1/2} . \tag{5}$$

The relative velocity due to different drift velocities is calculated by:

$$v_{dr} = [(u_1 - u_2)^2 + (w_1 - w_2)^2]^{1/2} , \tag{6}$$

where the radial drift velocity u_1 and the azimuthal velocity w_1 relative to the azimuthal gas velocity are given by eqs. (1) and (A10), respectively.

The turbulence-induced relative velocities have been calculated by Völk *et al.* (1980) using a Kolmogoroff power spectrum for the turbulence. For equal-sized particles the relative velocity can be approximated by (Völk *et al.*, 1980):

$$v_{tr} = 1.33\, v_t \frac{(t_e/\tau_t)^{1/2}}{1 + (t_e/\tau_t)} . \tag{7}$$

If the test-particle size a is larger than the background-particle size a_o the following expression is used:

$$\begin{aligned} v_{tr} &= \frac{2}{1 + (\tau_t/t_e)^{1/2}}\, v_t \qquad && t_e/\tau_t < 1 \\ &= v_t && t_e/\tau_t > 1 , \end{aligned} \tag{8}$$

where t_e is the test particle stopping time. For background-particle sizes less than about 1 cm this equation agrees within about 30% with the results of Völk *et al.* (1980) displayed in their Fig. 2.

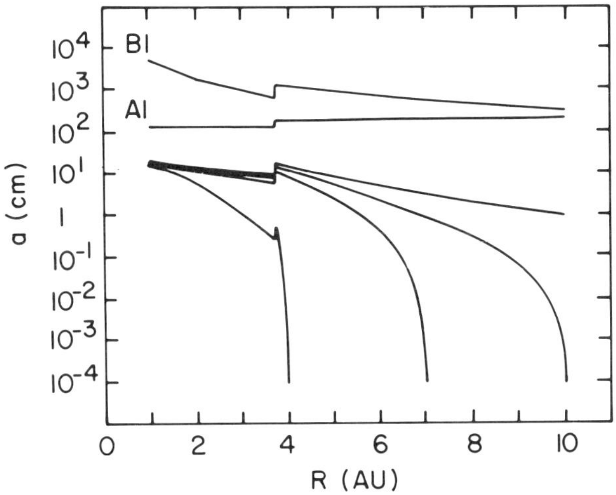

Fig. 2 Size of the text particle vs. radius using Weidenschilling's model. The background-particle size and the sticking probability are 1 μm and 0.1, respectively. The test particle starts out with 1 μm in size at 4, 7, and 10 AU and with 1 cm at 10 AU. The radial motion is approximated by the drift owing to gas drag. Curves A1 and B1 show how large particles must be to either: (A1) begin minimizing radial drift resulting from gas drag; or (B1) to decouple from the turbulence.

The total relative velocity is given by $v_{rel} = (v_{br}^2 + v_{dr}^2 + v_{tr}^2)^{1/2}$. By far the largest contribution to v_{rel} comes from turbulence. Brownian motion is negligible for particles larger than about 0.1 μm. For the above disk models v_{dr} remains smaller than v_{tr} for all particle sizes.

III. RESULTS

The size of the test particle and its radial position has been calculated by numerical integration of eqs. (1) and (4). The results are displayed in Fig. 2 using Weidenschilling's disk model. The size of the particle is plotted as a function of radial distance from the protosun for four different initial conditions. The particle starts with a size of 1 μm at 4, 7, and 10 AU and with 1 cm at 10 AU. The mean radial motion has been approximated by the radial drift from gas drag. The curve labeled A1 shows the particle size appropriate to the condition $2\pi\, t_e/T = 1$ (i.e., when the particle has the maximum inward drift due to gas drag). Curve B1 indicates the size appropriate to the condition $t_e/\tau_t = 1$. As long as the particle size is below curve B1 the particle is coupled to the turbulent motion. The calculations have been performed using a sticking probability Q = 0.1 and assuming a constant background particle size of 1 μm. Figure 3 shows the same calculations using Lin's disk. The numbers along one of the curves denote the time in years. At about 4 and 7 AU in Figs. 2 and 3, respectively, the temperature in the disk reaches 160 K. As mentioned before, we assume that the icy components evaporate at this point and the particle shrinks by about a factor of 2.

The results show that the particle grows rapidly from 1 μm to millimeter sizes in a few thousand years. The relative growth rate d*l*na/dt becomes smaller as the particle size increases. This is expected by eq. (4) in view of eqs. (8) and (2), since $t_e \propto a^{1/2}$ and thus $v_{tr} \lesssim a^{1/2}$ and da/dt

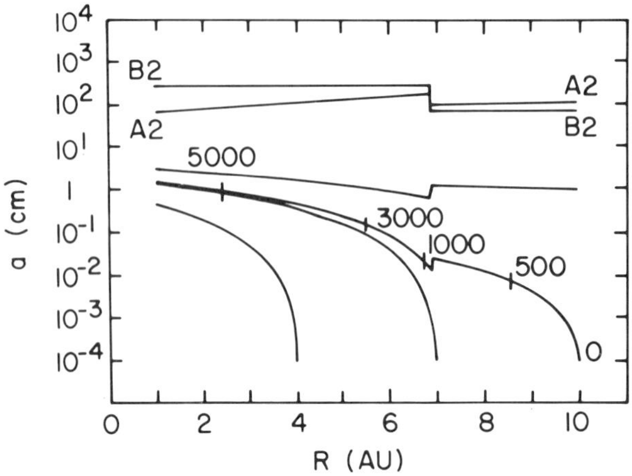

Fig 3. Same as Fig. 2 using Lin's model. The numbers denote the time in years.

$\propto a^{1/2}$. In addition, once the particle grows to centimeter sizes it acquires a large inward drift due to gas drag. This explains that all trajectories converge at 1 AU to sizes in the centimeter range. A one-cm-sized particle at 10 AU drifts inward rapidly and differs in size at 1 AU only slightly from a particle that starts out with 1 μm at 10 AU.

Since this calculation has not accounted properly for the random walk due to turbulence, the lower part of the curves should not be taken as actual trajectories; rather, they serve to define the timescale for growth up to centimeter sizes. Potentially, a small particle ($\lesssim$1 cm) can be turbulently convected radially outward or inward over distances of 10 AU in about 1000 yrs. So the actual trajectory for small particles is chaotic and variable. The situation changes for larger particles because the large inward velocity resulting from the gas drag prevents them from being mixed outward. The limiting size for which particles can still be transported outward is of the order of 1 cm as the following argument shows.

In a statistical ensemble description, turbulent mixing can be dealt with as diffusion with, according to the mixing-length theory, a diffusion coefficient identical to the turbulent viscosity $\upsilon = 1/3\ l_t v_t$. In a simple one-dimensional model, assuming that particles of a certain size are produced at a certain radius R_0 and have a systematic inward velocity u, then in the steady state the particle number density n for $R > R_0$ can be calculated by equating the flux outward due to turbulent mixing with the inward flux due to the velocity u: that is,

$$nu = -\upsilon \frac{dn}{dR} . \tag{9}$$

If u and υ are constant, which is only approximately true in the disk, then n is proportional to:

$$\exp(-\frac{u}{\upsilon} R) = \exp(-\frac{u}{v_t} \frac{3R}{l_t}). \tag{10}$$

The mixing length l_t is comparable to the disk height H. If particles are to diffuse outward any appreciable distance, say R, we required that $u/v_t \lesssim 1/30$, considering that $H/r \sim 1/10$. For the above disk models this condition corresponds to a particle size $a \lesssim 1$ cm. Since we neglected evaporation, recondensation of vapor, and growth of particles, this can only be a very rough estimate. But it is clear that such a limiting size must exist since the maximum inward drift velocity ($\sim 10^4$ cm/sec) is of the same order as the turbulent velocity ($1-3 \times 10^4$ cm/sec).

Figures 2 and 3 show that the 1 AU particles have not grown to sizes above curve A and B, which would have been necessary for decoupling from the turbulent motion and the gas drag. Because the growth rate of centimeter-sized particles is small and because they cannot be turbulently mixed outward, they are expected to drift inward rapidly (1 AU$/10^4$ cm/sec $= 48$ yrs) and evaporate at higher temperatures.

This result is quite independent of the disk parameters. The size of the particles at 1 AU for Weidenschilling's disk is somewhat larger than in Lin's model because the gas density and hence the growth rate are larger and because the average inward gas motion is smaller.

The calculations depend only weakly on the background-particle size (Fig. 4). The test particle has an initial size of 1 mm at 10 AU. For the upper curve we set the background-particle size equal to the test-particle size (i.e., we assumed that all particles evolved equally), whereas for the middle and lower curve we took a constant size of 1 mm and 1 μm, respectively. The final size of the particles at 1 AU differs only slightly. The difference is mainly due to a change in effective collisional cross section.

The main uncertainty in the calculations lies in the sticking probability Q. Greenberg *et al.* (1978) pointed out that gravity is ineffective for submillimeter-sized particles because the relative velocities resulting from Brownian motion alone exceed the gravitational escape velocity. Although the relative velocities from turbulence are not large enough to disrupt solids, they may disrupt grain aggregates. It is not yet clear whether particles stick at all upon colliding. It has been suggested that grains stick, for example, by electrostatic binding or cold welding. A reliable estimate of Q is not possible yet, but it might be very much less than unity.

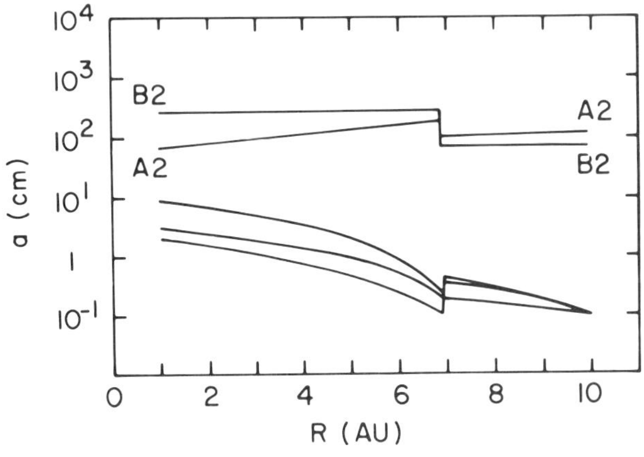

Fig. 4 Using Lin's model. The background-particle size is 1μm and 1 mm for the lowest curve and the middle one, respectively. It has been set equal to the test particle size for the upper curve. The particle starts out with 1 mm in size at 10 AU. The sticking probability is 0.1.

In Fig. 5 we varied Q from 1 to 0.01. The results show that even for Q = 1 the particle does not grow fast enough to escape being drawn into the evaporation zone. Note also that for a sticking coefficient of 0.01 we still expect particle growth from 1 μm to 1 mm in about 10,000 yrs.

IV. CONCLUSIONS

The calculations have shown that owing to turbulence-induced particle velocities, particles can grow rapidly by collisional agglomeration, provided only that they stick with reasonable efficiency upon collision. For a sticking probability Q = 0.1 the timescale for growing centimeter-sized objects from micron-sized dust is of the order 10^3 yrs for the innermost 10 AU of the disk. For Q = 0.01 it takes about 10^4 yrs to grow millimeter-sized particles. This is considerably less than the evolutionary timescale of the disk (10^5–10^7 yrs).

Since small particles are coupled to turbulent motion they are mixed throughout the disk. Particles larger than about 1 cm acquire a large inward drift velocity owing to gas drag and cannot be transported outward by turbulent mixing. In addition, the results show that they cannot grow fast enough into bodies large enough (>> 1 m) to decouple from the gas motion and stay on Kepler orbits. They are expected to drift rapidly toward the sun and evaporate, confirming similar conclusions (Weidenschilling, 1980; Völk, 1982; Morfill, 1983).

This result is quite independent of the background-particle size spectrum and the nebular parameters. For example, in a more massive, denser disk with a higher inward mass flux of 10^{-5} $M_{\odot}$/yr the growth rates are larger in general, but at the same time curves A2 and B2 in Fig. 3 are shifted upward because of a smaller value for t_e [see eq. (2)]. Hence, there is again a limiting size, about 10 cm–1 m, above which the particle growth is too slow and the inward drift due to gas drag too large to prevent destruction of the particle by drift into the evaporation zone.

So far, we assumed that the particles are more or less fluffy ice and dust balls. But this size-sorting process would apply equally well to denser chondrules. Since the limiting size of

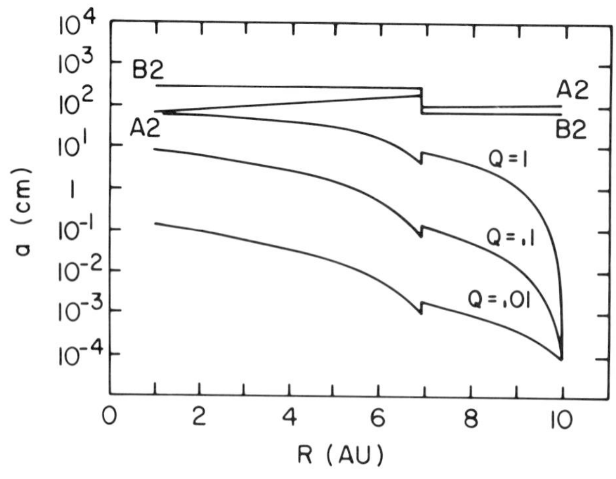

Fig. 5 Using Lin's model. Shown here is the dependence on the sticking probability Q. The background-particle size is equal to the test-particle size.

particles is a function of the nebular parameters, we expect different chondrule size distributions in different meteorites as is observed (e.g., Goswami, 1982), assuming that meteorites formed in different parts of the disk and that there is a certain time-lag ($\gtrsim 10^3-10^4$ yrs) between formation of chondrules and meteorite parent bodies.

The calculations as described offer no hope that large bodies ($\gtrsim$1 m) could aggregate in the turbulent disk, in agreement with the same conclusion reached by Morfill (1983) and Völk (1982). Sedimentation of particles toward the midplane of the disk is inhibited by turbulent stirring. It is not yet clear how large bodies and planetesimals or planets could form in such a scenario. To circumvent this very problem it has been proposed that turbulence is intermittent (Völk, 1982). Then through sedimentation a dense dust disk forms in about 3×10^3 yrs (Nakagawa *et al.*, 1981; Weidenschilling, 1980), which then breaks up into kilometer-sized objects because of a gravitational instability (Safranov, 1972; Goldreich and Ward, 1973).

Acknowledgments. This work was supported in part by NASA grant NSG-7361 and in part by the Robert A. Welch Foundation. Several discussions with G. Morfill were motivating factors for this study.

REFERENCES

Adachi I., Hayashi C., and Nakazawa K. (1976) The gas drag effect on the elliptical motion of a solid body in the primordial solar nebula. *Prog. Theor. Phys.* **56**, 1756–1771.

Cameron A. G. W. (1978) Physics of the primitive solar accretion disk. *Moon and Planets* **18**, 5–40.

Goldreich P. and Ward W. R. (1973) The formation of planetesimals. *Astrophys. J.* **183,** 1061–1061.

Goswami J. N. (1982) Size-distribution and morphology of chondrules in Dhajala (H-3) and Murchison (CM) chondrites (abstract). In *Papers Presented to the Conference on Chondrules and their Origins,* p. 22. Lunar and Planetary Institute, Houston.

Greenberg R., Hartmann W. K., Chapman C. R., and Wacker J. F. (1978) The accretion of planets from planetesimals. In *Protostars and Planets* (T. Gehrels, ed.), pp. 599–622. Univ. of Arizona Press, Tucson.

Lin D. N. C. (1981) Convection accretion disk model for the primordial solar nebula. *Astrophys. J.* **246**, 972–984.

Lin D. N. C. and Bodenheimer P. (1982) On the evolution of convective accretion disk models of the primordial solar nebula. *Astrophys. J.* **262**, 768–779.

Lynden-Bell D. and Pringle J. E. (1974) The evolution of viscous discs and the origin of the nebular variables. *Mon. Not. Roy. Astron. Soc.* **168**, 603–637.

Morfill G. (1983) Some cosmochemical consequences of a turbulent protoplanetary cloud. *Icarus* **53**, 41–54.

Nakagawa Y., Nakazawa K., and Hayashi C. (1981) Growth and sedimentation of dust grains in the primordial solar nebula. *Icarus* **45**, 517–528.

Podolak M. and Cameron A. G. W. (1974) Models of the giant planets. *Icarus* **22**, 123–148.

Safranov V. S. (1972) *Evolution of the Protoplanetary Cloud and Formation of the Earth and the Planets.* NASA TTF-677. 206 pp.

Völk H. J. (1982) Physical processes of relevance to the formation of the planetary system. In *Sun and Planetary System* (W. Fricke and G. Teleki, eds.), pp. 233–243. Reidel, Dordrecht.

Völk H. J., Jones F. C., Morfill G. E., and Röser S. (1980) Collisions between grains in a turbulent gas. *Astron. Astrophys.* **85**, 316–325.

Weidenschilling S. J. (1977) Aerodynamics of solid bodies in the solar system. *Mon. Not. Roy. Astron. Soc.* **180**, 57–70.

Weidenschilling S. J. (1980) Dust to planetesimals: Settling and coagulation in the solar nebula. *Icarus* **44**, 172–189.

Whipple F. L. (1964) The history of the solar system. *Proc. Natl. Acad. Sci.* **52**, 565–594.

APPENDIX A

This derivation follows closely the one by Weidenschilling (1977), but also includes the effect of a radial gas motion. Let the radial particle velocity be u and the azimuthal particle velocity be $v_g + w$, where v_g is the azimuthal gas velocity, which lags behind the Keplerian velocity v_ϕ by an amount:

$$\Delta v = v_\phi - v_g \simeq (\frac{\Delta_g}{2g})v_\phi \ . \tag{A1}$$

Δg is the "residual gravity" due to the pressure gradient:

$$\Delta g = \frac{1}{\rho}\frac{dP}{dR} \ , \tag{A2}$$

where ρ is the gas density.

The radial force balance equation is given by:

$$\frac{u-u_g}{t_e} + \frac{(v_g + w)^2}{R} - g = 0 \tag{A3}$$

with t_e and u_g denoting the particle stopping time ($t_e \equiv m v_{rel}/F_D$) and the radial gas velocity, respectively. The relative velocity, v_{rel}, between particle and gas is:

$$v_{rel} = [(u-u_g)^2 + w^2]^{1/3}. \tag{A4}$$

Keeping only terms of order $\Delta g/g$, eq. (A3) can be written as

$$\frac{u-u_g}{t_e} + \Delta g + \frac{2wv_\phi}{R} = 0 \ , \tag{A5}$$

since $v_g^2/R \equiv g + \Delta g$.

The transverse force equation is best expressed in terms of the torque acting on the particle, which is equal to the rate of change in specific angular momentum h of the particle:

$$\frac{dh}{dt} = -R\frac{w}{t_e}. \tag{A6}$$

Using $h = R(v_g + w)$ we get:

$$\frac{dh}{dt} = \frac{dR}{dt}(v_g + w) + R\frac{dR}{dt}\frac{d}{dR}(v_g + w) , \tag{A7}$$

since $u \equiv dR/dt$ is already first order in $\Delta g/g$ we can approximate $v_g + w$ by v_ϕ thus arriving at:

$$\frac{dh}{dt} = u\frac{v_\phi}{2} \tag{A8}$$

using $R\, dv_\phi/dR = -v_\phi/2$. Combining (A6) and (A8) leads to:

$$\frac{w}{t_e} + \frac{uv_\phi}{2R} = 0 \tag{A9}$$

or

$$w = -\frac{1}{2}u\left(\frac{2\pi t_e}{T}\right) , \tag{A10}$$

where T is the Keplerian orbital period. Substituting (A10) into (A5) finally gives:

$$u = (\Delta g t_e + u_g)(1 + \frac{4\pi^2 t_e^2}{T^2})^{-1} . \tag{A11}$$

For drag regimes where t_e is a function of v_{rel}, the value of u and w have been calculated interactively by assuming a trial value for t_e for each step, calculating u and w by eqs. (A10) and (A11) and v_{rel} by eq. (A4) and thus arriving at a new t_e. H converges rapidly.

Composition of Chondrules, Fragments and Matrix in the Unequilibrated Ordinary Chondrites Tieschitz and Sharps

Frank Wlotzka

Max-Planck-Institut für Chemie (Otto-Hahn-Institut), Abteilung Kosmochemie, Saarstrasse 23, D-6500 Mainz, Federal Republic of Germany

An electron microprobe study of the unequilibrated ordinary chondrites (UOC's) Tieschitz and Sharps shows that different populations of chondrules and fragments can be distinguished. Chondrules with clear glass (FeO 4.4%), spherical chondrules (FeO 7.1%), irregular chondrules (FeO 11.5%), and fragments (FeO 16.0%) form a sequence with increasing FeO content (all values from Tieschitz). Chondrules and fragments are coated by dark, fine-grained rims, the so-called black matrix. Its composition is not related to the enclosed particle; the mean is close to an average H-chondrite. The alkali-rich white matrix found in Tieschitz (50% albite, 25% nepheline in the norm) is a late deposit, perhaps formed by shock lithification. A genetic relation between chondrules and fragments is favored: chondrules were mechanically transformed into fragments and at the same time enriched in FeO. The FeO source could be the dark matrix material. This FeO distribution was established before agglomeration, because no *in situ* equilibration is found. Indications for this model are: an FeO-size relation, a radial increase in FeO in some fragments, and the constant average Fe/Mn ratio. The formation of chondrules from the fragment material by impact and reduction also seems possible. Secondary alterations have affected all components; they are mainly shown by the large variations in the Na/K ratio. Small, spherical Ca, Al, Ti-rich chondrules were also found. Besides olivine, they contain minerals not found in normal chondrules, namely spinel and Ti-augite. The average enrichment factors relative to Si and C1 are 12 for Al and Ti, and 4.7 for Ca.

1. INTRODUCTION

Unequilibrated ordinary chondrites (UOC's) are usually assigned to the H-, L-, or LL-group of uniform ordinary chondrites by their content of total iron. The meteorites Tieschitz (total Fe 24.6%; Hutchison *et al.*, 1981) and Sharps (total Fe 26.3%; Fredriksson *et al.*, 1969) were thus placed into the H-group, which has an average total iron content of 27.6% and a range of 24.5 to 30% (Mason, 1965), although their FeO contents of 13.5 and 13.6%, respectively, lie outside the H-group range (7–12% FeO, ave. 9.6%; Mason, 1965) and are more appropriate to the L-group (11.5–18% FeO, Mason, 1965; ave. 14.7%; Wlotzka and Jarosewich, 1977). The present study was undertaken in order to better understand this discrepancy and the relations of these "H"-group UOC's to the other, "equilibrated" (better called uniform), H-group chondrites. For this purpose a thin section study by electron microprobe was made. The broad-beam technique used for the bulk analysis of chondrules or fragments is not as accurate as chemical methods and does not yield trace elements. It is possible, however, to analyze *all* components in a given volume, including the matrix, which is difficult to achieve in the usual studies of separated chondrules or matrix material. It was hoped that this type of analysis would answer the following questions:

(1) Which constituent is responsible for the high mean FeO content of the meteorite and what is the contribution of the matrix to the bulk composition?

(2) Is there a uniform or "equilibrated" component corresponding in FeO to the H-, L-, or LL-group, or are there other populations of chondrules or fragments that can be distinguished by their composition?

Previous studies of chondrules from UOC's (Gooding *et al.*, 1980; Lux *et al.*, 1981) have not tried to compare the average composition of the chondrules to the bulk composition of the meteorite and discuss possible discrepancies. The reason was probably the limited number of analyzed chondrules, which did not allow significant conclusions. Therefore, we have attempted here to analyze a sufficiently large number of chondrules.

The meteorites studied are Tieschitz and Sharps. Tieschitz fell in 1878 in Czechoslovakia and was described in the same year by Tschermak (1878). Both are included in the survey of UOC's by Dodd *et al.* (1967). They give an unpublished analysis for Tieschitz by Wiik and determined the percent mean deviation (PMD) of the olivine composition as 45%. Several papers were published by Kurat (1969, 1970, 1971) with analyses of chondrule minerals, glasses, and the black matrix. Christophe Michel-Lévy (1976) described the black and the white matrix of Tieschitz. Hutchison *et al.* (1979, 1981) studied its petrography and mineral composition and gave a new chemical analysis. Sears *et al.* (1980) classified Tieschitz as a type 3.6 by the thermoluminescence sensitivity technique.

Sharps fell in Virginia, United States, in 1921. It was described by Watson (1923). A detailed petrographic description and analyses of its chondrules and lithic inclusions as well as a chemical analysis were given by Fredriksson *et al.* (1969). Dodd (1968, 1971) analyzed minerals and chondrules of Sharps. Dodd *et al.* (1967) give 37% for the PMD of the olivine composition. Sears *et al.* (1980) classified it as type 3.4.

2. RESULTS

a.) Chondrules and fragments

Figure 1 shows a thin section of Tieschitz at low magnification. Spherical chondrules can be distinguished from fragments, which have angular or square outlines. "Irregular chondrules" form a class in between "chondrules" and "fragments"; they show a rounded outline over more than half of their periphery. The rest may be irregular or straight, i.e., suggesting fragmentation or abrasion. Examples of these three types of particles are marked in Fig. 1. This definition is not the same as the distinction between "droplet" and "lithic" chondrules, which depends mainly on the internal structure of the particles (King and King, 1978). The internal structures of the fragments are the same as those known from chondrules, i.e., porphyritic, fine-grained, barred olivine, radiating pyroxene, etc. Also, all of them contain a glassy mesostasis as evidence for an igneous melt origin, the same as found in chondrules. The fragments are therefore most probably fragmented chondrules. In fact, these "fragments" fulfill the criteria for "chondrules" as defined by Dodd (1982).

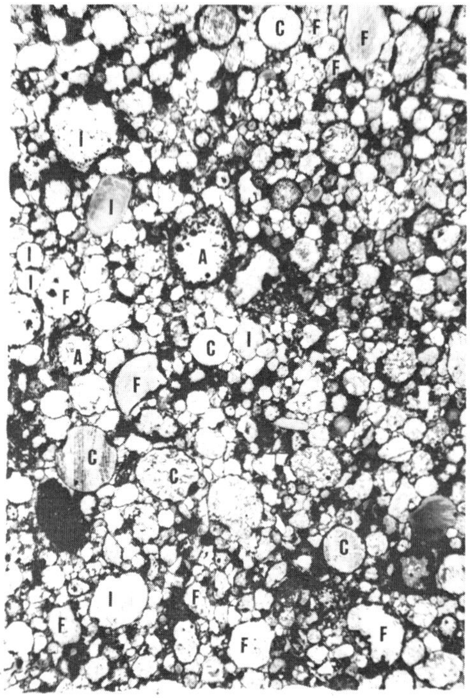

Fig. 1 Tieschitz, thin section in transmitted light. Length of section 16 mm. Examples of chondrules (C), irregular chondrules (I), and fragments (F) are marked. A = armoured chondrule.

In any case, the fragments described here from some UOC's should not be confused with the lithic fragments or clasts known from brecciated meteorites, especially the light-dark structured gas-rich chondrites (see, e.g., Wlotzka, 1963). These clasts have a chondritic texture in themselves and are usually larger in size than the chondrules of the host meteorite. The fragments found in Tieschitz and Sharps have the same size range as the chondrules in these meteorites and their internal texture is uniform in any given fragment. The reason for distinguishing fragments from chondrules in this paper lies mainly in the difference in composition as discussed below.

Chondrules and fragments were analyzed by electron microprobe with the broad-beam technique. The beam was placed on a cross-section as large as possible, avoiding metal and sulfide particles. In most cases, this was possible because metal and troilite occur outside the chondrules or at their periphery. Objects containing substantial amounts of metal or troilite throughout (which were very few) were not analyzed. Analyses showing sulfur were corrected by subtracting from the FeO value the amount of Fe to form FeS. The analysis was done by the energy dispersive method, using the correction procedures by Reed and Ware (1975). No further correction was applied for the fact that different element radiations come from different phases (e.g., McSween, 1977). This does not seem worthwhile here, because the analysis of random cross-sections instead of whole chondrules already introduces a certain error. The interpretation of the results will therefore be restricted to relatively large compositional differences.

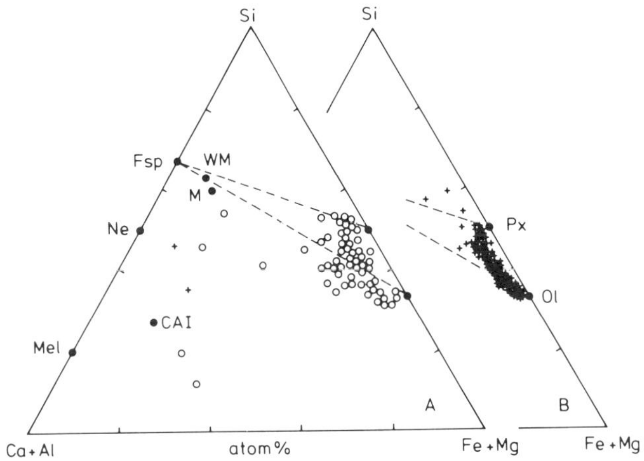

Fig. 2 **(a)** Composition of chondrules (o) and Ca,Al-rich fragments (+) in Tieschitz (atom.%). Px = pyroxene, Ol = olivine, Fsp = chondritic feldspar, Ne = nepheline, Mel = melilite, WM = white matrix, M = average of chondrule mesostases. CAI = Ca,Al-rich inclusion from Allende. **(b)** Composition of fragments in Tieschitz (atom.%). Note tighter clustering of points between Ol and Px.

The analyses plotted in Figs. 2 and 3 show that the compositions of chondrules and fragments are quite variable, in agreement with previous results (e.g., Gooding *et al.*, 1980; Grossman and Wasson, 1981; Lux *et al.*, 1981). Table 1 gives the mean compositions of chondrules and fragments. Chondrules with clear igneous glass, devitrified glass mesostases, irregular chondrules, and fragments form a sequence with increasing mean FeO content (Fig. 4). This holds for both Tieschitz and Sharps; the mean FeO contents and Fe/(Fe + Mg) ratios in these different groups are quite similar in these two meteorites (Table 2). MgO decreases with the increase of FeO. From the other elements, Al, Na, and K (i.e., the feldspar elements) are higher in the FeO-poor

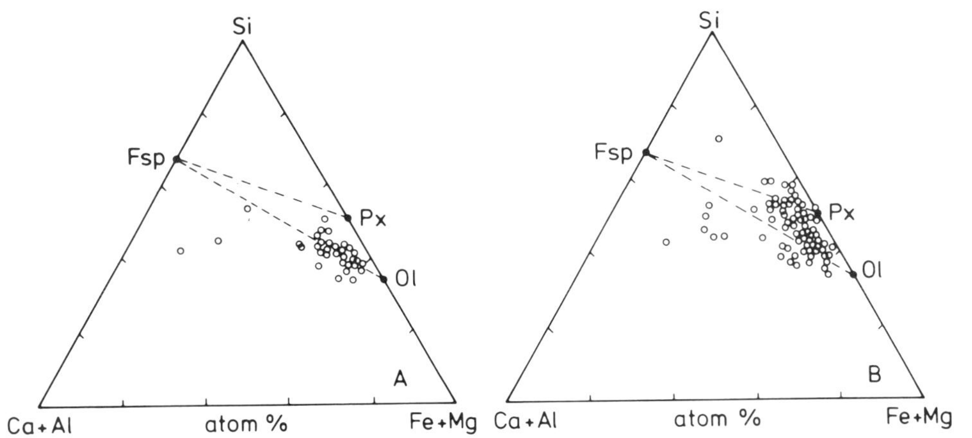

Fig. 3 Composition of chondrules with clear glass **(a)** and with devitrified glass **(b)**, in atom.%, from Sharps. Symbols as in Fig. 2. Fragments are not shown, because they plot very similar to normal chondrules.

Table 1. Mean composition of chondrules and fragments.

	Tieschitz				Sharps		
	Chondrules			Fragments	Chondrules		Fragments
	1	2	3		1	2,3	
	(10)*	(60)	(104)	(98)	(36)	(92)	(105)
SiO_2	43.9	47.1	46.7	44.9	46.3	51.1	49.0
Al_2O_3	7.66	5.20	4.13	3.26	5.38	3.67	2.96
Cr_2O_3	0.24	n.d.	n.d.	n.d.	0.41	0.59	n.d.
FeO	4.36	7.09	11.5	16.0	4.81	8.73	14.6
MnO	—	—	—	—	0.13	0.33	0.40
MgO	37.5	34.8	32.4	31.3	38.5	31.5	28.3
CaO	2.02	2.15	1.95	1.54	1.03	1.98	1.83
K_2O	—	—	—	—	0.50	0.26	0.16
Na_2O	3.39	1.88	1.42	1.02	2.71	1.45	1.29
Total	99.1	98.2	98.1	98.0	99.8	99.6	98.5

1. Chondrules with clear igneous glass.
2. Chondrules with devitrified glass.
3. Irregular chondrules, devitrified glass.

* Figures in parentheses give number analyzed.
n.d. Not determined.
— No average given, because many values are below detection limit.

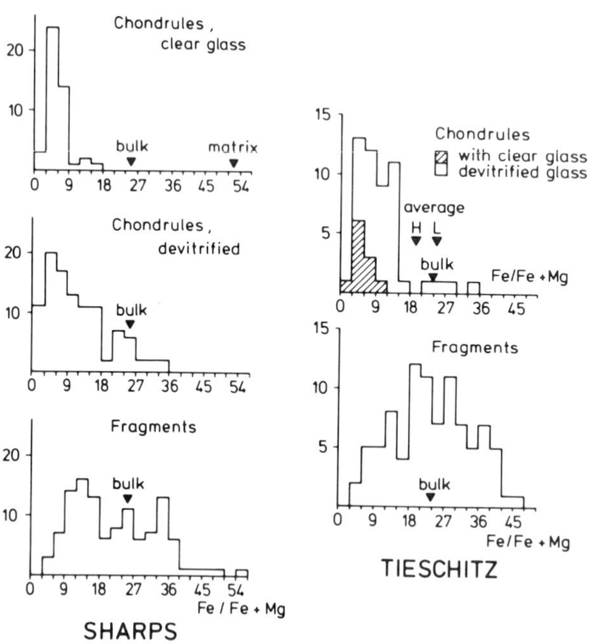

Fig. 4 Histogram of bulk Fe/(Fe + Mg) (mol.%) for chondrules and fragments from Tieschitz and Sharps.

Table 2. Mean FeO in chondrules and fragments.

		Chondrules with glass	Chondrules^{+}	Irregular Chondrules^{+}	Fragments	Bulk silicate
	Number analyzed	(10)	(60)	(104)	(98)	
Tieschitz	FeO%	4.4 ± 1	7.1 ± 4	11.5 ± 6	16.0 ± 7	16.5 (1)
	Fe/(Fe + Mg)(mol%)	6.1	10.3	16.6	22.3	23.8
	Number analyzed	(36)	(92)*		(105)	
Sharps	FeO%	4.8 ± 3	8.7 ± 6		14.6 ± 7	17.7 (2)
	Fe/(Fe + Mg)(mol.%)	6.6	13.5		22.4	25.6

(1) From Hutchison *et al.* (1981).
(2) From Fredriksson *et al.* (1969).
± Values are standard deviation.
$^{+}$ With devitrified glass.
* For Sharps, no distinction between spherical and irregular chondrules was made.

chondrule types. Dodd (1971) has found from a more limited number of analyses on Sharps that olivines in irregular chondrules were on the average richer in FeO than in spherical ones. Fredriksson *et al.* (1969) noted that olivines in chondrules with clear glass in Sharps tend to be very low in FeO.

Other plots of FeO with chondrule types did not show such clear differences. Chondrules and fragments differ in their relative abundance of textural types; fragments have a higher percentage of fine-grained and granular types. But these variations are not responsible for the different Fe/(Fe +Mg) distributions as shown in Fig. 4.

Figure 5 is a plot of Na vs. K in the chondrule mesostases, which is the main carrier of these elements inside the chondrules. The values scatter widely and for Tieschitz, most K-contents are very low with K/Na ratio well below the bulk ratio. Sharps shows the same scatter, but many K-contents lie above the K/Na bulk ratio.

In Fig. 6 the apparent diameter of chondrules and fragments is plotted against the FeO/(FeO + MgO) ratio. The correlation is weak and is not a linear one, but it is the same in both plots: the upper right of these plots is empty; i.e., there are no large, Fe-rich chondrules or fragments. In other words, small ones occur in the whole FeO/(FeO + MgO) range, low to high, whereas large ones tend to be Fe-poor.

b.) Black rims and white matrix

The occurrence of a "black" and a "white" matrix is a peculiarity of Tieschitz, first described by Christophe Michel-Lévy (1976) and also by Hutchison *et al.* (1979) and Ashworth (1980). The black "matrix" should really be called black rims, as it coats *each* constituent of the meteorite as thin, black rims (width about 5–20 microns), i.e., chondrules, fragments of chondrules and silicate

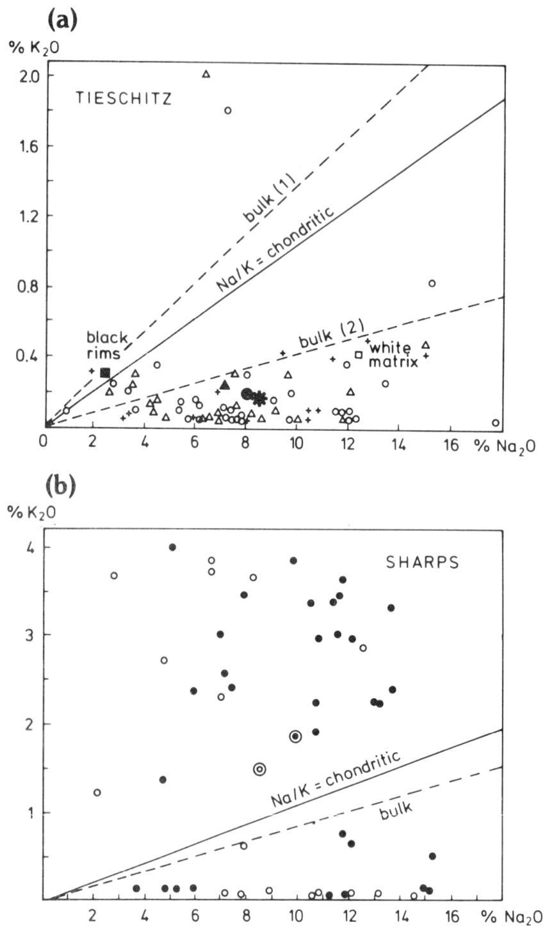

Fig. 5 Na_2O vs. K_2O (wt.%) in mesostases of chondrules and fragments. **(a)** Tieschitz, chondrules (open circles; average: filled circle), irregular chondrules (open triangles; average: filled triangle), fragments (crosses; average: star), black rims (filled square), white matrix (open square). Na/K bulk (1) from Hutchison *et al.* (1981), bulk (2) from Dodd *et al.* (1967). **(b)** Sharps, devitrified chondrules (open circles), chondrules with clear glass (filled circles). Encircled symbols are mean values. Bulk Na/K: Fredriksson *et al.* (1969). Note different scale for K_2O compared with the Tieschitz plot.

minerals, and also metal and troilite (Fig. 7). The outer surface of these rims is usually smooth, whereas the interior boundary is sharp and follows the shape of the enclosed particle, filling any indentations which might be present. Occasionally the black material is also found in a cavity completely inside a chondrule. This is probably not a real inclusion, however, but connected with the surface in the third dimension. In reflected light, the rims are medium grey, fine-grained, and occasionally contain resolvable specks of metal or troilite. Analyses by the broad-beam technique (Table 3) show the rims to be rich in iron. Rims are variable in composition, although not as variable as the chondrules themselves. No correlation was found between the composition of a given chondrule or fragment and the composition of its black rims. Thus rims around metal or troilite were not richer in Fe or Ni than others, and rims around Ca, Al-rich chondrules not enriched in Ca or Al.

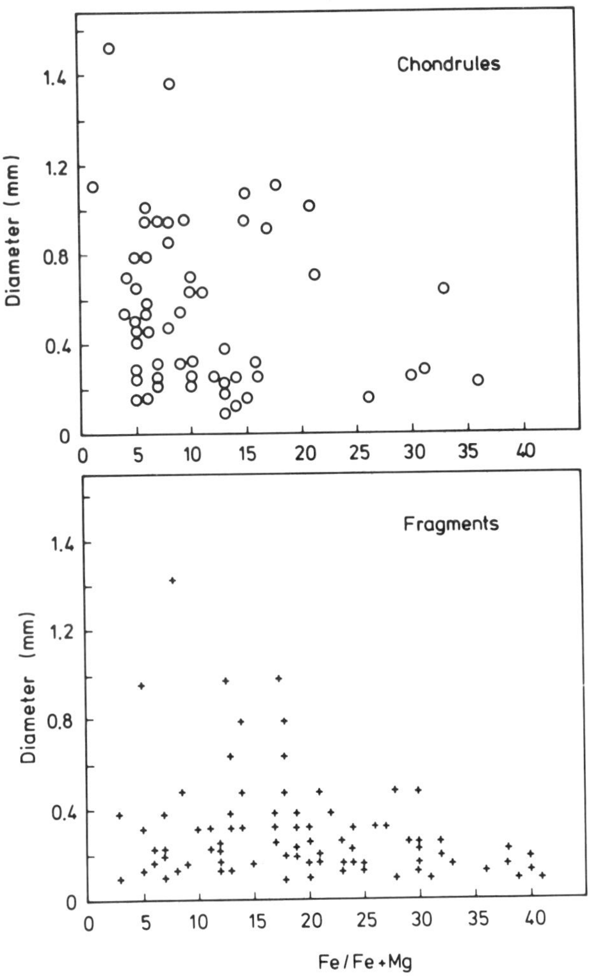

Fig. 6 Apparent diameter of chondrules and fragments from Tieschitz vs. Fe/(Fe + Mg) (mol.%).

The extensive study of chondrule rims by Allen *et al.* (1980) noted a greater variability in their composition, and often a high abundance of iron sulfides. In Tieschitz, this is not the case; no sulfide-rich rims were found, and sulfur averages were $0.37 \pm 0.15\%$ in the 20 analyses made. The reason for this discrepancy is probably a different definition of "rims." In this work on Tieschitz and Sharps, only "matrix rims" were analyzed, which are clearly separated from the chondrule bodies and appear to be secondary deposits. Metal and troilite-rich rims ("armoured chondrules") are different, usually more coarse-grained, and seem to be a part of the chondrule which they enclose. An example is shown in Fig. 1.

Chondrules and fragments in Tieschitz form a tightly fitting aggregate (Fig. 7). It seems remarkable that the black rims are preserved around all surfaces and that no clastic matrix is

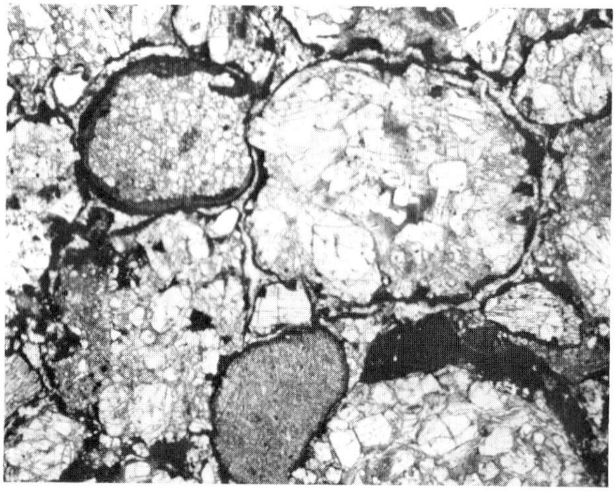

Fig. 7 Tieschitz, thin section in transmitted light. All chondrules and fragments are surrounded by dark rims, the "black matrix." In the interstices between chondrules and fragments, the transparent "white matrix" occurs. Note close fit between all chondrules and fragments; no clastic matrix is present. Length of section 1.3 mm.

present. Instead, the interstices are filled with the white matrix, a transparent fine-grained intergrowth of crystals of low birefringence. Its average composition is given in Table 3. Of all constituents of Tieschitz, the white matrix has the least variable composition; broad-beam analyses give a standard deviation of 6% (relative) for Na_2O and 18% (relative) for K_2O.

In Sharps the relations are less clear (Fig. 8). Black rims can be distinguished only around some chondrules. Their average composition (Table 3) is similar to that of the black rims in Tieschitz. No white matrix is found.

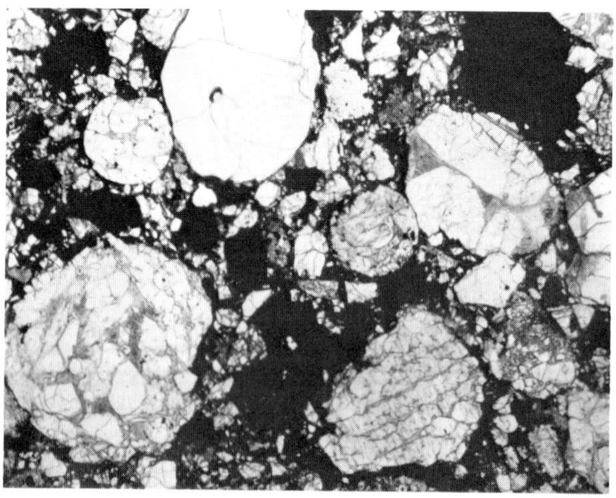

Fig. 8 Sharps, thin section in transmitted light, showing the gradation from spherical chondrules to irregular chondrules and fragments. Same magnification as Fig. 7.

Table 3. Composition of mesostases and matrices (in wt. %).

	Tieschitz			Sharps			White matrix	Black rims		Average H-chondrite*
	1	2	3	4	5	6	Tieschitz	Tieschitz	Sharps	
	(33)	(22)	(15)	(34)	(17)	(13)	(18)	(20)	(5)	
P_2O_5	0.11	0.41	0.23	0.1	0.30	0.50	—	0.30	0.28	0.26
SiO_2	55.8	56.3	54.4	58.5	64.4	59.8	56.6	36.7	34.1	34.6
TiO_2	0.35	0.23	0.27	0.52	0.35	0.46	0.06	—	—	0.1
Al_2O_3	17.3	16.9	19.2	18.0	13.2	15.3	19.3	3.81	2.60	2.01
Cr_2O_3	0.32	0.32	0.30	0.57	0.26	0.16	—	0.28	0.42	0.49
FeO	4.01	3.54	4.22	2.24	4.16	5.58	4.93	35.1 ± 5	38.5 ± 4	33.6
NiO	—	—	—	—	—	—	—	0.24	1.16	2.08
MnO	0.25	0.20	0.14	0.1	0.12	0.1	—	0.53	0.53	0.29
MgO	5.96	5.84	6.13	5.52	3.86	3.32	3.12	17.6	20.0	22.1
CaO	6.93	7.92	5.36	2.67	2.64	4.03	2.69	1.32	0.92	1.65
K_2O	0.20	0.25	0.19	1.85	1.48	1.63	0.44	0.29	0.23	0.09
Na_2O	8.00	7.21	8.26	9.88	8.45	8.83	12.2	2.49	0.81	0.80
S	0.20	0.23	0.51	0.11	0.23	0.15	0.22	0.37	0.21	1.93
Cl	—	—	—	—	—	—	0.1	0.3	0.15	—
Total	99.7	99.7	100.0	100.0	99.5	99.8	99.7	100†	100†	100
Fe/(Fe + Mg)	27.4	25.4	27.9	18.6	37.7	48.5	46.2	51.3	51.9	(20.3)
Na/K	35.7	25.8	33.2	4.8	5.1	4.8	24.8	7.7	3.15	7.9

— Not detected

* From Wlotzka and Jarosewich (1977), with total Fe calculated as FeO. 1 through 6: Mesostases in spherical chondrules (1), irregular chondrules (2), and fragments (3) in Tieschitz; in glass-containing chondrules (4), devitrified chondrules (5), and fragments (6) in Sharps. Numbers in parentheses are number of analyzed chondrules, fragments or matrix areas.

† Analyses totaling 85–90% normalized to 100%.

c.) Calcium-aluminum rich chondrules

Figure 9 shows the large spread in CaO and Al_2O_3 contents of chondrules and fragments in Sharps. For Tieschitz the relations are similar. Calcium and Al correlate close to the meteoritic ratio of 1.1 (Ahrens *et al.*, 1969) in devitrified chondrules and fragments, but not in the

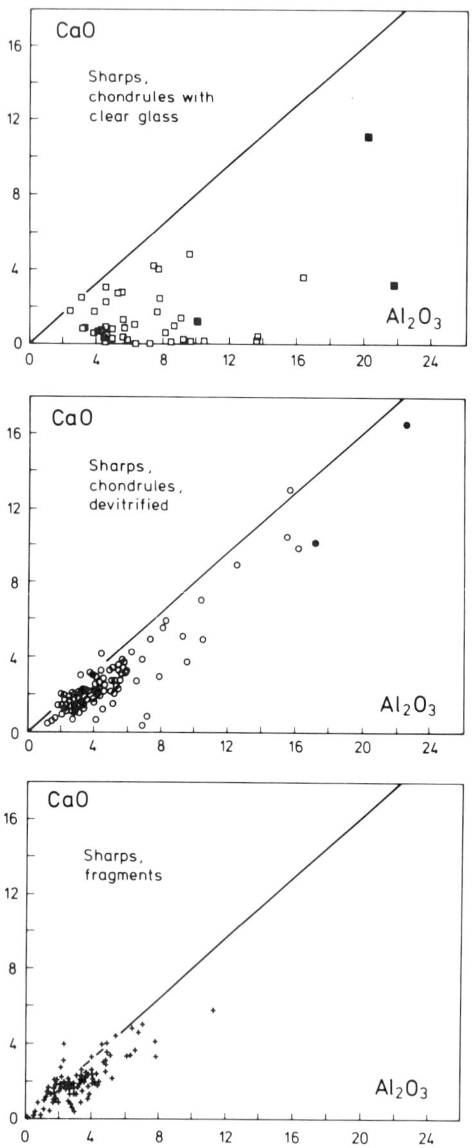

Fig. 9 CaO vs. Al_2O_3 (wt.%) for bulk chondrules and fragments from Sharps. The straight line gives the chondritic Ca/Al ratio (Ahrens *et al.*, 1969). Filled symbols are for Ca,Al-rich chondrules with Ti-augite and spinel.

glass-containing chondrules (see below). As the mesostasis is the main carrier of Ca and Al in the chondrules and fragments, the enrichment in these elements is often due to a high relative proportion of mesostasis in the analyzed cross-section. Sodium is often enriched together with Al, but because the Na content of the mesostases is rather variable (Fig. 5), Na and Al are not strongly correlated.

Besides these normal chondrules enriched in Ca and Al, special Ca, Al, Ti-rich chondrules also occur that are different in their mineralogy. They contain Ti-rich augite (fassaite), spinel, and olivine in an Al-rich mesostasis, usually a glass. Thus, they are similar to the Ca, Al-rich chondrules or inclusions found in carbonaceous chondrites, but they do not contain melilite. Eight such chondrules were found in Tieschitz, and several were found in Sharps; an example is shown in Fig. 10. They range in Al_2O_3 from 10 to 51% (ave. 26%), in CaO from 2.8 to 18,4% (ave. 9%), and in

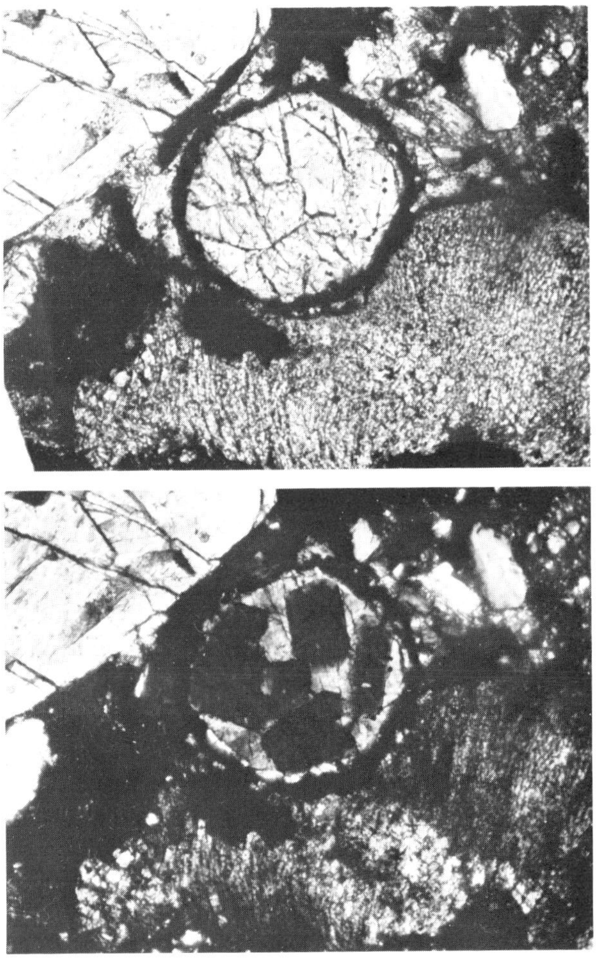

Fig. 10 Ca, Al-rich chondrule (apparent diameter 140 microns) in Tieschitz, in plain polarized light (top) and with crossed nicols (bottom). Isotropic grains (dark in bottom picture) are spinel. Dark rim around chondrule is "black matrix." Note concave contact surface of fine-grained fragment towards the chondrule.

TiO_2 from 0.24 to 6.6% (ave. 1.5%). The average enrichment factors (by weight) relative to C1 and Si are: for Al, 12; for Ca, 4.7; and for Ti, 12. These values are higher than the average refractory enrichment factor of 2.8 calculated by Grossman and Wasson (1981) for a refractory chondrule precursor component in Semarkona. Some of these chondrules are also enriched in Na (up to 14% Na_2O), as was already noted for some Ca, Al-rich inclusions in ordinary chondrites by Noonan *et al.* (1978).

Until now these Ca, Al-rich bodies were described in ordinary chondrites mainly as irregular inclusions (Noonan, 1975). Recently, Nagahara and Kushiro (1982) found chondrules similar to those described here in two Antarctic meteorites of type L3 and LL3. In Tieschitz all Ca,Al-rich chondrules are apparently rather small, 0.1 to 0.2 mm, i.e., at the lower end of the size range of normal chondrules. This is in contrast to their occurrence in the carbonaceous chondrite Allende, where they are much larger than the other chondrules (Mason and Taylor, 1982). A more detailed description of these special chondrules will be given elsewhere (Wlotzka, 1983).

3. INTERPRETATION

a.) Chondrules and fragments

Figures 2 and 3 show that the majority of chondrules and fragments plot inside the triangle defined by olivine, pyroxene, and feldspar. This agrees with an origin of chondrules from a granular precursor rock, which contained these three minerals (Suess, 1963), although it does not exclude other modes of origin. Fragments seem to be more "equilibrated" than chondrules, as they plot much closer together.

A main result is the lower mean FeO content and FeO/(FeO + MgO) ratio of chondrules compared to fragments (Table 2). The following interpretation will focus mainly on these relations and not on the origin of chondrules. It is important to note that the correlations found between FeO and chondrule shape and size are only statistical relations which refer to the *average* compositions and not to individual particles. There are chondrules rich in FeO and fragments poor in FeO. There are also small fragments in the Fe-rich black matrix that are very Fe-poor (Fig. 6). This excludes equilibration or diffusion process *in situ* after agglomeration of the meteoroid rock. Thus, the models all refer to pre-agglomeration processes that may have taken place either in the solar nebula or on the surface of a parent body:

Model 1: Chondrules and fragments have a separate origin and no genetic relation.

Model 2: Chondrules are the primary material; they are mechanically transformed into fragments and at the same time enriched in FeO.

Model 3: Chondrules are secondary, they are formed from FeO-rich rocks (now found as fragments); FeO is lost during chondrule formation by reduction to metal. The mechanism may be impact melting, as proposed by many authors (e.g., Urey, 1956; Fredriksson *et al.*, 1969; Dodd, 1971).

The following discussion will try to find arguments for or against one of these models:

In Model 1 chondrules and fragments could be formed by the same process, but at different times. Fragments should then be older than chondrules, thus having a higher chance of being fragmented before agglomeration or regolith consolidation. In most condensation scenarios (e.g., Wood, 1962; Grossman and Larimer, 1974), the opposite age sequence is produced, with Fe-free silicates condensing first.

Model 2 seems more appealing. The sequence—chondrules with clear glass–with devitrified glass–irregular chondrules–fragments—seems to be an evolutionary sequence; it agrees in the first two steps with the petrologic sequence by Van Schmus and Wood (1967). The envisioned sequence would be: (1) formation of chondrules from rapidly quenched melt droplets, resulting in silicate crystals in a supercooled residual melt, i.e., glass; (2) devitrification of the glass by mild thermal treatment; and (3) mechanical abrasion to form irregular chondrules and finally fragments. During all steps an enrichment with FeO takes place from the environment, be it a gas or a parent body surface. The fragmentation seems easier to achieve in a parent body regolith: the FeO source may be the Fe-rich fine-grained material, now present as the black matrix. Degree of alteration and FeO content will correlate, because both are acquired with time. This model would also satisfy the statistical nature of this correlation, as time can, but does not have to lead to fragmentation. This model has some similarity with the metamorphism model of Wood (1962); however, that model assumed an *in situ* equilibration between matrix and chondrules, whereas here the FeO enrichment takes place before agglomeration.

The first step in Model 2, namely the relation between chondrules with clear igneous glass and chondrules with devitrified glass, can be tested by a comparison of their bulk compositions. Figure 9 shows for Sharps that they are different in their Ca, Al-relations. Most of the glass-containing

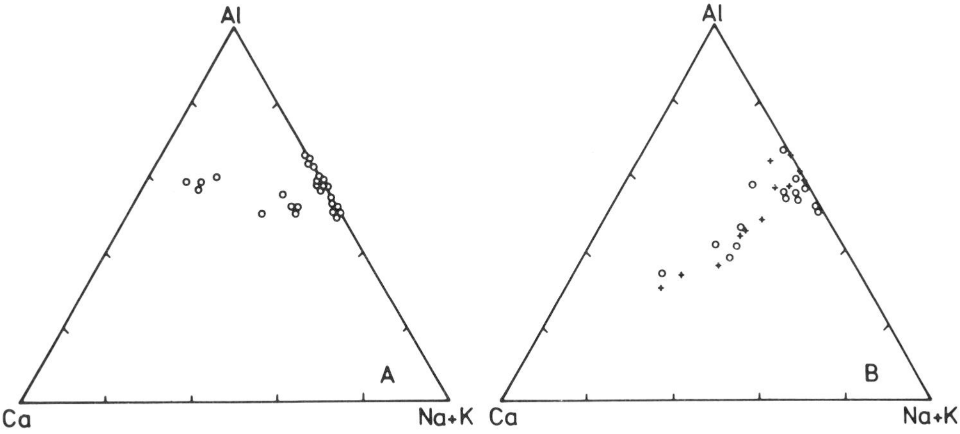

Fig. 11 Three-component plot for mesostases from Sharps (in wt.% of oxides). **(a)** clear glass; **(b)** devitrified glass in chondrules (o) and fragments (+).

chondrules are strongly depleted in Ca and their Ca/Al ratios are below the cosmic or primary ratio. Thus, they are probably not primitive in this sense, although they may be primitive in the petrologic sense.

This difference in bulk composition is also visible in the composition of the mesostases themselves. Figure 11 shows that in Sharps the glasses plot differently from the devitrified mesostases, and most of them are very Ca-poor. The higher Ca content of the devitrified glasses is probably due to the finely exsolved diopside, which is included in the analysis. The linear array of analysis points towards the Ca corner suggests this in Fig. 11, apparently a mixing line. A similar difference in composition between clear glasses and other mesostases was also noted for Tieschitz chondrules by Kurat (1971).

The compositional differences between chondrules with clear glass and chondrules with devitrified glass is the opposite of what one would expect from a condensation model. The more primitive, glass-containing chondrules should have cosmic Ca/Al ratios, whereas in the devitrified chondrules this ratio may have been changed by secondary processes during recrystallization. On the other hand, the low FeO content of the glass-containing chondrules could be in agreement with a primary origin, as the first condensing silicates should be Fe-poor (Wood, 1962; Grossman and Larimer, 1974).

An argument in favor of Model 2 is the reported radial increase in FeO toward the outside of several chondrules in Sharps (Dodd, 1971). Within several cases in this study, this increase was also found in Tieschitz chondrules and fragments. There is also the correlation between FeO content and size shown in Fig. 6. Both argue for a secondary FeO-enrichment from the outside. These observations are not clear and consistent enough, however, to make this argument very strong.

A difference in Ca-content of olivines from different chondrule types was reported by Dodd (1971) for Sharps and was used as an argument for different thermal histories. Such a difference could not be found here for chondrules and fragments in Tieschitz (Table 4). The data scatter too much for any firm conclusions to be drawn; better Ca measurements with a lower limit of detection may change this situation.

Table 4. Minor elements in Tieschitz olivine in wt.%.

		Chondrules		Fragments
CaO	(26)*	0.14 ± 0.07	(6)	0.16 ± 0.05
	(68)	<0.05	(30)	<0.05
MnO	(96)	0.22 ± 0.2	(36)	0.33 ± 0.2
FeO	(96)	13.0 ± 9	(36)	22.0 ± 9
Fe/Mn	ave.	59		68
Fe/Mn center of grains	(33)	48 ± 22	(17)	57 ± 27
Fe/Mn rim of grains	(21)	62 ± 28	(14)	80 ± 23

*Number of grains analyzed.

Model 3 implies the reduction of FeO during chondrule formation. The behavior of Mn may give a clue to this process. During reduction MnO would remain constant and the FeO/MnO ratio should decrease. An example of this is the relations in silicate inclusions of the Campo del Cielo iron meteorite. Here, reduction has caused the low FeO content of the olivines, e.g., 4.0% FeO in one inclusion, MnO has remained constant, 0.33%, resulting in the low Fe/Mn ratio of 12.1 (Wlotzka and Jarosewich, 1977). A normal chondritic Fe/Mn ratio is 42 as measured, for example, in olivines from the H5-chondrite Morro do Rocio (Wlotzka and Fredriksson, 1980 and unpublished data). No such difference is visible in Tieschitz chondrules compared to fragments. The average Fe/Mn ratios are similar and not far from the chondritic value mentioned above (Table 4). However, the individual Fe/Mn ratios vary widely; the range observed for olivines from chondrules is 9 to 125, from fragments 32 to 129. Often Mn varies in a crystal independently from Fe, but on the average crystal rims have a higher Fe/Mn ratio than the centers. Both values are slightly higher in the fragment than in the chondrule olivines (Table 4). The pyroxenes from Tieschitz have lower Fe/Mn than the olivines: the range is 1 to 27, and the average (10 values) is 14.6 ± 8. This olivine-pyroxene relation is the same as found for uniform chondrites (and terrestrial rocks); in Morro do Rocio the orthopyroxenes have an Fe/Mn of 23.6. Thus the Fe/Mn relations in Tieschitz do not seem to support a reduction model for chondrule formation, or at least not as a straightforward, simple mechanism.

b.) Black rims, white matrix, and Na,K-relations

The black rims are high in Fe: 27.3 ± 3.7% in Tieschitz, and 29.9 ± 3.1% in Sharps, which is remarkably close to the average total iron of H-chondrites (27.6%, see Table 3). The standard deviation of the mean for the 20 analyses for Tieschitz is small enough to allow a distinction from the average total iron of L-chondrites, 21.7% (Wlotzka and Jarosewich, 1977), although the field of the data points in Fig. 12 comes close to the L-chondrite composition point. This similarity, although not as close, also extends to other elements such as Si, P, Ca, Mn, and Cr; exceptions are Ni and S, which are depleted. The oxidation state of Fe in these rims is not known. Some of it may be metallic iron (minute metal particles being occasionally visible), but most of it is probably oxidized; Wood (1962) determined 30% magnetite in Tieschitz matrix. The actual microprobe analyses give totals of only 85 to 90%; the rest may be accounted for by porosity, water, and carbon. If all bulk C (0.1%, Hutchison *et al.*, 1981; 0.25%, Moore and Lewis, 1967) is assumed to reside in the black rims of Tieschitz (about 9 wt.% of the bulk, see below), a C-content of 1.1 to 2.8% can be calculated. This similarity in many elements agrees with the interpretation that the black rims may be a primary condensate, the low temperature matrix invoked by Wood (1962) and Anders (1964) as the carrier of volatile elements in ordinary chondrites.

In Tieschitz, the black rims are enriched in Al, Na, and K compared to H-chondrites. This can only partly be explained by infiltration by the alkali-rich white matrix (see Table 3) because the Na/K ratio of 7.7 in the rims is much lower than in the white matrix, 24.8, and close to the chondritic ratio of about 8. Furthermore, we find the same enrichment in the matrix of C1-chondrites (McSween and Richardson, 1977; also see Table 5). On the other hand, the black rims of Sharps, where no white matrix is found, are not enriched in Al and Na, only in K. The common feature of both meteorites is a higher K/Na ratio in the black rims than is found in the bulk meteorite (Tieschitz: 2.8 times; Sharps: 3.3 times), which would point to a K-enrichment process in these rims. It might be the formation of a K-rich sheet silicate, as it was found by Fredriksson *et al.* (1979) in the matrix of Bjurböle. A similar high K_2O content (0.28 K_2O, K/Na = 2.1 times chondritic) was found by Hoinkes *et al.* (1976) in black matrix rims in the L3-host of Dubrovnik.

As Fig. 5 shows, most chondrule and fragment mesostases in Tieschitz have very low K-contents and low K/Na ratios. Also, bulk Tieschitz seems to be inhomogeneous in Na and K, as the two analyses (Fig. 5 and Table 6) show. Minster and Allègre (1979) also noticed an inhomogeneity in Tieschitz with respect to Rb and Sr on a 0.5 g scale. The data obtained here for Na and K on a thin section agree better with the analysis by Wiik (see Fig. 5). The fractionation of Na from K cannot be a primary feature, as both condense at the same temperature (Fegley and Lewis, 1980). A redistribution of Na and K may have occurred through volatilization and recondensation during chondrule formation (see Kurat, 1971) or later high-temperature stages. Leaching may also have affected these elements (Kurat, 1969; Christophe Michel-Lévy, 1976). A similar K-depletion relative to Na was found in terrestrial impact "chondrules" from the Ries Crater (Graup, 1981). There is no systematic trend in the Na,K-contents and ratios between chondrules and fragments that would suggest different thermal histories. Tieschitz and Sharps behave in different ways: in the former, most mesostases are depleted in K and have high Na/K ratios, whereas in the latter, many are enriched in K (see Fig. 5) and the average Na/K ratios are low (Table 3).

This difference may be related to the occurrence of a white matrix in Tieschitz, but not in Sharps. This white matrix is rich in Al, Na, and K (Table 3) and its norm contains 50% albite and 25% nepheline. Christophe Michel-Lévy (1976) interpreted it as a hydrothermal deposit formed by leaching of the chondrules, whereas Hutchison *et al.* (1979) think of it as a liquid melt squeezed out of the chondrules during hot accretion. Its composition is indeed close to the average composition of the fragment and chondrule mesostases, as shown in Fig. 2.

A third possibility for the origin of the white matrix may be shock lithification, which may also have caused the tight-fitting texture of fragments and chondrules (Fig. 7). Allen *et al.* (1982)

Table 5. Element ratios (weight).

	Na/Si	K/Si	Al/Si	Ca/Si	Fe/Si
Tieschitz, black rims	0.106	0.015	0.116	0.047	1.56
Average H-chondrite (1)	0.036	0.005	0.065	0.073	1.61
C 1 matrix (2)	0.098	0.010	0.099	0.101	1.72

(1) See Wlotzka and Jarosewich (1977).
(2) See McSween and Richardson (1977).

have shown that rock powders shocked above 200 kbar develop an intergranular feldspar film, which transforms it into solid rock. According to Christophe Michel-Lévy (1976) and Ashworth (1980), however, no indications for post-accretionary shock were found in Tieschitz.

c.) Bulk composition and classification

Modal analysis of a thin section of Tieschitz (1750 points counted) gave the following composition: 74.8% silicates, 9.1% black rims, 2.5% white matrix, 9.5% metal, and 4.1% troilite (by weight). For this calculation the following densities were used: silicates 3.3, black rims 3, white matrix 2.6, metal 7.7, and troilite 4.7 (g/cm^3). The metal value confirms the low metal contents which were found by Wiik, 9.9% (see Dodd *et al.*, 1967), and by Hutchison *et al.* (1981), 12.6%, and which are below the range of equilibrated chondrites of the H-group: 16 to 20%, ave. 18.6% metal (Mason, 1965). Another integration gave the ratio 72% fragments to 28% chondrules by volume for Tieschitz, which was assumed to be the same ratio by weight. From these values the contribution of the different components to the bulk silicate composition of Tieschitz was calculated (Table 6). It shows that the black matrix contributes 24% of the total FeO. The summation to the bulk silicate composition agrees well with the two available chemical analyses, except that Al and Na are too high. This may result from sample inhomogeneity, as Na is also different in the two chemical analyses. But most of the discrepancy is probably due to the combined errors of the broad-beam microprobe technique and the point-counting.

It can be seen from Table 6 that chondrules plus fragments alone have an FeO/(FeO + MgO) ratio of 19.9 mol.%, which is near the average of the H-group [18.6 mol.% (Mason, 1965)],

Table 6. Bulk silicate composition of Tieschitz (in wt.%).

	(1) Chondrules + fragments	(2) Black matrix	(3) White matrix	(4) Bulk, 1 + 2 + 3	(5) Chemical analyses	(6)
SiO_2	40.0	4.16	1.72	45.9	47.4	44.3
Al_2O_3	3.19	0.43	0.59	4.21	3.11	2.89
FeO	12.2	3.98	0.15	16.3	16.7	21.1
MgO	28.1	2.00	0.09	30.2	30.1	28.6
CaO	1.47	0.15	0.08	1.70	2.20	1.94
Na_2O	1.07	0.28	0.37	1.72	0.43	1.14
Total	86.0	11.0	3.0	100	100	100
Fe/(Fe + Mg) mol. %	19.9	51.3	46.2	23.2	23.7	29.3

This table compares the bulk silicate composition (4) calculated from a summation of the components (1,2,3) with bulk chemical analyses (5,6 recalculated for the silicate portion). The abundance of the components was determined by point counting: 86 wt.% chondrules + fragments, 11 wt.% black matrix, 3 wt.% white matrix. It shows the large FeO contribution from the black matrix to total FeO. (5) is from Hutchison *et al.* (1981), (6) is from Dodd *et al.* (1967).

whereas the silicates plus matrix have FeO/(FeO + MgO) of 23.2 mol.% in the L-group range [ave. 25.0 mol.% (Wlotzka and Jarosewich, 1977)].

It was hoped to find a population of equilibrated chondrules or fragments in these chondrites that might correspond in their FeO/(FeO + MgO) ratio to either the H-, L-, or LL-group. A hint of such a component was found by Dodd (1968), who reported three equilibrated chondrules in Sharps (out of 25 analyzed) with homogeneous olivines of Fa 27–29, which would correspond to the LL-group. In this work no such component was found. Equilibrated chondrules and fragments with homogeneous olivines and/or pyroxenes of constant composition throughout do occur in both meteorites. But their Fe/(Fe + Mg) ratios range from 0 to 40 mol.% and no preferred compositions are present. Chondrule olivines again show lower mean Fa contents than fragment olivines.

A good argument for a relation of Tieschitz and Sharps to the H-group is the close similarity of the Fe-content of their black matrices to the average composition of H-chondrites, as discussed above. It would be interesting to see whether black matrix rims in L-group UOC's are also close to the average L-group composition and can be distinguished from the black matrix of Tieschitz and Sharps. The work of Huss *et al.* (1981) gives no indication for such a relation. However, Hoinkes *et al.* (1976) report black matrix rims around some chondrules in the L3-host of the L-chondrite Dubrovnik that have a bulk composition and Fe content (21.0% Fe) similar to the bulk composition and total iron of L-chondrites, 21.7% Fe (Wlotzka and Jarosewich, 1977). This relation between bulk and matrix composition would have consequences for determining the time of iron-silicate fractionation between the chondrite iron groups (see also Clayton *et al.*, 1981).

Thus the only clues found here for the classification of Tieschitz and Sharps into the H-group are their content of total iron and the composition of the black matrix. An additional argument for other type 3 chondrites was found in the content of siderophile trace elements by Sears (1982). However, for Tieschitz, the trace element content of the metal would place it in the L-group (see Rambaldi, 1977).

The two models proposed above for the genetic relations between chondrules, fragments, and matrix would lead to different results. If in Model 2 all chondrules were transformed into fragments and enriched in FeO (coming from the matrix), the resulting FeO/(FeO + MgO) ratio would be that of an L-chondrite. Some metal would then have to be lost from the system to bring total iron down into the L-group range. If in Model 3 chondrule formation with reduction moved the bulk composition towards lower FeO and higher metal contents, a normal H-chondrite composition may result. Because this H-chondrite classification is already inferred by total iron and matrix composition, Model 3 gets some support from this point of view. It is still not clear, however, whether such a genetic relationship between the unequilibrated chondrites and the higher petrologic types exists (see Wlotzka, 1981).

An oxidation-reduction relation between chondrules, fragments, and matrix is indicated by the FeO-SiO_2-MgO plots of Fig. 12 (these oxides account for more than 90% of the silicate portion of these meteorites). For both Tieschitz and Sharps, mixing lines can be drawn connecting the points of these three components, indicating addition or subtraction of iron. The direction of the

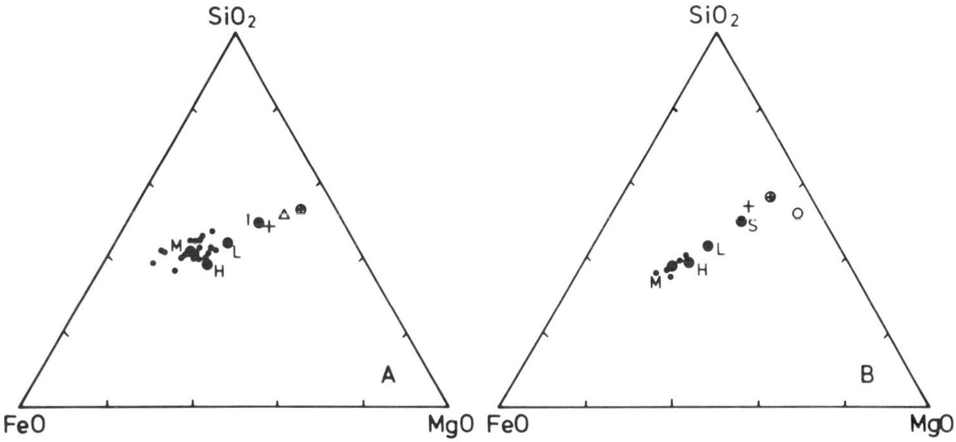

Fig. 12 Three-component plot (in wt.%) for mean compositions of chondrules with clear glass (o); chondrules with devitrified glass (⊕); irregular chondrules (△); and fragments (+), for Tieschitz **(a)** and Sharps **(b)**. M: black rims, average; small dots are single broad-beam analyses. H, L: average composition of H- and L-chondrites, respectively, with total iron as FeO (from Wlotzka and Jarosewich, 1977). T: bulk silicate of Tieschitz (from Hutchison *et al.*, 1981). S: bulk silicate of Sharps (from Fredriksson *et al.*, 1969).

iron movement is not known, of course. Figure 12 also shows that the average composition of the fragments is closer to the bulk composition of the meteorites than are the chondrules.

4. CONCLUSIONS

The results presented essentially confirm the findings by Dodd (1971) that shape and FeO-content of chondrules are correlated. Tieschitz and Sharps contain different populations of chondrules and fragments, which can be distinguished by their average FeO content and Fe/(Fe + Mg) ratios. There is also a small population of Ca, Al, Ti-rich chondrules, which are mineralogically distinct. The differences in average FeO among the various components were established before agglomeration of the meteorite. The most appealing model assumes a genetic relation between chondrules and fragments: Chondrules are the primary material; they are mechanically transformed into fragments and at the same time enriched in FeO. The FeO source may be the black matrix material. Indications for this relation are: the FeO-size relation, the radial enrichment of FeO in some chondrules and fragments, and the Fe/Mn relations.

The opposite relation, chondrules being formed from the fragment material by impact and reduction, also seems possible, although no direct indications for such a process were found. This process, however, may be able to transform UOC's like Tieschitz and Sharps into chondrites of normal H-group composition. A relation to the H-group is also indicated by the close similarity between the black matrix and the average H-group chondrite in bulk composition and, more importantly, total Fe content. If this black matrix is a primitive component, this would imply a metal-silicate fractionation between the H- and L-groups at a very early stage of chondrite evolution.

All components in Tieschitz and Sharps show secondary alterations, mainly visible in the Na,K-relations. These occurred either before or during the formation of the white matrix (in Tieschitz), which was deposited after accretion of the parent body.

Acknowledgments. I would like to thank the U.S. National Museum, Washington, D.C. and Dr. K. Fredriksson for the loan of excellent thin sections of Sharps and Tieschitz, and the Deutsche Forschungsgemeinschaft for financial support. Comments by J. L. Gooding and an anonymous reviewer helped to clarify certain points and to improve the manuscript.

REFERENCES

Ahrens L. H., von Michaelis H., Erlank A. J., and Willis J. P. (1969) Fractionation of some abundant lithophile element ratios in chondrites. In *Meteorite Research* (P. M. Millman, ed.), pp. 166–173. D. Reidel, Dordrecht.

Allen C. C., Jercinovic M. J., See T. H., and Keil K. (1982) Effects of target water on shock lithification (abstract). In *Lunar and Planetary Science XIII*, pp. 9–10. Lunar and Planetary Institute, Houston.

Allen J. S., Nozette S., and Wilkening L. L. (1980) A study of chondrule rims and chondrule irradiation records in unequilibrated ordinary chondrites. *Geochim. Cosmochim. Acta* **44**, 1161–1175.

Anders E. (1964) Origin, age and composition of meteorites. *Space Sci. Rev.* **3**, 583–714.

Ashworth J. R. (1980) Fine structure in H-group chondrites. *Proc. Roy. Soc. London* **A374**, 179–194.

Christophe Michel-Lévy M. (1976) La matrice noire et blanche de la chondrite de Tieschitz (H3). *Earth Planet. Sci. Lett.* **30**, 143–150.

Clayton R. N., Mayeda T. K., Gooding J. L., Keil K., and Olsen E. J. (1981) Redox processes in chondrules and chondrites (abstract). In *Lunar and Planetary Science XII*, pp. 154–156. Lunar and Planetary Institute, Houston.

Dodd R. T. (1968) Recrystallized chondrules in the Sharps (H3) chondrite. *Geochim. Cosmochim. Acta* **32**, 1111–1120.

Dodd R. T. (1971) The petrology of chondrules in the Sharps meteorite. *Contrib. Mineral. Petrol.* **31**, 201–227.

Dodd R. T. (1982) Objects we call chondrules (abstract). In *Papers Presented to the Conference on Chondrules and their Origins*, p. 15. Lunar and Planetary Institute, Houston.

Dodd R. T., Van Schmus W. R., and Koffman D. M. (1967) A survey of the unequilibrated ordinary chondrites. *Geochim. Cosmochim. Acta* **31**, 921–951.

Fegley B. and Lewis J. S. (1980) Volatile element chemistry in the solar nebula: Na, K, F, Cl, Br and P. *Icarus* **41**, 439–455.

Fredriksson K., Jarosewich E., and Nelen J. (1969) The Sharps chondrite—new evidence on the origin of chondrules and chondrites. In *Meteorite Research* (P. M. Millman, ed.), pp. 155–165. D. Reidel, Dordrecht.

Fredriksson K., Noonan A. F., and Nelen J. (1979) Bonding and reactions during accretion of ordinary chondrites (abstract). In *Lunar and Planetary Science X*, pp. 397–399. Lunar and Planetary Institute, Houston.

Gooding J. L., Keil K., Fukuoka T., and Schmitt R. A. (1980) Elemental abundances in chondrules from unequilibrated chondrites. *Earth Planet. Sci. Lett.* **50**, 171–180.

Graup G. (1981) Terrestrial chondrules, glass spherules and accretionary lapilli from the suevite, Ries Crater, Germany. *Earth Planet. Sci. Lett.* **55**, 407–418.
Grossman J. N. and Wasson J. T. (1981) The refractory component in Semarkona chondrules and the fractionation of refractory elements during the formation of ordinary chondrites (abstract). *) Dubrovnik: ein L3-6 Chondrit. Ann. Naturhist. Mus. Wien* **80**, 39–55.
Huss G. R., Keil K., and Taylor G. J. (1981) The matrices of unequilibrated ordinary chondrites: implications for the origin and history of chondrites. *Geochim. Cosmochim. Acta* **45**, 33–51.
Hutchison R., Bevan A. W. R., Agrell S. O., and Ashworth J. R. (1979) Accretion temperature of the Tieschitz, H3, chondritic meteorite. *Nature* **280**, 116–119.
Hutchison R., Bevan A. W. R., Easton A. J., and Agrell S. O. (1981) Mineral chemistry and genetic relations among H-group chondrites. *Proc. Roy. Soc. London* **A374**, 159–178.
King T. V. V. and King E. A. (1978) Grain size and petrography of C2 and C3 carbonaceous chondrites. *Meteoritics* **13**, 47–72.
Kurat G. (1969) The formation of chondrules and chondrites and some observations on chondrules from the Tieschitz meteorite. In *Meteorite Research* (P. M. Millman, ed.), pp. 185–190. D. Reidel, Dordrecht.
Kurat G. (1970) Zur Genese des kohligen Materials im Meteoriten von Tieschitz. *Earth Planet. Sci. Lett.* **7**, 317–324.
Kurat G. (1971) Die chemische Zusammensetzung von Gläsern und Chondrenmatrizes im Chondriten von Tieschitz. *Chemie der Erde* **30**, 235–249.
Lux G., Keil K., and Taylor G. J. (1981) Chondrules in H3 chondrites: textures, compositions and origins. *Geochim. Cosmochim. Acta* **45**, 675–685.
Mason B. (1965) The chemical composition of olivine-bronzite and olivine hypersthene chondrites. *Amer. Mus. Novitates* **2223**, 1–38.
Mason B. and Taylor S. R. (1982) Inclusions in the Allende meteorite. *Smithson. Contrib. Earth Sci.* **25**, 1–30.
McSween H. Y. (1977) Chemical and petrographic constraints on the origin of chondrules and inclusions in carbonaceous chondrites. *Geochim. Cosmochim. Acta* **41**, 1843–1860.
McSween H. Y. and Richardson S. M. (1977) The composition of carbonaceous chondrite matrix. *Geochim. Cosmochim. Acta* **41**, 1145–1161.
Minster J.-F. and Allègre C. J. (1979) ^{87}Rb-^{87}Sr chronology of H chondrites: constraint and speculations on the early evolution of their parent body. *Earth Planet. Sci. Lett.* **42**, 333–347.
Moore C. B. and Lewis C. F. (1967) Total carbon content of ordinary chondrites. *J. Geophys. Res.* **72**, 6289–6292.
Nagahara H. and Kushiro I. (1982) Calcium-aluminum-rich chondrules in the unequilibrated ordinary chondrites. *Meteoritics* **17**, 55–63.
Noonan A. F. (1975) The Clovis No. 1 meteorite and Ca,Al and Ti-rich inclusions in ordinary chondrites. *Meteoritics* **10**, 51–59.
Noonan A. F., Nelen J. A., and Fredriksson K. (1978) Ca-Al-Na-rich inclusions and aggregates in H-group and carbonaceous chondrites (abstract). *Meteoritics* **13**, 583.
Rambaldi E. (1977) Trace element content of metals from H- and LL-group chondrites. *Earth Planet. Sci. Lett.* **36**, 347–358.
Reed S. J. B. and Ware N. G. (1975) Quantitative electron microprobe analysis of silicates using energy-dispersive X-ray spectrometry. *J. Petrology* **16**, 499–519.
Sears D. W. G. (1982) Instrumental neutron activation analysis of type 3 ordinary chondrites (abstract). *Meteoritics* **17**, 279.
Sears D. W. G., Grossman J. N., Melcher C. L., Ross L. M., and Mills A. A. (1980) Measuring metamorphic history of unequilibrated ordinary chondrites. *Nature* **287**, 791–795.

Suess H. E. (1963) The Urey-Craig groups of chondrites and their states of oxidation. In *Isotopic and Cosmic Chemistry* (H. Craig, S. L. Miller, and G. J. Wasserburg, eds.), pp. 385–400. Elsevier North-Holland, Amsterdam.

Tschermak G. (1878) Der Meteorit von Tieschitz. *Mineral. Petrogr. Mitt.* **1**, 289.

Urey H. C. (1956) Diamonds, meteorites, and the origin of the solar system. *Astrophys. J.* **124**, 623–637.

Van Schmus W. R. and Wood J. A. (1967) A chemical-petrologic classification for the chondritic meteorites. *Geochim. Cosmochim. Acta* **31**, 747–765.

Watson T. L. (1923) The Sharps meteorite, Richmond County, Virginia. *Proc. U.S. Nat. Mus.* **64**, 1–4.

Wlotzka F. (1963) Über die Hell-Dunkel Struktur der urgashaltigen Chondrite Breitscheid und Pantar. *Geochim. Cosmochim. Acta* **27**, 419–429.

Wlotzka F. (1981) Classification of unequilibrated ordinary chondrites (abstract). *Meteoritics* **16**, 403–404.

Wlotzka F. (1983) Ca, Al-rich chondrules from the H3-chondrite Tieschitz (abstract). *Terra Cognita* **3**, 80.

Wlotzka F. and Fredriksson K. (1980) Morro do Rocio, an unequilibrated H5-chondrite (abstract). *Meteoritics* **15**, 387–388.

Wlotzka F. and Jarosewich E. (1977) Mineralogical and chemical compositions of silicate inclusions in the El Taco, Campo del Cielo, iron meteorite. *Smithson. Contrib. Earth Sci.* **19**, 104-125.

Wood J. A. (1962) Metamorphism in chondrites. *Geochim. Cosmochim. Acta* **26**, 739-749.

Morphological Features of Pore Spaces in Chondrules

Marek Żbik and Bruno Lang

Institute of Hydrogeology and Engineering Geology and Department of Chemistry, Warsaw University, Żwirki i Wigury 93, 02-089 Warsaw Poland

Morphological features of porosity of chondrules were studied with the scanning electron microscope at magnifications from 50 to 10,000 times. Microscopy of polished thin sections was also applied. The distinction was made between open and closed porosity components, which were found to belong to five types: (1) joints, (2) fissures, (3) intercrystalline pores, (4) vesicles, and (5) microcraters. Vesicles are the most important porosity components in chondrules, which lack matrices. The abundances of vesicles depends upon the volatile content. Chondrules are hypothesized to be older than matrices and therefore are thought to have a double origin derived from impact events. These origins are: (1) impact-molten rocky material of the meteorite parent body and (2) material from the ejecta cloud of dust and condensed vapor from the target.

INTRODUCTION

Porosity of meteorites, particularly porosity of chondrules, has not yet attracted the wide interest of students working in the field. The scarcity of pertinent data is in contrast to the volume of study of chondrule textures, as recently stimulated by the effort devoted to a reasonable classification of primary textural types.

One example of the use of porosity data other than bulk, in correlation with the anisotropy of magnetic susceptibility, was shown by Hamano and Yomogida (1982). A valuable study of fissures in the Jilin meteorite (Chang, 1981) is too specific to be extended to other meteorites. Olsen (1981) studied vugs in chondrites and showed that porosity as vugs can be used as a diagnostic tool to identify shock events of varying intensity.

Żbik (1982) presented a classification of morphological features of meteoritic porosity and made a distinction between open and closed porosity components. This classification contains genetic aspects. Thus, porosity is regarded as a consequence of structural peculiarities, originated by the action of stress-induced mechanical forces. Other effects are the flow of matter and phase transitions such as melting and freezing, evaporation and condensation.

In this paper we extend the approach of Żbik, but with no pretention to generalizations for all chondrites, beause we studied only a limited number of chondrules from six chondrites.

METHODS

The six chondrites studied were Bremervörde (H3), Beardsley (H5), Faith (H5), Gnadenfrei (H5), Krymka (LL3), and Allende (CV3). A Tesla scanning microscope (SEM) was used for photomicrographs, with magnifications being from 50 to 10,000 times. The separated and non-separated pores in chondrules and those exposed by breakage or sectioning were observed on the original surface layer. SEM examination was followed by the preparation of polished thin sections to assess the mineral compositions and textures of chondrules, using microscopic techniques.

POROSITY COMPONENTS

The morphology of pores was classified into five types:

(1) *Joints:* highly anisometric voids produced by the action of mechanical stresses and appearing as localized ruptures without displacement.

(2) *Fissures:* highly anisometric hollow spaces, produced by the action of mechanical stresses (stretching), traversing the interiors of chondrules and occasionally producing displacements.

(3) *Intercrystalline pores:* isometric and anisometric voids due to the spatial orientation of adjacent crystals.

(4) *Vesicles:* isometric and anisometric void spaces evolved from inclusions, both gaseous and gas-liquid.

Fig. 1 Angular vesicles on a pyroxene surface in a chondrule from Beardsley.

(5) *Microcraters:* negative surficial concave morphological forms on chondrules, regarded as a primary texture.

CLOSED PORES—VESICLES

Vesicles in chondrules seem to be ubiquitous and their shapes vary from oval to angular or irregular. Angularly-shaped vesicles are observed in pyroxene in a chondrule from Beardsley (Fig. 1). Such pores are typical and sizes range from submicron to a few microns.

With regard to sizes we follow the classification of pores by Sergeyev *et al.* (1980). Accordingly, the sizes adopted are:

<0.1 μm for ultrapores;
0.1–10 μm for micropores;
10–1000 μm for mesopores; and
>1000 μm for macropores.

The angular vesicles in the chondrule from Beardsley are ultra- and micropores.

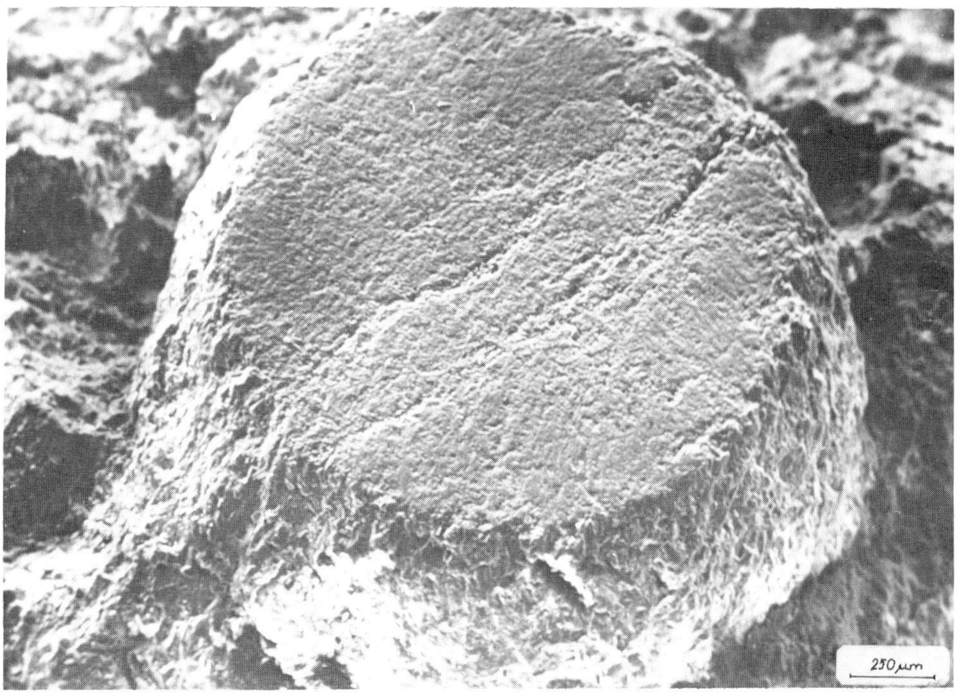

Fig. 2 Abraded chondrule from Beardsley. Zones with porous structure are displayed.

The structure of the interior of a chondrule (Fig. 2) appears to be compact, because at this magnification, the pores are indiscernible. However, the polished surface is traversed by a narrow zone that is quite distinct. At greater magnification it exhibits a porous nature (Fig. 3). The abundance of vesicles in this zone suggests that the fissures were filled with viscous vein-forming glassy material before lithification. At the same time, the void spaces were filled around the matrix-forming chunks. The sizes of vesicles, 1–10 μm, are those of micropores.

Similar veins of cementing glassy material were found in a chondrule from Faith. The walls of a vesicle (Fig. 4) might have traces of matter either from rapid cooling or from impact reheating. Fine bubble-like micropores, irregular in shape, were found in apparently recrystallized glassy material filling a large (6 mm) chondrule in Faith (not shown).

Our conclusion is that the moderate porosity of chondrules is due mainly to closed bubble-like fine pores called micropores. This can probably be attributed to the low abundance of volatiles in chondritic material.

The contribution of volatiles to total porosity shows a distinct increase in carbonaceous chondrites. This can be observed in a chondrule-bearing section of Allende (Fig. 5). Vesicles are much larger than in ordinary chondrites, reaching the size of mesopores (10–200 μm). The

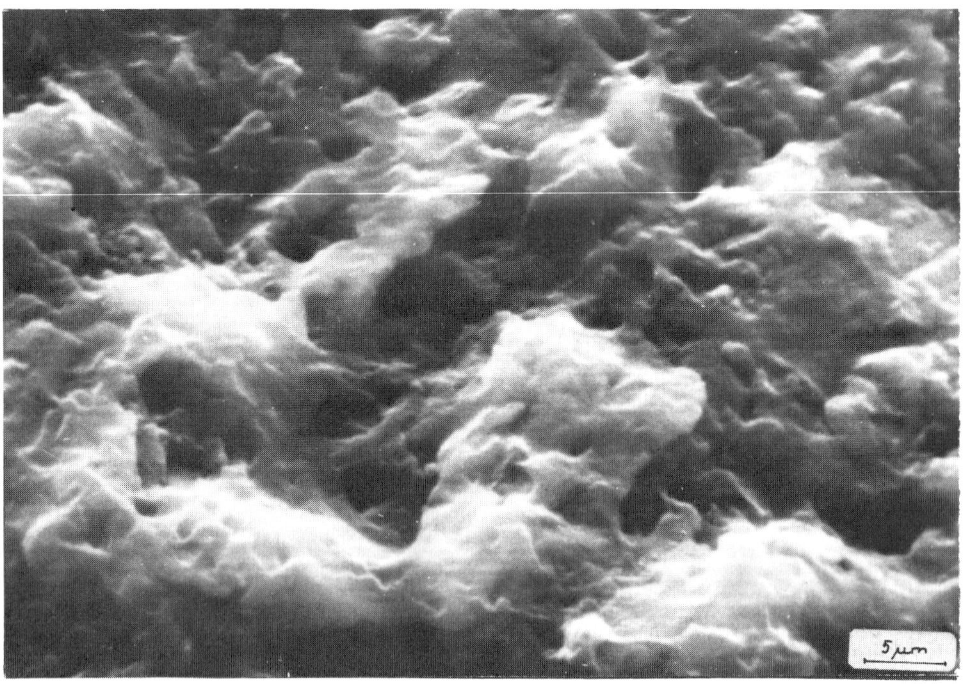

Fig. 3 Chondrule in Faith embedded in matrix in a polished thin section. Fine vesicles are displayed.

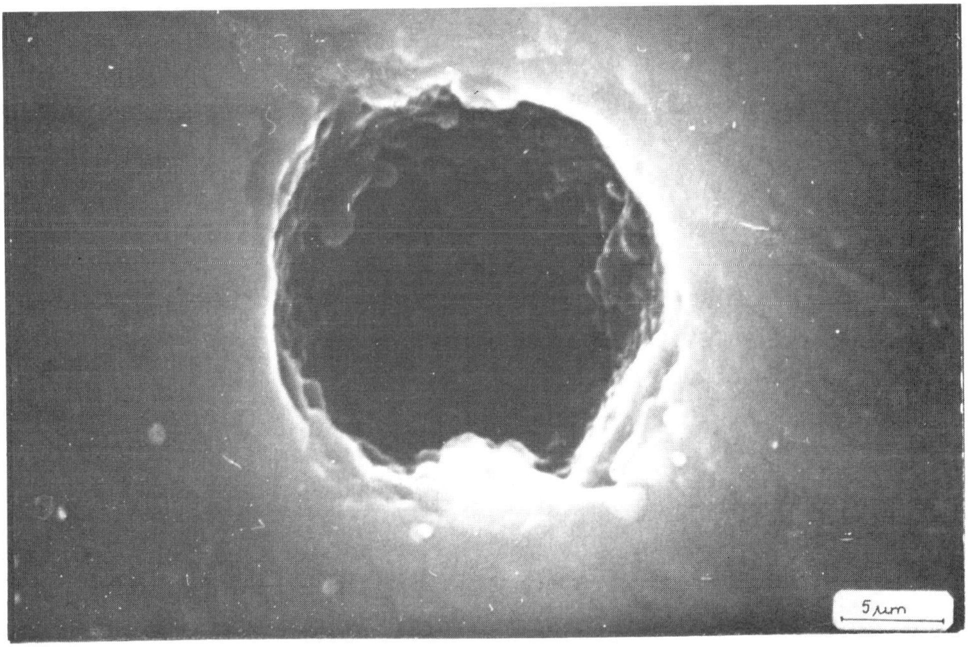

Fig. 4 One vesicle in a chondrule from Faith, enlarged.

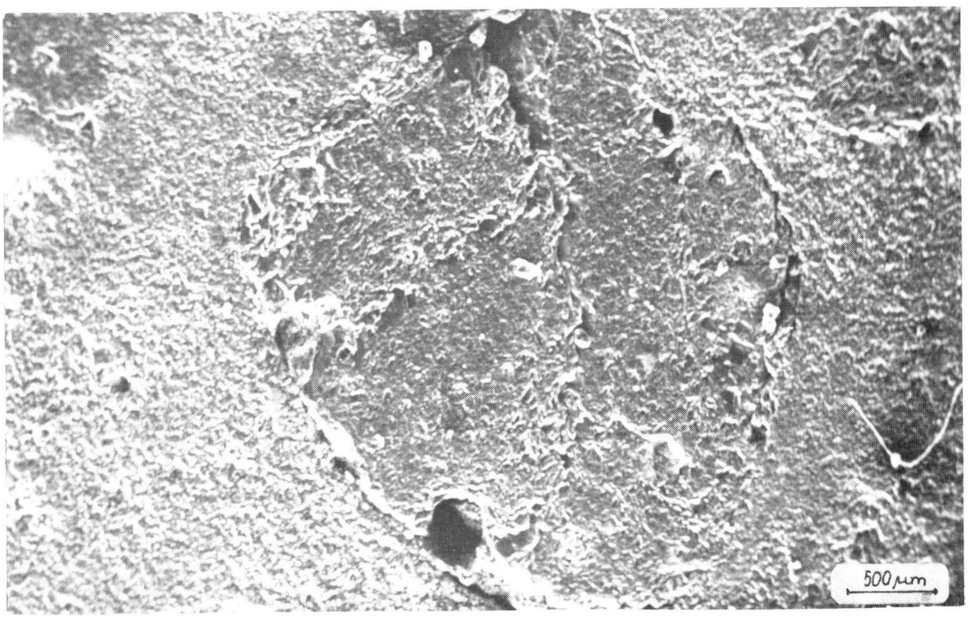

Fig. 5 Section of a chondrule in Allende. Fissures and vesicles of various sizes are displayed.

regularity of shape of these vesicles argues for their formation in molten material. Jagged walls, an amorphous appearance, and lack of flow structure suggests slow cooling. A bubble that formed under the thin cover of the outermost shell of the chondrule (Fig. 5) is possibly a morphological depression of that layer. It is shown at larger magnification in Fig. 6. Similar morphological features were observed by Rode *et al.* (1979) in lunar chondrules. Because the vesicle under the surface of the chondrule is unfilled by matrix material, we assume that the chondrule formed outside the matrix. It looks like it formed in a regolith-like environment with low pressure during lithification. Bubble-like pores attracted the attention of King (1976, p. 20) and Christophe Michel-Lévy (1982). We hypothesize that the structure of the regolithic environment on the surface of meteoritic parent bodies, the alleged site of accumulation of chondrules, had to be very loose.

One can see that there are no vesicles in the matrix material of Allende and no amorphous structures of joints and intercrystalline pores (Fig. 7, 8). The material looks like it has not been affected by metamorphism or heating and melting. Jagged walls of bubble-like vesicles (Fig. 9) appear to argue for low temperature degassing of components in the chondrules.

Apart from vesicles, intercrystalline pores in chondrules are of the closed type. Due to the random orientation of adjacent crystals (Fig. 10), resulting in disorder of their packing, they are less characteristic and thus of minor importance.

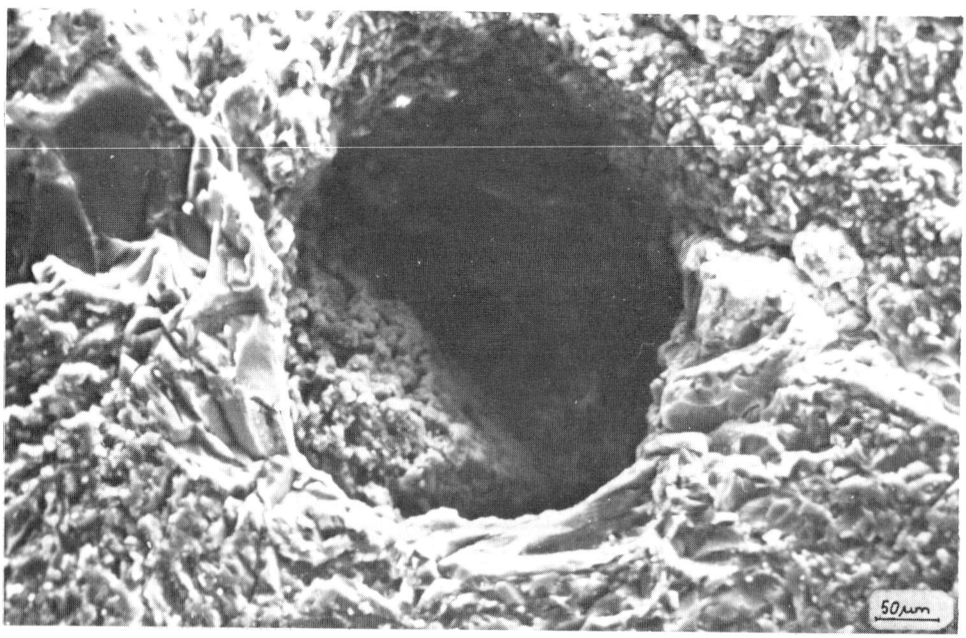

Fig. 6 Large vesicle in chondrule in Fig. 5, enlarged.

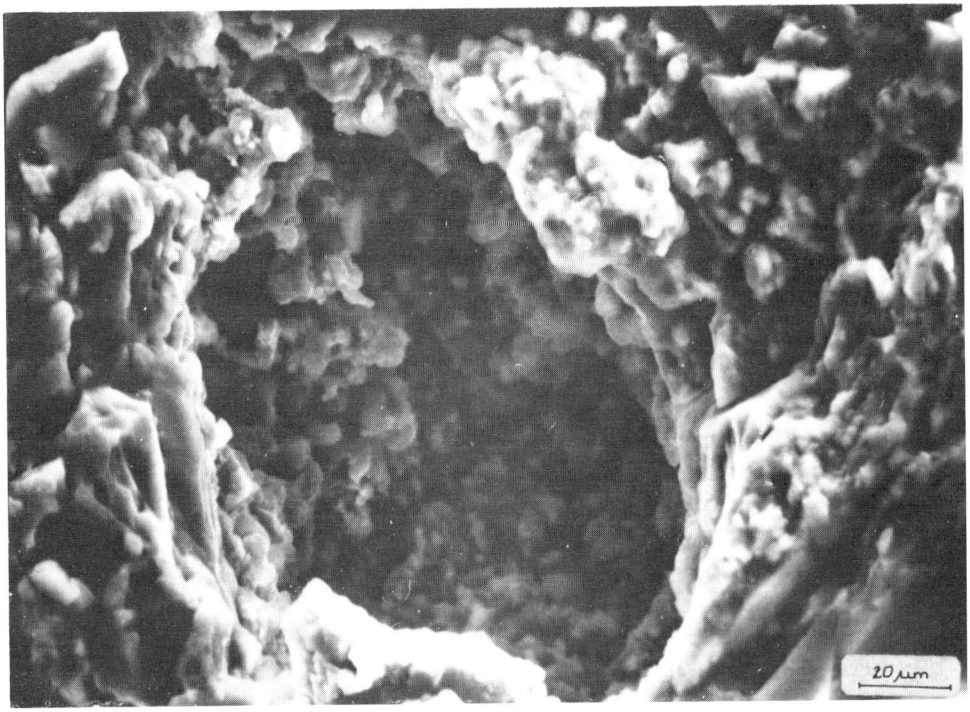

Fig. 7 Vesicle in chondrule from Fig. 5.

OPEN TYPE POROSITY

Joints and fissures in chondrules (Figs. 1, 2, 5) contribute to effective porosity; i.e., they form a network of interconnected void spaces open to penetration. Their occurrence can be explained with mechanical stress pulses of thermal origin, or with the effects of propagation of impact-produced shock waves. Some of the fissures were generated prior to ultimate lithification. Such a case can be observed in Fig. 5, where the fissure crossing the chondrule ends at the border, leaving the bed of matrix material uninjured. It should be emphasized that the pore spaces in the outermost shell of this chondrule are empty (i.c., not filled with matrix). We hypothesize that the loose structure, low hydrostatic pressure of the matrix material and the mild action of gravity forces produced the barrier. Thus the patterns of porosity in chondrules prove that they may have evolved earlier than the matrix material.

The width of the fissures qualifies them as mesopores and macropores, while the characteristic sizes of joints (as seen in Fig. 1) match the micropores and mesopores. Fissures in chondrules from Faith may have been generated prior to lithification of the matrix, and were filled with cementing glassy material. Such a textural pattern was also observed in a chondrule from Krymka.

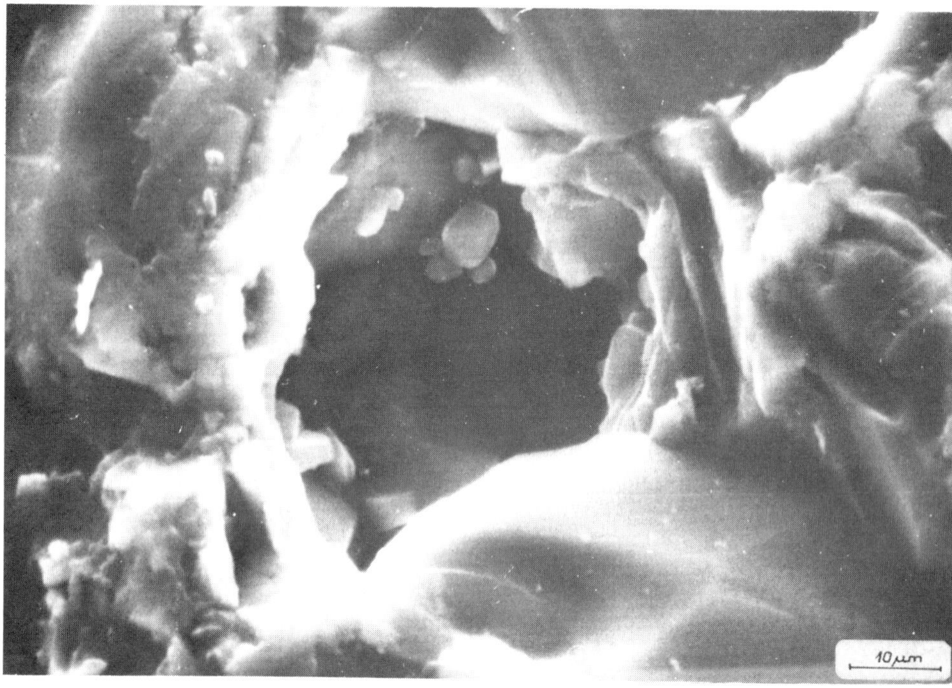

Fig. 8 Intercrystalline pore (closed type) in the interior of the chondrule in Fig. 5.

MICROCRATERS

The porosity of chondrules can be related to morphological depressions such as craters and microcraters. The porous aspect of chondrule microcraters does not exclude them as textural features, as suggested by Gooding and Keil (1981), and their subsequent consideration as main primary textural types of chondrules.

As observed by Rode *et al.* (1979), such depressions resemble those on the surface of lunar chondrules, which are assumed to be impact-produced. The difference between lunar and meteoritic microcraters is the apparent erosion of morphological patterns in the lunar soil; they appear to be preserved in meteoritic chondrules. Microcraters were also identified in glassy particles from howardites (Noonan *et al.*, 1974).

Gooding and Keil (1981) suggested that microcraters formed as plastic, semi-molten material. However, we see another way to form microcraters—the migration of a bubble from the inner part of a chondrule towards the surface of a droplet of molten material. Such an example can be seen in Fig. 5 for the chondrule from Allende, where the subsurface vesicle is close to its conversion into a concave crater-like surficial feature.

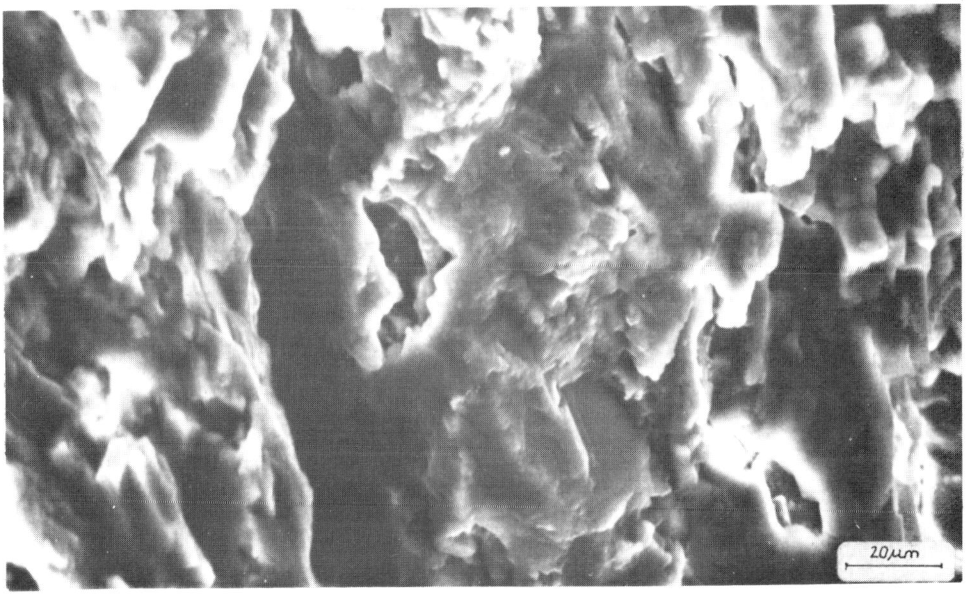

Fig. 9 Irregularly-shaped fine vesicles in Allende.

Fig. 10 Surface of the chondrule in Fig. 1. Intercrystalline pores are the result of secondary minerals.

CONCLUDING REMARKS

This review of the components of porosity of chondrules argues that there is a difference between those in carbonaceous chondrites as compared to ordinary chondrites. In the latter, open porosity components such as intergranular pores and open channels prevail, while closed porosity components are abundant in chondrules. With closed porosity, a previous episode involving droplets of molten material is recorded. Lithification of chondrules incorporated into matrices means a new period of their history, in common with the matrix. Fissures traversing chondrules and matrices indicate the latter.

We suggest that chondrules are possibly of double origin—both derived from impact events. One origin is as droplets of impact-produced molten material from their parent bodies. The other origin is as a mixture of dust and "condensate" from the impact ejecta. The growth of chondrules in both matters, in a cloud of impact ejecta, seems plausible. Collision, breakage, melting, and mixing of the material on the surface of meteorite parent bodies should lead to the observed diversity of structural and textural features. An impact model seems to be in agreement with constraints, as advanced by Taylor *et al.* (1982) and Gooding (1982).

Acknowledgments. We gratefully acknowledge the interest and friendly suggestions of Prof. Barbara Grabowska-Olszewska, and the extremely helpful criticism and remarks of Martin Prinz. Elbert King is acknowledged for providing the sample of Allende, while Carleton B. Moore provided the samples of Faith and Beardsley. We are indebted to Prof. M. Sachanbiński for samples of Bremervörde and Gnadenfrei. The work was supported by Grant No. 06.7-1.5.5 from the Center for Space Research (Poland).

REFERENCES

Chang S. (1981) Fissures in the Jilin meteorite. *Mem. Natl. Inst. Polar Res., Spec. Issue No. 20* (T. Nagata, ed.), pp. 285–291. National Institute of Polar Research, Tokyo.

Christophe Michel-Lévy (1982) A special case of associated chondrules in Allende: Snapshot of a punching-out process (abstract). In *Papers Presented to the Conference on Chondrules and their Origins*, p. 12. Lunar and Planetary Institute, Houston.

Gooding J. L. (1982) Origin of chondrules in ordinary chondrites by incomplete melting of geochemically fractioned solids (abstract). In *Papers Presented to the Conference on Chondrules and their Origins*, p. 20. Lunar and Planetary Institute, Houston.

Gooding J. L. and Keil K. (1981) Relative abundances of chondrule primary textural types in ordinary chondrites and their bearing on conditions of chondrule formation. *Meteoritics* **16**, 17–43.

Hamano Y. and Yomogida K. (1982) Magnetic anisotropy and porosity of Antarctic chondrites. *Mem. Natl. Inst. Polar Res., Spec. Issue No. 25* (T. Nagata, ed.), pp. 281–290. National Institute of Polar Research, Tokyo.

King E. A. (1976) *Space Geology: An Introduction*. Wiley, N. Y. 349 pp.

Noonan A. F., Rajan R. S., and Chodos A. A. (1974) Microprobe analyses of glassy particles from howardites. *Meteoritics* **9**, 385–386.

Olsen E. (1981) Vugs in ordinary chondrites. *Meteoritics* **18**, 45–59.

Rode O. D., Ivanov A. V., Nazarov M. A., Cimbalnikova A., Jurek K., and Hejl V. (1979) *Atlas of Photomicrographs of the Surface Structure of Lunar Regolith Particles.* Academia, Praha.

Sergeyev Y. M., Grabowska-Olszewska B., Osipov V., and Sokolov V. N. (1980) The classification of microstructure of clay soils. *J. Microscopy* **120**, pt. 3, 237–260.

Taylor G. J., Scott E. R. D., and Keil K. (1982) Cosmic setting for chondrule formation (abstract). In *Papers Presented to the Conference on Chondrules and their Origins*, p. 58. Lunar and Planetary Institute, Houston.

Żbik M. (1982) Pore spaces in stony meteorites. *Bull. de l'Acad. Pol. des Sciences* **30**, 59 65. Série des Sciences de la Terre.

SELECTED REFERENCES

The following list of references has been compiled over the past several years by Debbie Taylor and Elbert King, Department of Geosciences of the University of Houston. This list includes papers that specifically refer to the origins of chondrules, assume an origin of chondrules, or provide some of the fundamental data from which ideas about the origins of chondrules have been developed. Doubtless, there have been many *omissions in this rather extensive literature, but the compilers hope that the list will serve as a nucleus of relevant papers or researchers in this field. Additional references can, of course, be found in the standard texts. For the older literature, the work "A Bibliography on Meteorites," edited by Harrison Brown and published by the University of Chicago Press in 1953, continues to be especially useful and complete.*

Adib D. and Liou J. G. (1979) The Naragh meteorite: A new olivine-bronzite chondrite fall. *Meteoritics* **14**, 257–272.

Afiattalab F. and Wasson J. T. (1980) Composition of the metal phases in ordinary chondrites: Implications regarding classification and metamorphism. *Geochim. Cosmochim. Acta* **44**, 431–446.

Alexander E. C., Jr. and Manual O. K. (1969) Xenon in chondrules and matrix material of chondrites. *Geochim. Cosmochim. Acta* **33**, 298–301.

Alfvén H. (1971) Apples in a spacecraft. *Science* **173**, 522–525.

Allen J., Nozette S., and Wilkening L. L. (1979) Chondrule rims: Composition and texture (abstract). In *Lunar and Planetary Science X*, pp. 27–29. Lunar and Planetary Institute, Houston.

Allen J. S., Nozette S., and Wilkening L. L. (1980) A study of chondrule rims and chondrule irradiation records in unequilibrated ordinary chondrites. *Geochim. Cosmochim. Acta* **44**, 1161–1175.

Anonymous (1978) Mineralogy and petrology of Kirin meteorite and its formation and evolution. *Scientia Sinica* **21**, 805–822.

Arrhenius G. and Alfvén H. (1971) Fractionation and condensation in space. *Earth Planet. Sci. Lett.* **10**, 253–267.

Ashworth J. R. (1980) Chondrite thermal histories: Clues from electron microscopy of orthopyroxene. *Earth Planet. Sci. Lett.* **46**, 167–177.

Ashworth J. R. (1981) Fine structure in H-group chondrites. *Proc. Roy. Soc. London* **A374**, 179–194.

Aurell J. (1969) Los meteoritos condritos. *Inst. Invest. Geol. Univ. Barcelona,* Número Especial **25**, 157–164.

Berwerth F. (1901) Über die Struktur der Chondritischen Meteorsteine. *Zentralblatt Min. Geol.* **21**, 641–647.

Bevan A. W. R. and Axon H. J. (1980) Metallography and thermal history of the Tieschitz unequilibrated meteorite: Metallic chondrules and the origin of polycrystalline taenite. *Earth Planet. Sci. Lett.* **47**, 353–360.

Bevan A. W. R., Axon H. J., Scott E. R. D., and Rajan R. S. (1981) Comments on "metallic minerals, thermal histories and parent bodies of some xenolithic ordinary chondrite meteorites"; and reply. *Geochim. Cosmochim. Acta* **45**, 1957–1958.

Bhandari N., Shah V. B., and Wasson J. T. (1980) The Parsa enstatite chondrite. *Meteoritics* **15**, 225–233.

Binns R. (1967) Structure and evolution of non-carbonaceous chondritic meteorites. *Earth Planet. Sci. Lett.* **2**, 23–28.

Binns R. A. (1968) Cognate xenoliths in chondritic meteorites: Examples in Mezo-Madaras and Ghubara. *Geochim. Cosmochim. Acta* **32**, 299–317.

Blander M. (1982) Condensation of chondrules (abstract). In *Papers Presented to the Conference on Chondrules and their Origins,* pp. 1–2. Lunar and Planetary Institute, Houston.

Blander M. and Abdel-Gawad M. (1969) The origin of meteorites and the constrained equilibrium condensation theory. *Geochim. Cosmochim. Acta* **33**, 701–716.

Blander M. and Fuchs L. H. (1975) The CAI and chondrules—solid or liquid condensates? (abstract). *Meteoritics* **10**, 365.

Blander M. and Katz J. L. (1967) Condensation of primordial dust. *Geochim. Cosmochim. Acta* **31**, 1025–1034.

Blander M., Keil K., Planner H. N., and Nelson L. S. (1973) The origin of chondrules: Experimental investigation of metastable liquid silicates (abstract). *Meteoritics* **8**, 15.

Blander M., Planner H. N., Keil K., Nelson L. S., and Richardson N. L. (1976) The origin of chondrules: experimental investigation of metastable liquids in the system Mg_2SiO_4-SiO_2. *Geochim. Cosmochim. Acta* **40**, 889–896.

Borgstrom L. H. (1904) The Shelburne meteorite. *J. Roy. Aston. Soc. Canada,* 69–94.

Bostrom K. and Frederiksson K. (1966) Surface conditions of the Orgueil meteorite parent body as indicated by mineral associations. *Smithson. Misc. Coll.* **151**, No. 3.

Bosscha J., Jr. (1887) Über den Meteorit von Karang-Modjo oder Magetan auf Java. *Neues Jahrb. Mineral. Geol. Beilage Bd.* **5**, 126–144.

Bott M. H. P. (1982) *The Interior of the Earth: Its Structure, Constitution, and Evolution.* Edward Arnold Ltd., London. 403 pp.

Bournon J. L. C. de (1802) Mineralogical description of the various stones said to have fallen upon the Earth. *Phil. Trans. Roy. Soc. London* **92**, 180–186.

Boynton W. V. and Hill D. H. (1982) Rare-earth and other trace elements in rim and interior portions of a peculiar Allende chondrule (abstract). In *Lunar and Planetary Science XIII,* pp. 63–64. Lunar and Planetary Institute, Houston.

Boynton W. V., Hill D. H., and Wilkening L. L. (1982) Trace elements in rim and interior of chondrules from Chainpur (abstract). In *Papers Presented to the Conference on Chondrules and their Origins*, p. 3. Lunar and Planetary Institute, Houston.

Bradley J. G., Huneke J. C., and Wasserburg G. J. (1978) Ion microprobe evidence for the presence of excess 26 Mg in an Allende anorthite crystal. *J. Geophys. Res.* **83**, 244–254.

Brezina A. (1885) Die Meteoritensammlung des k. k. mineralogischen Hofcabinetes in Wien am 1. Mai 1885. *Jahr. Geol. Reichsanstalt Wien* **35**, 151–276.

Bridge T. E. (1969) Possible origin of chondrules and meteorites (abstract). *Meteoritics* **4**, 262–263.

Brigham C., Murrell M. T., and Burnett D. S. (1982) SiO_2-rich chondrules in ordinary chondrites (abstract). In *Papers Presented to the Conference on Chondrules and their Origins*, p. 4. Lunar and Planetary Institute, Houston.

Brownlee D. E., Bates B., and Beauchamp R. H. (1982) Meteor ablation spherules as chondrule analogs (abstract). In *Papers Presented to the Conference on Chondrules and their Origins*, p. 5. Lunar and Planetary Institute, Houston.

Bunch T. E. and Chang S. (1980) An alternative origin of Allende CAI inclusion rims, or a correlation between the early solar system and a British steel furnace (abstract). In *Papers Presented to the 43rd Annual Meteoritical Society Meeting*, p. 39. Lunar and Planetary Institute, Houston.

Bunch T. E. and Chang S. (1980) Carbonaceous chondrites—II, carbonaceous chondrite phyllosilicates and light element geochemistry as indicators of parent body processes and surface conditions. *Geochim. Cosmochim. Acta* **44**, 1543–1577.

Bunch T. E., Chang S., Frick U., Neil J. M., and Moreland G. (1979) Carbonaceous chondrites—I, Characterization and significance of carbonaceous chondrite (CM) xenoliths in the Jodzie howardite. *Geochim. Cosmochim. Acta* **43**, 1727–1742.

Bunch T. E., Quaide W., Prinz M., Keil K., and Dowty E. (1972) Lunar ultramafic glasses, chondrules and rocks. *Nature Phys. Sci.* **239**, 57–59.

Butler J. C., King E. A., Jr., and Carman M. F. (1971) Grain size frequency distribution of lunar samples and estimation of lunar surface ages (abstract). *Meteoritics* **6**, 254–255.

Butler J. C., King E. A., and Carman M. F. (1972) Size frequency distributions and petrographic observations of Apollo 15 samples (abstract). In *The Apollo 15 Lunar Samples*, (J. W. Chamberlain and C. Watkins, eds.), pp. 45–47. Lunar and Planetary Institute, Houston.

Caffee M. W., Hohenberg C. M., Hudson B., and Swindle T. D. (1982) I-Xe ages of individual Bjürbole chondrules (abstract). In *Lunar and Planetary Science XIII*, pp. 75–76. Lunar and Planetary Institute, Houston.

Caffee M. W., Hohenberg C. M., and Swindle T. D. (1982) Isotopic measurements in individual chondrules (abstract). In *Papers Presented to the Conference on Chondrules and their Origins*, pp. 6–8. Lunar and Planetary Institute, Houston.

Cameron A. G. W. (1966) The accumulation of chondritic material. *Earth Planet. Sci. Lett.* **1**, 93–96.

Cameron A. G. W. (1973) Accumulation processes in the primitive solar nebula. *Icarus* **18**, 407–450.

Cameron A. G. W. (1982) Chondrule-related processes in the primitive solar nebula (abstract). In *Papers Presented to the Conference on Chondrules and their Origins*, pp. 9–11. Lunar and

Planetary Institute, Houston.
Cameron A. G. W. and Fegley M. B. (1982) Condensation events near the total evaporation front within the primitive solar nebula (abstract). In *Lunar and Planetary Science XIII*, pp. 81–82. Lunar and Planetary Institute, Houston.
Christophe Michel-Lévy M. (1975) A few observations on the Tieschitz H3 chondrite (abstract). *Meteoritics* **10**, 381–383.
Christophe Michel-Lévy M. (1981) Some clues to the history of the H-group chondrites. *Earth Planet. Sci. Lett.* **54**, 67–80.
Christophe Michel-Lévy M. (1982) A special case of associated chondrules in Allende: snapshot of a punching-out process? (abstract). In *Papers Presented to the Conference on Chondrules and their Origins*, p. 12. Lunar and Planetary Institute, Houston.
Christophe Michel-Lévy M. and Lévy C. (1972) Mineralogical aspects of Apollo XIV samples: lunar chondrules, pink spinel bearing rocks, ilmenites (abstract). In *Lunar Science III*, pp. 136–138. The Lunar Science Institute, Houston.
Cirlin E. H. and Housley R. M. (1982) On the distribution of Zn in Allende chondrules, inclusions, and matrix (abstract). In *Lunar and Planetary Science XIII*, pp. 104–105. Lunar and Planetary Institute, Houston.
Clarke R. S., Jr., Jarosewich E., Mason B., Nelen J., Gomez M., and Hyde J. R. (1970) The Allende, Mexico, meteorite shower. *Smithson. Contrib. Earth Sci.* **5**, 1–53.
Clayton D. D. (1980) Chemical energy in cold-cloud aggregates: the origin of meteoritic chondrules. *Astrophys. J.* **239**, L37–L41.
Clayton D. D. (1980) Internal chemical energy: Origin of chondrules (abstract). In *Lunar and Planetary Science XI*, pp. 162–164. Lunar and Planetary Institute, Houston.
Clayton D. D. (1982) The initial chemical state of precondensed matter (abstract). In *Papers Presented to the Conference on Chondrules and their Origins*, p. 13. Lunar and Planetary Institute, Houston.
Clayton D. D. and Wieneke B. (1982) Dust growth in turbulent solar disk (abstract). In *Papers Presented to the Conference on Chondrules and their Origins*, p. 14. Lunar and Planetary Institute, Houston.
Clayton R. N., Mayeda T. K., Gooding J. L., Keil K., and Olsen E. J. (1981) Redox processes in chondrules and chondrites (abstract). In *Lunar and Planetary Science XII*, pp. 154–156. Lunar and Planetary Institute, Houston.
Clayton R. N., Mayeda T. K., and Olsen E. J. (1982) Oxygen isotopes in ordinary chondrites: What is the origin of the iron groups? (abstract). *Meteoritics* **17**, 198–199.
Cohen E. (1878) Über den Meteoriten von Zsadány, Temesvar Comitat, Banat. *Verhandl. Naturhist.-med. ver. Heidelberg* **2**, Heft **2**, 154–163.
Cohen E. (1903) *Meteoritenkunde*, Heft **II.** E. Schweizerbart, Stuttgart. 302 pp.
Crabb J., Lewis R. S., and Anders E. (1980) I-Xe ages of carbonaceous chondrites (abstract). In *Lunar and Planetary Science XI*, pp. 174–176. Lunar and Planetary Institute, Houston.

Das Gupta S. P., Charabarti C., Sen Gupta P. R., and Dube A. (1978) A statistical analysis of chondrules in Chainpur meteorite (abstract). *Meteoritics* **13**, 434–435.

Das Gupta S. P., Sen Gupta P. R., Dube A., Sen Gupta N. R., and Das Gupta D. R. (1978) The Dhajala meteorite. *Mineral. Mag.* **42**, 493–497.

Daubrée A. (1867) Contribution a Panatomie des Météorites. *Compt. Rend. Acad. Sci.* **65**. Paris.

Daubrée G. A. (1879) *Etudes Synthétiques de Géologie Experimentale,* pp. 473–730. Paris.

Daubrée G. A. (1893) Observation sur les conditions que paraissent avoir préside à la formation des météorites. *Compt. Rend.* **116**, 345–347.

de Gasparis A. A., Fredriksson K., and Brenner P. (1975) Composition of individual chondrules in ordinary chondrites (abstract). *Meteoritics* **10**, 390–392.

Desnoyers C. (1980) The Niger (I) carbonaceous chondrite and implications for the origin of aggregates and isolated olivine grains in C2 chondrites. *Earth Planet. Sci. Lett.* **47**, 223–234.

Dews J. R. (1966) The isotopic composition of lithium in chondrules. *J. Geophys. Res.* **71** 4011–4020.

Dodd R. T. (1968) Recrystallized chondrules in the Sharps (H-3) chondrite. *Geochim. Cosmochim. Acta* **32**, 1111–1120.

Dodd R. T. (1969) Petrofabric analysis of a large microporphyritic chonrdrule in the Parnallee meteorite. *Mineral. Mag.* **37**, 230–237.

Dodd R. T. (1971) Calcium and aluminum as thermal tracers in chondritic olivine (abstract). *Meteoritics* **6**, 261.

Dodd R. T. (1971) The petrology of chondrules in the Sharps meteorite. *Contrib. Mineral. Petrol.* **31**, 201–227.

Dodd R. T. (1976) Accretion of the ordinary chondrites. *Earth Planet. Sci. Lett.* **30**, 281–291.

Dodd R. T. (1978) Compositions of droplet chondrules in the Manych (L-3) chondrite and the origin of chondrules. *Earth Planet. Sci. Lett.* **40**, 71–82.

Dodd R. T. (1978) The composition and origin of large microporphyritic chondrules in the Manych (L-3) chondrite. *Earth Planet. Sci. Lett.* **39**, 52–66.

Dodd R. T. (1981) Chondritic meteorites: a changed and changing view. *Nature* **290**, 189–190.

Dodd R. T. (1981) *Meteorites: A Petrologic-Chemical Synthesis.* Cambridge University Press. 368 pp.

Dodd R. T. (1982) Objects we call chondrules (abstract). In *Papers Presented to the Conference on Chondrules and their Origins,* p. 15. Lunar and Planetary Institute, Houston.

Dodd R. T. and Teleky L. (1967) Preferred orientation of olivine crystals in porphyritic chondrules. *Icarus* **6**, 407–416.

Dodd R. T. and Van Schmus W. R. (1971) Dark-zoned chondrules. *Chem. Erde* **30**, 59–69.

Dodd R. T. and Walter L. S. (1970) Chemical variations among chondrules from the Chainpur meteorite (abstract). *Meteoritics* **5**, 193–194.

Dodd R. T. and Walter L. S. (1972) Chemical constraints on the origin of chondrules in ordinary chondrites. In *L'origine du Système Solaire* (H. Reeves, ed.), p. 293. Centre National de

Recherche Scientifique, Paris.

Doelter C. and Hussak E. (1884) Ueber die Einwirkung geschmolzener Magmen auf verschiedene Mineralien. *Neues Jahrb. Mineral.* **59**, 18–44.

Doss B. (1892) Über den Meteoriten von Misshof in Kurland und die Ursachen der Schallphonomenen bei Meteoritenfallen im Allgemeinen. *Neues Jahrb. Mineral. Geol. Paleontol.* **I**, 71–113.

Drever H. I. (1960) Immiscibility in the picritic intrusion at Igdlorssuit, West Greenland. *Proc. 21st Int. Geol. Congr., Section 13*, pp. 47–58.

Dufrénoy P. A. and Biot E. (1841) Hote sur la composition de l'aérolithe de Chateay-Renard et remarque suivant. *Compt. Rend.* **13**, 47–53.

Dugan J. P., Jr. (1978) Oxygen isotope geochemistry of the Allende carbonaceous chondrite and its implications for the origin of the Solar System. Master's thesis, Northern Illinois Univ., De Kalb. 118 pp.

Ehmann W. D. and McKown D. (1967) Silicon abundances in meteoritic chondrules. *Geochim. Cosmochim. Acta* **31**, 1975–1985.

El Goresy A., Ramdohr P., and Rambaldi E. R. (1981) Sulfide-rich chondrule rims in Bishunpur (L3) chondrite: Types, textures, mineralogy and genetic implications (abstract). *Lunar and Planetary Science XII,* pp. 254–256. Lunar and Planetary Institute, Houston.

Englehardt W. von, Arndt J., Muller W. F., and Stöffler D. (1971) Evidence of meteorite impacts found in lunar soil and breccias of the Apollo landing sites (abstract). *Meteoritics* **6**, 267.

Englehardt W. von, Arndt J., Stöffler D., and Scheider H. (1972) Apollo 14 regolith and fragmental rocks, their compositions and origin by impacts. *Proc. Lunar Sci. Conf. 3rd,* pp. 753–770.

Englund E. (1969) Experimental studies of the origin and thermal metamorphism of chondrules in chondritic meteorites. Master's thesis, Univ. of Vermont, Burlington. 33 pp.

Evenson N. M., Carter S. R., Hamilton P. J., O'Nions R. K., and Ridley W. I. (1979) A combined chemical petrological study of separated chondrules from the Richardson meteorite. *Earth Planet. Sci. Lett.* **42**, 223–236.

Fallick A. E. and Pillinger C. T. (1979) Meteoritics—Life begins at seventy. *Nature* **281**, 527–528.

Farrington O. C. (1915) *Meteorites, their Structure, Composition, and Terrestrial Relations.* Lakeside Press, Chicago. 233 pp.

Fegley B. and Kornacki A. S. (1982) Spinel-rich chondrules and the condensation of metastable liquids in the solar nebula (abstract). In *Papers Presented to the Conference on Chondrules and their Origins,* p. 16. Lunar and Planetary Institute, Houston.

Feierberg M. A., Lebofsky L. A., and Larson H. P. (1981) Spectroscopic evidence for aqueous alteration products on the surfaces of low-albedo asteroids. *Geochim. Cosmochim. Acta* **45**, 971–981.

Fermor L. L. (1938) Garnets and their role in nature. *Indian Assoc. Adv. Sci., Spec. Publ. No. 6*, pp. 87–91.

Fermor L. L. (1938) On khoharite, a new garnet, and on the nomenclature of garnets. *Records Geol. Survey India* **73**, 145–156.

Fesenkov V. G. (1952) The nature and possible origin of the meteorites, zodiacal light, and the asteroids. *Prob. of Cosmogony* **1**, 89–127. JPRS #24,735.

Fish R. A., Goles G. G., and Anders E. (1960) The record in the meteorites—III. On the development of meteorites in asteroidal bodies. *Astrophys. J.* **132**, 243–258.

Fodor R. V. and Keil K. (1973) Composition and origin of lithic fragments in L- and H-group chondrites (abstract). *Meteoritics* **8**, 366–367.

Fodor R. V., Keil K., Prinz M., Ma M. -S., Murali A. V., and Schmitt R. A. (1980) Clast-laden melt-rock fragment in the Adams County, Colorado, H5 chondrite. *Meteoritics* **15**, 41–62.

Fodor R., Keil K., and Gomes C. B. (1977) Studies of Brazilian meteorites; IV, origin of a dark-colored, unequilibrated lithic fragment in the Rio Negro chondrite. *Rev. Bras. Geocienc.* **7**, 45–57.

Foullon H. V. (1883) Über die mineralogische und chemische Zusammensetzung des am 16. Februar 1883 bei Alfianello gefallenen Meteorsteines. *Sitzber. Akad. Wiss. Math-naturw. Kl. Wien* **88**, No. 1, 433–443.

Fredriksson K. (1963) Chondrules and the meteorite parent bodies. *Trans. N. Y. Acad. Sci.*, Ser. 2, **25**, 756–769.

Fredriksson K. (1969) The Sharps chondrite—new evidence on the origin of chondrules and chondrites. In *Meteorite Research* (P. M. Millman, ed.), p. 155–165. Reidel, Dordrecht.

Fredriksson K. (1982) Chondrites are not metamorphic rocks (abstract). In *Papers Presented to the Conference on Chondrules and their Origins*, pp. 17–18. Lunar and Planetary Institute, Houston.

Fredriksson K. (1982) Chondrule compositions in different type chondrites (abstract). *Lunar and Planetary Science XIII,* pp. 233–234. Lunar and Planetary Institute, Houston.

Fredriksson K., Dube A., and Milton D. (1973) Micro-breccias, impact glasses and spherules from Lonar Crater, India; Lunar analogs (abstract). *Meteoritics* **8**, 34.

Fredriksson K., Dube A., Milton D. J., and Balasundaram M. S. (1973) Lonar Lake, India: an impact crater in basalt. *Science* **180**, 862–864.

Fredriksson K., Nelen J., and Melson W. G. (1970) Petrography and origin of lunar breccias and glasses. *Proc. Apollo 11 Lunar Sci. Conf.,* pp. 419–432.

Fredriksson K., Nelen J., Melson W. G., Henderson E. P., and Anderson C. A. (1970) Lunar glasses and micro-breccias: properties and origin. *Science* **167**, 664–666.

Fredriksson K., Nelen J., and Noonan A. (1971) Lunar chondrules (abstract). *Meteoritics* **6**, 270–271.

Fredriksson K., Nelen J., Noonan A., and Kraut F. (1972) Apollo 14: glasses, breccias, chondrules (abstract). In *Lunar Science III*, pp. 280–282. The Lunar Science Institute, Houston.

Fredriksson K., Noonan A., and Nelen J. (1973) Meteoritic, lunar and lonar impact chondrules. *The Moon* **7**, 475–482.

Fredriksson K., Noonan A. F., Nelen J., and Beauchamp R. H. (1979) Bonding and reactions during accretion of ordinary chondrites—I. Bjürbole (abstract). In *Lunar and Planetary Science X,* pp. 397–399. Lunar and Planetary Institute, Houston.

Fredriksson K., O'Keefe J., and Brenner P. (1978) The bulk composition of individual chondrules (abstract). *Meteoritics* **13**, 464.

Frederiksson K. and Reid A. M. (1965) A chondrule in the Chainpur meteorite. *Science* **149**, 856–860.

Fredriksson K. and Ringwood A. E. (1963) Origin of meteoritic chondrules. *Geochim. Cosmochim. Acta* **27**, 639–641.

Fruland R. M., King E. A., and McKay D. S. (1978) Allende dark inclusions. *Proc. Lunar Planet. Sci. Conf. 9th*, pp. 1305–1330.

Fujii N. (1982) Constraints on the heating and cooling process of chondrule formation by "lightning" (abstract). In *Papers Presented to the Conference on Chondrules and their Origins*, p. 19. Lunar and Planetary Institute, Houston.

Fujimaki H., Matsu-Ura M., Aoki K., and Sunagawa I. (1981) Ferro-pseudobrookite-silica mineral-albite-chondrule in the ALH-77015 chondrite (L3). In *Proceedings of the Sixth Symposium on Antarctic Meteorites* (T. Nagata, ed.), pp. 119–123. National Institute of Polar Research, Tokyo.

Fujimaki H., Matsu-Ura M., Sunagawa I., and Aoki K. (1981) Chemical compositions of chondrules and matrices in the ALH-77015 chondrite (L3). In *Proceedings of the Sixth Symposium on Antarctic Meteorites* (T. Nagata, ed.), pp. 161–174. National Institute of Polar Research, Tokyo.

Funaki M., Nagata T., and Momose K. (1981) Natural remanent magnetizations of chondrules, metallic grains and matrix of an Antarctic chondrite, ALH-769. In *Proceedings of the Sixth Symposium on Antarctic Meteorites* (T. Nagata, ed.), pp. 300–315. National Institute of Polar Research, Tokyo.

Gehrels T. (1979) *Asteroids.* Univ. of Arizona Press, Tucson. 1181 pp.

Gluyas J. G. (1978) Meteorite studies: petrography and origin of chondrules. *J. Geol. Soc. - Sheffield Univ.* **7**, 123–125.

Gooding J. L. (1979) Petrogenetic properties of chondrules. Ph.D. dissertation, Univ. of New Mexico, Albuquerque. 392 pp.

Gooding J. L. (1982) Origin of chondrules in ordinary chondrites by incomplete melting of geochemically fractionated solids (abstract). In *Papers Presented to the Conference on Chondrules and their Origins*, p. 20. Lunar and Planetary Institute, Houston.

Gooding J. L., Fukuoka T., Keil K., and Schmitt R. A. (1979) Refractory and siderophile element variations among chondrules: Evidence for primary compositional differences (abstract). In *Lunar and Planetary Science X,* pp. 443–445. Lunar and Planetary Institute, Houston.

Gooding J. L. and Keil K. (1981) Estimated viscosities and implied thermal histories of chondrule droplets (abstract). In *Lunar and Planetary Science XII,* pp. 353–354. Lunar and Planetary Institute, Houston.

Gooding J. L. and Keil K. (1981) Relative abundances of chondrule primary textural types in ordinary chondrites and their bearing on conditions of chondrule formation. *Meteoritics* **16**, 17–43.

Gooding J. L., Keil K., Fukuoka T., and Schmitt R. A. (1980) Elemental abundances in chondrules from unequilibrated chondrites; evidence for chondrule origin by melting of pre-existing materials. *Earth Planet. Sci. Lett.* **50**, 171–180.

Gooding J. L., Keil K., Fukuoka T., and Schmitt R. A. (1980) The origin of chondrules as secondary objects: Evidence from chemical-petrological heterogeneities (abstract). In *Lunar and Planetary Science XI,* pp. 345–347. Lunar and Planetary Institute, Houston.

Gooding J. L., Mayeda T. K., Clayton R. N., Keil K., Fukuoka T., and Schmitt R. A. (1982) Oxygen isotopic compositions of chondrules in unequilibrated chondrites: further petrological interpretations (abstract). In *Lunar and Planetary Science XIII,* pp. 271–272. Lunar and Planetary Institute, Houston.

Goswami J. N. (1982) Refractory spherules in CM chondrites and chondrule forming processes (abstract). In *Papers Presented to the Conference on Chondrules and their Origins*, p. 21. Lunar and Planetary Institute, Houston.

Goswami J. N. (1982) Size-distribution and morphology of chondrules in Dhajala (H-3) and Murchison (CM) chondrites (abstract). In *Papers Presented to the Conference on Chondrules and their Origins*, p. 22. Lunar and Planetary Institute, Houston.

Goswami J. N. and Macdougall J. D. (1980) Irradiation features in different inclusion types in CM chondrites (abstract). In *Lunar and Planetary Science XI,* pp. 354–356. Lunar and Planetary Institute, Houston.

Graup G. (1981) Terrestrial chondrules, glass spherules, and accretionary lapilli from the Suevite, Ries Crater, Germany. *Earth Planet. Sci. Lett.* **55**, 407–418.

Griffin W. L. and Jorgensen K. A. (1979) The Grefsheim (Norway) meteorite: a new L5 chondrite. *Meteoritics* **14**, 117–120.

Grossman J. N. (1980) Interrelationships of petrography, mineralogy, and chemistry in Chainpur chondrules (abstract). In *Papers Presented to the 43rd Annual Meteoritical Society Meeting,* p. 136. Lunar and Planetary Institute, Houston.

Grossman J. N. (1982) Mechanisms for the depletion of metal in chondrules (abstract). In *Papers Presented to the Conference on Chondrules and their Origins*, p. 23. Lunar and Planetary Institute, Houston.

Grossman J. N. (1982) The abundance and distribution of moderately volatile elements in Semarkona chondrules (abstract). In *Lunar and Planetary Science XIII,* pp. 289–290. Lunar and Planetary Institute, Houston.

Grossman J. N., Kracher A., and Wasson J. T. (1979) Volatiles in Chainpur chondrules. *Geophys. Res. Lett.* **6**, 597–600.

Grossman J. N., Kracher A., Kallemeyn G. W., and Wasson J. T. (1978) Chemical-petrographic study of chondrules (abstract). In *Lunar and Planetary Science IX,* pp. 422–423. Lunar and Planetary Institute, Houston.

Grossman J. N., Kracher A., and Wasson J. T. (1979) Compositional and petrographic constraints on the origin of Chainpur chondrules (abstract). In *Lunar and Planetary Science X,* pp. 464–466. Lunar and Planetary Institute, Houston.

Grossman J. N. and Wasson J. T. (1981) Compositional study of chondrules from the highly unequilibrated (LL3.0) Semarkona chondrite (abstract). In *Lunar and Planetary Science XII,* pp. 371–373. Lunar and Planetary Institute, Houston.

Grossman J. N. and Wasson J. T. (1982) Evidence for primitive nebular components in chondrules from the Chainpur chondrite. *Geochim. Cosmochim. Acta* **46**, p. 1081–1099.

Grossman L. and Larimer J. W. (1974) Early chemical history of the solar system. *Rev. Geophys. Space Phys.* **12**, 71–101.

Grossman L. and Steele I. M. (1976) Amoeboid olivine aggregates in the Allende meteorite. *Geochim. Cosmochim. Acta* **40**, 149–155.

Gumbel C. W. (1875) Über die Beschaffenheit des Steinmeteoriten von Fall am 12. Februar 1875 in der Grafschaft Iowa N.-A. *Sitzungsber. Math.-Phys. Kl. Bayer. Akad. Wiss. München* **5**, 313–330.

Gumbel C. W. and Schwager (1878) Über die in Bayern gefundenen Steinmeteoriten. *Sitzungsber. Math.-Phys. Kl. Bayer Akad. Wiss. München* **8**, 14–72.

Haggerty S. E. (1974) Apollo 17 orange glass: textural and morphological characteristics of devitrification. *Proc. Lunar Sci. Conf. 5th*, pp. 193–205.

Haggerty S. E. and McMahon B. M. (1979) Magnetite-sulfide-metal complexes in the Allende meteorite. *Proc. Lunar Planet. Sci. Conf. 10th*, pp. 851–870.

Hahn O. (1880) *Die Meteorite (chondrite) und ihre Organismen.* Verlag der H. Lauppschen Buchhandlung, Tubingen. 56 pp.

Hahn O. (1883) Organische Einschlüsse in Meteoriten. *Z. Deut. Geol. Ges.* **35**, 636.

Hamilton P. J., Evenson N. M., and O'Nions R. K. (1979) Chronology and chemistry of Parnallee (LL-3) chondrules (abstract). In *Lunar and Planetary Science X*, pp. 494–496. Lunar and Planetary Institute, Houston.

Hartmann W. K. and Wilkening L. L. (1980) Chondrule-sized spherules from an explosion crater (abstract). In *Papers Presented to the 43rd Annual Meteoritical Society Meeting*, p. 141. Lunar and Planetary Institute, Houston.

Heide F. (1934) *Kleine Meteoritenkunde*. Springer-Verlag, Berlin. 120 pp.

Herrwerth I., Jessberger E. K., Muller N., and Kirsten T. (1982) ^{40}Ar-^{39}Ar dating of individual chondrules (abstract). In *Papers Presented to the Conference on Chondrules and their Origins*, p. 24. Lunar and Planetary Institute, Houston.

Hertogen J. (1982) Arguments in favor of a volatilization-recondensation model for chondrule formation (abstract). In *Papers Presented to the Conference on Chondrules and their Origins*, p. 25. Lunar and Planetary Institute, Houston.

Hewins R. H. (1982) Dynamic crystallization experiments as constraints on chondrule origins (abstract). In *Papers Presented to the Conference on Chondrules and their Origins*, p. 26. Lunar and Planetary Institute, Houston.

Hewins R. H. and Klein L. C. (1980) Cooling histories of chondrules in the Manych (L-3) chondrite (abstract). In *Papers Presented to the 43rd Annual Meteoritical Society Meeting*, p. 140. Lunar and Planetary Institute, Houston.

Hewins R. H., Klein L. C., and Fasano B. V. (1981) Conditions of formation of pyroxene excentroradial chondrules. *Proc. Lunar Planet. Sci. 12B*, pp. 1123–1133.

Heymann D. (1982) Stellar chemical processes (1) (abstract). In *Papers Presented to the Conference on Chondrules and their Origins*, pp. 27–28. Lunar and Planetary Institute, Houston.

Hirt W. H. and Kieffer S. W. (1978) Barred olivine chondrules: an optical crystallographic study and model for their formation. *EOS (Trans. Amer. Geophys. Union)* **59**, 1122.

Housley R. M. (1981) Oxidation of Allende parent material (abstract). In *Lunar and Planetary Science XII*, pp. 477–478. Lunar and Planetary Institute, Houston.

Housley R. M. (1981) SEM, optical, and Mossbauer studies of submicrometer chromite in Allende. *Proc. Lunar Planet. Sci. 12B*, pp. 1069–1077.

Housley R. M. (1982) On the interpretation of magnetism in Allende chondrules (abstract). In *Lunar and Planetary Science XIII*, p. 341. Lunar and Planetary Institute, Houston.

Housley R. M. and Cirlin E. H. (1982) On the composition and state of Allende chondrule precursor material (abstract). In *Papers Presented to the Conference on Chondrules and their Origins*, p. 29. Lunar and Planetary Institute, Houston.

Housley R. M. and Cirlin E. H. (1982) On the origin of Allende chondrules and mafic inclusions (abstract). In *Lunar and Planetary Science XIII*, pp. 342–343. Lunar and Planetary Institute, Houston.

Hughes D. W. (1977) A comparison between the mass distribution indices of chondrules and cometary meteoroids. *Earth Planet. Sci. Lett.* **33**, 428–436.

Hughes D. W. (1978) Chondrule mass distribution and the Rosin and Weibull statistical functions. *Earth Planet. Sci. Lett.* **39**, 371–376.

Hughes D. W. (1980) The dependence of chondrule density on chondrule size. *Earth Planet. Sci. Lett.* **51**, 26–28.

Hunt E. H. and Muller P. (1979) Comments on light and dark spots in the equatorial regions of Jupiter. *Planet. Space Sci.* **27**, 1127–1129.

Huss G. R., Keil K., and Taylor G. J. (1981) The matrices of unequilibrated ordinary chondrites: Implications for the origin and history of chondrites. *Geochim. Cosmochim. Acta* **45**, 33–51.

Hutchison R. and Bevan A. W. R. (1982) Conditions and time of chondrule accretion (abstract). In *Papers Presented to the Conference on Chondrules and their Origins*, pp. 30–32. Lunar and Planetary Institute, Houston.

Hutchison R., Bevan A. W. R., Agrell S. O., and Ashworth J. R. (1979) Accretion temperature of the Tieschitz, H3, chondritic meteorite. *Nature* **280**, 116–119.

Hutchison R., Bevan A. W. R., Agrell S. O., and Ashworth J. R. (1980) Thermal history of the H-group of chondritic meteorites. *Nature* **287**, 787–790.

Hutchison R., Bevan A. W. R., Easton A. J., and Agrell S. O. (1981) Mineral chemistry and genetic relations among H-group chondrites. *Proc. Roy. Soc. London* **A374**, 159–178.

Jeremine E. (1954) Eucrite de Chaves (Portugal). *Bol. Soc. Geol. Port.* **11**, 127–138.

Jeremine E. and LeLubre M. (1952) Sur la meteorite d'Ourbari. *Geochim. Cosmochim. Acta* **2**, 217–228.

Jessberger E. K., Dominik B., Staudacher T., and Herzog G. F. (1980) ^{40}Ar-^{39}Ar ages of Allende. *Icarus* **42**, 380–405.

Juan V. C., Chen J. C., Huang C. K., Chen P. Y., and Wang Lee C. M. (1972) Petrology and chemistry of some Apollo 14 lunar samples. *Proc. Lunar Sci. Conf. 3rd*, pp. 687–705.

Kashkarov L. L., Genaeva L. I., and Lavrukhina A. K. (1980) Track studies of Weston meteorite chondrules (abstract). In *Lunar and Planetary Science XI*, pp. 530–532. Lunar and Planetary Institute, Houston.

Kashkarov L. L., Genaeva L. I., Tarasov L. S., and Baryshnikova G. V. (1982) Radiation history of lunar microbreccias and lithic chondrules from Weston meteorite by track data (abstract). In *Lunar and Planetary Science XIII*, pp. 375–376. Lunar and Planetary Institute, Houston.

Keidel W. (1965) Untersuchungen am Meteoriten von Borkut und anderen Chondriten über Form, Aufbau, und Entstehung der Chonden. *Beit. Mineral. Petrogr.* **11**, 487–506.

Keil K. and Fodor R. V. (1973) Composition and origin of lithic fragments in LL-group chondrites (abstract). *Meteoritics* **8**, 394–396.

Keil K. and Fredriksson K. (1964) The iron, magnesium, and calcium distribution in coexisting olivines and rhombic pyroxenes of chondrites. *J. Geophys. Res.* **69**, 3487–3515.

Keil K., Gomes C. B., and Dealmeid F. F. (1978) (PT) Meteorites in Paranaiba, Mato Grosso: Review (meeting abstract). *An. Acad. Bras. Cienc.* **50**, 114.

Keil K., Kurat G., Prinz M., and Green J. A. (1972) Lithic fragments, glasses and chondrules from Luna 16 fines. *Earth Planet. Sci. Lett.* **13**, 243–256.

Keil K., Mason B., Wiik H. B., and Fredriksson K. (1964) The Chainpur meteorite. *Amer. Mus. Novit.* **2173**, 1–28.

Keil K., Prinz M., Planner H. N., Skaggs S. R., Dowty E., Nelson L. S., Richardson N. L., and Blander M. (1973) A qualitative comparison of textures in lunar chondrules and CO_2 laser-formed synthetic chondrule-like spherules. *Univ. of New Mexico Dept. of Geol. and Inst. of Meteoritics, Spec. Publ. No. 7*. 16 pp.

Kerridge J. F. (1970) Condensation in the primordial solar nebula (abstract). *Meteoritics* **5**, 205.

Kieffer S. W. (1974) Glass produced by shock waves in terrestrial and lunar rocks and implications regarding the origin of chondrules (abstract). *Meteoritics* **9**, 360–362.

Kieffer S. W. (1974) Minimum impact velocities required for impact melting of irregular particles of bronzite, dunite, and SiO_2. *EOS (Trans. Amer. Geophys. Union)* **55**, 1142.

Kieffer S. W. (1975) Droplet chondrules. *Science* **189**, 333–340.

Kieffer S. W. (1975) Glass production by shock waves in terrestrial and lunar rocks and implications regarding the origin of chondrules (abstract). *Meteoritics* **30**, 89–90.

Kimura M. and Yagi K. (1980) Crystallization of chondrules in ordinary chondrites. *Geochim.*

Cosmochim. Acta **44**, 589–602.

Kimura M., Yagi K., and Onuma K. (1980) Petrology of the Yamato chondritic meteorites from Antarctica with special reference to the origin of the chondrules (abstract). *Int. Geol. Congr. 26, Resumes* **3**, 1248.

King E. A. (1976) *Space Geology: An Introduction.* Wiley and Sons, N.Y. 349 pp.

King E. A. (1977) The lunar regolith: physical characteristics and dynamics. *Phil. Trans. Roy. Soc. London* **A285**, 273–278.

King E. A. (1980) Multi-zoned chondrules: a newly recognized particle type from ordinary chondrites (abstract). In *Papers Presented to the 43rd Annual Meteoritical Society Meeting*, p. 138. Lunar and Planetary Institute, Houston.

King E. A. (1982) Iron loss from fluid drop chondrules by partial evaporation (abstract). In *Papers Presented to the Conference on Chondrules and their Origins*, p. 33. Lunar and Planetary Institute, Houston.

King E. A. (1982) Refractory residues, condensates and chondrules from solar furnace experiments (abstract). In *Lunar and Planetary Science XIII*, pp. 389–390. Lunar and Planetary Institute, Houston.

King E. A. (1982) Refractory residues, condensates and chondrules from solar furnace experiments. *Proc. Lunar Planet. Sci. Conf. 13th*, in *J. Geophys. Res.* **87**, A429–A434.

King E. A. (1982) Spattering: The dominant process of fluid drop chondrule formation (abstract). In *Papers Presented to the Conference on Chondrules and their Origins*, p. 34. Lunar and Planetary Institute, Houston.

King E. A., Butler J. C., and Carman M. F. (1971) Apollo 14 sample results: chondrule-like bodies, grain size analyses and estimation of lunar surface ages (abstract). *Geol. Soc. Amer. Abstr. with Prog.* **3**, 623.

King E. A., Butler J. C., and Carman M. F. (1972) Chondrules in Apollo 14 breccias and estimation of lunar surface exposure ages from grain size analyses (abstract). In *Lunar Science III*, pp. 449–451. The Lunar Science Institute, Houston.

King E. A., Jr., Butler J. C., and Carman M. F. (1972) Chondrules in Apollo 14 samples and size analyses of Apollo 14 and 15 fines. *Proc. Lunar Sci. Conf. 3rd*, pp. 673–686.

King E. A., Butler J. C., and Carman M. F. (1972) The lunar regolith. *Proc. 24th Int. Geol. Congr., Section 15,* pp. 58–63.

King E. A., Jr., Carman M. F., and Butler, J. C. (1972) Chondrules in Apollo 14 samples: implications for the origin of chondritic meteorites. *Science* **175**, 59–60.

King T. V. V. and King E. A. (1978) Grain size and petrography of C2 and C3 carbonaceous chondrites. *Meteoritics* **13**, 47–72.

King T. V. V. and King E. A. (1979) Size frequency distributions of fluid drop chondrules in ordinary chondrites. *Meteoritics* **14**, 91–96.

King T. V. V. and King E. A. (1980) Accretionary dark rims in unequilibrated chondrites (abstract). In *Lunar and Planetary Science XI*, pp. 557–559. Lunar and Planetary Institute, Houston.

King T. V. V. and King E. A. (1981) Accretionary dark rims in unequilibrated chondrites. *Icarus* **48**, 460–472.

Klein C. (1906) Studien über Meteoriten, vorgenommen auf Grund des Materials der Sammlung der Universität Berlin. *Abhandl. Akad. Wiss. Berlin*, pp. 1–141.

Klein L. C., Fasano B., and Hewins R. H. (1980) Flow behavior of droplet chondrules in the Manych (L-3) chondrite. *Proc. Lunar Planet. Sci. Conf. 11th*, pp. 865–878.

Kluger F. and Weinke H. H. (1980) Chondrule formation by impact: the cooling rate (abstract). In *Papers Presented to the 43rd Annual Meteoritical Society Meeting*, p. 142. Lunar and Planetary Institute, Houston.

Koeberl C. (1982) Chondrules and the isotopic view of interstellar comets (abstract). In *Papers Presented to the Conference on Chondrules and their Origins*, p. 35. Lunar and Planetary Institute, Houston.

Korotkova N. N., Lavrukhina A. K., Kashkarov L. L., and Dmitriev N. B. (1979) (RS) A pre-compaction irradiation of olivine chondrule crystal from meteorite Allende by low-energetic cosmic rays. *Geokhimiya* **1979**, 420–426.

Korotkova N. N., Lavrukhina A. K., and Perelygin V. P. (1979) Iron group cosmic ray tracks in the Allende olivine: evidence of complex exposure history. *Proc. Lunar Planet. Sci. Conf. 10th*, pp. 907–920.

Krinov E. L. (1960) *Principles of Meteoritics.* Pergamon, N. Y. 535 pp.

Kurat G. (1967) Einige Chondren aus dem Meteoriten von Mezo-Madaras. *Geochim. Cosmochim. Acta* **31**, 1843–1857.

Kurat G. (1967) Zur Entstehung der Chondren. *Geochim. Cosmochim. Acta* **31**, 491–502.

Kurat G. (1969) The formation of chondrules and chondrites and some observations on chondrules from the Tieschitz meteorite. In *Meteorite Research* (P. M. Millman, ed.), pp. 185–190. Reidel, Dordrecht.

Kurat G. (1973) The Lance chondrite: Further evidence for the complex development of chondrites (abstract). *Meteoritics* **8**, 51–52.

Kurat G. (1982) Geochemistry of Chainpur chondrules, evidence for evaporation and recondensation (abstract). In *Papers Presented to the Conference on Chondrules and their Origins*, p. 37. Lunar and Planetary Institute, Houston.

Kurat G. (1982) Impact origin of chondrules (abstract). In *Papers Presented to the Conference on Chondrules and their Origins*, p. 36. Lunar and Planetary Institute, Houston.

Kurat G., Fredriksson K., and Nelen J. (1969) Der Meteorit von Siena. *Geochim. Cosmochim. Acta* **33**, 765–773.

Kurat G., Keil K., and Prinz M. (1972) A "chondrite" of lunar origin: textures, lithic fragments, glasses and chondrules (abstract). In *Lunar Science III*, pp. 463–465. The Lunar Science Institute, Houston.

Kurat G., Keil K., and Prinz M. (1974) Rock 14318: a polymict lunar breccia with chondritic texture. *Geochim. Cosmochim. Acta* **38**, 1133–1146.

Kurat G., Keil K., Prinz M., and Nehru C. E. (1971) Chondrules of lunar origin (abstract). *Meteoritics* **6**, 285–286.

Kurat G., Keil K., Prinz M., and Nehru C. E. (1972) Chondrules of lunar origin. *Proc. Lunar Sci. Conf. 3rd*, pp. 707–721.

Kurat G., Kracher A., and Keil K. (1980) Luna 16 microbreccia lithic fragments: samples of an early basalt filling of Mare Fecunditatis. *Tschermaks Mineral. Petrogr. Mitt.* **27**, 41–66.

Lambert P. (1982) Fluid inclusions in chondrules and history of the meteorite hosts (abstract). In *Papers Presented to the Conference on Chondrules and their Origins*, p. 38. Lunar and Planetary Institute, Houston.

Lang B., Stakheav Y. I., Baryshnikova G. V., and Lavrukhina A. K. (1975) Mass/size frequency distribution of chondrules from four L-chondrites (abstract). *Meteoritics* **10**, 437–439.

Lange D. E. and Larimer J. W. (1973) Chondrules: An origin by impacts between dust grains. *Science* **182**, 920–922.

Lanoix M., Strangway D. W., and Pearce G. W. (1978) Paleointensity determination from Allende chondrules (abstract). In *Lunar and Planetary Science IX*, pp. 630–632. Lunar and Planetary Institute, Houston.

Lanoix M., Strangway D. W., and Pearce G. W. (1978) The primordial magnetic field preserved in chondrules of the Allende meteorite. *Geophys. Res. Lett.* **5**, 73–76.

Lasaulx A. V. (1882) Über die Vermehrung der Meteoritensammlung des mineralogischen Museums. *Sitzber. Niederrh. Ges. Natur. u. Heilk. Bonn.* **39**, 100–110.

Lavrukhina A. K. (1973) On the origin of meteorites. *Acta Geophys. Pol.* **21**, 185–201.

Lavrukhina A. K. and Korotkova N. N. (1981) Chemical composition of Allende chondrite olivine (abstract). In *Lunar and Planetary Science XII*, pp. 601–603. Lunar and Planetary Institute, Houston.

Leitch C. A. and Smith J. V. (1981) Enstatite chondrites II: Genesis (abstract). In *Lunar and Planetary Science XII*, pp. 610–612. Lunar and Planetary Institute, Houston.

Leitch C. A. and Smith J. V. (1981) Mechanical aggregation of enstatite chondrites from an inhomogeneous debris cloud. *Nature* **290**, 228–230.

Leitch C. A. and Smith J. V. (1982) Complex chemistries of 24 individual Indarch (EH4) chondrules released by HCl treatment (abstract). In *Papers Presented to the Conference on Chondrules and their Origins*, p. 39. Lunar and Planetary Institute, Houston.

Levi F. A. (1982) Non-spherical chondrules: Possible mechanisms of formation connected with dynamic anisotropy (abstract). In *Papers Presented to the Conference on Chondrules and their Origins*, p. 40. Lunar and Planetary Institute, Houston.

Levi-Donati G. R. (1970) Micro-chondrules. *Meteoritics* **5**, 33–42.

Levi-Donati G. R. (1975) A qualitative comparison of textures in type 4-chondrites chondrules and laser-prepared synthetic chondrule-like spherules (abstract). *Meteoritics* **10**, 442–444.

Levi-Donati G. R., Maras A., and Sighinolfi G. P. (1980) An H4-6 Chondrite: Motta Di Conti. *Meteoritics* **15**, 211–223.

Levi-Donati G. R., Shima M., and Sighinolfi G. P. (1976) Brazilian meteorites: the Mafra, Santa Catarina State, chondrite. *Meteoritics* **11**, 29–41.

Levin B. Y. (1958) Über den Ursprung der Meteoriten. *Chem. Erde* **19**, 286–295.

Levin B. Y. and Slonimsky G. L. (1957) Title unknown. *Doklady Akad. Nauk. S.S.S.R.* **113**, 62.

Levin B. Y. and Slonimsky G. L. (1958) Question of the origin of meteoritic chondrules. *Meteorika* **16**, 30–36.

Lewis R. S., Hertogen J., Alaerts L., and Anders E. (1979) Isotopic anomalies in meteorites and their origins—V. Search for fission fragment recoils in Allende sulfides. *Geochim. Cosmochim. Acta* **43**, 1743–1752.

Lofgren G. (1971) Devitrified glass fragments from Apollo 11 and Apollo 12 lunar samples. *Earth Planet. Sci. Lett.* **9**, 225–231.

Lofgren G. (1982) The importance of heterogeneous nucleation for the formation of microporphyritic chondrules (abstract). In *Papers Presented to the Conference on Chondrules and their Origins*, p. 41. Lunar and Planetary Institute, Houston.

Lumpkin G. R. (1980) Nepheline and sodalite in a barred olivine chondrule from the Allende meteorite. *Meteoritics* **15**, 139–147.

Lux G., Keil K., and Taylor G. J. (1980) Metamorphism of the H-group chondrites: implications from compositional and textural trends in chondrules. *Geochim. Cosmochim. Acta* **44**, 841–855.

Lux G., Keil K., and Taylor G. J. (1981) Chondrules in H3 chondrites: textures, compositions and origins. *Geochim. Cosmochim. Acta* **45**, 675–685.

McCall G. J. (1973) *Meteorites and their Origins*. Wiley and Sons, N.Y. 352 pp.

McKeever S. W. and Sears D. W. (1980) Meteorites that glow. *Sky and Telescope* **60**, 14–16.

McMahon B. M. and Haggerty S. E. (1980) Experimental studies bearing on the magnetite-alloy-sulfide association in the Allende meteorite: Constraints on the conditions of chondrule formation. *Proc. Lunar Planet. Sci. Conf. 11th*, pp. 1003–1025.

McMahon B. M., Haggerty S. E., and Bence R. (1980) Experimental studies bearing on magnetite-alloy associations in the Allende meteorite (abstract). In *Lunar and Planetary Science XI*, pp. 711–713. Lunar and Planetary Institute, Houston.

McMahon B. M., Tompkins L. A., and Haggerty S. E. (1981) Bulk compositional variations, viscosities, and the effects of oxidation and sulfide abundances on the formation of chondrules in the Allende meteorite (abstract). In *Lunar and Planetary Science XII*, pp. 697–698. Lunar and Planetary Institute, Houston.

McNaughton N. J. and Swart P. K. (1982) In, out and about meteorites. *Nature* **297**, 453–454.

McSween H. Y., Jr. (1977) Chemical and petrographic constraints on the origin of chondrules and inclusions in carbonaceous chondrites. *Geochim. Cosmochim. Acta* **41**, 1843–1860.

McSween H. Y., Jr. (1977) On the nature and origin of isolated olivine grains in carbonaceous chondrites. *Geochim. Cosmochim. Acta* **41**, 411–418.

McSween H. Y., Jr. (1979) Are carbonaceous chondrites primitive or processed? A review. *Rev. Geophys. Space Phys.* **17**, 1059–1078.

McSween H. Y., Jr. (1982) Chondrites chronicle solar system. *Geotimes* **27**, 20–21.

McSween H. Y. (1982) Chondrule populations in carbonaceous chondrites (abstract). In *Papers Presented to the Conference on Chondrules and their Origins*, p. 46. Lunar and Planetary Institute, Houston.

Macdougall J. D. (1981) Refractory spherules in the Murchison meteorite: are they chondrules? *Geophys. Res. Lett.* **8**, 966–969.

Macdougall J. D. (1982) Refractory chondrules in the CM meteorites (abstract). In *Papers Presented to the Conference on Chondrules and their Origins*, p. 42. Lunar and Planetary Institute, Houston.

MacPherson G. J. and Grossman L. (1982) Chondrules and aggregates (abstract). In *Papers Presented to the Conference on Chondrules and their Origins*, pp. 43–44. Lunar and Planetary Institute, Houston.

Makowsky A. and Tschermak G. (1978) Bericht über den Meteoritenfall bei Tieschitz in Mahren. *Denkschr. Akad. Wiss. Wien.* **39**, Abt. 14, 187–202.

Malysheva T. V., Tobelko K. I., Khramov D. A., and Matveeva O. A. (1979) On the problem of the origin of enstatite chondrites. *Proc. Lunar Planet. Sci. Conf. 10th*, pp. 977–988.

Martin P. M. and Hughes D. W. (1980) Meteoritic chondrules and the Weibull function. *Earth Planet. Sci. Lett.* **49**, 175–180.

Martin P. M. and Mills A. A. (1976) Size and shape of chondrules in the Bjurbole and Chainpur meteorites. *Earth Planet. Sci. Lett.* **33**, 239–248.

Martin P. M. and Mills A. A. (1978) Size and shape of near-spherical Allegan chondrules. *Earth Planet. Sci. Lett.* **38**, 385–390.

Martin P. M. and Mills A. A. (1980) Preferred chondrule orientations in meteorites. *Earth Planet. Sci. Lett.* **51**, 18–25.

Masaitis V. L., Raikhlin A. I., and Mashchak M. S. (1980) Immiscibility of impact melts (abstract). In *Lunar and Planetary Science XI*, pp. 677–679. Lunar and Planetary Institute, Houston.

Mason B. (1960) Origin of chondrules and chondritic meteorites. *Nature* **186**, 230–231.

Mason B. (1960) The origin of meteorites. *J. Geophys. Res.* **65**, 2965–2970.

Mason B. (1962) *Meteorites*. Wiley and Sons, N.Y. 274 pp.

Matsomoto Y., Hayashi M., Daishi M., and Miura Y. (1981) On homogeneity of the Yamato-75110 chondrite. In *Proceedings of the Sixth Symposium on Antarctic Meteorites* (T. Nagata, ed.), pp. 106–118. National Institute of Polar Research, Tokyo.

Matsui T., Yomogida K., and Hamano Y. (1982) Physical properties of ordinary chondrites and implications of their accretion and consolidation processes (abstract). In *Papers Presented to the Conference on Chondrules and their Origins*, p. 45. Lunar and Planetary Institute, Houston.

Menninga C., Rancitelli L. A., and Beauchamp R. H. (1973) Chemical composition of 36 individual chondrules from the Allende meteorite (abstract). *Meteoritics* **8**, 56.

Merrill G. P. (1888) On the San Emigdio meteorite. *Proc. U. S. Nat. Mus.* **11**, 161–167.

Merrill G. P. (1915) Researches on the chemical and mineralogical composition of meteorites. *Proc. Nat. Acad. Sci.* **1**, 429–431.

Merrill G. P. (1920) On chondrules and chondritic structure in meteorites. *Proc. Nat. Acad. Sci.* **6**, 449–472.

Merrill G. P. (1921) On metamorphism in meteorites. *Geol. Soc. Amer. Bull.* **32**, 395–414.

Merrill G. P. (1928) Concerning the origin of metal in meteorites. *Proc. U. S. Nat. Mus.* **73**, Art. 21.

Merrill G. P. (1929) The story of meteorites. In *Minerals from Earth and Sky*, pp. 1–75. Smithsonian Science Series, vol. 3, pt. 1.

Methot R. L., Noonan A. F., Jarosewich E., deGasparis A. A., and Al-Far D. M. (1975) Mineralogy, petrology and chemistry of the Isna (C3) meteorite. *Meteoritics* **10**, 121–131.

Meunier S. (1883) Essai d'application de la théorie cyclonique de M. Faye a L'histoire de Météorites primitires. *Compt. Rend.* **96**, 866–869.

Meunier S. (1884) Encyclopédie Chimique, publiée sous la direction de M. Fremy. *Météorites*, book 2 (Dunod, ed.), Paris. 532 pp.

Miura Y. (1982) Compositional variations of plagioclase, whitlockite, apatite, and chromite in Yamato-75135.93 chondrite (abstract). In *Lunar and Planetary Science XIII*, pp. 526–527. Lunar and Planetary Institute, Houston.

Miura Y. and Matsumoto Y. (1981) Classification of several Yamato-75 chondrites (III). In *Proceedings of the Sixth Symposium on Antarctic Meteorites* (T. Nagata, ed.), pp. 53–68. National Institute of Polar Research, Tokyo.

Mueller G. (1964) Interpretation of micro-structures in carbonaceous meteorites. In *Advances in Organic Geochemistry* (U. Colombo and G. D. Hobson, eds.), pp. 119–140. Pergamon, Oxford.

Mueller G. (1966) Significance of inclusions in carbonaceous meteorites. *Nature* **210**, 151–155.

Mueller G. (1966) Title unknown. *Meteoritika* **27**, 3.

Mueller G. (1968) Significance of microchondrules of olivine in type I carbonaceous chondrites. *Nature* **218**, 1239–1240.

Mueller G. (1969) Genetical interrelations between ureilites and carbonaceous chondrites. In *Meteorite Research* (P. M. Millman, eds.), pp. 505–517. Reidel, Dordrecht.

Mueller G. (1971) Morphology of sprays from the moon and elsewhere (abstract). *Meteoritics* **6**, 294–295.

Muller W. F. and Wlotzka F. (1982) Mineralogical study of the Leoville meteorite (CV3): macroscopic texture and transmission electron microscopic observations (abstract). In *Lunar and Planetary Science XIII*, pp. 558–559. Lunar and Planetary Institute, Houston.

Murrell M. T. and Burnett D. S. (1982) Actinide chemistry in chondrites (abstract). In *Lunar and*

Planetary Science XIII, pp. 560–561. Lunar and Planetary Institute, Houston.
Nagahara H. (1981) Evidence for secondary origin of chondrules. *Nature* **292**, 135–136.
Nagahara H. (1981) Petrology of chondrules in ALH-77015 (L3) chondrite. In *Proceedings of the Sixth Symposium on Antarctic Meteorites* (T. Nagata, ed.), pp. 145–160. National Institute of Polar Research, Tokyo.
Nagahara H. (1982) Effect of heating temperature on the texture of chondrules with special reference to the porphyritic chondrules (abstract). In *Papers Presented to the Conference on Chondrules and their Origins*, p. 47. Lunar and Planetary Institute, Houston.
Nagahara H. and Kushiro I. (1982) Calcium-aluminum-rich chondrules in the unequilibrated ordinary chondrites. *Meteoritics* **17**, 55–63.
Nagahara H. and Kushiro I. (1982) Genetic relation of chondrules, inclusions and isolated olivines in ALH-77307 (CO3) chondrite (abstract). In *Lunar and Planetary Science XIII*, pp. 562–563. Lunar and Planetary Institute, Houston.
Nagahara H. and Kushiro I. (1982) Petrology of chondrules, inclusions and isolated olivine grains in ALH-77307 (CO3) chondrite. *Mem. Natl. Inst. Polar Res., Spec. Issue No. 25* (T. Nagata, ed.), pp. 66–77. National Institute of Polar Research, Tokyo.
Nagle J. S. (1982) Textural and structural evidence of regolithic and non-regolithic processes in chondrites (abstract). In *Lunar and Planetary Science XIII*, pp. 574–575. Lunar and Planetary Institute, Houston.
Naney M. T., Crowl D. M., and Papike J. J. (1976) The Apollo 16 drill core: statistical analysis of glass chemistry and the characterization of high alumina-silica poor (HASP) glass. *Proc. Lunar Sci. Conf. 7th*, pp. 155–184.
Nagata T. and Funaki M. (1981) The composition of natural remanent magnetization of an Antarctic chondrite, ALHA 76009 (L6). *Proc. Lunar Planet. Sci. 12B*, pp. 1229–1241.
Nayak V. K. (1972) Glassy objects (impactite glasses?). A possible new evidence for meteoritic origin of the Lonar Crater, Maharashtra State, India. *Earth Planet. Sci. Lett.* **14**, 1–6.
Nelen J., Noonan A., and Fredriksson K. (1972) Lunar glasses, breccias and chondrules. *Proc. Lunar Sci. Conf. 3rd*, pp. 723–737.
Nelen J., Noonan A., and Fredriksson K. (1974) Al-Ca-rich chondrules in the Coolidge chondrite (abstract). *Meteoritics* **9**, 384–385.
Nelson L. S., Blander M., Skaggs S. R., and Keil K. (1972) Use of a CO_2 laser to prepare chondrule-like spherules from supercooled molten oxide and silicate droplets. *Earth Planet. Sci. Lett.* **14**, 338–344.
Nelson L. S., Skaggs S. R., Blander M., and Keil K. (1970) Laboratory preparation of chondrule-like spherules with a CO_2 laser: Origin of chondrules (abstract). *Meteoritics* **5**, 213–214.
Niemeyer S. and Lugmair G. W. (1981) Ti isotopes in Allende inclusions and unequilibrated chondrite chondrules (abstract). In *Lunar and Planetary Science XII*, pp. 768–770. Lunar and Planetary Institute, Houston.
Nininger H. H. (1952) *Out of the Sky*. Dover, N.Y. 336 pp.

Noonan A. F. (1980) Chondrules in the Kapoeta and Bununu howardites (abstract). In *Papers Presented to the 43rd Annual Meteoritical Society Meeting*, p. 139. Lunar and Planetary Institute, Houston.

Noonan A. F. and Nelen J. A. (1976) A petrographic and mineral chemistry study of the Weston, Connecticut, chondrite. *Meteoritics* **11**, 111–130.

Noonan A. F., Fredriksson K., Jarosewich E., and Brenner P. (1976) Mineralogy and bulk, chondrule, size-fraction chemistry of the Dhajala, India, chondrite (abstract). *Meteoritics* **11**, 340–343.

Notsu K., Onuma N., and Nagasawa N. (1978) High temperature heating of the Allende meteorite. *Geochim. Cosmochim. Acta* **42**, 903–907.

O'Keefe J. A. (1976) Chondrules as volcanic ash. *Meteoritics* **11**, 347–348.

Olsen E. (1982) Melt droplet chondrule in Murchison: Evidence for slow cooling before quench (abstract). In *Papers Presented to the Conference on Chondrules and their Origins*, p. 48. Lunar and Planetary Institute, Houston.

Olsen E. and Grossman L. (1978) On the origin of isolated olivine grains in Type 2 carbonaceous chondrites. *Earth Planet. Sci. Lett.* **41**, 111–127.

Onuma N., Nishida N., Ohtsuka Y., Kimura M., and Yanai K. (1979) Origin and evolution of chondrules based on Na/Al dispersion and convergence in Yamato-74 ordinary chondrites. *Mem. Natl. Inst. Polar Res., Spec. Issue No. 15* (T. Nagata, ed.), pp. 206–214. National Institute of Polar Research, Tokyo.

Osborn T. W. and Schmitt R. A. (1971) Elemental abundances in meteoritic chondrules (abstract). *Meteoritics* **6**, 297–298.

Osborn T. W., Smith R. H., and Schmitt R. A. (1973) Elemental composition of individual chondrules from ordinary chondrites. *Geochim. Cosmochim. Acta* **37**, 1909–1942.

Osborn T. W., Warren R. G., Smith R. H., Wakita H., Zellner D. L., and Schmitt R. A. (1974) Elemental composition of individual chondrules from carbonaceous chondrites, including Allende. *Geochim. Cosmochim. Acta* **38**, 1359–1378.

Paque J. M. and Stolper E. (1982) Dynamic crystallization experiments on an average Ca-Al-rich inclusion composition (abstract). In *Papers Presented to the Conference on Chondrules and their Origins*, p. 49. Lunar and Planetary Institute, Houston.

Partsch P. (1843) *Die Meteoriten oder von Himmel gefallen Stein-und Eisenmassen im KK. Hof-Mineralien-Kabinette zu Wien*. Kaulfuss, Witiuel, Prandel u. Comp., Vienna. 175 pp.

Patchett P. J. (1980) Isotopic fractionation during accretion of Allende: SR double-spike data from olivine chondrules (abstract). In *Lunar and Planetary Science XI*, pp. 865–867. Lunar and Planetary Institute, Houston.

Patchett P. J. (1980) Sr isotopic fractionation in Allende chondrules: A reflection of solar nebular processes. *Earth Planet. Sci. Lett.* **50**, 181–188.

Planner H. N. (1982) A chondrule fragment containing silica spheroids (abstract). In *Papers Presented to the Conference on Chondrules and their Origins*, p. 50. Lunar and Planetary Institute, Houston.

Planner H. N. and Keil K. (1982) Evidence for the three-stage cooling history of olivine-porphyritic fluid droplet chondrules. *Geochim. Cosmochim. Acta* **46**, 317–330.

Podolak M. and Cameron A. G. W. (1975) Possible formation of meteoritic chondrules and inclusions in the precollapse Jovian protoplanetary atmosphere (abstract). *Meteoritics* **10**, 97.

Prinz M., Fodor R. V., and Keil K. (1977) Comparison of lunar rocks and meteorites: implications to histories of the Moon and parent meteorite bodies. *NASA SP-370*, Vol. 1, pp. 183–199.

Prinz W. (1885) Les météorites tombées en Belgique et les météorites en général. *Ciel et Terre* **5**, No. 21, 498–505.

Proust L. (1806) Über einen Meteorstein, der im J. 1773 unweit Sigena. im Arogonien herabgefallen ist, und über die Meteorsteine uberhaupt. *Ann. Physik.* **24**, 261–292.

Rajan R. S. (1974) On the irradiation history and origin of gas-rich meteorites. *Geochim. Cosmochim. Acta* **38**, 777–788.

Rambaldi E. R. (1981) Relict grains in chondrules. *Nature* **293**, 558–561.

Rambaldi E. R. (1982) Dusty metal in chondrules: Implication for their origin (abstract). In *Papers Presented to the Conference on Chondrules and their Origins*, p. 51. Lunar and Planetary Institute, Houston.

Rambaldi E. R. (1982) New evidence of relict grains in chondrules of highly unequilibrated ordinary chondrites (abstract). In *Lunar and Planetary Science XIII*, pp. 638–639. Lunar and Planetary Institute, Houston.

Rambaldi E. R. and Wasson J. T. (1980) The origin of chondrule rims in the Bishunpur (L3) chondrite (abstract). In *Papers Presented to the 43rd Annual Meteoritical Society Meeting*, p. 137. Lunar and Planetary Institute, Houston.

Rambaldi E. R. and Wasson J. T. (1981) Metal and associated phases in Bishunpur, a highly unequilibrated ordinary chondrite. *Geochim. Cosmochim. Acta* **45**, 1001–1015.

Rambaldi E. R. and Wasson J. T. (1982) Fine, nickel-poor Fe-Ni grains in the olivine of unequilibrated ordinary chondrites. *Geochim. Cosmochim. Acta* **46**, 929–939.

Rambaldi E. R., Sears D. W., and Wasson J. T. (1980) Si-rich Fe-Ni grains in highly unequilibrated chondrites. *Nature* **287**, 817–820.

Ramsay W. and Borgström L. H. (1902) Der Meteorit von Bjurböle bei Borga. *Bull. Comm. Geol. Finlande, No. 12.* 28 pp.

Rao A. S. P. (1982) Clues to the genesis of meteoritic chondrules (abstract). In *Papers Presented to the Conference on Chondrules and their Origins*, p. 52. Lunar and Planetary Institute, Houston.

Rasmussen K. L. and Wasson J. T. (1982) A new lightning model for chondrule formation (abstract). In *Papers Presented to the Conference on Chondrules and their Origins*, p. 53. Lunar and Planetary Institute, Houston.

Reichenbach C. V. (1860) Meteoriten in Meteoriten. *Ann. Physik.* **111**, 353–386.

Reid A. M. and Fredriksson K. (1967) Chondrules and chondrites. In *Researches in Geochemistry* (P. H. Abelson, ed.), pp. 170–203. Wiley, N.Y.

Reid A. M., Williams R. J., Gibson E. K., Jr., and Fredriksson K. (1974) A refractory glass chondrule in the Vigarano chondrite. *Meteoritics* **9**, 35–45.

Rénard A. F. (1897) Recherches sur le mode de formation des météorites pierreuses (chondrites). *Bull. Soc. Geol. Belg. Bruxelles* **11**, 61–65.

Reusch H. (1886) Om meteorstene. *Naturen* **10**, 129–137.

Reynolds J. H. and Jeffery P. M. (1980) Rare gases in individual meteor ablation spherules (abstract). In *Lunar and Planetary Science XI*, pp. 920–922. Lunar and Planetary Institute, Houston.

Reynolds J. H., Lumpkin G. R., and Jeffery P. M. (1980) Search for 129Xe in mineral grains from Allende inclusions: An exercise in miniaturized rare gas analysis (abstract). In *Lunar and Planetary Science XI*, pp. 923–925. Lunar and Planetary Institute, Houston.

Richardson S. M. (1978) Chemical alteration and mechanical degradation in C2 chondritic meteorites. *EOS (Trans. Amer. Geophys. Union)* **59**, 314–315.

Richardson S. M. (1981) Alteration of mesostasis in chondrules and aggregates from three C2 carbonaceous chondrites. *Earth Planet. Sci. Lett.* **52**, 67–75.

Richardson S. M. and McSween H. Y., Jr. (1978) Textural evidence bearing on the origin of isolated olivine crystals in C2 carbonaceous chondrites. *Earth Planet. Sci. Lett.* **37**, 485–491.

Richardson S. M. and Willson L. A. (1980) Cosmic contamination: elemental clues to the sun's birth. *Astronomy* **8**, 6–22.

Ringwood A. E. (1959) On the chemical evolution and densities of the planets. *Geochim. Cosmochim. Acta* **15**, 257–283.

Ringwood A. E. (1961) Chemical and genetic relationships among meteorites. *Geochim. Cosmochim. Acta* **24**, 159–197.

Ringwood A. E. (1966) Genesis of chondritic meteorites. *Rev. Geophys.* **4**, 113–175.

Rinne F. (1895) Über rhombischen Augit als Kontaktproduct, chondrenartige Bildungen aus künstlichen Schmelzen und über Konkretionen in Basalten. *Neues Jarhb. Mineral. Geol. Palaeontol.* **2**, 229–246.

Rinne F. (1897) Kugelrunde Eiskrystalle und Chondren von Meteoriten. *Neues Jahrb. Mineral. Geol. Palaeontol.* **1**, 259–261.

Rose G. (1863) Beschreibung und Einteilung der Meteoriten auf Grund der Sammlung im mineralogischen Museum zu Berlin. *Abhandl. Akad. Wiss. Phys.-Math. Kl. Berlin.* 161 pp.

Rosman K. J. R., De Laeter J. R., and Gorton M. P. (1980) Cadmium isotope fractionation in fractions of two H3 chondrites. *Earth Planet. Sci. Lett.* **48**, 166–170.

Roy S. K. (1957) The problems of the origin and structure of chondrules in stony meteorites. *Fieldiana Geol.* **10**, No. 31, 383–396.

Rubin A. E., Scott E. R. D., Keil K., and Okada A. (1982) Microchondrules: Their occurrence in new kinds of type 3 chondrites and their bearing on the origin of chondrules (abstract). In *Lunar and Planetary Science XIII*, pp. 663–664. Lunar and Planetary Institute, Houston.

Rubin A. E., Taylor G. J., Keil K., and Nelson G. (1981) The Correo and Suwanee Spring meteorites: two new ordinary chondrite finds. *Meteoritics* **16**, 9–12.

Salisbury W. W. and Fernald D. F. (1970) Experimental chondrule formation (abstract). *Meteoritics* **5**, 220.

Schmitt R. A., Smith R. H., Ehmann W. D., and McKowan D. (1967) Silicon abundances in meteoritic chondrules. *Geochim. Cosmochim. Acta* **31**, 1975–1985.

Schmitt R. A., Smith R. H., and Goles G. G. (1965) Abundances of Na, Sc, Cr, Mn, Fe, Co, and Cu in 218 individual chondrules via activation analyses, 1. *J. Geophys. Res.* **70**, 2419–2444.

Schmitt R. A., Smith R. H., and Olehy D. A. (1968) Rare earth abundances in meteoritic chondrules. In *Origin and Distribution of the Elements* (L. H. Ahrens *et al.*, eds.), pp. 273–282. Pergamon, Oxford.

Scott E. R. D. (1981) Meteorites—Solar system fossils. *Geotimes* **26**, 22–23.

Scott E. R. D., Taylor G. J., and Keil K. (1982) Origins of ordinary and carbonaceous type 3 chondrites and their components (abstract). In *Lunar and Planetary Science XIII*, pp. 704–705. Lunar and Planetary Institute, Houston.

Scott E. R. D., Taylor G. J., Rubin A. E., Keil K., and Kracher A. (1982) Origin of chondrules by preaccretionary melting of silicate matrix material (abstract). In *Papers Presented to the Conference on Chondrules and their Origins*, p. 54. Lunar and Planetary Institute, Houston.

Shima M., Murayama S., and Okada A. (1979) Chemical composition, petrology, and mineralogy of the Shibayama chondrite found in Shibayama-Maghi, Sanbu-Gun, Chiba-Ken, Japan. *Meteoritics* **14**, 317–330.

Simon S. B. and Haggerty S. E. (1979) Petrography and olivine mineral chemistry of chondrules and inclusions in the Allende meteorite. *Proc. Lunar Planet. Sci. Conf. 10th*, pp. 871–883.

Simon S. B. and Haggerty S. E. (1980) Bulk chemistry of a diverse chondrule suite in the Allende meteorite. *Proc. Lunar Planet. Sci. Conf. 11th*, pp. 901–927.

Smith J. V. (1980) Origin of O- and E-chondrites by cooling, sorting, aggregation, condensation, metamorphism, and brecciation of debris from planetesimal collisions (abstract). In *Lunar and Planetary Science XI*, pp. 1054–1056. Lunar and Planetary Institute, Houston.

Snellenburg J. W. (1978) A petrographic grid applicable to chondrule formation (abstract). In *Lunar and Planetary Science IX*, pp. 1080–1082. Lunar and Planetary Institute, Houston.

Sonett C. P. (1979) On the origin of chondrules. *Geophys. Res. Lett.* **6**, 677–680.

Sonett C. P. (1982) An electrical mechanism for chondrule formation (abstract). In *Papers Presented to the Conference on Chondrules and their Origins*, p. 55. Lunar and Planetary Institute, Houston.

Sorby H. C. (1864) On the microscopical structure of meteorites. *Proc. Roy. Soc. London* **13**, 333–334.

Sorby H. C. (1864) On the microscopical structure of meteorites. *Phil. Mag.* **28**, 157–159.

Sorby H. C. (1877) On the structure and origin of meteorites. *Nature* **15**, 495–498.

Sparks M. H. and Sears D. W. G. (1982) Mechanism for metamorphism-induced TL changes in UOC: Evidence from separated Dhajala chondrules (abstract). In *Lunar and Planetary Science*

XIII, pp. 754–755. Lunar and Planetary Institute, Houston.

Strangway D. W. (1978) The record of magnetic fields in the early solar system. *Moon and Planets* **18**, 273–279.

Suess H. E. (1949) Zur Chemie der Planeten—und Meteoritenbildung. *Z. Electrochem.* **53**, 237–241.

Suess H. E. (1963) The Urey-Craig groups of chondrites and their states of oxidation. In *Isotopic and Cosmic Chemistry* (H. Craig, S. L. Miller, and G. J. Wasserburg, eds.), pp. 385–400. North-Holland, Amsterdam.

Suess H. E. (1982) Can chondrules form from a gas of solar composition? (abstract). In *Papers Presented to the Conference on Chondrules and their Origins*, p. 56. Lunar and Planetary Institute, Houston.

Sugiura N., Lanoix M., and Strangway D. W. (1979) Magnetic fields of the solar nebula as recorded in chondrules from the Allende meteorite. *Phys. Earth Planet. Inter.* **20**, 342–349.

Sugiura N. and Strangway D. W. (1978) Magnetic studies on the Allende meteorite. In *Papers Presented to the Conference on Origins of Planetary Magnetism*, p. 115. Lunar and Planetary Institute, Houston.

Sugiura N. and Strangway D. W. (1982) Early magnetic fields recorded in individual chondrules from the Allende meteorite (abstract). In *Papers Presented to the Conference on Chondrules and their Origins*, p. 57. Lunar and Planetary Institute, Houston.

Taylor G. J., Scott E. R. D., and Keil K. (1982) Cosmic setting for chondrule formation (abstract). In *Papers Presented to the Conference on Chondrules and their Origins*, p. 58. Lunar and Planetary Institute, Houston.

Taylor S. R. (1975) *Lunar Science: A Post-Apollo View*. Pergamon, N.Y. 372 pp.

Tomeoka K. and Buseck P. R. (1982) An unusual Fe- and O-rich layered material in chondrules and aggregates of carbonaceous chondrites (abstract). In *Lunar and Planetary Science XIII*, pp. 803–804. Lunar and Planetary Institute, Houston.

Tschermak G. (1874) Die Trummerstructur der Meteoriten von Orvinio und Chantonnay. *Sitzungsber. Akad. Wiss. Wien* **70**, 459–472, plus 2 plates.

Tschermak G. (1875) Die Bildung der Meteoriten und der Vulcanismus. *Sitzungsber. Akad. Wiss. Wien Math.-Naturwiss. Kl.* **71**, Abt. **2**, 661–673.

Tschermak G. (1878) Der Meteoritenfall bei Tieschitz in Mahren (Zweiter Bericht). *Sitzungsber. Akad. Wiss. Wien* **78**, 580–582.

Tschermak G. (1882) Uber die Meteoriten von Mocs. *Sitzungsber. Akad. Wiss. Kl. Wein* **85**, Abt. **3**, 195–209, plus 1 plate.

Tschermak G. (1883) Beitrag zur Classification der Meteoriten. *Sitzungsber. Akad. Wiss. Wien* **88**, Abt. **1**, 347–371.

Tschermak G. (1885) *Die mikroskopische Beschaffenheit der Meteoriten, erläutert durch photographische Abbildungen*. E. Schweizerbart'sche Verlagshandlung, Stuttgart. 24 pp. [English trans., Wood J. A. and Wood E. M. (1964) *Smithson. Contrib. Astrophys. No. 4*. 137 pp.]

Tsuchiyama A. and Nagahara H. (1981) Effects of precooling thermal history and cooling rate on the texture of chondrules: a preliminary report. *Proceedings of the Sixth Symposium on Antarctic Meteorites* (T. Nagata, ed.), pp. 175–192. National Institute of Polar Research, Tokyo.

Tsuchiyama A., Nagahara H., and Kushiro I. (1980) Experimental reproduction of textures of chondrules. *Earth Planet. Sci. Lett.* **48**, 155–165.

Tsuchiyama A., Nagahara H., and Kushiro I. (1981) Volatilization of sodium from silicate melt spheres and its application to the formation of chondrules. *Geochim. Cosmochim. Acta* **45**, 1357–1367.

Tsuchiyama A., Nagahara H., and Kushiro I. (1982) Conditions of chondrule formation: Experimental reproduction of texture and volatilization of sodium from chondrules (abstract). In *Papers Presented to the Conference on Chondrules and their Origins*, p. 59. Lunar and Planetary Institute, Houston.

Urey H. C. (1952) *The Planets.* Yale University Press, New Haven. 245 pp.

Urey H. C. (1956) Diamonds, meteorites and the origin of the solar system. *Astrophys. J.* **124**, 623–637.

Urey H. C. (1958) The early history of the solar system as indicated by the meteorites. *Proc. Chem. Soc. (London)*, pp. 67–78.

Urey H. C. (1959) Primary and secondary objects. *J. Geophys. Res.* **64**, 1721–1737.

Urey H. C. (1961) Criticism of Dr. B. Mason's paper on the origin of meteorites. *J. Geophys. Res.* **66**, 1988–1991.

Urey H. C. (1967) On the origin of chondrules and chondrites (abstract). *Meteoritics* **3**, 129–130.

Urey H. C. (1967) Parent bodies of the meteorites. *Icarus* **7**, 350–359.

Urey H. C. and Craig H. (1953) The composition of the stone meteorites and the origin of the meteorites. *Geochim. Cosmochim. Acta* **4**, 36–82.

Van Schmus W. R. (1967) Polymict structure of the Mezö-Madaras chondrite. *Geochim. Cosmochim. Acta* **31**, 2027–2042.

Van Schmus W. R. (1969) Classification of types 3 and 4 carbonaceous chondrites. In *Meteorite Research* (P. M. Millman, ed.), pp. 480–491. Reidel, Dordrecht.

Van Schmus W. R. and Wood J. A. (1967) A chemical-petrological classification for the chondritic meteorites. *Geochim. Cosmochim. Acta* **31**, 747–765.

Wadsworth M. E. (1883) Meteoritic and terrestrial rocks. *Science* **1**, 127–130.

Wahl W. (1910) Beitrage zur Chemie der Meteoriten. *Z. Anorg. Allgem. Chem.* **69**, 52–96.

Wahl W. (1952) The brecciated stony meteorites and meteorites containing foreign fragments. *Geochim. Cosmochim. Acta* **2**, 91–117.

Walter L. S. (1969) The major element composition of individual chondrules of the Bjurbole meteorite. In *Meteorite Research* (P. M. Millman, ed.), pp. 191–205. Reidel, Dordrecht.

Walter L. S. and Dodd R. T. (1972) Evidence for vapor fractionation in the origin of chondrules. *Meteoritics* **7**, 341–352.

Walter L., Dodd R., and Smidinger P. (1973) Sampling model of chondrule compositional variations (abstract). *Meteoritics* **8**, 449–450.

Wark D. A. (1978) Early solar system stratigraphy: The condensation sequence from Allende to chondrites, and the origin of chondrules (abstract). In *Lunar and Planetary Science IX*, pp. 1208–1210. Lunar and Planetary Institute, Houston.

Wark D. A., Wasserburg G. J., and Lovering J. F. (1979) Structural features of some Allende coarse-grained Ca-Al-rich inclusions: Chondrules within chondrules (abstract). In *Lunar and Planetary Science X*, pp. 1292–1294. Lunar and Planetary Institute, Houston.

Wänke H., Baddenhausen H., Palme H., and Spettel B. (1974) On the chemistry of the Allende inclusions and their origins as high temperature condensates. *Earth Planet. Sci. Lett.* **23**, 1–7.

Warren R. G., Osborn T. W., and Schmitt R. A. (1971) Chemical composition of 61 individual Allende chondrules (abstract). *Meteoritics* **6**, 321–322.

Wasilewski P. (1981) New magnetic results from Allende C3(V). *Phys. Earth Planet. Sci. Inter.* **26**, 134–148.

Wasilewski P. J. and Saralker C. (1981) Stable NRM and mineralogy in Allende: Chondrules. *Proc. Lunar Planet. Sci. 12B*, pp. 1217–1227.

Wasson J. (1972) Formation of ordinary chondrites. *Rev. Geophys. Space Phys.* **10**, 711–759.

Wasson J. (1975) *Meteorites*. Springer-Verlag, Berlin. 316 pp.

Wasson J. T. and Grossman J. N. (1982) Progress in the compositional characterization of chondrules (abstract). In *Papers Presented to the Conference on Chondrules and their Origins*, p. 60. Lunar and Planetary Institute, Houston.

Weinland D. F. (1881) Korallen in Meteorsteinen. *Das Ausland* **54**, No. 16, 301–303, No. 26, 501–508.

Whipple F. L. (1966) Chondrules: suggestion concerning the origin. *Science* **153**, 54–56.

Whipple F. L. (1970) Aerodynamic processes for selectively accumulating chondrules (abstract). *Meteoritics* **5**, 228.

Whipple F. L. (1972) On certain aerodynamic processes for asteroids and comets. In *From Plasma to Planet* (A. Elvius, ed.), pp. 211–232. Wiley and Sons, N.Y.

Wlotzka F. (1969) On the formation of chondrules and metal properties by "shock melting." In *Meteorite Research* (P. M. Millman, ed.), pp. 174–184. Reidel, Dordrecht.

Wlotzka F. (1981) Classification of unequilibrated ordinary chondrites (abstract). *Meteoritics* **16**, 403–404.

Wlotzka F. (1982) Composition of chondrules, fragments and matrix in Tieschitz (abstract). In *Papers Presented to the Conference on Chondrules and their Origins*, p. 61. Lunar and Planetary Institute, Houston.

Wlotzka F. and Palme H. (1982) A refractory-rich "basaltic" chondrule from Ornans (abstract). In *Papers Presented to the Conference on Chondrules and their Origins*, p. 62. Lunar and Planetary Institute, Houston.

Wlotzka F. and Wark D. A. (1982) The significance of zeolites and other hydrous alteration products in Leoville Ca-Al-rich inclusions (abstract). In *Lunar and Planetary Science XIII*, pp. 869–870. Lunar and Planetary Institute, Houston.

Wood J. A. (1958) Silicate meteorite structures and the origin of the meteorites. *Smithson. Astrophys. Obs. Tech. Rept. No. 10.*
Wood J. A. (1962) Chondrules and the origin of terrestrial planets. *Nature* **194**, 127–130.
Wood J. A. (1962) Metamorphism in chondrites. *Geochim. Cosmochim. Acta* **26**, 739–749.
Wood J. A. (1963) On the origin of chondrules and chondrites. *Icarus* **2**, 152–180.
Wood J. A. (1963) Physics and chemistry of meteorites. In *The Solar System*, vol. 4 (B. M. Middlehurst and G. P. Kuiper, eds.), pp. 337–401. Univ. of Chicago Press, Chicago.
Wood J. (1967) Chondrites: Their metallic minerals, thermal histories, and parent planets. *Icarus* **6**, 1–49.
Wood J. A. (1968) *Meteorites and the Origin of Planets.* McGraw-Hill, N.Y. 117 pp.
Wood J. A. (1981) The interstellar dust as a precursor of Ca,Al-rich inclusions in carbonaceous chondrites. *Earth Planet. Sci. Lett.* **56**, 32–44.
Wood J. A. and McSween H. Y. (1977) Chondrules as condensation products. In *Relationships between Comets, Minor Planets, and Meteorites* (E. Anders, A. H. Delsemme, and T. Gehrels, eds.), pp. 366–373. Proceedings of I. A. U. Colloquim 39, Lyon, France, 1976.
Wood J. A. and Motylewski K. (1979) Meteorite research. *Rev. Geophys. Space Phys.* **17**, 912–925.
Yang J. and Epstein S. (1982) The concentration and isotopic composition of hydrogen and carbon in meteorites (abstract). In *Lunar and Planetary Science XIII*, pp. 885–886. Lunar and Planetary Institute, Houston.
Zavaritskii A. N. (1948) On the structure of crystalline chondrites. *Meteoritika* **4**, 50–70.
Żbik M. and Lang B. (1982) Morphological features of pore spaces in chondrules (abstract). In *Papers Presented to the Conference on Chondrules and their Origins*, p. 63. Lunar and Planetary Institute, Houston.
Zhang P. (1978) (CH) Investigation on the material constitution of Jilin meteorite shower and discussion on its significance to the study of origin and evolution of the earth. *Sci. Geol. Sin.* **1978**, 113 –133.
Zook H. A. (1981) On a new model for the generation of chondrites (abstract). In *Lunar and Planetary Science XII*, pp. 1242 –1244. Lunar and Planetary Institute, Houston.
Zook H. A. (1982) Planetoids to chondrules? (abstract). In *Papers Presented to the Conference on Chondrules and their Origins*, p. 64. Lunar and Planetary Institute, Houston.

SUBJECT INDEX

AUTHOR INDEX